# How the **Sun God** Reached America

## c.2500 BC

Revised Edition

**A Guide to Megalithic Sites**

**Dr. Reinoud M. de Jonge**
**Jay Stuart Wakefield**

For my wife, Tineke, and my three sons, Rogier, Michiel and Tjako,
who encouraged me to write
Reinoud de Jonge

And for my father Richard,
who is always seeking a reasonable explanation
Jay Stuart Wakefield

---

"Over and beyond mere living, the human spirit adds and creates what is better than what was before" -- R. Roefield

---

cover: Dr. de Jonge explaining the main petroglyph inside megalithic Cairn T, Loughcrew, Ireland, July 8, 1998

# Table of Contents

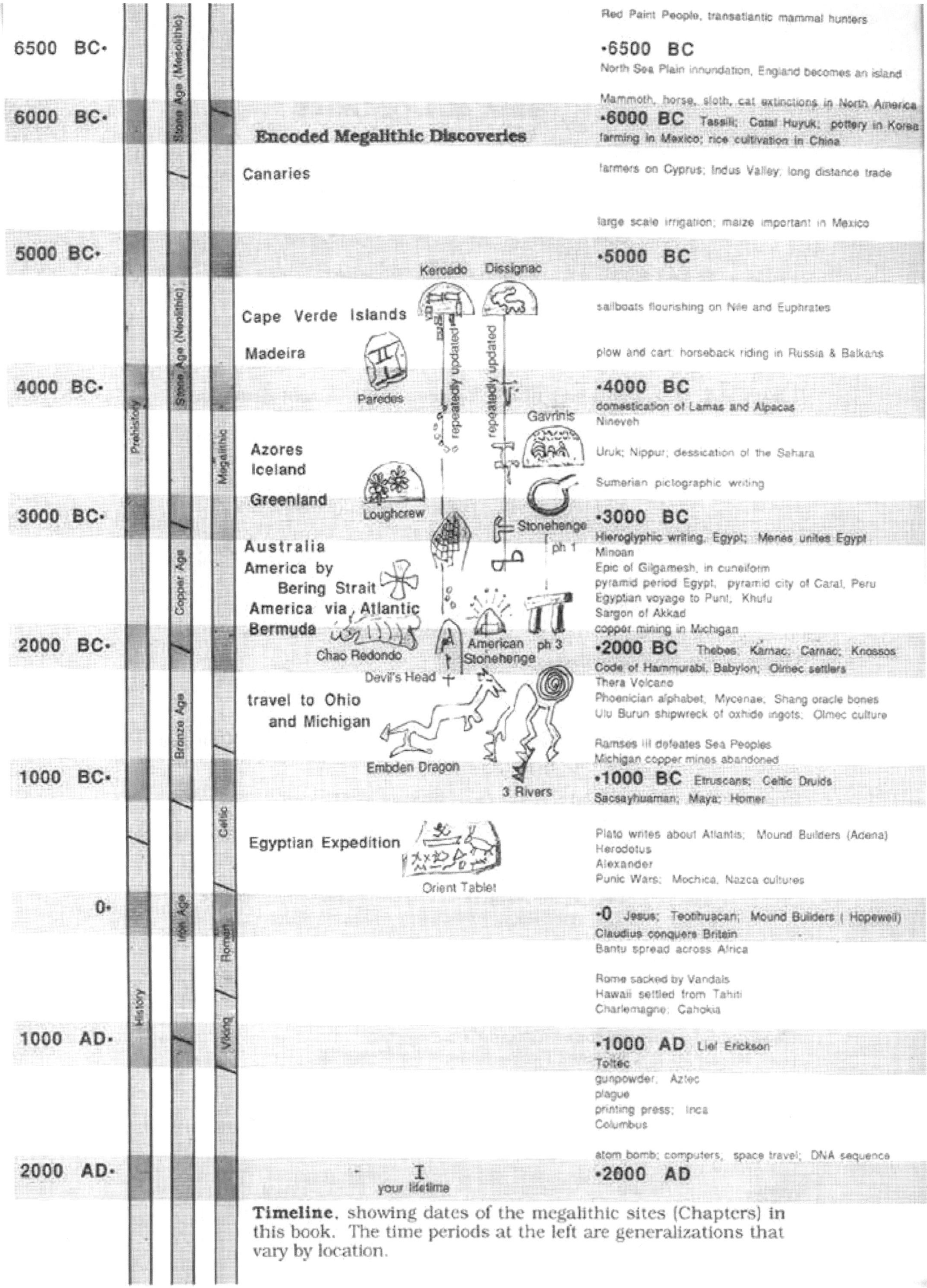


**Timeline**, showing dates of the megalithic sites (Chapters) in this book. The time periods at the left are generalizations that vary by location.

# Introduction

Dr. R.M. de Jonge. drsrmdejonge@hotmail.com
J.S. Wakefield, jayswakefield@yahoo.com

- **Do ancient stones record the quest to find the other side of the world?**
- **Is Stonehenge a monument for the discovery of America?**
- **Is there a place in New Hampshire where sea captains were trained for ancient transatlantic voyages?**

## The Decipherment of Megalithic Petroglyphs and Monuments

We have discovered that ancient "art" contains geographical and mathematical pictographs (picture drawings). These petroglyphs (rock drawings) tell stories of sailing exploration and of the discovery of islands in the Oceans. We are learning what early people believed and achieved, putting light on events in late prehistory, prior to the invention of Egyptian hieroglyphics or other written languages by man. This new evidence explains why prehistoric people were motivated to explore the Atlantic Ocean, and how they were able to discover America and exploit its resources long before it has been thought possible. **Figure 1** (next page) gives a general overview of the locations of the ancient sites in the 14 chapters of this book. The **Timeline** on the opposite page shows these sites in historical order, and the discoveries that are associated with them. These discoveries extend known history thousands of years into the past.

This is the story of the exploration of the world by people who left behind their records in rock. We do not know who they were, but they are defined as **"megalithic"** because they left behind "big rocks". Some of these rocks are carved, leaving **"petroglyphs"**, or "rock carvings". This period is called "pre-historic" because it is before "history", which is what we know about the past from written records. The **"megalithic period"** starts c.6000 BC ("about" 6,000 BC) in the Mediterranean, and along the western coasts of Europe. This was 4,000 years after the last Ice Age, at the start of the "Neolithicum", or "new stone age", as warmer weather in Europe and developing agriculture brought increasing human populations. This period is called "late prehistory". We thought that because they were unable to write their spoken (phonetic) languages, they left no written records. But we will show you how they have left their stories encoded in numbers in their petroglyphs and monuments.

Human-constructed piles of rocks, are what we call **"monuments"**. These are found around the world, dating from different times in different places, but they mostly date from 6000 BC to 1500 BC in Europe. Their origin seems to have been in the Small Mediterranean Sea, between southern Italy and the Strait of Gibraltar. The oldest megalithic monuments are on the islands and underwater in this area, and the oldest megalithic petroglyphs depict this inland sea. Slowly, this culture expanded along the coasts of Europe, and then worldwide. In Scandinavia, the end of this culture was c.1500 BC. This is a long span, and it is remarkable that the cultural

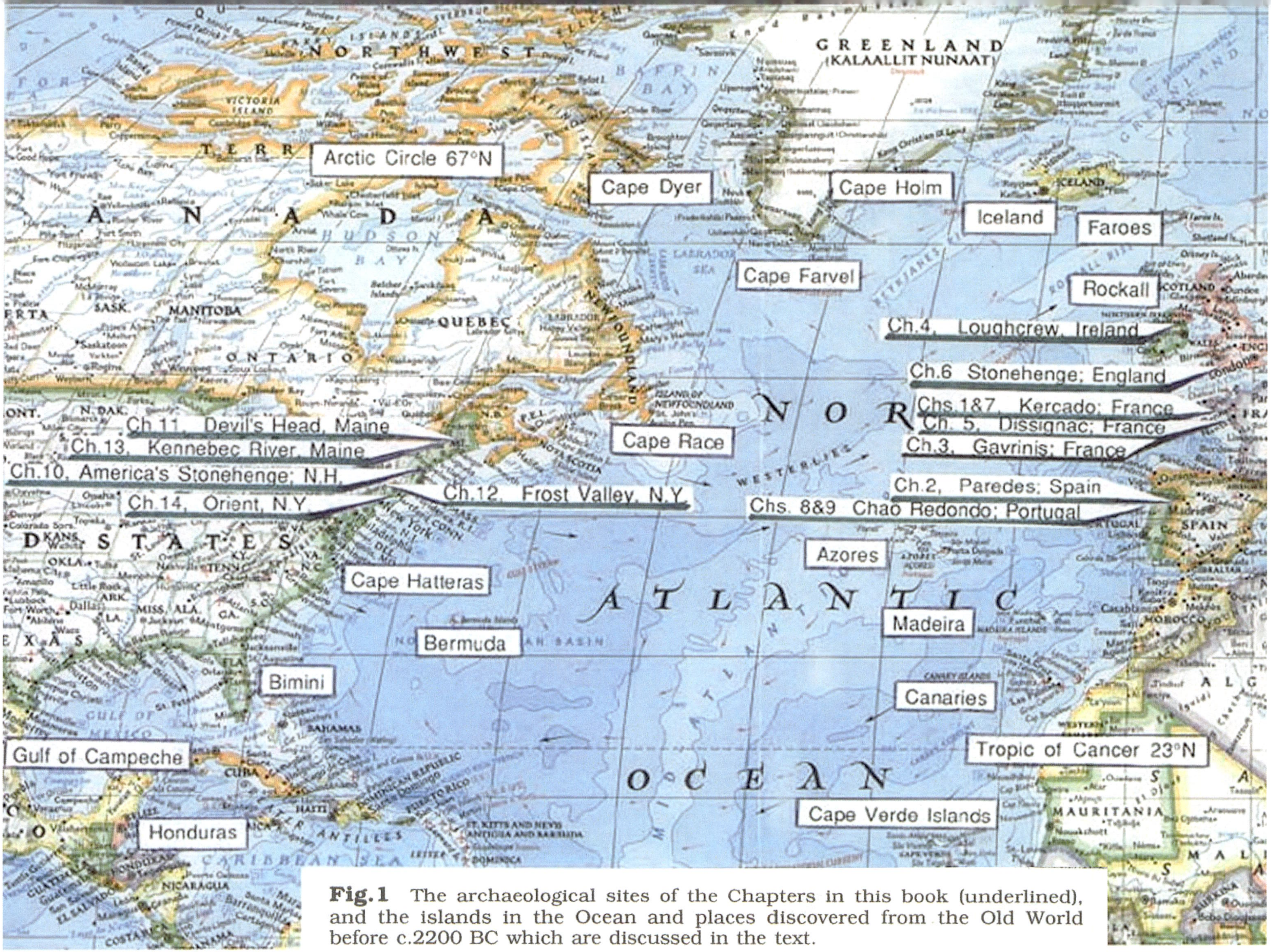


**Fig.1** The archaeological sites of the Chapters in this book (underlined), and the islands in the Ocean and places discovered from the Old World before c.2200 BC which are discussed in the text.

tradition we are studying was consistent for such a long period. The earliest megalithic monuments in Europe, made of very big stones, clearly are holy sites, built for eternity. Archaeologists have studied them with precision, but their true meaning has remained unclear. In the central interiors they have almost always found skeletons, and for that reason, they have called these constructions "megalithic graves". In many monuments, a passage runs to a chamber, so these have been called "passage graves". However, the number of skeletons found is always very small, as in medieval European cathedrals, which have passages, chambers and a few skeletons. The cathedrals have within them references to the origin of religion: e.g. Jerusalem, Bethlehem, and Rome. As we shall later explain, European passage graves contain references to the origin of the Sun religion in Egypt. The solar alignments of these megalithic chambers, with their annual displays of rays of midwinter sun down the passages onto glyphs at the inner end, have been shown to be cardinal features of these monuments.

The "**SunGod**" religion characterized the megalithic culture, and evidence shows a major ideologic connection with the culture of Egypt. In all the monuments and petroglyphs of the megalithic people, there is evidence of a great admiration for Egypt, which was the great center of civilization at the far end of the Mediterranean Sea. These passage graves appear to be temples of the Sun religion. In the literature (Ref.2), Egyptian civilization becomes visible about 3400 BC. Before the Archaic Period, which started with King Menes (c.3000 BC), the sun year of 365 days had been officially accepted. Sun temples and sanctuaries for the SunGod Ra had appeared. During the 3rd dynasty of King Djoser (c.2700 BC), the first pyramid, a monument of 60 meters was built, after earlier mastabas (brick structures over underground vaults). The sun city Heliopolis was flourishing with its huge Temple of the Sun. During the whole history of Egypt, the East was considered the "empire of the living", and the West was considered the "empire of the dead", with SunGod Ra traveling around the Earth in his sunboat on his daily journey.

By **"decipherment"**, we mean that by counting lines in a petroglyph, or by counting the number of stones at a site, or by observing the angles between stones, we have found consistently intelligible data with predictive power, that has provided interpretations of the meanings of these sites, and of megalithic prehistory in general. We know from reading the literature on this subject, that megalithic people developed a sophisticated understanding of astronomy and spherical mathematics. Early in the Megalithic Period, they knew the world was round, and that they knew only half of it. The curious monument in England near Stonehenge called "Robin Hood's Ball" (Ref.4, c.4000 BC) is an enormous earthen model of half the earth. They were, of course, curious to know what was in the other half of the world, on the other side of the ocean, where the SunGod went every evening, and had theories that it was "paradise" or "the land of the dead". The stories of these explorations are told in this book as a series of discoveries of the islands in the Atlantic. Of course, the islands close to shore were found first, starting with the Canaries, off Africa. It appears that each of these discoveries in the ocean was celebrated as the finding of the new western home of the SunGod. Once the New World itself was explored, its valuable metal resources were a powerful motivation for long-distance trade and colonization.

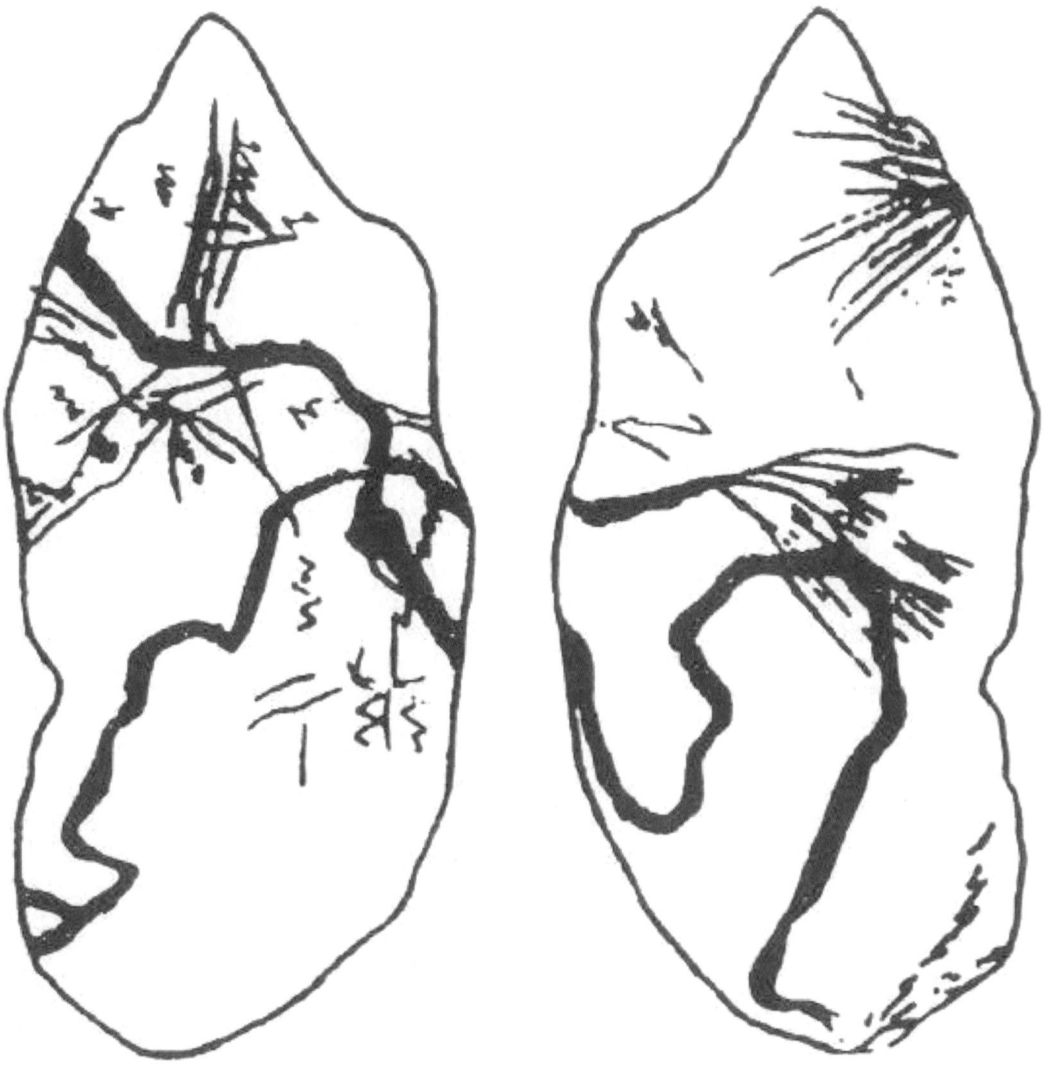

**Fig.2** Inscription on both sides of a stone found in the Grotto of Romanelli, eastern peninsula of southern Italy (Ref.6), c.6000 BC.

It is no coincidence that the "Megalithic Period" ended when the entire world had been explored and had become fully known. This took them about **four thousand years** to accomplish. Unfortunately, the geographic knowledge of antiquity was later lost through natural disasters, cultural conflicts, and victories of ignorance and faith over learning, such as the library burnings in Troy and Alexandria, and was not to be rediscovered for thousands of years.

**Dr. Reinoud de Jonge** is a Dutch physical chemist, who teaches chemistry and physics, and studies megalithic art as a hobby. That is Reinoud, holding the flashlight in Cairn T at Loughcrew, Ireland four years ago, in the cover photo. His enthusiasm is infectious, as he exclaims, "Now I can read many of these inscriptions, because I now know how they were thinking!"

He studied all the available megalithic inscriptions of Western Europe with a mathematical eye, asking "what evidence can be found within the "art" itself"? This has been possible because all known inscriptions in Europe have been carefully recorded, and published (Ref.1), and so many sites have been dated. He has discovered that while megalithic people did not have a phonetic written language as we do, they were able to record their stories on stone with mathematical symbols, creating the first written stories in the world. This expansion of "cognitive archaeology" is bringing us an enormous cache of information about megalithic life and ideology. Reinoud's book with Dr. IJzereef, De Stenen Spreken, was published in Dutch in 1996 (Ref.5), and subsequently he has developed other materials, including this book with Jay Wakefield. This work is indebted to the work of many other multi-disciplinary archaeological "amateurs", such as Fell, Bailey, da Silva Ramos, Mallery, Lockyer, Hawkins, Thom, Schliemann, Fawcett, Heyerdahl, and so many others who are slowly unraveling the last few thousand years of prehistory.

The decipherments done for this book were laborious, and proceeded in a step-by-step series of surprising discoveries over a period of ten years. These decipherments are all based on the discovery that megalithic people made geographic petroglyphs, and encoded latitudes in their petroglyphs and monuments. At the Tumulus of Kercado (in Western France) and Stonehenge (in South-Central England), latitudes are encoded in the number of stones. At Dissignac (France), they are encoded in the symbols of the petroglyph, while at America's Stonehenge (in New Hampshire, USA), they are encoded in the angles between the central axis, and the peripheral stones of the site. As Reinoud puts it, it has involved "learning to think like they did". As a result, we have been able to construct the **"de Jonge Rules of Decipherment"**, which appear in a later chapter. These will help teach you what we have learned, and will enable you to contribute to this study by deciphering new sites by yourselves.

## An example of a geographic petroglyph, from South Italy

Dr. de Jonge has discovered that a great proportion of megalithic inscriptions are geographic. Most are coastal maps. In the early neolithic, before 4000 BC, seafaring was limited to coastal navigation. The earliest inscriptions depict the coasts of the Mediterranean, Europe, and Africa. Figure 2 is a magnificent example of a very old geographical inscription

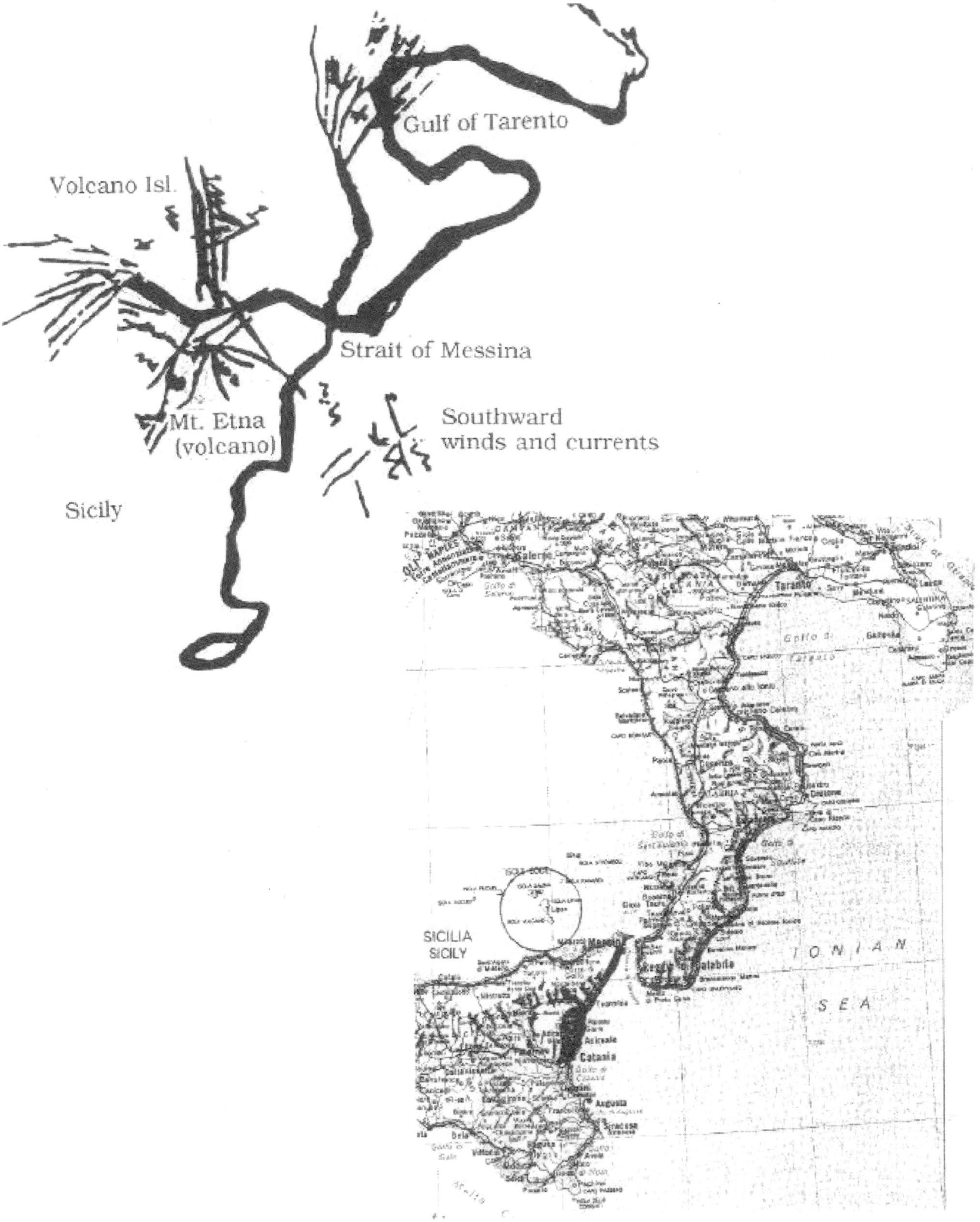


**Fig.3** Representation of the complete inscription of the stone by the authors, with modern map of southern Italy.

(Ref.6), on a stone from the Grotto of Romanelli, on the eastern peninsula of southern Italy, which has been dated at approximately 6000 BC. The "art" on this fairly flat stone wraps around it, so the two sides are shown in the drawing. When you compare this with a modern map of Italy (Fig.3), you can see the inscription is a coastal map of southern Italy, with the "boot" of Italy to the right, and the Island of Sicily on the left, a pattern you would get if you were to roll the stone like an old seal cylinder (this appears to be an early map version of those later found in great quantities in both Mesoamerica and the Middle East (Ref.15). In total, more than 1200 kilometers (800 miles) of coastline have been engraved accurately. You can even see the Island of Vulcano in the Stromboli Islands, Mount Etna, and perhaps a depiction of the wind and currents that flow strongly to the south through the Strait of Messina, where Roman galleys always had to be rowed when traveling northward. In view of the nature of the carving, and considering the other artifacts found in the Grotto of Romanelli, the inscription probably dates back to the early Neolithicum, c.6000 BC (Ref.6). If this turns out to be correct, this map is one of the oldest coast maps ever found in Europe.

Sailors today can appreciate that rivers, lakes, and oceans have been the highways of the world. Colleagues in Seattle recently paddled their kayaks to Alaska, and then from Alaska to Russia, and they tell of others doing so each year. Many of us read Heyerdahl's books (Refs.7,8) as teenagers, and understand his findings (and those of Tim Severin of the Brendan Voyage, Ref.9), that "primitive" sailors were in many ways (including foods, clothing, training) better equipped for ocean voyaging in small craft than modern sailors. The oceans, never having been commercially fished, and unpolluted, would have astonished us with their sealife. Bailey (Ref.10) quotes from Roman historian Strabo: "the ancients made longer journeys, both by land and sea, than men of later times". When making a chart, a sailor will be particularly interested in points, bays, islands, and heights, so it should not surprise us when we see a disproportionate emphasis on these in megalithic "art".

## An example of a geographic petroglyph, from Rhode Island

Another example of a geographic inscription, although more recent, is visible on the back of an oil lamp (Fig.4), dredged out of the Taunton River near Dighton Rock in the 1970's (Ref.13). It is a sandstone oil lamp, 6 by 9 inches (15x23cm). The "face" drawing is a stylized chart of Narragansett Bay, Rhode Island (USA), where the lamp was found. As the home of the "America's Cup", this area is still an important sailing area, as it has been for thousands of years. The "mouth" of the face, which is the mouth of Narragansett Bay, leads into three waterways running to the north. East of Newport the Sakonnet River leads into the Taunton River (the finding spot). In the middle, west of Newport and east of some larger islands, is the route to the City of Providence, and the Blackstone and other rivers. The western entrance runs north from Narragansett Pier, and side branches to Wickford, E. Greenwich, Apponaug and Warwick are shown on the lamp. Although hard to estimate, this oil lamp could date as late as the Middle Ages, when Indians caught whales along the East Coast. You will find other geographic inscriptions, if you keep in mind that they are not uncommon among prehistoric petroglyphs.

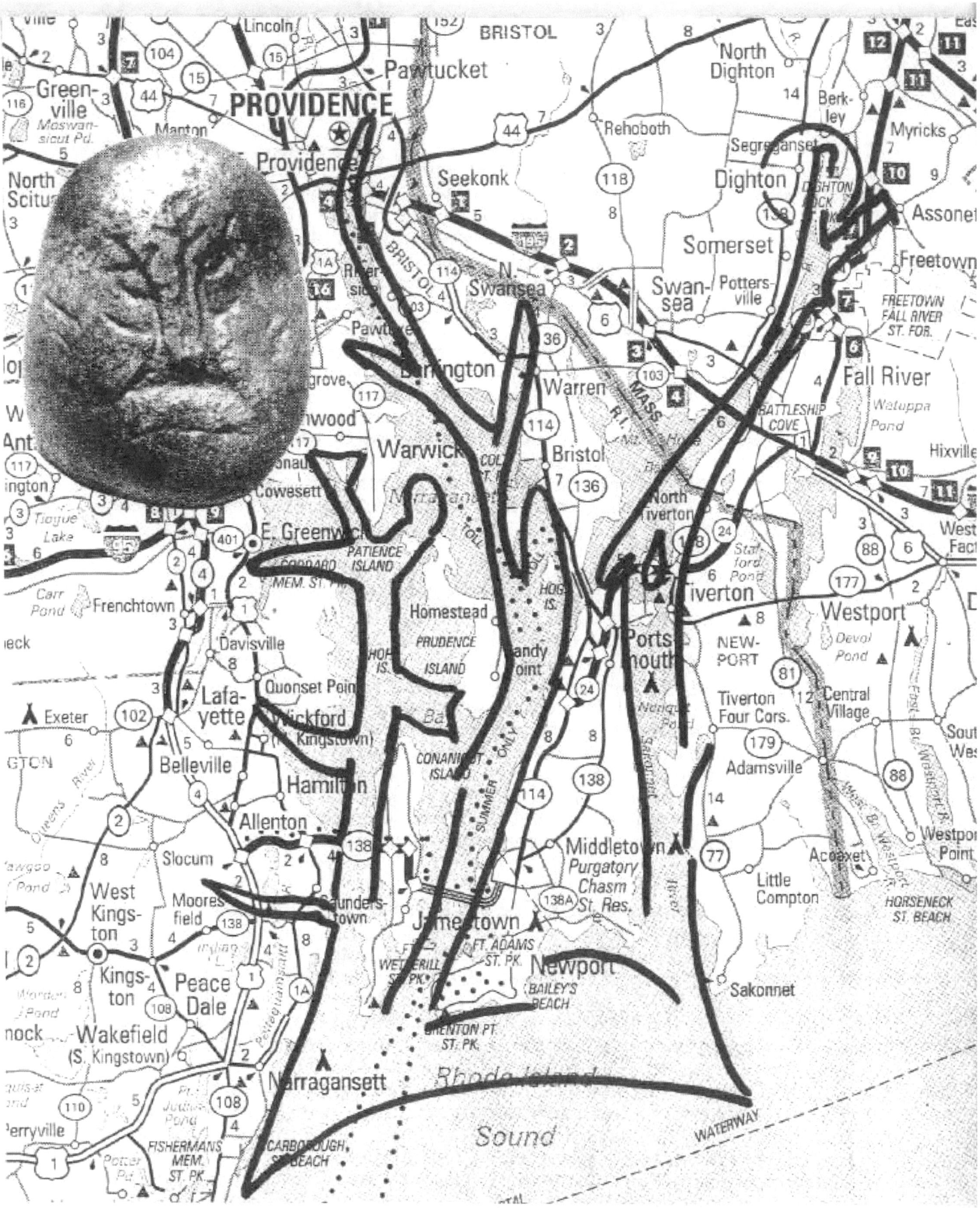


**Fig.4** Geographic petroglyph of Narragansett Bay on the backside of an oil lamp dredged from the Taunton River, near Dighton Rock, Massachusetts, USA, (sandstone, height 9", 500-1500 AD, map scale 10cm=14mi, Ref.13).

## An example of a petroglyph of the North Atlantic Ocean

Figure 5 is a photo taken by the authors, with a drawing (Ref.1), of one of the inscriptions in a 78 foot-long tomb called "Les Pierres Plates", now on the oceanfront of the Gulf of Morbihan, Locmariaquer, Brittany, France. This is one of many so-called "mother goddess figures" with "multiple breasts", as shown in Fig.7. The "breasts" of the archaeological literature probably stem from the Christian belief that sex is sinful, and therefore worth being preoccupied about. These are actually all sailing charts of the North Atlantic Ocean, and the "breasts" are islands, with distance lines (dl) of exploration around them.

Dr. de Jonge calls a repeating feature of many inscriptions "distance lines", or "**dl**" for short (when dls are used in multiples of 10 he calls them **DL**s). One dl is the distance of one degree of latitude on the surface of the earth, which corresponds to a distance of 40,000km (the circumference of the earth), divided by 360 degrees (the circle angle of the earth), which equals 111.11 kilometers. On the top of Figure 5 we recognize the southern half of the island of Greenland, the most western area of the then-known world (c.2700 BC). Below it we see an early facsimile of the Mid-Atlantic Ridge, which divides the ocean in two. In the eastern half of the ocean, four island groups were engraved, all of them discovered long before. Below, you see the Cape Verde Islands, around which the sea was poorly investigated (single line), because of the strong winds and sea currents from the northeast. Above it you see the Canary Islands, which belong to Africa (the island inscription is open on the right side). Next are the Azores Islands, which are thought to belong to the other side of the ocean (their symbol is open on the left side). On top is Iceland, important because of its large size. Four island groups are inscribed on the west side of the ocean too. At the time this was inscribed, the people indulged in fantasies about groups of islands in the western half of the ocean as counterparts of islands known to exist in the eastern half. They did not know anything about them, but just assumed a symmetric distribution of land and sea.

Until you read this book, the other lines in Figure 5 will not be easy to understand. The three spaces between the lines below Greenland mean that the coastal waters were investigated there over a distance of 3 degrees of latitude, or 3dl in all directions (3°= 3dl= 333km). The first line offshore at the right side marks a sailing distance of half a "big distance line" (0.5DL= 5dl= 5°= 555km). The second line shows the limit of all the coastal waters ever investigated (2 x 0.5DL= 10dl= 1111km). The inscription shows that the width of the ocean was estimated to be 4DL= 4444km, which happens to be quite a realistic figure. As we will explain later, the inscription was made between the discovery of Greenland (c.3300 BC) and the discovery of America (c.2500 BC) (Ref.5). All the details indicate the probable date for this petroglyph is c.2700 BC.

## Megalithic use of Latitudes and Distance Lines

After thousands of years of experience with their astronomic observatories, neolithic people had learned that a one-degree change in height of the sun, moon, or star would be caused by moving some distance further north or south, so that **one degree of latitude** (called a "moira" by Egyptians) equalled

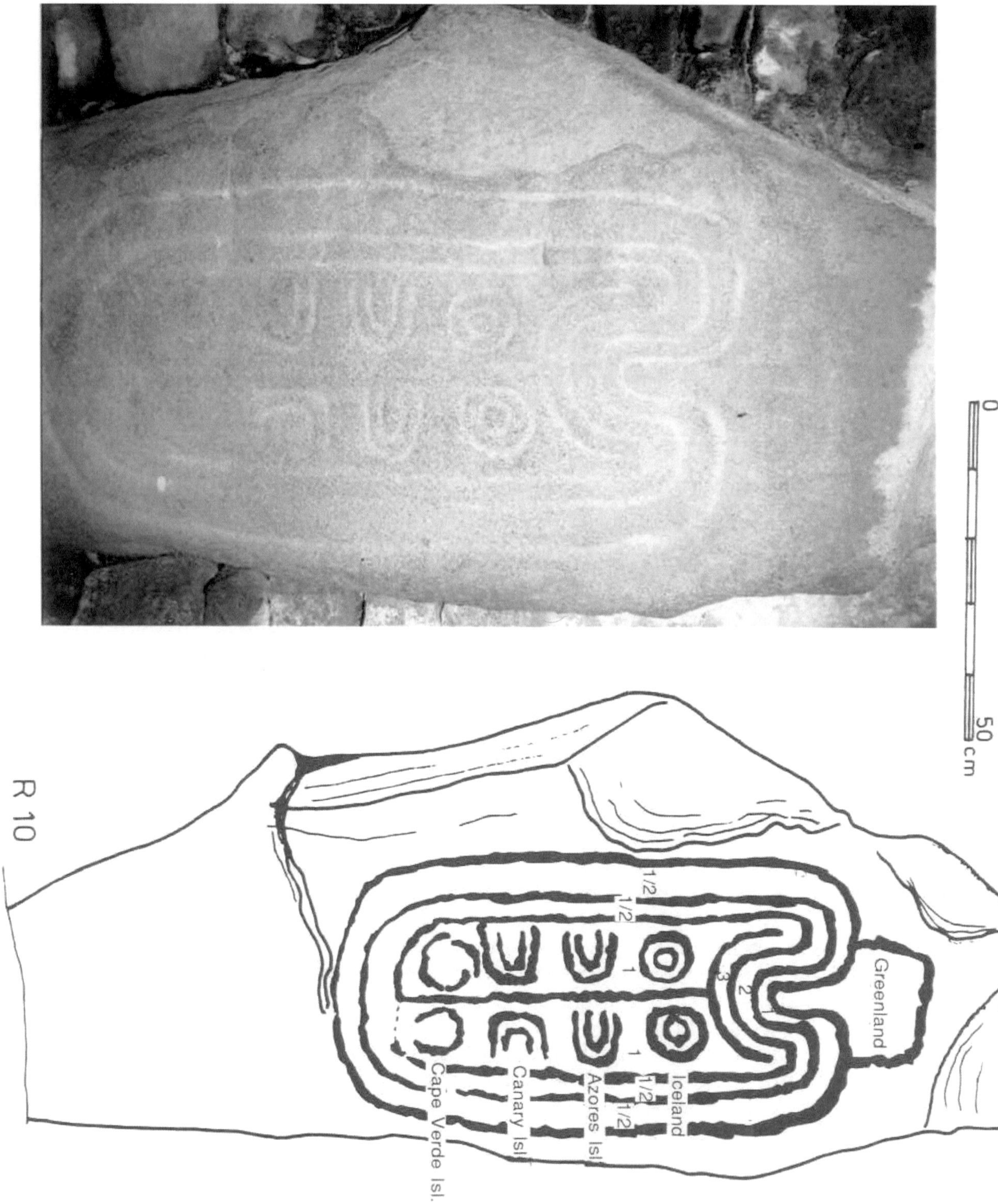


**Fig.5** Photo by authors and drawing with geographic meanings of megalithic inscription on one of the upright stones (R10) of the passage grave of Les Pierres Plates, on the peninsula of Locmariaquer, Gulf of Morbihan, south Brittany, France (Ref.1, c.2700 BC).

60 Egyptian miles. This distance changes a bit as you move north, due to a slight bulge of the Earth at the equator, but not much. Their accuracy is discussed by Tomkins (Ref.11): "computing from the third millenium BC Egyptian text, the mean length of a degree of latitude in Egypt would be 110,832 meters, while the modern estimate is 110,800 meters".

Modern english statute miles are derived, but a little different, from very ancient mile measures ("some Greek temples are in English feet"-Ref.11). We have primarily used the French meter (1km= 0.6214 statute miles), which is derived from a curved meridian of the earth, and, like the foot, not related to time, as were many ancient measures. Thomas Jefferson was a surveyor, and read extensively in classical works in their original languages. He understood that ancient measures were often related to time, which is why he opposed and prevented the planned US adoption of the decimal system by Congress. The ancients knew that "the speed of the rotation of the vault of heaven is one degree every four minutes. The Egyptians (found...) it was expedient to divide the circumference of the earth not only into 360 degrees, but also into 24 hours. According to the second system, a degree is equal to 4 minutes of time, and a minute of degree is equal to 4 seconds of time" (Ref.11). English Nautical Miles (NM) (= 1.1515 statute miles, and = 1.853km) are used for distances at sea, because it is the length of one minute of arc of any great circle of the earth (Ref.12). **Therefore if we were to use 60 nautical miles (NM) for a dl (600 for a DL), we would be using a modern unit of measure equivalent to 60 Egyptian miles, both equal to the distance of 1°.** It is amazing that the ancient Egyptian mile is equal to the modern Nautical Mile. Apparently Jefferson should have recommended adoption of the Nautical Mile. We suggest future research be based in Nautical (Egyptian) miles, but we have not seen others do this. We have used kilometers (for European sites) and miles (for American sites) because so many readers are accustomed to visualizing distances in these units.

Tomkins and Dr. Stecchini, a Professor of the History of Science, explain that calculations of **longitude** between fixed points were easy in Egyptian prehistory: "because every observable star comes to the meridian of every place on the globe once in 24 hours, the interval which elapses between the same star coming upon the meridian of two different places is the difference in longitude of the two places" (Ref.11). In other words, the time interval traveled by the sun between two places gives the distance between them, though doing this at sea required a way to keep accurate time at sea, not accomplished until the 17th century AD. To simplify navigation for sailing, as latitudes and longitudes form nearly perfect squares at near-equitorial latitudes, it made sense to use a dl or DL grid in all directions. Thus, for example, to show distance explored around an island, the megalith builders drew multiple rings of one dl around it. So we find that most of the "mother figures" with "breasts" are actually primitive charts with islands surrounded by encircling distance line (dl) rings of exploration, as shown in Figure 5.

## Petroglyphs of the Ocean carved after the Discovery of America

Dr. de Jonge has found that all "mother goddess figures" at sites dated after c.2500 BC show features of the American coast. For example, this is shown in Fig.6, a beautiful late inscription (Ref.1) on a support stone in the Ile Longue dolmen also in the Gulf of Morbihan, Brittany, dated c.2200 BC. It is

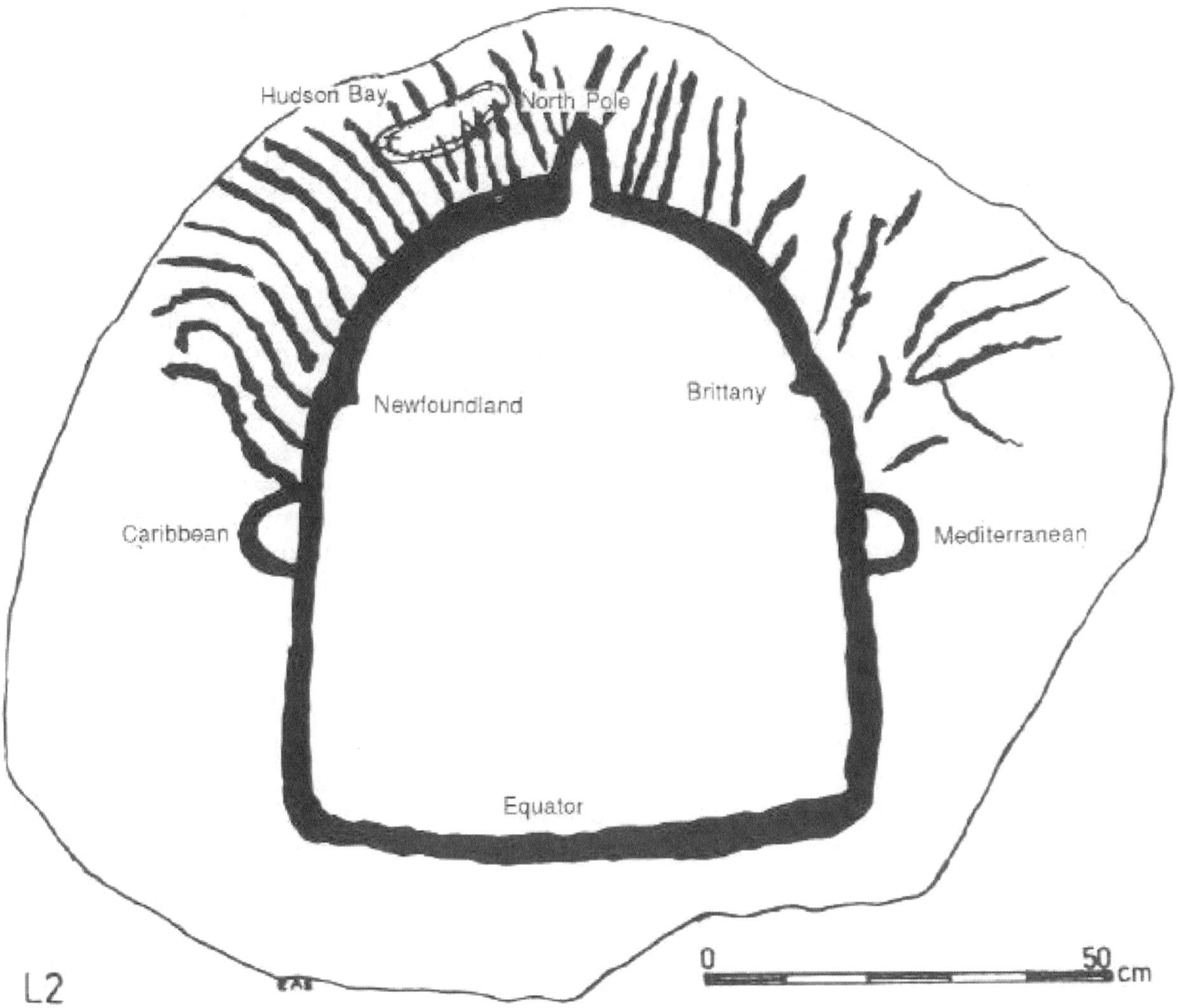


**Fig.6** Megalithic inscription on one of the upright stones (L2) of the passage grave of Isle Longue, Locmariaquer, Brittany (Ref.1, after the discovery of America, c.2200 BC).

a stylized inscription of the North Atlantic Ocean (Ref.5). We recognize the mainland of Europe and Africa on the right, the equator below, and the other side of the ocean on the left. On the top, the North Pole is indicated. Note the inward "points" on each side of the Atlantic, indicating Brittany and Newfoundland, both at 47°N. The bays which look like ears on the figure are the Mediterranean Sea, and the Gulf of Mexico/Caribbean Sea. The lines feathering out above them indicate that land is found everywhere around the north of the ocean. As a consequence, navigation of the north is made easier than first thought. Above the inscription at the left side, a natural oval depression in the stone can be identified. Although not important, it nicely indicates Hudson Bay. From a comparison of all the details with the features of petroglyphs at other sites, it can be estimated that this carving dates from c.2200 BC. Other megalithic inscriptions of the ocean, found in a variety of locations in Brittany, are shown in Fig.7. Some of them show various islands, and some show dl and/or DL. When you have finished reading this book, you will probably be able to "tell the story" about each of these figures.

## How the SunGod reached America

This book challenges the accepted paradigm that diffusion of culture across oceans before Columbus was not possible. Not a single hand-worked mineshaft, corbelled solar-oriented building, hilltop fortress, pyramid, temple mound, coin, or any of the other of thousands of constructions, artifacts or inscriptions in the Americas is accepted as credible evidence of pre-Columbian contact with Europe, with the exception of the Viking camp unearthed in Newfoundland. Because of the professional insistence that the sun temples and civilizations of the Americas arose in isolation, we have drafted this publication for the public.

Efforts to uncover and document the past need the political help that can be provided by widespread public interest. We have all been reading about the Taliban blowing up ancient Buddha statues and smashing museum collections in Afghanistan. The largest libraries in both the Old World and the New World were burned on purpose by man. Unfortunately, even today, in the field we are studying, important evidence is being destroyed by people who do not want their turfs, beliefs or rights challenged. In New York State, archaeologists are reportedly being paid to sign Declarations of Non-Significance so stone chambers can be destroyed by developers on housing lots; the US Army Corps of Engineers has dumped 30 tons of rock on the Kennewick Man site on the Columbia River (Washington State, USA), because the Indians "already know their history"; the Brazilian Navy reportedly has dumped tons of mud on a shipwreck full of amphorae (dated by experts to the 3rd century AD, and pinpointed to the Roman City of Kouass, on the Moroccan coast, Ref.14), and confiscated the jars, preferring their history of being discovered by the Portuguese; in New Zealand, caves of non-Polynesian/Melanesian skeletons and artifacts have reportedly been bulldozed shut by authorities, so as not to offend Maori rights, and so on. It is our belief that self-serving contemporary motivations should not be allowed to prevent research, or control humanity's knowledge of its history.

Each **Chapter** is about a prehistoric site, and they stand alone, in chronological order. The chapters can be serialized from this book with our written permission, and credit to this publication. They may prove useful as

**Fig.7** Megalithic petroglyphs of the North Atlantic Ocean from various locations in Brittany (Ref.5). The top left one is very old (c.5000 BC), the bottom right one is more recent (c.2000 BC).

tourist guides at some of the sites. We find, however, that the chapters are strongly supportive of one another, like instruments in a symphony. Only by listening to all of them together do we hear the story of the global sailing effort, the quest of prehistoric man to explore and understand the scope of his world, and exploit its resources.

## References


1. Twohig, E. Shee, The Megalithic Art of Western Europe, Clarendon Press, Oxford, 1981
2. Old World Civilizations/The Rise of Cities and States, Weldon Owen Pty Limited, Australia, 1995
3. Balfour, M., Megalithic Mysteries, Dragon's World, 1992 (ISBN 1-85028-163-7)
4. Richards, J., Stonehenge, English Heritage, 1992
5. Jonge, R. M. de, and IJzereef, G. F., De Stenen Spreken, Kosmos Z&K, Utrecht/Antwerpen, 1996 (ISBN 90-215-2846-0) (Dutch, pg.152)
6. Grote Encyclopedie van de Mens uit de Oertijd, J. Jelinek, 1994 (Dutch, pg. 458)
7. Heyerdahl, T., The RA Expeditions, George Allen & Unwin, London, 1971
8. Heyerdahl, T., The Tigris Expedition, George Allen & Unwin, London, 1983
9. Severin, T., The Brendan Voyage, Hutchinson, London, 1978
10. Bailey, J., Sailing to Paradise, Simon & Schuster, 1994
11. Tompkins, P., Secrets of the Great Pyramid, Harper Colophon Books, Harper & Row, New York, 1971 (ISBN 0-06-09-0631-6)
12. Mixter, G., Primer of Navigation, 4th Edition, D. Van Nostrand Co., Princeton, 1960
13. Cahill, R., New England's Ancient Mysteries, Old Saltbox Publishing, Salem, Mass., 1993 (ISBN 0-9626162-4-9)
14. Feats and Wisdom of the Ancients, Time-Life Books, Alexandria, Va., 1990 (ISBN 0-8094-7675-4)
15. Irwin, C., Fair Gods and Stone Faces, St. Martin's Press, New York, 1963 (pg.188)
16. Cunliffe, B., Facing the Ocean, The Atlantic and its Peoples 8000 BC - AD 1500, Oxford Press, 2001 (ISBN 0-19-924019-1)

**Fig.1** The entrance of the tumulus of Kercado, Brittany, with at the left side menhir #47 encoding its latitude at 47°N. (Photo by authors)

# The Discovery of the Cape Verde Islands

## The Decipherment of Encoded Site Design

(The Tumulus of Kercado, Brittany, c.4500 BC)

Dr. R.M. de Jonge, drsrmdejonge@hotmail.com
J.S. Wakefield, jayswakefield@yahoo.com

### Introduction

The Tumulus of Kercado (Figs.1,2) is one of the oldest structures located in Carnac, the megalithic center of Brittany (France) (Refs.1,2,14). This tumulus, dated in archaeological literature at c.4500 BC (Ref.6), consists of a passage grave having a length of c.11m (Fig.3), covered with a mound of earth with a radius of about the same length (c.12m). At the south side of the tumulus is a stone semicircle at a distance of 5 to 10m.

The monument is located rather close to the sea, and the passage with the important burial chamber points in the WNW direction, showing an interest in the ocean. The monument and the inscriptions inside reveal geographic meanings related to crossing the Atlantic Ocean (Refs.3,4). Previous work at other megalithic sites have shown us that these structures are "mission churches" for the spreading of archaic (or Egyptian) king theory (Refs.3,4), and usually show connections with Sungod beliefs. Also, many monuments have astronomical and calendrical meanings (Refs.6,10).

# The Passage Grave (c.4500 BC):

## The Discovery of the Cape Verde Islands

The passage contains nine (9) upright menhirs and five (5) big coverstones (Fig.3, and Fig.4 with notations). These 14 stones (9+5=14) encode the latitude of Cape Verde at 14°N, the westernmost point of known continental land. The burial chamber contains eight (8) upright menhirs and one enormous coverstone, for a total of 9 stones (8+1=9), encoding the 9 Cape Verde Islands (see Fig.3-5). So the Cape Verde Islands had been discovered!

The corridor sidewalls contain 4 southern and 5 northern menhirs, corresponding to the 4 southern and 5 northern Cape Verde Islands. The corridor has 5 coverstones (see Fig.3 again), showing the shortest sailing distance from Cape Verde to these islands, 5dl, or 555km (5dl= 5 distance lines= 5 degrees of latitude= 555km). The burial chamber has a big eastern part with 6 menhirs, which resembles the shape of this archipelago (see Fig.6). The corridor and this eastern part of the chamber contain together 15 menhirs (9+6=15), showing the latitude of Brava, the SW Cape Verde Island at 15°N. The western side of the burial chamber has two upright menhirs. The SW menhir (Fig.5, stone C4) represents the SW island of Brava, and the NW menhir represents the NW island of Santo Antao. In total, the tumulus contains 17 (9+8=17) upright menhirs, encoding the latitude of Santo Antao at 17°N. Santo Antao became the westernmost island of the then known world.

**Fig.2** The entrance of the burial chamber of Kercado with its 9 enormous stones, symbolizing the discovery of the 9 Cape Verde Islands (c.4500 BC). (Photo by authors)

The Tumulus of Kercado was built at 47°N when the Nile Delta at 30°N was the center of one of the most important cultures on earth. Subtracting the 30°N of the Nile Delta from this 47°N again reveals the 17°N of the Santo Antao discovery. This latitude is also confirmed by the passage axis of 17° (Fig.5). So this tumulus has been built here at 47°N to commemorate the discovery of the Cape Verde Islands, in c.4500 BC, and in particular the westernmost island of Santo Antao at 17° N. These Islands have replaced the old known Canary Islands as the westernmost islands in the then known world. On the islands of Boa Vista and Santo Antao, their northwest capes are both labeled "Ponta do Sol" (bottom of Fig.6), the Cape of the Sun, perhaps still revealing a connection with the Sunreligion.

The passage (Fig.5) is constructed of 14 big stones, and the burial chamber is made of 9 big stones, for in total 23 stones, encoding the latitude of the Tropic of Cancer, 23°N. On midsummer day the sun is at a right angle above this latitude, and then the slow northerly movement of the sun turns around and the sun begins moving back to the south. In honor of their Sungod it appears these people wanted to cross the ocean at this holy latitude (Ref.3). Diodorus Siculus (c.50 BC) has said, that the godkings (substitutes of the SunGod) and heroes ruled Egypt for fifteen hundred years before Menes, the first king of the First Dynasty, who reigned about c.3000 BC (Ref.10, pg.337). The axis of the burial chamber makes an angle of 23° (Fig.5), which points 23°ESE, directly from Kercado to the Nile Delta, confirming the importance of this area and the 23°N latitude of the Tropic of Cancer, and the religious meanings of the monument.

A "fitting stone" is found between the two westernmost menhirs and the enormous coverstone of the burial chamber (Fig.4, cross section). It indicates that the small islet of Santa Luzia was not included in the count of 9 islands. It also adjusts the latitude of Cape Verde to a more accurate 14.5°N, rather than the 14°N we calculated above.

Between the standing menhirs of the passage grave are N-S and E-W axes (Fig.5), and other lines. These other lines make certain angles, which confirm the meaning of the monument in detail. This way of explaining Kercado is not unique, as most megalithic monuments, whether passage graves, stone circles, or alignments, can be explained in a similar way.

The monument of Kercado shows that the real start of the attempts to cross the Atlantic Ocean in the late Neolithic began with the discovery of the Cape Verde Islands c.4500 BC, because the accepted date for this tumulus, which encodes these islands, is c.4500 BC. The inscriptions of Dissignac, Brittany, have revealed that America was later discovered via the Aleutian Islands, south of the Bering Sea (Ref.3). Other inscriptions, such as those at Chao Redondo, and especially the monuments of Stonehenge and America's Stonehenge, show that a hundred years later (c.2500 BC) the discovery of America via the Atlantic was being celebrated. Thus in the megalithic age, the attempts to cross the Atlantic from Europe lasted about 2000 years. This

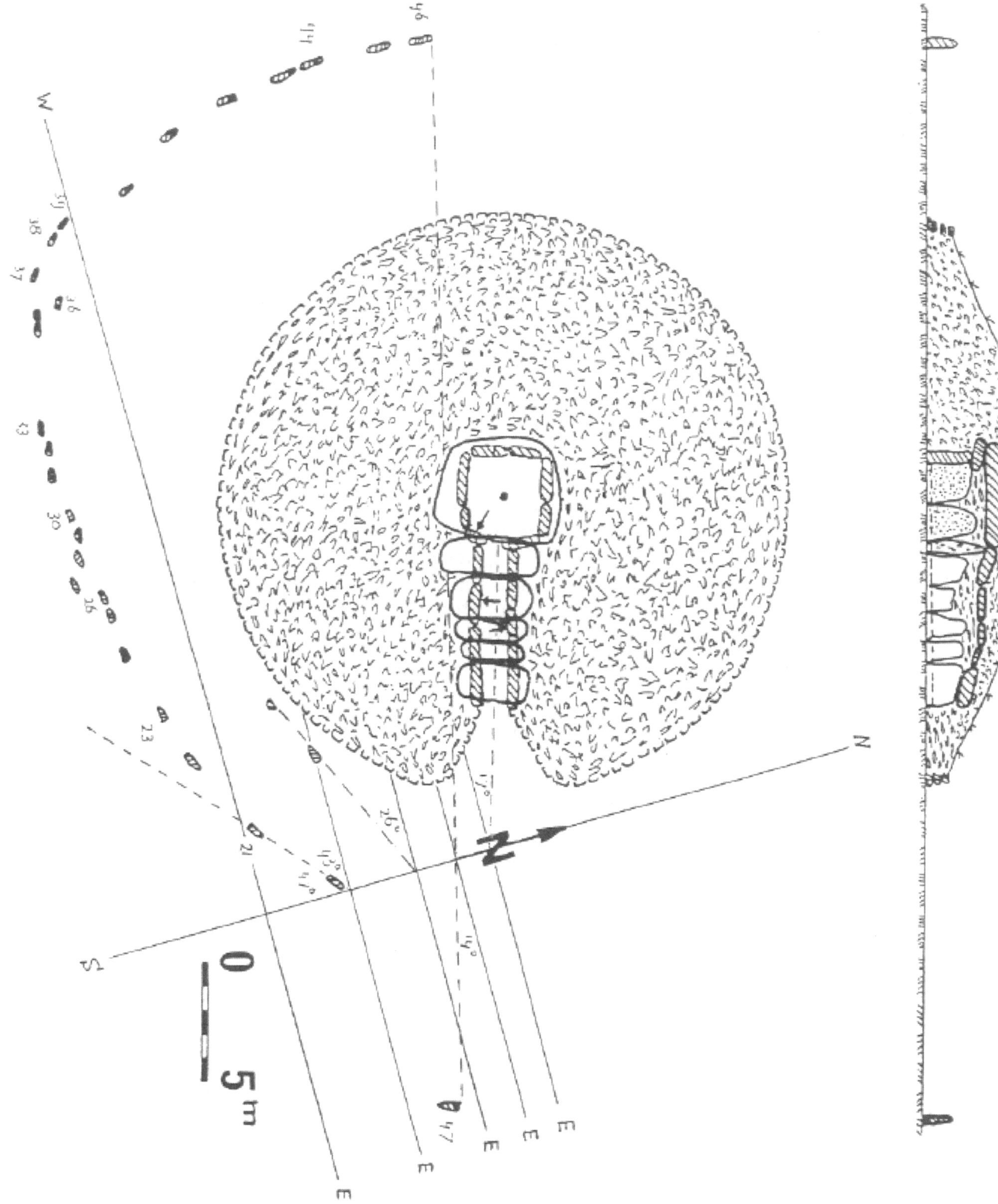

**Fig.3** Groundplan of the tumulus of Kercado, Brittany, with cross section (Refs.1,2). The 17 upright menhirs correspond to the latitude of Santo Antao, the westernmost island of the then known world, at17°N (c.4500 BC). The 30 stones of the semicircle correspond to the latitude of the Nile Delta, the center of the Northern Egyptian Empire, at 30°N (c.3500 BC).

is a time span similar to the length of the Christian Era in which we are presently living. Following that, a great deal of worldwide exploration and trading occurred in the following 2000 years. Overall, the megalithic period of five thousand years, from 6000 BC to 1000 BC was characterized by a desire to find the other side of the earth that they knew should be there. The megalithic period ended when this desire had been fulfilled, and the entire world had been explored. Unfortunately these achievements would be largely forgotten except in the mythology of the Americas, where stories recorded the arrival of the "Four Becabs", "Viracocha", "Quetzalcoatl", and other white cultural heroes from across the sea (Refs.10,11).

# The Stone Semicircle (c.3500 BC):

## The Discovery of Madeira and the Azores

The ancient circular earth mound symbolizes the spherical planet earth, in particular the then known part of it, which is the "Old World". As a consequence, the stone semicircle symbolizes the edge of it, and the discovered islands in the western ocean. As shown in Fig.3, these stones are smaller than those of the passage grave. They are placed at the south side of the tumulus, because they indicate degrees of latitude of important land points between the Cape Verde Islands and Kercado, south of the monument. Three of these stones are not placed in the semicircle (#18, #19 beside the monument, & #47 out in front). They were moved to these places later in time, as we shall explain later.

In his popular book Megalithic Brittany (Ref.14), Aubrey Burl states "these stones are the remains of a ring", but does not present evidence for that, perhaps falling for the pattern-recognition tendency of the human mind to complete circles. We are using the stone pattern as reported from other sources (Refs.1,2), and working with the evidence as it appears.

Count the stones, starting from the entrance as shown in Fig.4, following the daily movement of the sun toward the west. This seems in agreement with the spreading of the Sunreligion. Adding the 2 stones close to the tumulus, and the first 2 stones of the semicircle to the 17 standing menhirs of the passage grave, brings us to stone #21 (17+2+2= 21), labeled in Fig.4. This stone represents Cape Blanco, at 21°N, from where the Cape Verde Islands may be reached by sailing with the wind and current at your back (Ref.3). The count of the stones encodes the latitudes of all the places they are trying to remember.

The second stone of the next group, #23 (21+2= 23), provides the latitude of the Tropic of Cancer, and the center of the Southern Egyptian Empire, at 23°N, which was the center of the Sunreligion at that time. Note that from here on, the stones run due west, because this is the latitude people want to use to cross the ocean. The third stone of the next group, #26, provides the latitude of the center of the United Egyptian Empire at 26°N, this stone representing the greatest civilization on earth, emerging in this time period (c.3500 BC). The first stone of the next group (#27), has been placed at the

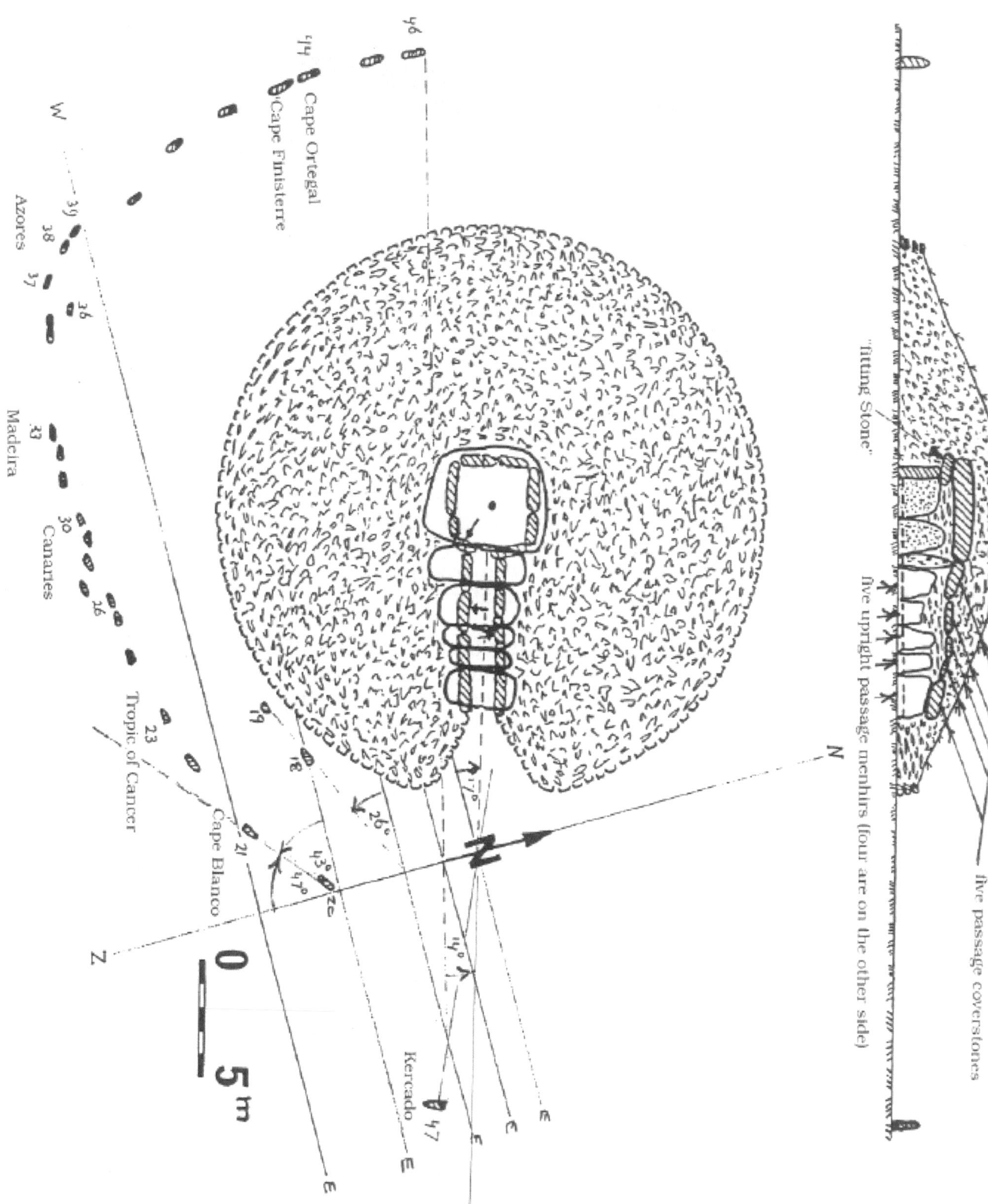


**Fig.4** Same as Fig.3, with notations added by authors. The 30 stones of the semicircle describe the latitudes of the discovered islands of the eastern Atlantic - the Cape Verde Islands, Madeira, and the Azores - in historical order.

outside of the circle, because at this latitude, 27°N, the ocean was known by them for a considerable distance offshore, because of the presence of the Canary Islands at 28°N and 29°N (menhirs #28 and #29). These islands had already been discovered c.6000 BC (Refs.3,4). The fourth and last stone of this group, #30, provides the latitude of the important Nile Delta, at 30°N, the center of the Northern Egyptian Empire.

The third stone of the next group, #33, provides the latitude of the Islands of Madeira, at 33°N. Madeira, only 450km north of the Canaries, and 700km offshore, was discovered c.4100 BC. Note that stones of the first part of the semicircle are placed at increasing distances from the mound, which means that islands were discovered at increasing distances offshore.

The third stone of the next group, #36, has been placed at the inside of the circle. As a consequence, it does not refer to a land point, but to the Strait of Gibraltar, at 36°N. It is the important entry and exit of the Mediterranean. The three stones of the next group, the East, Central, and West Azores, at #37, #38, and #39 represent the 3 island groups of the Azores, at 37°N, 38°N, and 39°N respectively. The Azores were discovered c.3600 BC, 900km from Madeira and 1400km offshore. The stones of the Azores are the furthest away from the earth mound. This means that at that time, the Azores were the westernmost islands of the then known world. They would remain the westernmost known land, the western abode of the Sungod, for the next 1100 years. That is why the 3-circle, 3-spiral, three groups of three island symbolism so pervades "megalithic art". Note also that the monument of Kercado shows the discovery of all these islands in historical order. In numeric pictograms in a stone cairn at Loughcrew, Ireland, dated 250 years later, this serial story of discovery has been repeated in the same order, including islands in the Upper North.

Now the semicircle of stones turns sharply toward the north (Fig.4). Obviously they had not succeeded in crossing the ocean from the Azores! The fourth (#43) and fifth stone (#44 ) of the next group provide the latitudes of Cape Finisterre at 43°N, and Cape Ortegal at 44°N, both on the NW coast of Spain. From this coast they did not succeed either! The third stone of the last group, #47, is presently located east of the monument, and encodes the latitude of Kercado itself in Brittany at 47°N (or including the fitting stone, at 47.5°N). The line through the first two stones, #20 and #21, makes an angle of 47°, confirming this important latitude. However, sailing west from Kercado they did not discover new land either!

Note that the semicircle which was added to the thousand-year-old Tumulus contains in total 30 new stones, in deference to the important latitude of the Nile Delta at 30°N. In Egypt, they want the people in the west to try to cross the ocean. The passage of this Kercado tumulus contains 5 big coverstones, which when combined with the 47 stones on the ground on this site, add up to 52, encoding the latitude of Dunmore Head, at 52°N, the southwest point of Ireland, and the westernmost point of known Western

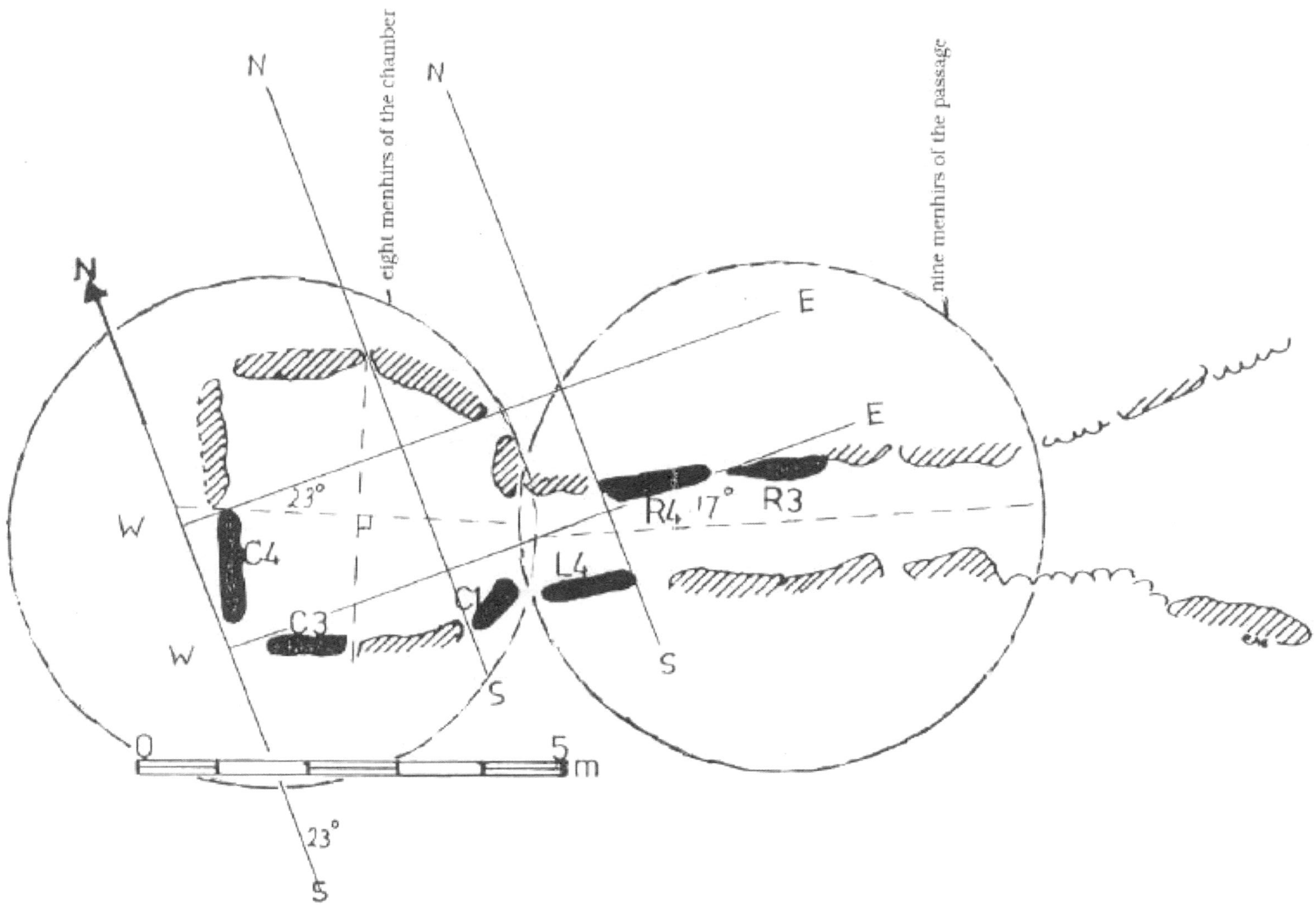


**Fig.5** Groundplan of the passage grave of Kercado (Ref.6). The passage axis points 17°WNW, encoding the discovery of Santo Antao at 17°N, and the chamber axis points 23°WNW, referring to the Sunreligion. Between the menhirs are the NS and EW axes. The dark menhirs contain geographical petroglyphs at the inside.

Europe. (That 52 is also the total number of stones used at the nearby Tumulus of Gavrinis is no coincidence.) The ocean had not yet been explored to the west from Ireland.

Finally, the burial chamber is covered by one enormous coverstone, totaling the stones on the site to 53, for the latitude of the coast of western Ireland at 53°N. This is the promising area, where explorations will focus next, as told in the numeric pictographs of Loughcrew, Ireland.

The stone semicircle dates from c.3500 BC, because it is between the discovery of the Azores c.3600 BC (shown at Gavrinis), and the later discovery of the Faeroes and Iceland at c.3400 BC (shown at other monuments, including Loughcrew). The semicircle has been added to the old monument because they know and respect that the 9 large menhirs of the burial chamber have for a millenium represented the 9 islands discovered near Cape Verde. Now, the Azores have been discovered, which also consist of nine islands. Although the Azores discovery is frequently represented in other "passage graves", you can see why it was important to update this early Kercado monument. By adding the semicircle on the outside, they could continue to honor the holy Cape Verde 3-island group in the ocean, which had just been reinforced by the discovery of another nine islands in three groups, the Azores, a reinforcing double symbolism of the western abode of the SunGod.

# The Three Moved Stones (c.2200 BC):

## The Discovery of America

From the evidence given above, it appears that the monument of Kercado was changed again, updating it again, after the discovery of America, c.2500 BC (Refs.3,8-12). Let us see what happens in the groundplan of Fig.7, if we suspect that the two menhirs (#18 & #19), that formerly might have started the semicircle were moved to their current positions near the mound, and the last menhir #47 was moved from the west side of the circle to its present position far in the east.

## The Route to Central America

The old passage grave layout was used again to add the new knowledge of the sailing route to Central America, updating and therefore continuing to honor and respect the ancient monument of the ancestors. After the discovery of America via the Aleutian Islands (c.2600 BC), the Cape Verde Islands became important for the crossing of the Atlantic Ocean (Refs.1,2). The passage of the tumulus (Fig.5) is built of 9 upright menhirs, and the big eastern part of the chamber is built of 6 upright menhirs, together 15 stones, corresponding to the latitude of Brava, the small southwestern island at 15°N. The SW menhir at the west side of the chamber represents Brava. This island became the start of the Southern Crossing. The burial chamber contains one enormous coverstone, which now would encode the latitude of the Island of St. Paul, at 1°N. It is the first island on the other side, after the Southern Crossing of the Atlantic. In accord with the Sunreligion, it is the

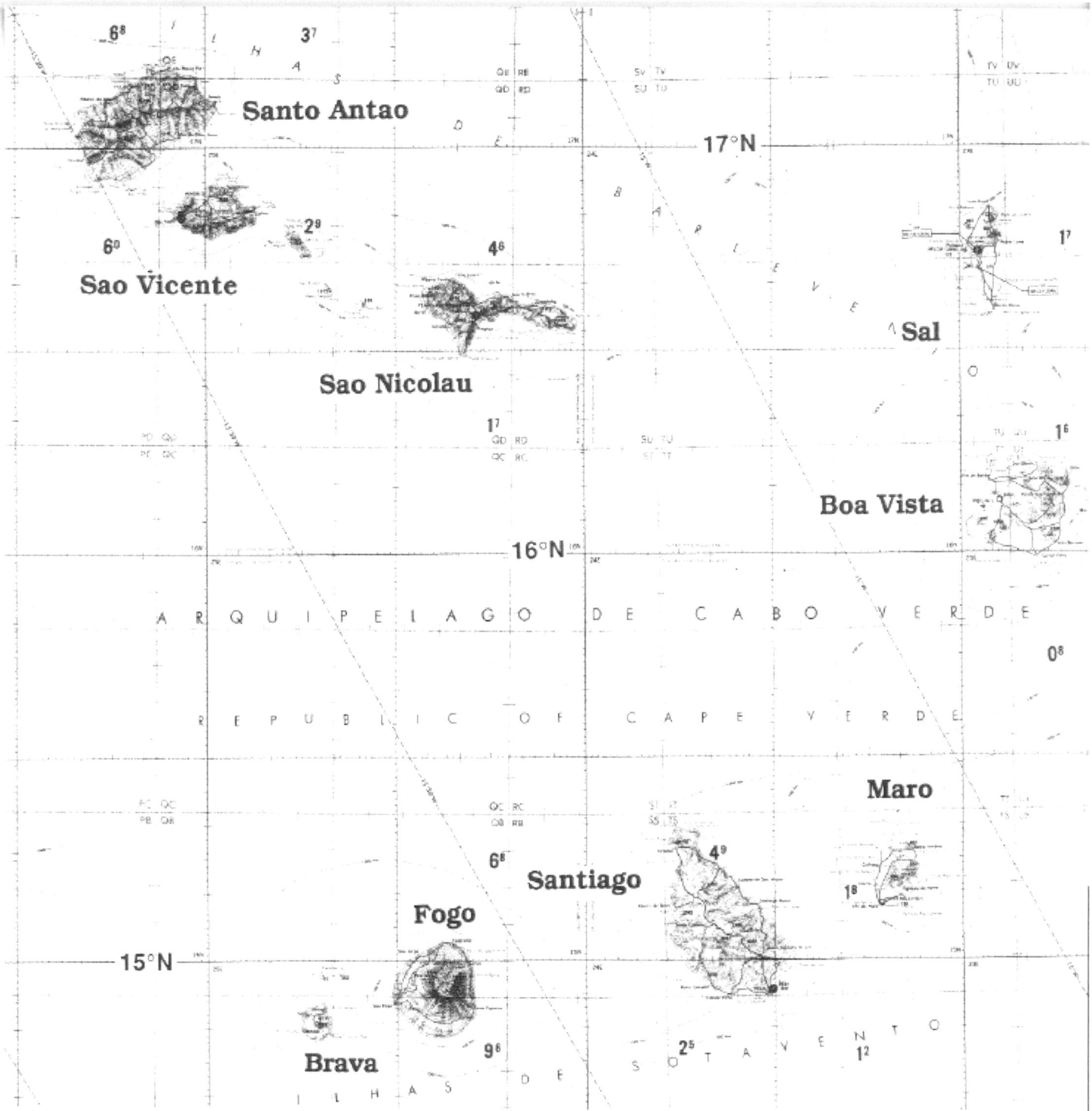


**Fig.6** Map of the Cape Verde Islands, c.600km west of North Africa, at 16°N (photoreduced air chart). The eastern part of the chamber of Kercado resembles the shape of this archipelago. North shores of Boa Vista and Santo Antao enlarged below.

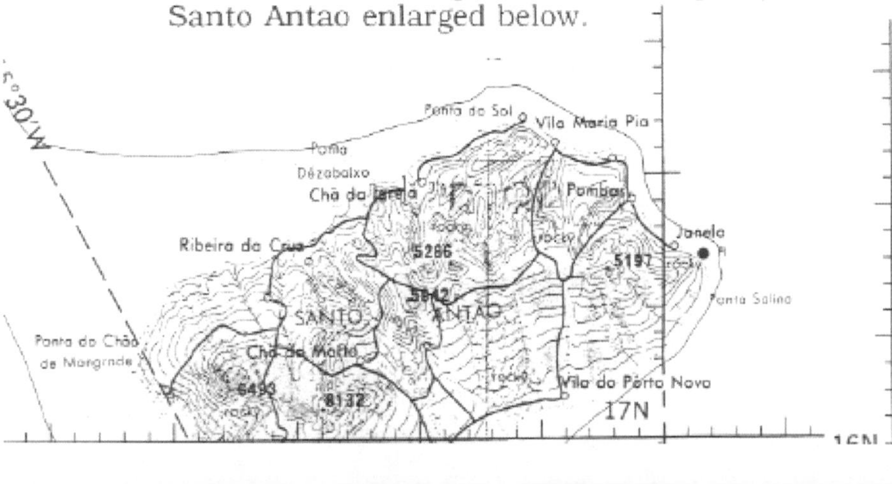


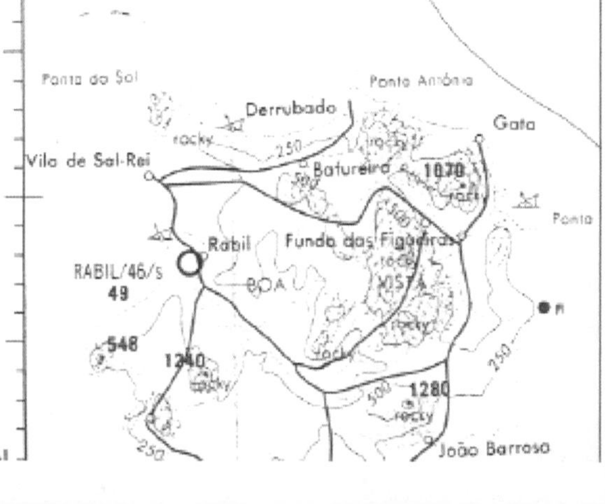

passage of the monument that is symbolic for the crossing to the Other Side. The south side of the passage contains 4 standing menhirs, corresponding to the latitude of the two islands of Fernando de Noronha, at 4°S. More important, the passage also contains 5 coverstones, corresponding to the latitude of Cape Sao Roque, the NE point of Brazil, at 5°S. But often sailors cut the corner, and sailed more to the north on the coast of South America. The north side of the passage contains 5 standing menhirs, encoding the latitude of the protruded coast of French Guyana at 5°N.

The 15 menhirs encoding the latitude of Brava now also represent Cape Gracias a Dios, the NE Cape of Honduras, at 15°N too. Together with the menhir that represents Brava, they total 16 menhirs, corresponding to the latitude of the north coast of Honduras at 16°N. These are the two degrees of latitude of the culture along the north coast of Honduras.

In total, the passage grave contains 17 upright menhirs. Together with the first menhir at the edge of the tumulus (#18 in Fig.7), the total of 18 menhirs represents the latitude of the south point of the Gulf of Campeche, at 18°N, which is encoded in Stonehenge too. Adding the next menhir along the edge gives the next latitude in this area, 19°N. These are the two degrees of latitude of the Olmec civilization around the Gulf of Campeche, which orthodox archaeology claims arose suddenly and independently from nothing to sun-worshipping step-pyramids at c.2300 BC. The two stones are placed close to the earth mound, because of the importance of this place. The line through these two stones makes an angle of 26°, referring to the latitude of the United Egyptian Empire at 26°N, showing that contact with the government of Egypt was important.

The 17 upright menhirs of the passage grave correspond to the latitude of the center of the United Central American Empire, halfway between the north coast of Honduras and the Gulf of Campeche, at 17°N (Refs.10-13). So, this empire extended over 5 degrees of latitude, which is strongly confirmed in the famous monument of Stonehenge in England.

## The Route back from Central America

The new positions of the three moved menhirs (#18, #19, & #47) have been chosen with great care. They are in accord with the ancient system of symbolic representation, and yet update the monument, while keeping damage to the original semicircle to a minimum. While originally, the tumulus was constructed to commemorate the discovery of the Cape Verde Islands in c.4500 BC, and the stone semicircle was added after the discovery of the Azores c.3600 BC. The monument was even later modified to accomodate the encoding of much later information about sailing routes to the Americas at c.2200 BC. This remarkable over-encoding, or updating of monuments over amazingly long spans of time is also shown in the inscriptions of Dissignac, Brittany, the construction of Stonehenge in England, and to a lesser extent the petroglyphs of Loughcrew, Ireland, and the design of America's Stonehenge in New Hampshire.

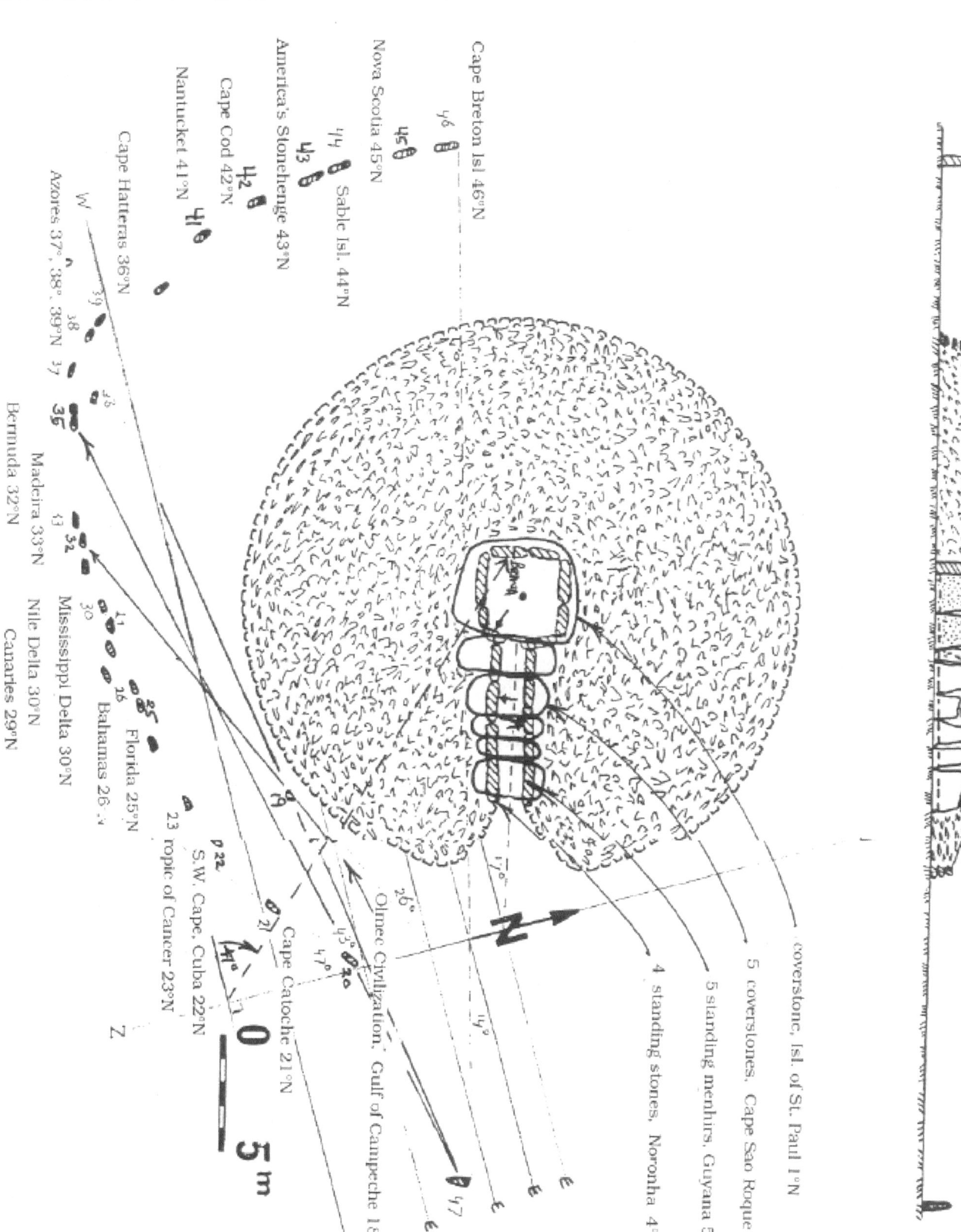


**Fig.7** Groundplan of Kercado. The stones #18, #19, and #47 were moved in c.2200 BC to update the monument a second time. The passage grave describes the sailing route from Africa to Central America. The 17 upright menhirs and the 30 stones of the semicircle indicate latitudes of important places around the Atlantic.

Because of the governing winds and currents, it is impossible to return by sailing via the Southern Crossing. For that reason the start of the stone semicircle describes the return route in Central America to the north (see Fig.7). The second and last menhir (#21) of the first group provides the latitude of Cape Catoche, the NE point of Yucatan, at 21°N (17+2+2=21). This is the point of departure for crossing to Cuba. The first menhir of the next group, #22, gives the latitude of the SW cape of Cuba, 22°N (21+1= 22). Both stones, parallel to the menhirs close to the mound (that is, 18 & 19), also provide the length of the crossing, c.2dl, (=2°of latitude) = 222km. The next stone, #23, provides the latitude of the north coast of Cuba, 23°N (22+1=23), and the Tropic of Cancer, important in the Sunreligion.

Sailing from the north coast of Cuba we cross the sea to the peninsula of Florida at 25°N (see stone #25, Fig.7). Stone #26 is now the latitude of Bimini and Great Abaco in the Bahamas at 26°N. So now we see the stone circle is embellished with new meanings, fitting in, and sometimes overlapping the older ones. When leaving Central America, the stone semicircle now provides important new waypoints on the return routes from North America to the Old World. In the course of the second millennium BC, these became important as a point of departure for crossing the ocean via Bermuda (1300km).

Stones of the next group, including #29, continue to provide the latitude of the two eastern Canary Islands at 29°N, again of some importance, for a safe return route to the NW coast of Africa. The last stone of the group, #30, gives the latitude of the south coast of the U.S., with mouth of the Mississippi Delta, in particular, at 30°N. which is also the latitude of the Nile Delta at 30°N, revealing how the SunGod himself uses a doubling of physical characteristics in encoding important places. The petroglyph in Chao Redondo, North Portugal (c.2200 BC), and the monument of Stonehenge in England both also show that the Mississippi became an important goal of voyages.

Sight lines are used to encode new known places. A line through stones #18 and #19 (in Fig.7) points to menhir #32 in the next group, reveals the latitude of Bermuda, 1000km from the American Coast, at 32°N. Line 18-19 makes an angle of 26°, referring to the center of the United Egyptian Empire, at 26°N. So, thanks to the great efforts of the government of Egypt, Bermuda had already been discovered by c.2200 BC! For a long time this small island was important for the return of ships from Central America as seen at American Stonehenge. The fact that Bermuda is clearly indicated proves that the positions of the three moved menhirs were firmly established only after the discovery of Bermuda in c.2200 BC.

A line drawn in Fig.7 between menhir #47 and stone #20 points to menhir #35 of the next group, providing the latitude of Cape Hatteras South, at 35°N. Compared to Bimini, this place is even closer to Bermuda (1000km), and for that reason, and probably for better winds, was often a point of

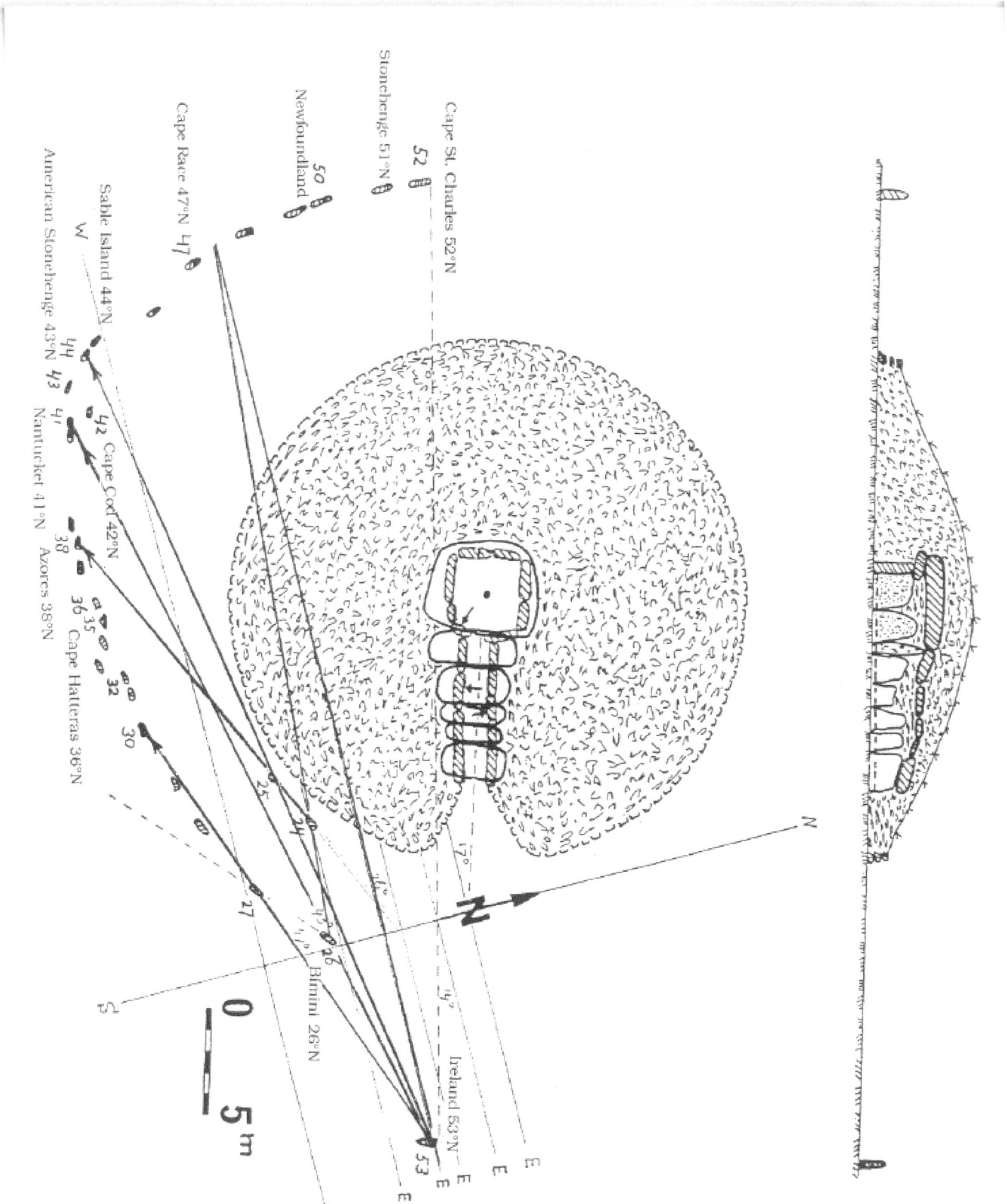


**Fig.8** Groundplan of Kercado, with stones re-numbered by the authors, starting with the total 23 menhirs of the Tumulus (not just the 17 vertical stones used before). Here the moved stones are now #24, #25, and #53. The 23 menhirs of the passage grave encode the latitude of the Tropic of Cancer, symbol of the Sunreligion, at 23°N. The 23 menhirs and the 30 stones of the semicircle indicate latitudes of important places around the Atlantic a second time.

departure for Bermuda. The next stone, #36, gives the latitude of Cape Hatteras North, at 36°N, important for the direct crossing to the Azores. In Chao Redondo (North Portugal) is a beautiful petroglyph of these sailing routes (c.2200 BC).

Line 47-19 points to menhir #38, representing the Central Azores at 38°N. These islands form the famous archipelago in the middle of the Atlantic Ocean. The most important megalithic center in the New World, America's Stonehenge, NH, shows that the Azores played a predominate role in oceanic crossings after the discovery of America c.2500 BC.

From stone #39, the circle of stones bends strongly to the north, because here are the departure places for the "Northern Crossings" to the Old World. The stone #41 encodes the latitude of Nantucket Island, at 41°N. This is the first important departure place, oriented on Newfoundland, for crossing the ocean via the Azores. The next group contains three important stones. They represent Cape Cod at 42°N, America's Stonehenge and Cape Sable, the south cape of Nova Scotia, at 43°N, and Sable Island at 44°N, respectively. The stone #44 has a more westerly position, because the island is more than 150km offshore. From all of these places, the route to the Azores passed just south of Cape Race, Newfoundland, though from Sable Island, they may have passed further to the south. The reason for this is that the shortest distance across the Atlantic is from Cape Race to the Azores (c.2000km). This will be especially clear to you on Great Circle Sailing Chart, where the projection is not Mercator, but focused upon a square of latitude and longitude in the center of the chart. The result is that a straight line drawn between any two points on the chart represents a great circle across the Earth's curved surface, and is therefore the shortest possible track line between the points. Before the discovery of Bermuda, c.2200 BC, these were the only points of departure for the crossing of the ocean to the Old World. In the centuries after the discovery of Bermuda, these northern routes continued to be used from northern ports, because sailing south against the Gulfstream was difficult, as well as a longer route. The newer routes from Cape Hatteras or the Bahamas to Bermuda were convenient for ships sailing from the Caribbean, and were probably more comfortable by being warmer.

The last menhir is stone #47, now located far in the east. Originally, before the discovery of America, this stone was located in the west, above stone #46. As a consequence, it clearly represented Cape Race at Newfoundland (new-found-land), at 47°N, because this important place should have a stone. Attempts to cross the North Atlantic Ocean started here. However, later on, departure places were chosen at more southern locations, and, as discussed, after the discovery of Bermuda, even Cape Hatteras and the Bahamas became points of departure. This stone #47 also represents Kercado in Brittany, at 47°N (or, including the "fitting stone", at 47.5°N). A line drawn through the locations of the last two menhirs, #46 and #47, makes an angle of 14° (Fig.7), referring to the latitude of Cape Verde, the westernmost point of

**Fig.9** Megalithic Tumulus of Knowth, Ireland, undergoing professional archaeological exploration and reconstruction during the summer of 1999. Following redevelopment and beautification of the site, it probably will be added to the adjacent Irish National Heritage tourist complex of Newgrange, which has also been substantially reconstructed.

known continental land, at 14°N. This was celebrated in the design of the passage grave for the Cape Verde Islands (c.4500 BC), celebrated in the island encodings of the stone semicircle (c.3500 BC), and the Cape Verdes served as the start of the sailing route encoding when the three stones were moved (c.2200 BC)!

The old passage grave of Kercado possesses 17 upright menhirs and 6 coverstones, together 17+6= 23 big stones, corresponding to the latitude of the Tropic of Cancer, 23°N (c.4500 BC). We should not forget, that this passage grave was a "mission church" of the archaic king theory, of the divine descendance of kings from the SunGod! The total number of stones in the central monument also corresponds to the latitude of the Southern Egyptian Empire at 23°N, the center of the SunGod religion (c.3500 BC). At 2200 BC, this is still the center of the Sunreligion, and also in Central America at midsummer day, at the holy 23° of latitude, the sun is at right angles above the Tropic of Cancer. The Central American construction of Sun-Temples, more than 2000 years after the erection of this monument of Kercado (c.4500 BC), shows continued belief in the SunGod for a very long period of time, and expansion of the ideas to the New World. (Ironically, this Divine Right of Kings was later reversed by the Americans through the separation of church and state, an idea which has spread around the world in the reverse direction.)

## Another Route Back from Central America

In the Kercado groundplans we have been working with so far, the stones of the semicircle have been numbered starting with the 17 upright menhirs of the tumulus. In Figure 8, let us look at the result if we number the semicircle stones starting with the total 23 menhirs of the Tumulus. This reveals another encoded route back from Central America.

From the north coast of Cuba at the Tropic of Cancer (23°N), we start to cross the sea to the peninsula of Florida. The total number of Menhirs of the old passage grave and the two stones near the edge of the tumulus provide the latitude of the south point of Florida, at 23°+2= 25°N. The two stones also give the length of the crossing, c.2dl= 2° of latitude= 222km.

The first stone of the semicircle gives the latitude of Bimini and the Abacos in the Bahamas, as well as the center of the United Egyptian Empire at 26°N. There are several indications in other sites, including America's Stonehenge in New Hampshire, that in the course of the second millennium BC that these northern Bahama Islands, across the Gulfstream from Florida, became the most popular departure point for Bermuda when sailing from Central America.

The new alignment line 53-27 points to the last stone, #30, of the next group, indicating the latitude of the mouth of the Mississippi River, and the Nile River, both deltas at 30°N. The Nile Delta was the economic center of the Northern Egyptian Empire. The Mississippi Delta was often an important goal of voyages, from where Poverty Point, fortified Aztalan, the City of Cahokia with its huge pyramid mounds, and other cities later developed upriver.

The second and last stone of the next group, #32, encodes the latitude of Bermuda, 1000km from the American coast, at 32°N. For a long time Bermuda was important for the return of small vessels to Europe from Central America. Again, the fact that Bermuda is clearly indicated in the semicircle shows that the positions of the three moved menhirs were established only after the discovery of Bermuda c.2200 BC.

The last two stones of this group, #35 and #36, provide the latitudes of Cape Hatteras South and North, at 35°N, and 36°N, respectively. Other petroglyphs, such as Chao Redondo (North Portugal) illustrate these sailing routes.

Line 24-25 points to menhir #38, indicating the latitude of the Central Azores at 38°N. The three stones of this group refer to the three island groups of the Azores, at 37°N, 38°N, and 39°N, respectively. They form the famous archipelago in the middle of the Atlantic Ocean. The line through the two menhirs along the edge of the tumulus cuts the stone group of the Azores at an angle of 26°, encoding the United Egyptian Empire at 26°N, and commemorating the discovery of the Azores c.3600 BC through the efforts of the government of Egypt.

Line 53-26 points to the second stone of the next group, #41, representing Nantucket Island at 41°N (Fig.8). The third and last stone, #42, provides the important latitude of Cape Cod at 42°N. At times, these were important places of departure for the "Northern Crossing" to the Azores.

The first stone of the next group, #43, is the furthest away from the mound of Kercado. It provides the latitude of America's Stonehenge (Ref.7) in New Hampshire at 43°N, and of Cape Sable, the south cape of Nova Scotia ("New Scotland"), both at 42°+1= 43°N. A visit to America's Stonehenge always makes it easier to cross the ocean safely! Line 53-25 points to the second stone, #44, providing the latitude of Sable Island, c.2dl= 222km offshore, at 42+2= 44°N. In the centuries before and just after the discovery of Bermuda, c.2200 BC, these were the most important points of departure for the crossing of the ocean to the Old World. The sailing distance over open sea to the Azores was much shorter from Bermuda than other southern departure points, Cape Hatteras, or the Bahamas. (All these latitudes can be seen again in the site angles at America's Stonehenge.)

Previously, menhir #53 on the east side of the tumulus, (which was #47 in Fig.7), represented Cape Race and Kercado in Brittany itself at 47°N. This big, isolated menhir is the moved menhir, which symbolizes Kercado, at the other side of the ocean. The east-west line running through #53 runs between stones #47 and #48 in the semicircle, and line 24-26 intersects there. With this method of counting the stones, the latitude of menhir #53 is 47+1/2= 47.5°N, which is correct. So again, this menhir represents the monument where we stand at the moment! This is an encoded confirmation that we are counting the stones of the semicircle in the right manner. Stone #53 is furthest away from the earth mound, which means it is the last placed stone in the monument, to accommodate the new lands discovered at such large distances. As a consequence, this stone is important for the last story of the monument.

From the stones #46 and #47 on, the semicircle bends off strongly to the north, because by these later dates, the higher latitudes are of less importance for the crossing the ocean. Stone #47 provides the latitude of Cape Race on Newfoundland, at 47°N. It is the easternmost point of North America, at the same latitude of Kercado, and the sailing distance from Cape Race to the Azores is only c.2000km. Yet, as said before, there are strong indications that sailors often preferred to sail via Sable Island, at 44°N.

The third stone of the next group, #50, gives the big latitude of 50°, which connects the NE coast of Newfoundland with Cornwall in South England, both at 47+3= 50°N (Fig.8). Above this latitude the ocean narrows. A long time ago, the megalith builders discovered Iceland and Greenland up there
c.3300 BC (Refs.3-6). However, they did not succeed in reaching North America at that time, as explained by the picture writing at Loughcrew.

The first stone of the next group, #51, provides the latitude of Stonehenge in South England (Ref.5), the monument built to commemorate the discovery of America, at 50+1= 51°N. This is also the latitude of the southern Aleutian Islands, where America was first discovered, c.2600 BC, as shown in the Dissignac petroglyphs. The second stone of this group, #52, provides the latitude of Cape St. Charles, the east cape of the mainland of North America, and of Dunmore Head, Ireland, the westernmost point of West Europe, at 50+2= 52°N. It is also the latitude of the mouth of Belle Isle Strait, of some importance for people who want to cross to Europe via the Upper North, and via Greenland and Iceland, because they can sail in more protected waters for a while west of Newfoundland, while sailing to the north. But this is a crossing in fog, cold, and icebergs, which is to be disuaded.

The last stone of the group, #53, has been placed far in the east, c.2200 BC. It represents the latitude of Ireland, 53°N, at the other side of the Atlantic. For crossing the ocean from Cape Race, Ireland is an alternative for the Azores. The direct sailing distance is about 50% longer, but all the way, the wind and current are off your stern as you ride the Gulfstream.

## Discussion

If these encodings seem too complicated to you, remember the length of time they had to think about it, compared to you. Like our computer chips, these monuments have seemed magical because of things going on below our level of understanding. It now appears that the multi-layered encodings of these stones constitute mneumonic devices that enabled accumulations of knowledge of the world to be passed down through many generations. These monuments should therefore be seen as knowledge-reservoirs, or libraries of prehistory. We have found recordings of their geographic explorations related to the spreading of the Sunreligion. Perhaps others may be able to find other kinds of information encoded.

We are fortunate that this monument of Kercado is still in excellent shape. The Romans are reported to have pulled much of Stonehenge apart in order to interfere with its Druidic use as a "school of learning". Figure 9 shows photos of the monument of Knowth, in Ireland, when we visited in 1999. Professionals were exploring the monument with a steamshovel. It might be viewed as unfortunate to have disturbed a valuable site to this extent.

Because of the mana, or power, of these "cairns", they have been used to bury people in, or store cremated remains. Most are solar-oriented, so the rays of the SunGod enter once a year down the passage, and light up symbols carved on the back walls. Perhaps this impregnates the Earth Goddess, and brings the Spring. Probably for this reason, the Celts felt it was productive to have sex inside. These are peripheral uses to the primary role of these sites as libraries of knowledge for those trained to use them. They are mnemonic devices, not buildings with shelves of books, but nevertheless they are tools designed to assist the human memory in prehistory. They memorialize the achievements of these people, so long forgotten.

# References


1. Briard, J., The Megaliths of Brittany, Gisserot, 1991
2. Batt, M., and others, Au Pays des Megalithes, Carnac-Locmariaquer, Jos, 1991 (French)
3. Jonge, R.M. de, and IJzereef, G.F., De Stenen Spreken, Kosmos Z&K, Utrecht/Antwerpen, 1996 (ISBN 90-215-2846-0) (Dutch)
4. Jonge, R.M. de, and IJzereef, G.F., "Exhibition: The Megalithic Inscriptions of Western Europe" (1996)
5. Richards, J., Stonehenge, English Heritage, 1992
6. Twohig, E. Shee, The Megalithic Art of Western Europe, Clarendon Press, Oxford, 1981
7. Lambert, J.D., America's Stonehenge, An Interpretive Guide, Sunrise Publications, Kingston, N.H., 1996 (ISBN 0-9652630-0-2)
8. Ferryn, P., "5000 Years Before Our Era: The Red Men of the North Atlantic", NEARA Journal, Vol. XXXI, No. 2 (1997)
9. Fell, B., America BC, Pocket Books, Simon & Schuster, 1994
10. Bailey, J., Sailing to Paradise, Simon & Schuster, 1994
11. Thompson, G., American Discovery, Misty Isles Press, Seattle, 1994
12. Mallery, A., The Rediscovery of Lost America, The Story of the Pre-Columbian Iron Age in America, Dutton, NY, 1951 (ISBN 0-525-47545-1)
13. Peterson, F.A., Ancient Mexico, (1959)
14. Burl, A., Megalithic Brittany, Thames and Hudson, GDR 1985

# The Discovery of Madeira and Rockall

(The Tablet of Paredes, Galicia, Spain, c.4100 BC)

Dr. R.M. de Jonge, drsrmdejonge@hotmail.com
J. S. Wakefield, jayswakefield@yahoo.com

## Introduction

In 1936, a beautifully inscribed stone ( Fig.1), was found two meters from a megalithic tomb at Paredes (East Galicia, NW Spain). It is now in the Provincial Museum at Lugo (Ref.1). The flat granite stone has a length of 31.5cm, a width of 21cm, and a thickness of 3-4cm. The surface areas of both sides are well made and polished. However, one side is smoother, and more nicely engraved. On the basis of these two features, it can be concluded that it must be the front side. Previously, this stone has been described as an anthropomorphic stelae, or humanoid figure (Ref.1), but as is the case for most megalithic inscriptions, it has a geographic meaning.

## The Front Side: The Discovery of Madeira

At the right side of the stone (Fig.2) a carved line can be identified, representing a coastal map of the west part of the Iberian Peninsula (NW Spain and Portugal, Ref.2). The line shows a good image of the real coastline when compared with a contemporary map (Figs.3&4). It includes the area in the north of Iberia, where this stone was found, as we should expect. In the center of the inscription a horizontal line runs from right to left across the whole width. This line can only be the 40th latitude line, which divides Iberia in two, more or less equal parts. The peninsula is situated between the Strait of Gibraltar at 36°N in the south, and the Bay of Biscay at 44°N in the north, or roughly between 35°N and 45°N. As a consequence, the long lower and upper lines that run horizontally and parallel with each other represent the 30th and 50th latitude lines, as shown in Fig.4.

Southwest of the Iberian Peninsula the long-known Canary Islands are located at about 28.5°N. One of the easternmost islands, Fuerte Ventura, is visible from the shore of Africa on a clear day (c.100 km), and the archipelago is first shown in megalithic petroglyphs before c.5500 BC (Ref.2). Both the Canaries and their latitude of 28.5°N are represented by the lower edge of the stone. Due north of the main island of Tenerife,(shown as the carved notch at the bottom of the stone), and at a distance of 1.5dl (=166km), the small Selvages Islands are situated. The northerly route towards them, but especially these islands themselves at exactly 30°N, are clearly indicated on the stone. These are situated at the crossing of the vertical line and the 30th latitude line. North of them the vertical line is divided into three separate parts, which are clearly visible. Each part represents a degree of latitude. Due north of the Selvages Islands (at 30°N), the southern-most island of the small Desertas Islands (belonging to Madeira, top map, Fig.5) can be reached, at a sailing distance of 255km (c.2°= 2 moira, or 222km). In the inscription, this southernmost island at

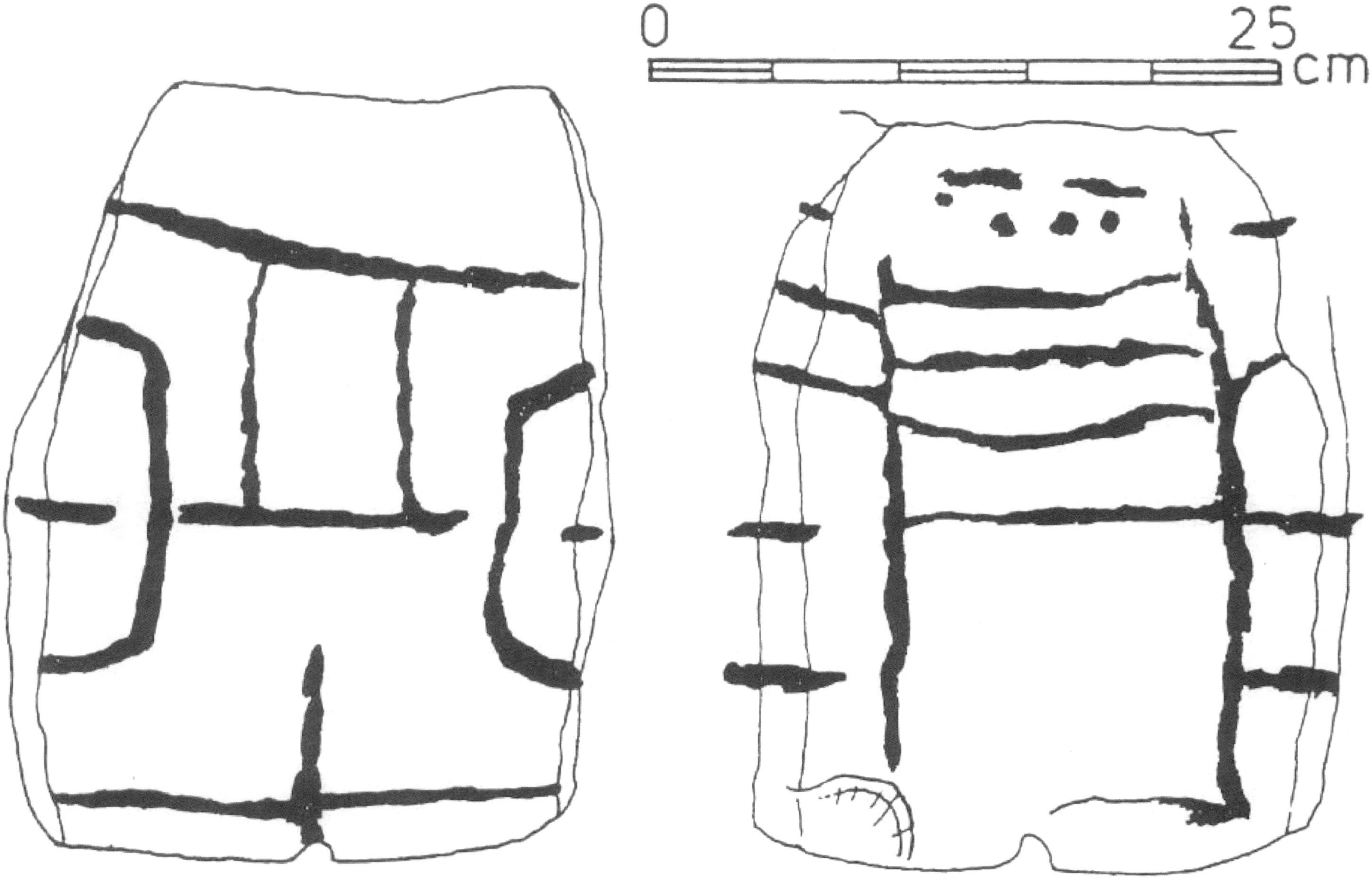


**Fig. 1** The Tablet of Paredes, Galicia, NW Spain (Ref. 1). **Left**, the front side, showing the discovery of Madeira at 33°N (c.4200 BC). **Right**, the back side, showing the discovery of Rockall at 90-33 =57°N (c.4100 BC).

30+2=32°N has been clearly indicated. About 1° (111km) further north, the main island of Madeira is located at 32+1=33°N. Madeira is situated 3° north of the Selvagens Islands (30°N), at the latitude of 30+3=33°N. Madeira, and the smaller island of Porto Santo, are indicated by the tip of the vertical line. Since this is the center and focus of the stone, it suggests that the inscription was made because of the discovery of these islands of Madeira.

In the inscription, the distance between the west coast of Iberia and the meridian through Madeira appears to be approximately one half of a big distance line, or 1/2 DL =5°of latitude =555km. And indeed, the islands of Madeira are located 600km west of the Iberian Peninsula. However, we think that at the time this inscription was made (c.4200 BC), such a long distance could not be easily covered over open sea, because we see no evidence for it. In the sailing route from the Canary Islands to Madeira just described, the longest sailing distance over open sea was 255km, which is about 2°, or 222km. The first big vertical DL is shown off the coast of Iberia about 1/4 DL, or 2.5°, or 277km. This indicates that the waters around the Iberian Peninsula (and elsewhere in the surroundings) were completely explored over this distance. After the discovery of Madeira, the seas to the east and west were also explored to a distance of 2.5°, but the rest of the ocean was still largely unknown.

The inscription not only reveals what was known about the ocean, but also reveals the subjects people were thinking about. They speculated that there would be an unknown opposite side to the ocean, a kind of mirror image of the Iberian Peninsula, as shown at the left side of the stone. This mirror-imaging is found in many other megalithic petroglyphs created prior to the actual discovery of the other side in 2500 BC. If you look at a lot of megalithic petroglyphs, you will see that all inscriptions after c.2500 BC show actual features of the other side (Ref.2). In this petroglyph at Paredes, it is clear that they assumed that with the discovery of Madeira, they were halfway across to the fantasized other side of the ocean. Note, that the distance between Madeira and the supposed opposite side equals 1/2DL= 5° of latitude, or 555km. This means, that after the discovery of the Cape Verde Islands at the same distance offshore, they were still not able to cover a longer distance over open sea. The spacing of the distance lines shows they estimate the width of the Atlantic Ocean to be about 1/4 +1/2 +1/4= 1DL, or 10° of latitude, or 1111km. It took these megalithic people almost two thousand years more to discover that the true width of the ocean was closer to 4DL, or 40° of latitude, or 4444km.

Note that the Iberian Peninsula is shown over a breadth of 1/4DL. Paredes, where this stone was found, is about 255km from the west coast, so this carving site is within the scope of the petroglyph, which is to be expected, in our experience. Also note that the geographical position of Madeira in the inscription has been indicated too much to the north by about 3° (3dl), compared with the Strait of Gibraltar, at 36°N. They could read latitudes

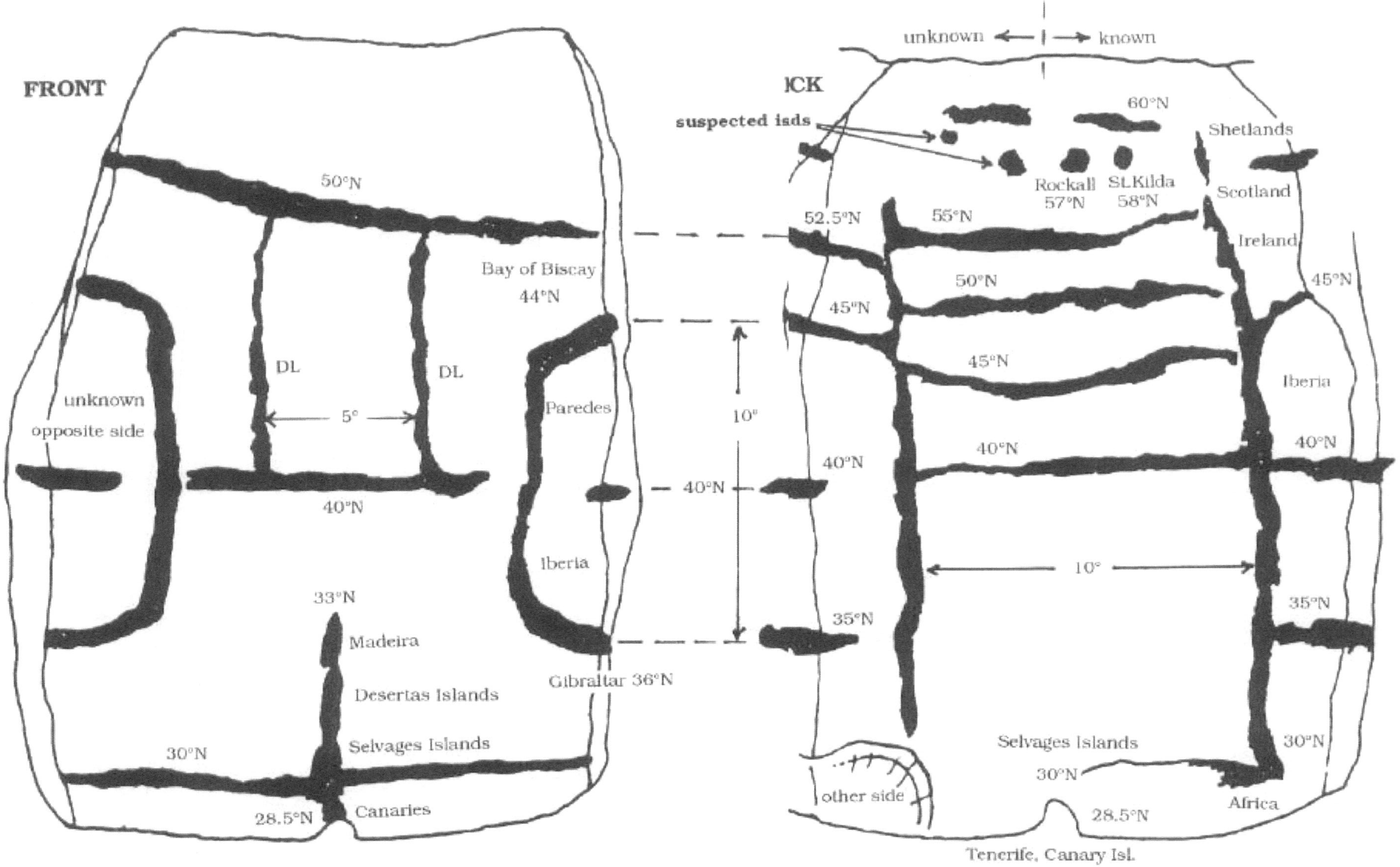


**Fig. 2** Fig.1, with geographic meanings. At›th sides, Right, the west portion of Iberia, and Left, the fantasized otheiide of the ocean. Shown horizontally, are latitude lines, and vertically, ċance lines. The width of the ocean is estimated to be about 1DL =1111knc.4150 BC).

well enough to know that Madeira was at 33°N, as also shown, so this error must have been done on purpose, and for political reasons! This probably is related to their association of the westernmost islands with the western abode of the SunGod (Refs.2,9). They consider the newly discovered archipellago of Madeira as Iberian Islands, not belonging to Africa.

The petroglyphs of Dissignac and Gavrinis (Brittany) and of Loughcrew (Ireland) show that Madeira was discovered after the Cape Verde Islands (c.4500 BC), but before the Azores (c.3600 BC) (Refs.1-4). The details at these sites indicate that the most probable date for this inscription at Paredes, and so for the discovery of Madeira is c.4200 BC. The megalith builders considered them always as consisting of two islands, Madeira, and Porto Santo, because they are often shown as twin symbols in petroglyphs and monuments. Officially, Madeira was indicated for the first time on a map from Florence, Italy, in 1351 AD, with the name Isola di Legname, or "Wood Island", which is still the meaning of the word Madeira.

## The Backside: The Discovery of Rockall

The latitude lines on the front side of the Paredes Tablet and the drawing of the Iberian Peninsula, and the imagined opposite side continue to run around the side edges of the tablet to the back side (Figs.1,2). So you can see the latitude lines corresponding to 35°N, 40°N, and 45°N running right to left, indicating the Iberian Peninsula, the imagined opposite side, and the 40° latitude line a second time. Near the bottom of the stone, the 30th latitude line runs from the mainland of Africa to the Selvages Islands (30°N, thin line), just above the cut notch at the center of the bottom edge of the stone (Tenerife, Canary Islands, 28.5°N). At 4100 BC, people thought these islands were in the middle of the ocean.

Slightly different from the front side, the width of the ocean at the back side is estimated to be 1DL =1111km at all latitudes, a width equal to the height of the Iberian Peninsula, 10° (=45°-35°). As will be discussed later, and as shown in other petroglyphs (esp. Kercado), these megalithic geographers are using a grid system of squares over a spherical surface, like netting laid on a beach ball, so latitude DLs equal longitude DLs, with distortions along all the edges. Note in the left corner of the carving, the far side of the ocean has been carved with a semi-circular inscription (the thin lines in the lower left). This has been added here, because they would prefer to cross the ocean and find land at the latitude of the Nile Delta, at 30°N, in honor of the SunGod.

However, the inscriptions on the backside of the tablet do not relate to the south in particular, but more to the north. We should not forget that this stone was found in Paredes, East Galicia, which is in the NW part of Spain. Obviously, they are interested in the northwestern direction. In the center of the stone, between the well-known continental land and the supposed other side, four latitude lines across the ocean are shown, corresponding to 40°, 45°, 50°, and 55°N, respectively (see also Fig.6). Between the last two

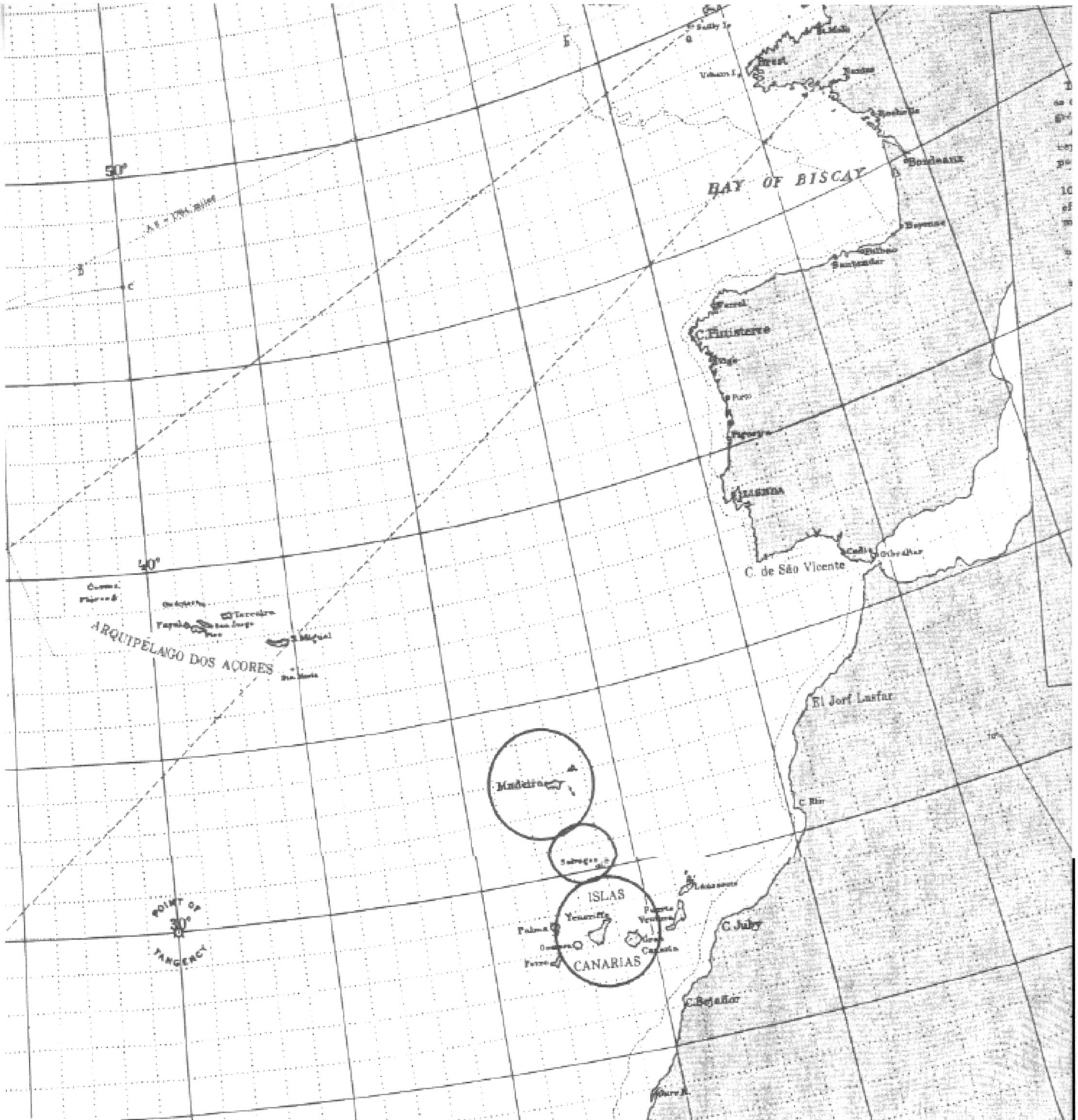


**Fig. 3** Great Circle Sailing Chart of the ocean around Madeira, southwest of the Iberian Peninsula (Ref.8).

lines of 50° and 55°N, Ireland is located (vertically carved piece of line). This island belongs together with Iberia to the westernmost areas of Europe. Dunmore Head in Ireland, at c.52.5°N, is the westernmost point of the whole of Western Europe. It is quite possible that from that coast the opposite side could be reached. For that reason, at the left side of the stone, the corresponding latitude line has been engraved separately (extra line at 52.5°N (Fig.2).

Above Ireland, at the upper right side of the stone, a vague, vertical piece of line is shown. It is Scotland, the Hebredes, and the Orkneys, which string out to the Shetlands at the 60°N latitude line. At the top center of the glyph, the 60th latitude has been indicated by two separate pieces of line, right and left, for the known and unknown sides of the ocean. Below the right one, and at the right side, is a small dot. This represents the Islet of St. Kilda, discovered west of the Hebrides, c.5 sq km in area, at 58°N, 70km offshore (Figs.5,6). According to an inscription in Barnenez, North Brittany (Refs.1,6), St. Kilda had been discovered c.300 years prior to the Paredes Tablet, c.4400 BC. A bigger dot was carved a bit lower and further west.

This is the excitingly western islet of Rockall (Figs. 5,6), at 57°N and 320km west of St. Kilda. Note that the sailing distance between these islands of 320km is a little larger than the sailing distance to Madeira of 277km on the other side of the tablet, a lengthening of the sailing voyages of exploration. Note that the discovery at 57°N was at the complementary latitude of Madeira, at 90°-33°=57°N, which may have seemed magical to them. Near Stonehenge in South England is Vespasian Camp, which dates from the Iron Age. Nearby, a Long Barrow is situated, which is oriented on the old monument of Robin Hoods Ball (c.4000 BC, Refs.2,5). These two monuments make an angle of 45°NW, pointing to this islet of Rockall. Thus this Long Barrow was made because of the discovery of Rockall too, at about the same date as the Paredes tablet (c.4100 BC).

The inscriptions right and left at both sides of the Paredes Tablet are almost symmetrical. The two other dots at the left side below the 60th latitude line indicate the possibility of two other islands in the western part of the ocean. These "suspected" islands provideed the megalith builders with hope for a northern sailing route to the opposite side of the Atlantic. The whole backside of this tablet has been engraved because of the discovery of Rockall, 320km west of St. Kilda. This stone from the Iberian Peninsula shows that Rockall was discovered shortly after the discovery of Madeira (in c.4200 BC), but well before the discovery of the Azores (c.3600 BC). Details suggest that Rockall had been discovered about a century later, or c.4100 BC.

The old monument of Barnenez, overlooking the English Channel on the coast of Brittany at 49°N (c.4700-4400 BC), consists of an old eastern cairn with 5 passage graves, and a more recent western cairn with 6 passage graves (Fig.7). Because of its age, huge size, and northern location on the coast, it is one of the most interesting megalithic monuments in Europe.

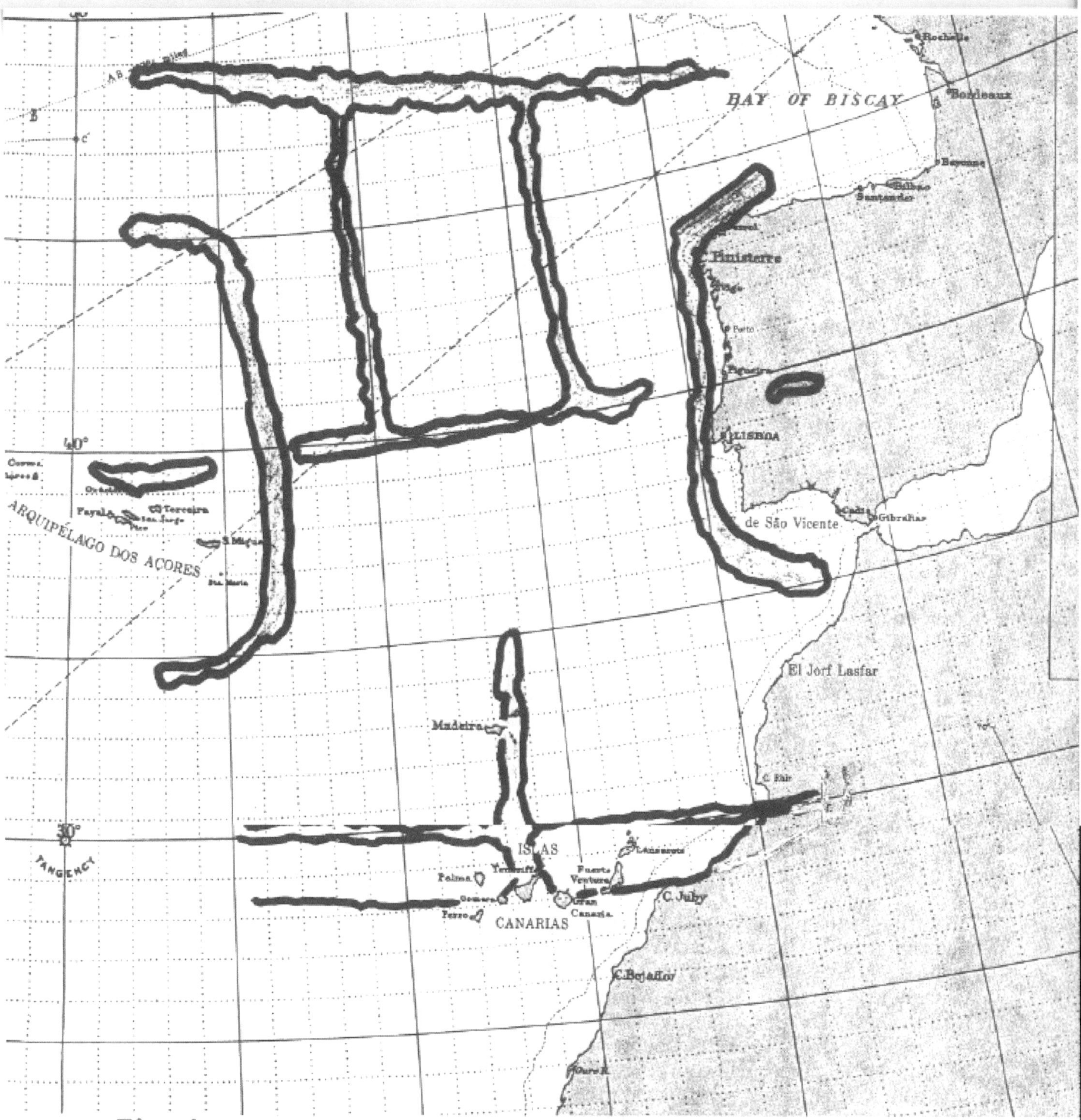


**Fig. 4** The front side of the Tablet of Paredes drawn on the Great Circle Sailing Chart of the ocean around Madeira, southwest of the Iberian Peninsula (Ref.8).

The eleven (5+6) passage graves and their angles of construction are related to the megalithic sea explorations to the west over the 11° from 49°N to 60°N (see Fig.6). The petroglyphs in these chambers deal primarily with the crossing of the English Channel, and explorations of the British Isles.

Here we are interested only in the most recent inscription, which shows the discovery of Rockall, c.4100 BC (see the lower portion of Fig.7). On the right side are carved 5 "crossing signs". This symbol is a cross-section, or depth/height profile, which was originally meant to show the crossing of a deep place, or channel. Together, they are used here to form a western shoreline, describing the coast at each degree of latitude. The lower two form the west coast of Ireland at 54°N and 55°N, and the upper three represent the west coast of Scotland at 56°N, 57°N, and 58°N. Via the 58th latitude (St. Kilda), the small island of Rockall (the 6th crossing sign, left) was discovered, at a little over 57°N. The petroglyph was inscribed in the youngest and most western chamber, because Rockall was found 3° to the west c.4100 BC, three hundred years after the western cairn was built. However, because of the discovery of Rockall at 57°N, the 8th passage grave D of Barnenez (at 49°N) became important. Grave D then symbolized the latitude of Rockall at 49°+8 =57°N. In the book of Towhig (Ref.1) we read: "Near the entrance to (passage grave) D a sterile layer was found, on top of which was a secondary deposit which included beaker pottery, a barbed and tanged arrowhead, a copper dagger, four transverse arrowheads, some "retouchoirs and dolerite axes." In this passage grave more artifacts were found than in all the other graves of Barenez put together.

## Discussion

On one side of the Paredes stone the discovery of Madeira was the central focus, and on the other, Rockall was the focus. The discovery of Madeira was definitely more important than Rockall to people on the Iberian Peninsula, where the Tablet was found. The discovery of Rockall had probably taken place somewhat later in time. For these reasons, and the finish of the stone, mentioned earlier, the Madeira side of the stone is the front side. The Tablet of Paredes (c.4200-4100 BC) shows the oldest inscriptions of big (horizontal) latitude lines and big (vertical) distance lines that we have seen. The big latitude lines (DL) correspond to 10° of latitude, equal to distance of 1111km, or 600 nautical miles. Older examples of big latitude lines may be found, but we are sure that examples of big distance lines of this early date are very rare.

On the Tablet of Paredes the discovered islands of Madeira and of Rockall are placed in the middle, which is presumed to be halfway across the ocean. As a consequence, they estimated the width of the ocean to be c.1DL =10°of latitude, =1111km (c.4150 BC). Fig.8 shows two other granite tablets of similar size, which are made a millennium later (Ref.1). At the left side is the tablet of Parafita, Portugal. On both sides are stylized petroglyphs of the North Atlantic Ocean. At the left top of the front side we see the discovery of South Greenland (c.3200 BC). It is presumed again to be halfway across

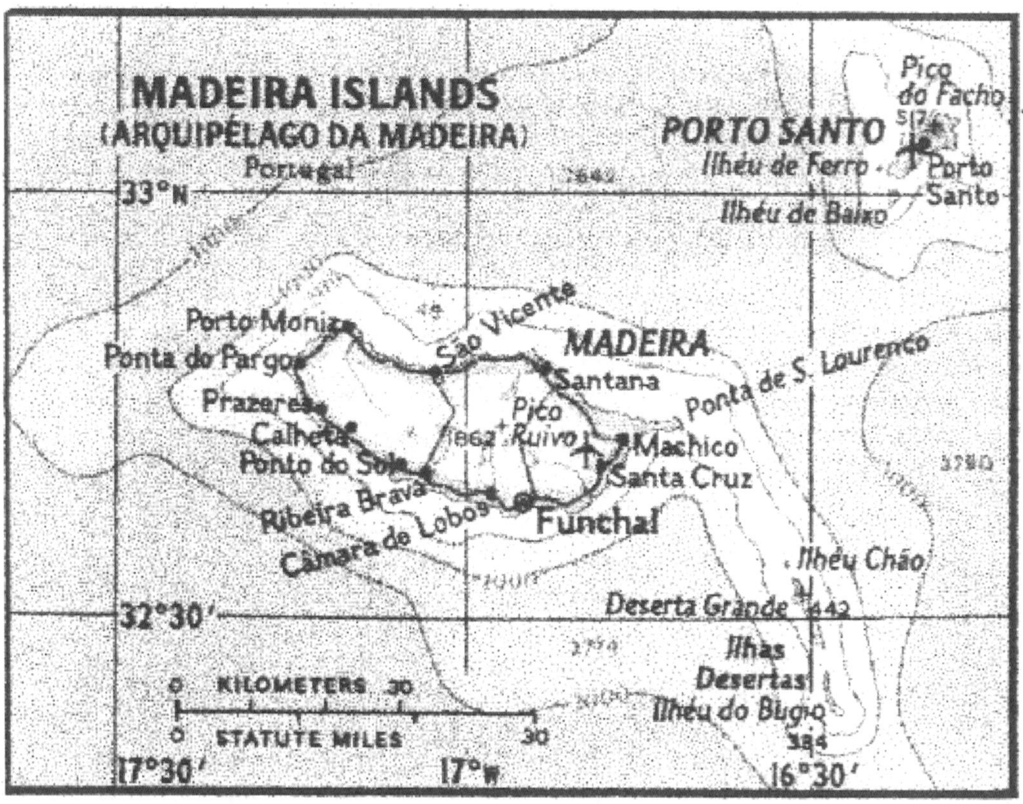


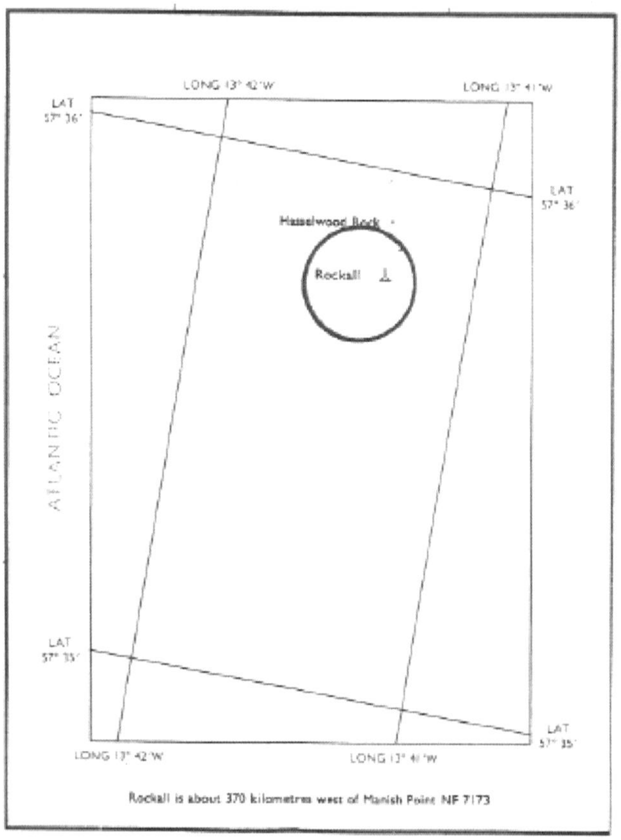

**Fig. 5** Maps of the Islands of Madeira (33°N, above) (Ref.10), and of St. Kilda (right, 58°N, Ref.11), and Rockall (57°N) (left, 57°N, Ref.12).

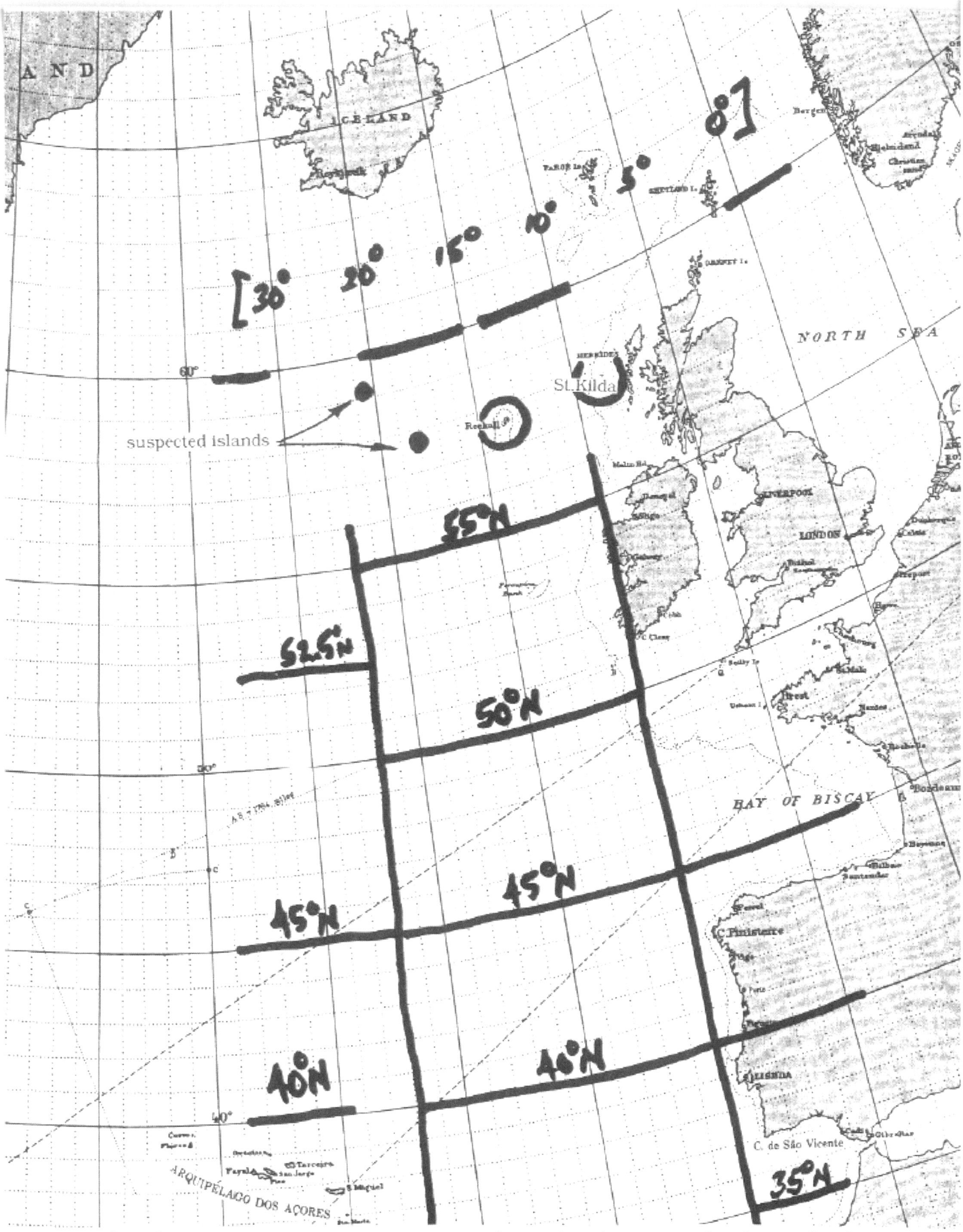


**Fig. 6** The back side of the Tablet of Paredes drawn on the Great Circle Sailing Chart of the ocean along Western Europe (Ref.8). In the north the discovered islands of St. Kilda (58°N) and Rockall (57°N).

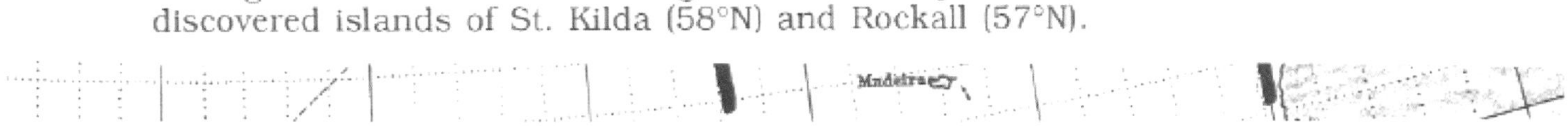

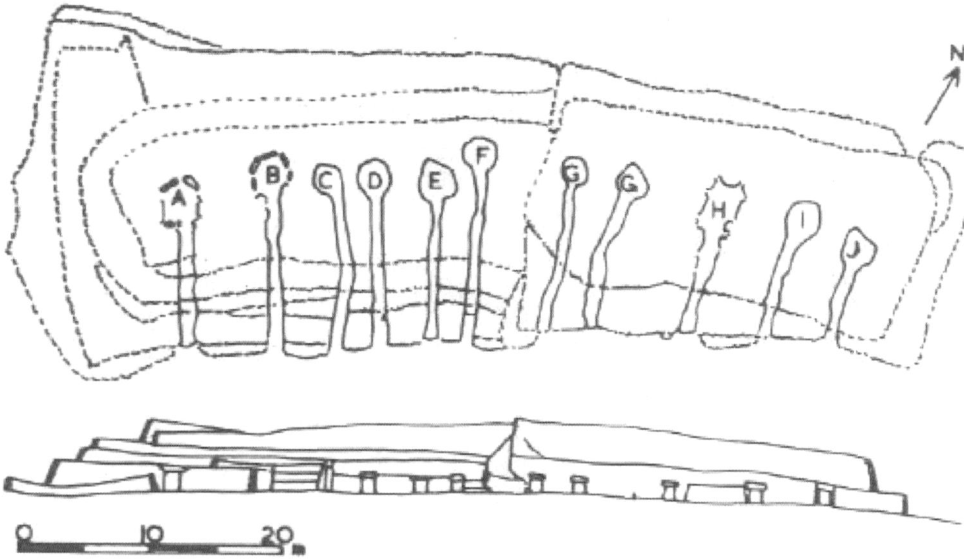


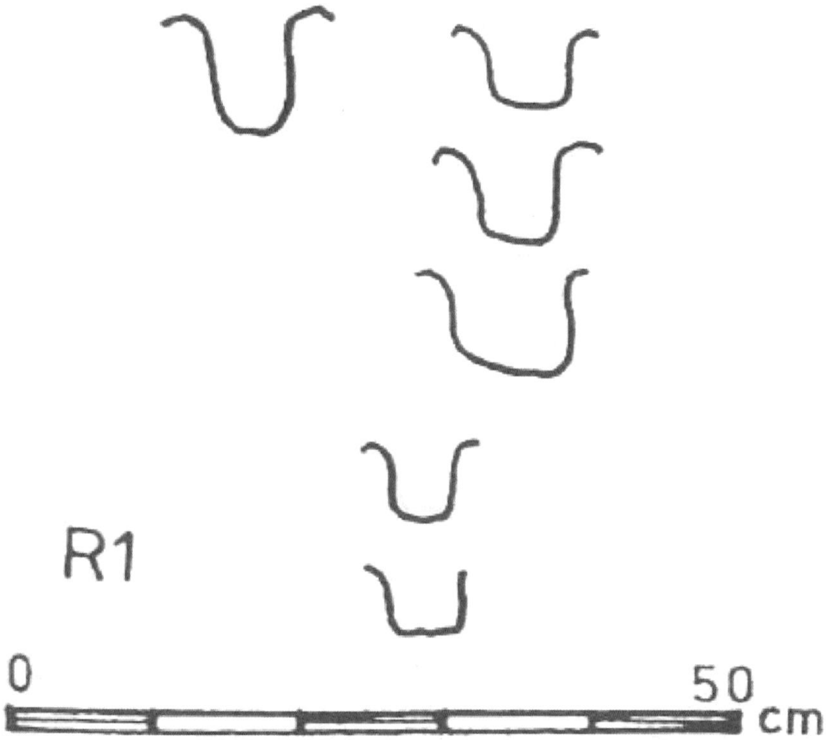


**Fig. 7 Top**: groundplan and front view of the two Cairns of Barenez, with 5+6 =11 passage graves (c.4700-4400 BC) (Ref.1).
**Bottom**: petroglyph on stone R1 of passage grave A. The upper left "crossing sign" shows the discovery of Rockall and its latitude of 57.5°N off the coasts of Ireland and Scotland (right).

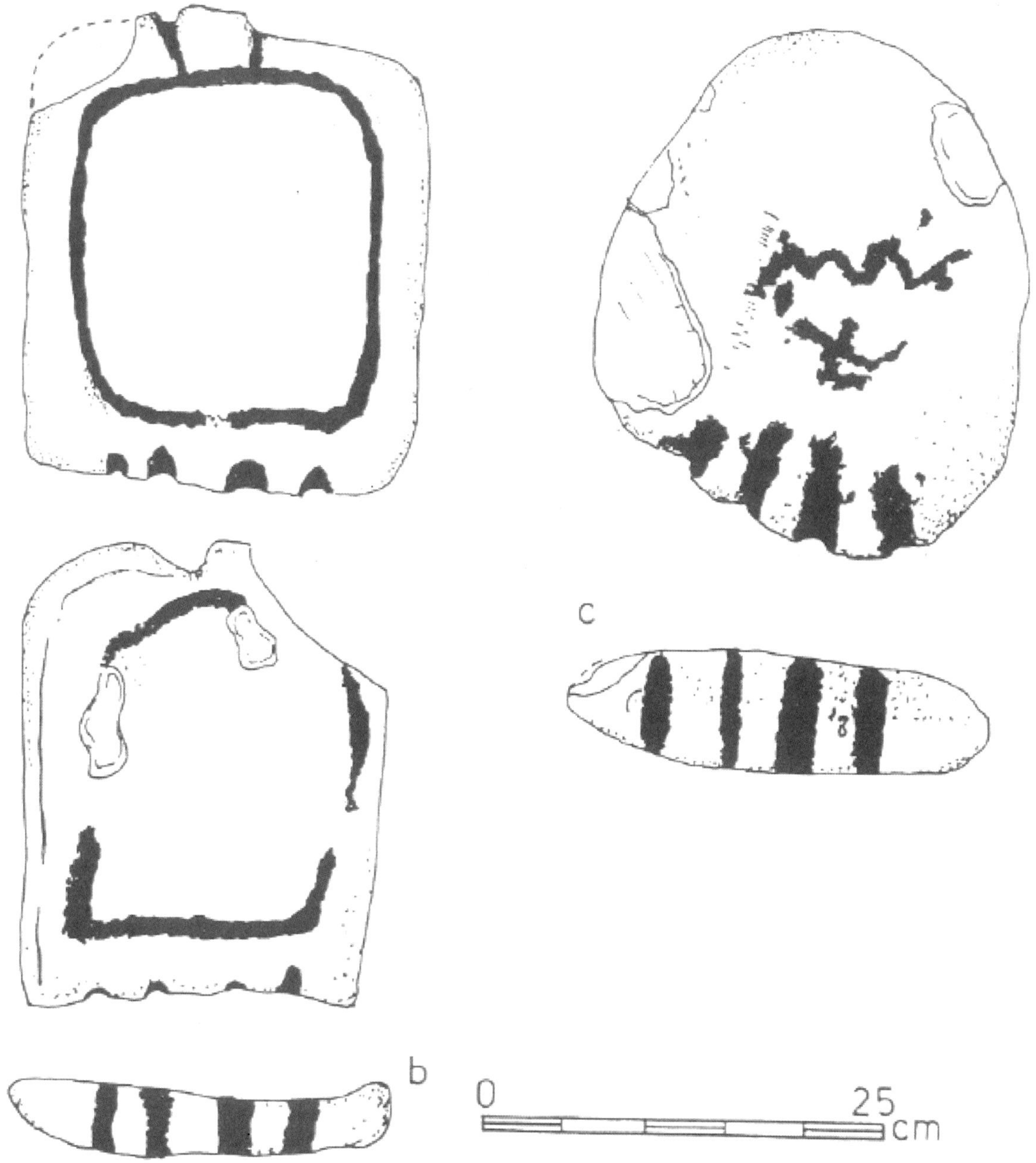


**Fig. 8** **Left**: the tablet of Parafita (Alijo, Portugal), showing petroglyphs of the North Atlantic Ocean, with the discovered south point of Greenland (top).
**Right**: the tablet of Newgrange (Ireland), also showing the ocean. On both stones, the width of the ocean (5 spacings) is estimated to be 5DL =5555km (c.3200 BC) (Ref.1).

the ocean. As a consequence, they estimated the width of the ocean to be 5DL =50°of latitude =5555km (5 spacings on the bottom of the stone). On the right side is the tablet of Newgrange, Ireland. The passage grave of Newgrange has 60 upright menhirs, corresponding to the latitude of Cape Farvel (South Greenland), at 60°N (Ref.7). The monument and the tablet are also made because of the discovery of Greenland, c.3200 BC. Again, Greenland is presumed to be halfway across the ocean (the surface area of the tablet), and they estimated the width of the ocean to be 5DL =50°of latitude =555km (5 spacings on the bottom of the stone). Because of this pessimistic view, and because of difficulty in their explorations (see our Story of Loughcrew), people gave up their efforts to cross the Atlantic Ocean (c.3200 BC), although they were less than 1DL =10°of latitude =1111km from the coast of North America!

# References


1. Twohig, E. Shee The Megalithic Art of Western Europe, Clarendon Press, Oxford, 1981.
2. Jonge, R.M., de, and IJzereef, G.F., De Stenen Sprecken, Kosmos Z&K, Utrecht/Antwerpen, 1996 (ISBN 90-215-2846-0) (Dutch)
3. Briard, J., The Megaliths of Brittany, Gisserot, 1991
4. Le Roux, C-T., Gavrinis, Gisserot, 1995 (French)
5. Richards, J., Stonehenge, English Heritage, 1992
6. Giot, P.R., Barnenez, Ouest France, 1991 (ISBN 2-7373-0933-6) (French)
7. Towhig, E. Shee, "Irish Megalithic Tombs", Shire Archaeology, 1990.
8. Portion of Great Circle Sailing Chart of North Atlantic Ocean, 43rd Edition, Oct. 1944, rev. 3/30/70, Defense Mapping Agency, Hydrographic Topographic Center, Washington, D.C. 20315
9. Whishaw, E.M., Atlantis in Spain, (1928 title: Atlantis in Andalucia), Adventures Unlimited Press, Illinois, 1997 (ISBN 0-932813-22-4)
10. National Geographic CD Map Collection, Madeira
11. National Geographic Map of British Isles, July 1958
12. British Ordinance Survey Map of 1967

# The Discovery of the Azores
(The Tumulus of Gavrinis, Brittany, c.3600 BC)

Dr. R.M. de Jonge, drsrmdejonge@hotmail.com
J.S. Wakefield, jayswakefield@yahoo.com

## Introduction
The Tumulus of Gavrinis is situated on the small islet of Gavrinis in the Gulf of Morbihan, close to Carnac, the megalithic center of Brittany (France). It is a mound of stones, having a diameter of c.55 meters, about 8 meters high. Inside the mound is a passage grave (Fig.1), famous for its beautiful petroglyphs. The first account of exploration was in 1831, and in 1875 the Musee des Antiquities Nationales made casts of all the decorated slabs. In 1980, it was thoroughly investigated by French archaeologist Le Roux (Refs.1-4). The old passage grave (c.3600 BC, Ref.4) is a "mission church" for the spreading of archaic God/King theories from Egypt (Refs.5,6).

The whole monument is close to the sea, and the passage with the important grave chamber points to the northwest (Fig.2). As we shall show, the monument and the inscriptions inside have a geographic meaning, related to a desire of these people to cross the ocean. The passage has a total length of c.15 meters. The access to the inside is easy, because the passage is almost 2 meters high and more than a meter wide. The chamber is about 2 meters wide. The 29 upright granite menhirs along the sides of the passage are big. On the average, they are about a meter wide, half a meter thick, and 2 meters high. There are also 10 coverstones and 13 groundstones (see Fig.2, right). The 23 western upright stones are engraved with petroglyphs over their entire surfaces (the black stones in Fig.2, left, see also Figs. 6-11). The omitted stone, R7, is of white quartz, special to the SunGod and astronomic events, as we shall explain. We shall show that the number and layout of the stones, and the glyphs on them, encode data important to the builders, in accord with the geomathematics and SunGod religion of the time.

## Egypt
The whole passage grave has 29 upright stones, corresponding with the latitude of the Nile Delta, at 29°N. The 23 upright stones of the passage (and the 23 decorated western stones of the whole monument) encode both the direction of Gavrinis to the Nile Delta, 23°ESE, the center of the Northern Egyptian Empire, and also the latitude of the Tropic of Cancer, which crosses the River Nile at 23°N. This point, at the center of the Southern Egyptian Empire, is 29°ESE of Gavrinis, encoded by 29 upright stones. In the time period of the construction of this monument, Egypt became the most important civilization on earth. This orientation upon Egypt is further evidence for our claim that Gavrinis should be considered a "mission church" for the spreading of Egyptian King theory (Refs.5,6).

**Fig.1** Photo of the passage grave of Gavrinis (Ref.8), looking out along the south wall (entry doorway is on the left side of the photo). Together with the protecting heap of stones, it forms the Cairn of Gavrinis. It is a kind of "mission church" for the spreading of the archaic God/King theory (c.3600 BC).

## The Canary Islands

The passage grave of Gavrinis is located at the west coast of Western Europe, and in addition, it points to the west. We start by looking for something in the west, in the stone counts and angles in the design of the monument. The last two big upright stones of the passage (L11 and R12) are wider than the other stones (Fig.2). These are the so-called portal stones, where one enters the holy grave chamber. For this reason the first part of the passage contains 23-2= 21 stones, corresponding with the latitude of important Cape Blanco at 21°N, which is the first western cape of Africa before you sail for the islands of Cape Verde.

The whole passage contains 23 stones, encoding the latitude of the point where the Tropic of Cancer leaves the continent of Africa, at 23°N. On midsummer day, the sun is at a right angle above the Tropic of Cancer, and that day the slow northern movement of the sun turns into a southern movement. They believed in the SunGod, and that the Southern Egyptian Empire near later Karnak at 23°N was the center of that religion. In honor of the SunGod, the 23 westerly stones are completely engraved. In fact, at this latitude, people wanted to cross the ocean. Counting the menhirs in the direction of the daily movement of the sun, as the SunGod wishes it, stone 23 (R7) should be an important stone, and it is pure white quartz, and not engraved (Fig.2, left).

The whole passage grave contains 29 upright stones, corresponding with the latitude of the (northern) Canary Islands at 29°N. Until the Cape Verde Islands were discovered, these were the westernmost islands of the then known world. The grave chamber is built of 6 upright stones and one coverstone, corresponding with the 6+1=7 Canary Islands. The two western stones C3 and C4 symbolize the two westerly islands, Ferro and Palma. At the end of an "intermittently paved" road on the northern tip of the northern island, La Palma, at Fuente de la Zarza are large megalithic inscriptions of "concentric arcs and circles" on the cliff face similar to those of Gavrinis, telling of discoveries and islands further north and west in the ocean (Ref.10).

## Madeira

The monument has huge coverstones (Fig.2). You know something important is going on when so much work is involved to move such big stones. The passage contains 23 upright stones (and the monument has 23 decorated western stones), and there are 10 coverstones, altogether (in both cases) 23+10= 33 stones, corresponding with the latitude of the Islands of Madeira, discovered at 33°N (c.4200BC). The last two stones of the passage are wider portal stones. They represent the two islands of Madeira. These two islands of Madeira turn out to be very important, now the stepping- stones to other islands.

## The Discovery of the Azores

By all the details of its site, orientation, layout, and stone sizes, the passage

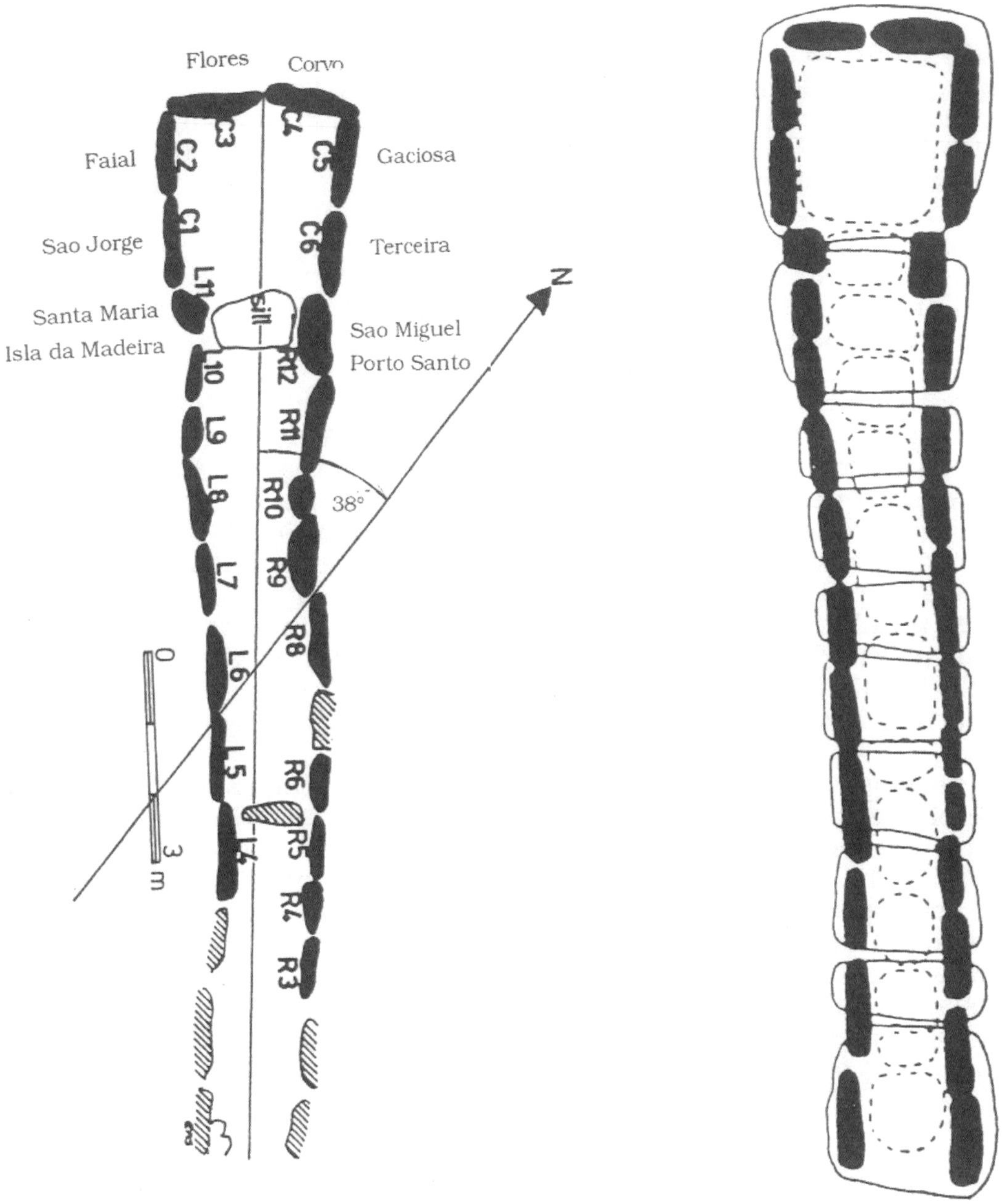


**Fig.2 Left**, groundplan of the passage grave of Gavrinis (Ref.4). The 23 western upright stones (black, except R7) are engraved. They encode the latitude of the Holy Tropic of Cancer at 23°N. The head axis is between the stones C3,C4, and L1,R1, and the NS axis is between L5,L6 and R8,R9. **Right**, the less accurate groundplan is stylized (Ref.3), but showing the coverstones (white), and the groundstones (dotted lines). There are 29 upright stones, and 10 coverstones, encoding the latitude of the West Azores, just discovered at 39°N (c.3600 BC).

grave of Gavrinis was carefully designed and built to commemorate the discovery of the Azores, the far western home of the SunGod in c.3600 BC.The upright stones are the most important ones, and the two islands of Madeira form the "gate" for an important discovery. The passage grave contains 29 upright stones and 7 smaller coverstones, together 29+7= 36 stones, encoding the latitude of the Strait of Gibraltar, at 36°N. This is the important entry and exit of the Mediterranean Sea. But the 36 stones also correspond to the initial sailing direction from Madeira to Santa Maria, the easternmost island of the Azores, 36°NW! The 29 upright stones provide the terminal sailing direction in the neighborhood of Santa Maria, 29°WNW. The difference between the sailing directions is due to the curvature of the Earth. The 8 eastern groundstones (No.8 is a large one, see Fig.2) provide the sailing distance from Madeira to Santa Maria: 8dl= 8°of latitude= 888km.

In addition to the 7 small coverstones, there are 3 large ones. These 3 symbolize the 3 island groups of the Azores, just discovered. The 13-8= 5 western groundstones provide the sailing distance within the Azores: 5dl= 5° of latitude= 555km. At the entrance of the passage grave is the first large coverstone. The 29 upright stones, plus the 7 small coverstones, plus the first huge coverstone= 29+7+1= 37, or the latitude of Santa Maria at 37°N. This first coverstone has been placed on top of 3 upright stones, in further honor of the discovery of the three island groups of the Azores.

The second large coverstone is at the entrance of the chamber. If we add this stone to all the upright stones of the monument and coverstones of the passage, the result is 37+1= 38, which encodes the latitude of the Central Azores, 38°N. This coverstone leans on 5 upright stones, corresponding to the 5 islands of the Central Azores.

The last huge coverstone is located on top of the holy grave chamber. As a consequence, the total number of stones of the whole passage grave now results in 38+1,= 29 total upright stones +10 total capstones= 39, encoding the latitude of the especially important West Azores at 39°N. This coverstone leans on the two portal stones, now corresponding with the two islands of the East Azores, Santa Maria (L11) and Sao Miguel (R12). See how the sizes and shapes of these islands on the groundplan (Fig.2) compare with the layout of the islands on the small map (Fig.3). The coverstone also leans on the 2 western stones, now corresponding to the 2 islands of the West Azores, Flores (C3) and Corvo (C4). It also leans on the 4 other islands of Sao Jorge (C1), Fayal (C2), Terciera (C6), and Graciosa (C5) (see map, Fig.3). The large coverstone itself represents the main island of the Central Azores, Pico, which has the largest mountain of the archipelago at 2351m.

The head axis of Gavrinis is situated between the end stones C3 and C4, and between the entrance stones L1 and R1 (Fig.2). This head axis makes an angle of 38° with the NS axis, confirming the latitude of the Azores, and of Pico and the Central Azores in particular, at 38°N.

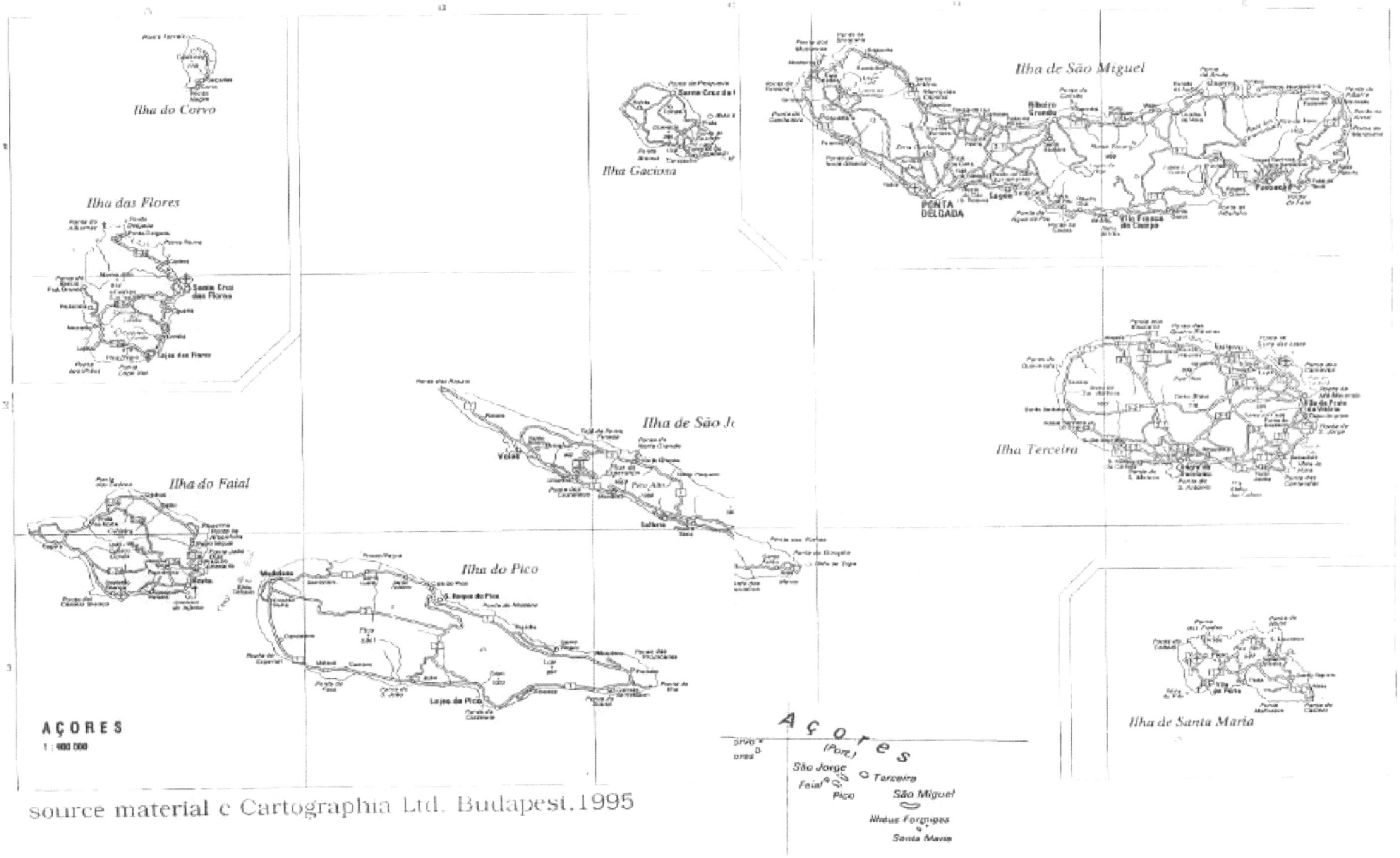


source material c Cartographia Ltd. Budapest.1995

**Fig.3** Map of the Azores in the middlae ocean, 38°N, Ref.13). The East Azores (2 islands), the Central Azorésislands), and the West Azores (2 islands).

## Africa and Europe

Besides the Azores, other areas in Africa and Europe continue to be of importance in the attempts to cross the ocean. The grave chamber has 6 upright stones and one coverstone, together 6+1= 7 stones, encoding the latitude of Sherbro Island, at 7°N, the far SW point of North Africa.

Halfway up the passage, between the stones L5,6 and R8,9 is the NS axis of the monument (Fig.2). West of this NS axis are 16 upright stones, encoding the latitude of the Cape Verde Islands, at 16°N. Until the recent discovery of the Azores, these were the western-most islands in the known world.

We have learned to look carefully for the latitude of the site itself in megalithic site designs, because in our experience it was always encoded. As noted before, the 8th groundstone is surprisingly large. The 39 stones of the passage grave and the first 8 groundstones total 39+8= 47 stones, here encoding the latitude of Gavrinis in Brittany (now France), at 47°N. This was the most important center of megalithic culture of Western Europe.

Finally, the 29 upright stones, plus the 10 capstones and 13 groundstones together equal a total of 52 stones (29+10+13= 52). These encode the latitude of Dunmore Head in Ireland, at 52°N. The head axis of the passage grave makes an angle of 52°NW, confirming this latitude (Fig.4) This is the westernmost point of Europe. If it turns out to be impossible to cross the ocean from the Azores at 38°N, then the attempts will be continued here, on the complementary latitude of 90-38= 52°N (which is what later happened!). A large number of lines can be drawn between the upright stones. These lines make special angles with the NS axis and with the head axis of the monument. If this is done in a careful, systematic manner (see Fig.4), the story above will be confirmed again by the angles. In addition, other information about the early history of the attempts to cross the Atlantic Ocean will be found.

## The Petroglyphs on the Sill

The most important petroglyph of Gavrinis (Fig.5) was not engraved on one of the upright menhirs, but on the sill, the groundstone between the portal stones L11 and R12 at the entrance of the grave chamber (Fig.2). This was done, because it was the first time that a discovery had been made far in the ocean, from Madeira (L11), and Porto Santo (R12). The surface area of the stone is the ocean, between the equator (the bottom edge of the stone) and the 50th latitude line near Brittany (the top edge of the stone). We are looking over the waters to the north as we enter the grave chamber. At the right side, to the east, we see the Canary Islands, and behind them Madeira. In the middle, are the Cape Verde Islands, and at the left side, to the west, are the just discovered Azores. The semi-circular lines are distance lines, each of them corresponding to one degree of latitude. Around the Canaries and Madeira the coastal waters have been explored over 4+4= 8dl= 8°of latitude= 888km. The coastal waters of the Cape Verde Islands were only explored over 6dl= 6°of latitude= 666km, possibly due to the strong

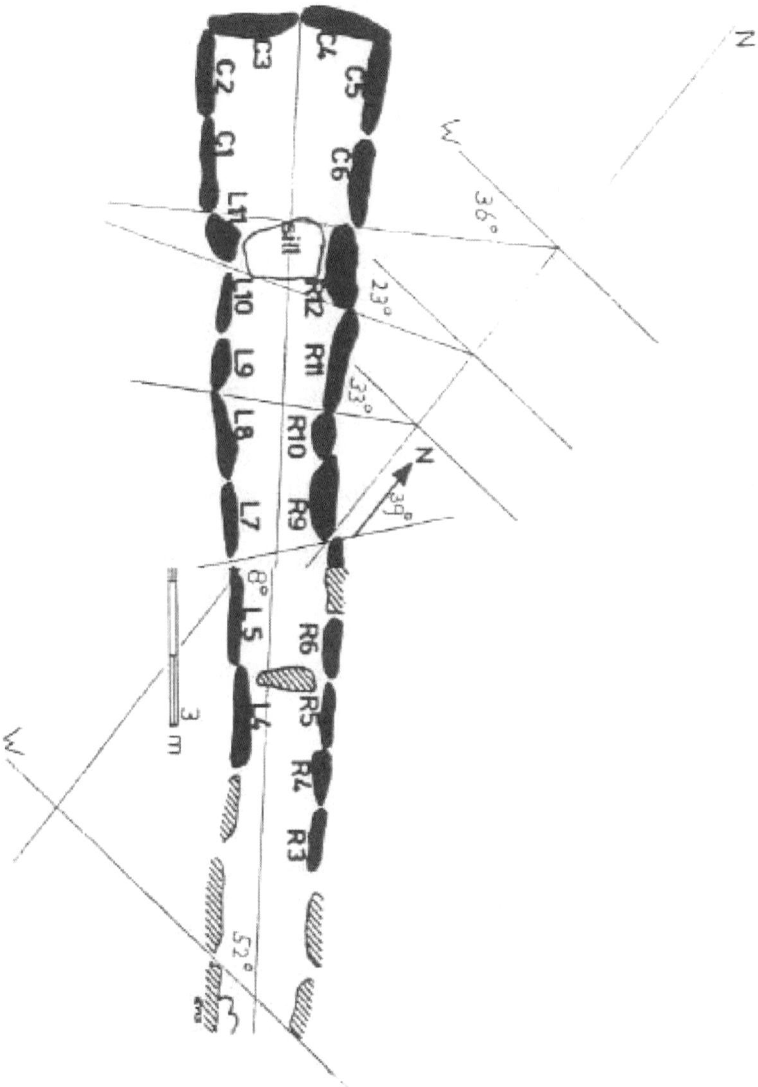


**Fig.4** Groundplan of Gavrinis (Ref.4), showing some lines between the menhirs that make important angles with the NS axis (c.3600 BC).

westerly currents and offshore winds in the area. Around the Azores the coastal waters are shown as having been explored over 8dl= 8°of latitude= 888km. This is equal to the sailing distance from Madeira, necessary for the discovery of the Azores. At the far right side, we see the 6 important latitudes (6 spaces of distance) between Madeira and the West Azores (39°-33°= 6°=6dl). At the left side of Fig.5 we see a man with a lobster (still representing Cancer in astrology), because people want to cross the ocean at the latitude of the Tropic of Cancer (Refs.5,6). Below, on the edge of the stone, we see 9 equilateral triangles, corresponding to the 7 Canary Islands and the 2 islands of Madeira (7+2= 9). These triangles or pyramids symbolize death. On the (northern) edge above, there are 9 crosses, corresponding to the 9 more northerly islands of the Azores, just discovered. These Xs or rays symbolize light, hope, and life (Ref.16).

Because of the 39 stones of the passage grave, and the 12 groundstones of the passage, the sill stone is number 51 (39+12=51), encoding the latitude of 51°N. At this sacred latitude the rise of the midsummer sun (the appearance of the god Horus) is at right angles with the southernmost rise of the full moon (the appearance of the god Osiris). This religious notion would form the basis of the later famous Stonehenge in England, at 51°N. But the sill stone (#51) also shows the complementary angle of 90°-51°= 39°, confirming again the latitude of the just discovered West Azores, at 39°N. For this reason, the 9th cross on the upper edge of the stone (Fig.5), symbolizing the westernmost island of the West Azores, Flores, has been placed slightly apart.

The quality of this petroglyph (Fig.5) is so high, it must have been made by a master. For this reason, it is worthwhile to look at it in greater detail. The total number of distance lines around the islands is 4+4+6+8=22. Behind the sailing man (with the lobster) around the Azores we see a small added piece of an extra distance line (the small hump), allowing the master carver to now have a total of 22+1=23 lines here. This number now corresponds to the Southern Egyptian Empire and the holy latitude of the Tropic of Cancer, at 23°N. At the far right side, six spaces are shown, now bringing a total of 23+6= 29, corresponding to the latitude of the Northern Egyptian Empire, and the Canaries at 29°N. Below, 9 triangles are shown, now making a total of 29+9= 38, corresponding to the latitude of the Azores, just discovered at 38°N. Above, 9 crosses are shown, making for a total of 38+9= 47, encoding the latitude of the monument of Gavrinis, at 47°N!

## The Petroglyphs of Gavrinis

The megalithic petroglyphs on the upright stones of Gavrinis are the most beautiful ones of Western Europe. They are engraved with hard, stone chisels in the pick-dressed, granite menhirs, more than a centimeter deep (Refs.1,4). Almost all of them have a geographic meaning, but are meant for decoration. The parallel lines are distance lines, each of them corresponding to 1dl= 1°of latitude= 111km. These distance lines show the explored coastal waters around islands, sometimes also around capes and

**Fig.5** Petroglyph on the sill stone, with engravings on both ends (Ref.4), with author's numbers and labels. We are facing the ocean to the north. Right, the Canary Islands, with behind them Madeira, in the middle the Cape Verde Islands, and left, the Azores, just discovered. The number of distance lines show exactly the explored coastal waters around the islands. At the left side above is a man with a cancer, because people want to cross the ocean at the Tropic of Cancer at 23°N, in honor of the SunGod. On the stone end below are 9 equilateral triangles symbolizing the 7 Canary Islands plus the 2 islands of Madeia, and on the edge above are nine Xs, symbolizing the 9 discovered islands of the Azores (Gavrinis, Brittany, c.3600 BC).

along the shores of the Atlantic Ocean. If there are about 8 distance lines, the existing situation is described in the neighborhood of Madeira and the Azores. If there are more than 8 distance lines, people indulge in fancies of what the future may bring. The petroglyphs are applied on the 23 western upright stones of the passage grave because people want to cross the ocean on the Tropic of Cancer, at 23°N, because that latitude is the track important to the SunGod. We will follow the inscriptions (from Ref.4) along the walls of the passage grave, and will provide the reader an understanding of the clear ones.

On stone **L4** (Fig.6), we see in the middle, the two islands of the West Azores, Flores and Corvo, with left and right the distance lines (dl) of the explored coastal waters.

On stone **L5** (Fig.6), we see at the right a square, which is a stylized inscription of the lower half of the North Atlantic Ocean. In the middle the Azores are placed, because they are convinced they have reached halfway to the opposite side, which is correct. In the "cartouche" of the Azores is a U sign with a vertical stroke in the center, again a symbol of "halfway". The rest of the inscriptions are probably heroic exploits or accomplishments of individual sailing expeditions.

On stone **L6** (Fig.6), we see at the bottom the two islands of Madeira. At the left side a boat with 23 rowers (the holy number), that sails to the west along the 3 island groups of the Azores. At the right side is the Strait of Gibraltar, the door or gate to the ocean, with distance lines. At the top of the stone is a square pattern of latitude and distance lines in the west. The rest is details of their trips.

On stone **L7** (Fig.6), we see the two islands of the West Azores, Flores and Corvo. Behind the main island of Flores is a small islet (Ponta Delgada), which they have found.

On stone **L8** (Fig.7), we see below the 2 islands of Madeira, and above them the 9 islands of the Azores.

On stone **L9** (Fig.7), are 18 axeheads, symbolizing the 18 islands of the Canaries, Madeira and the Azores (7+2+9=18), and corresponding to the approximate sailing direction from the Strait of Gibraltar to the Central Azores (shown in Dissignac), and from the East Azores to the West Azores, 18°WNW. Low on the stone are distance lines around Cape Bojador, at 27°N. This special 9th stone is symbolic for the 9 islands of the Azores.

On stone **L10** (Fig.7), we see below the 2 easterly Canary Islands, above them at the right side the 2 islands of Madeira, and at the top, the 9 islands of the Azores. On the right edge of the stone the distance from Madeira to the West Azores is shown, c.14dl= 1555km.

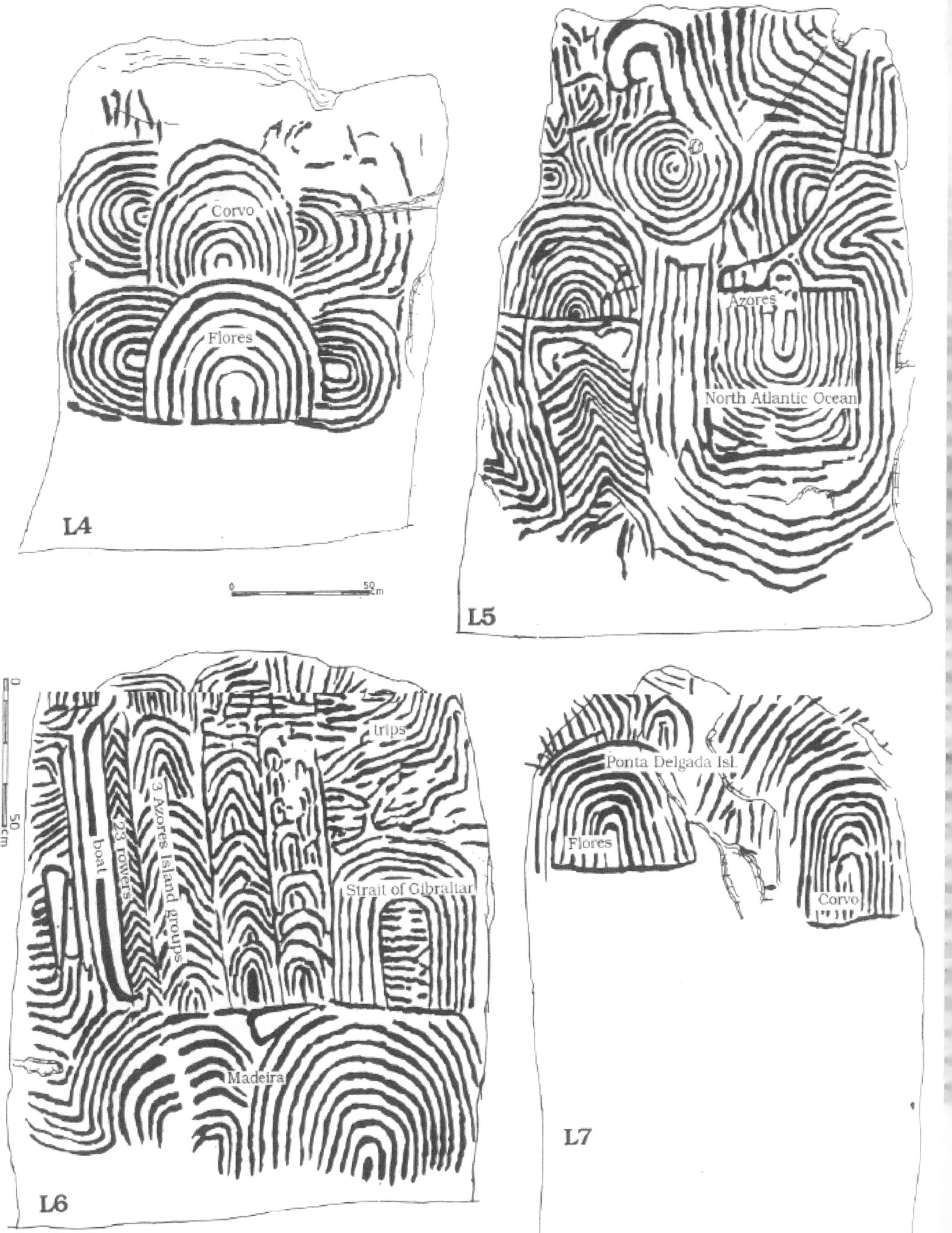


**Fig.6** Four of the 23 fully engraved western menhirs of Gavrinis, c.3600 BC (Ref.4).

On stone **L11** (Fig.7), it seems that the Central Azores, left, have the two Western Azores (square shaped) on top of them (more westerly), with many explorations mapped out to the north of these islands.

On stone **R3** (Fig.8), there appear to be expeditions mapped out around the 5 islands of the Central Azores.

On stone **R4** (Fig.8), we see below, the 2 islands of Madeira, and above the 3 island groups of the Azores, with in the middle the main island of Pico. Left, south of the Azores, the endless ocean (Ref.4).

On **R5** (Fig.8), in the middle, the West Azores, with behind them, in the west, the endless ocean around the 39°N latitude line.

On **R6** (Fig.8) are the West Azores, with hope of other islands in the southwest direction.

Stone **R7** (not shown) is a white quartz stone, special to the builders, because it honored the SunGod. Like Newgrange, the entrance is exactly aligned to the mid-winter sunrise (Ref.3).

On **R8** (Fig.9), we see below, the easternmost Canary Island, with right of it the sailing route over 8dl from Madeira to the Azores, then the sailing route over 6dl from Africa to Madeira, and finally the internal sailing route over c.5dl from Santa Maria to the West Azores. Above it the 2 islands of Madeira. On top, right, the 9 islands of the Azores, and left, a hope for discovery of other islands in the south, in the direction of the Tropic of Cancer.

On **R9** (Fig.9), we see below, in the middle, Madeira, with behind it the 2 islands of the East Azores. Above it the Central Azores and the West Azores. (The humps are imitations of the sill stone.) Left and right of this column we see again the 5+4= 9 islands of the Azores. Along the edges of the stone are the less important, other islands of the ocean. On the left edge of the stone, and with a thin line, 2 1/2 "big" distance lines have been engraved, corresponding with 2.5 DL= 25° of latitude= 25x1111km= 2777km. Over this distance the ocean has been explored, west of the coast of Iberia, at the level of the West Azores. This corresponds exactly with 8dl= 8° of latitude= 888km west of the West Azores, as shown in these drawings, and is in agreement with the technical abilities of the megalithic people at the time of the building of Gavrinis. After the discovery of America (c.2500 BC), a later visitor of the passage grave has added a thick line to the carving, updating it. The new distance of 3 1/2 DL= 35°of latitude= 3888km corresponds to the approximate distance of the shortest crossing of the ocean from Newfoundland to the Canary Islands, or to Gavrinis (Brittany).

On **R10** (Fig.9), we see at the right side the 7 islands of the Central and West Azores, with at the left, the explored ocean over 5 latitude lines, to the

**Fig.7** Four of the 23 fully engraved western menhirs of Gavrinis, c.3600 BC (Ref.4).

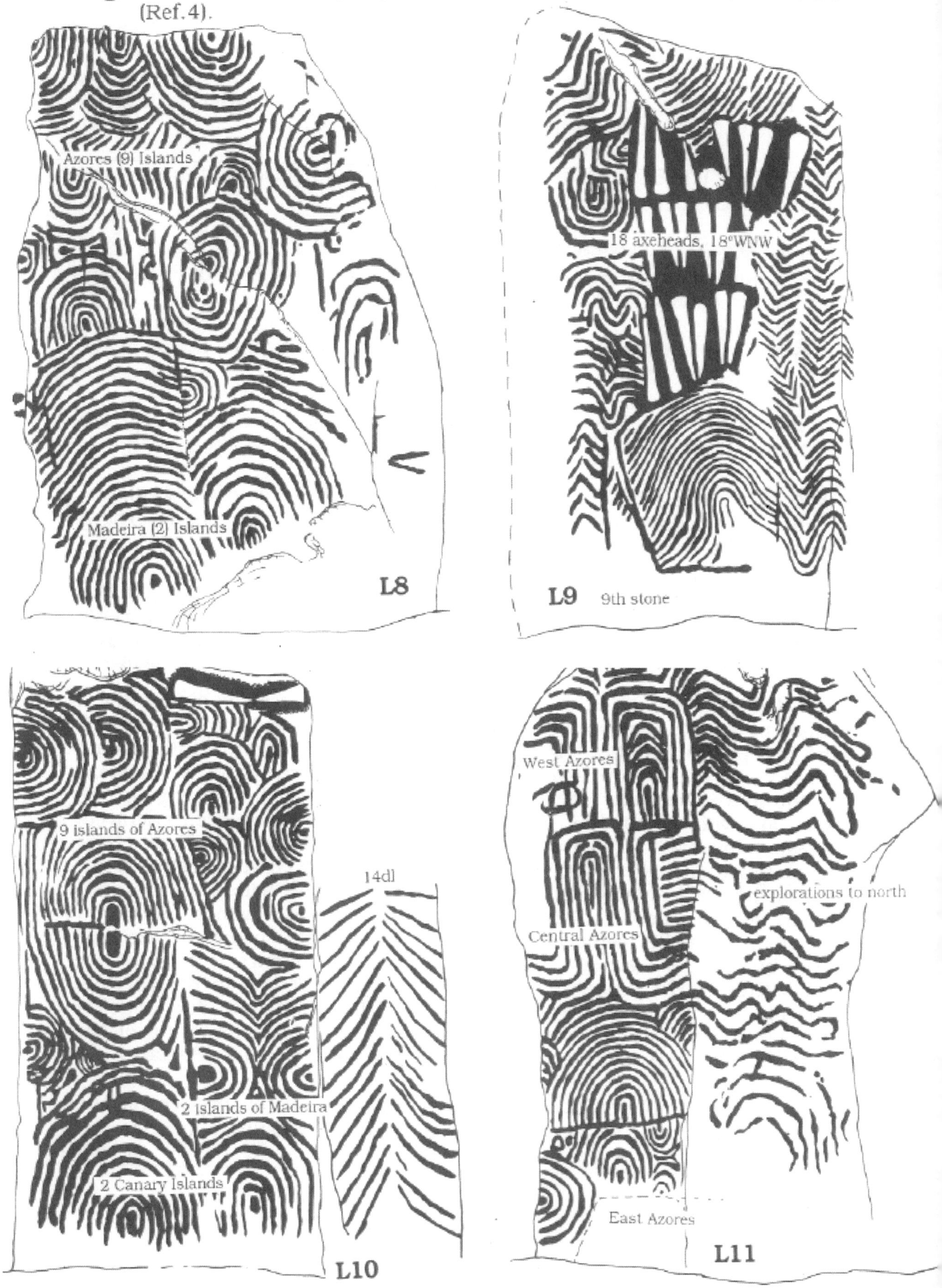

latitude of Madeira at 33°N (38°-5°=33°). The 4 axeheads correspond with the 4 island groups of Madeira and the Azores.

**R11** (Fig.10), shows sailing explorations around some islands of the Azores.

On **R12** (Fig.10), we see in the middle the route from Madeira to the East Azores in the endless ocean, and on top, the route from the East Azores to the West Azores, both in the endless ocean. At the left side, explorations to the south, in the direction of the Tropic of Cancer.

On stone **C1** (Fig.10), the first engraved stone at the west side of the grave chamber, probably is the most important one of the passage grave. In the middle we see 3 very deep, internally connected, cup-shaped hollows (c.15cm deep!), with some protruding stone, like handles, between. The 3 holes are the 3 island groups of the Azores. They form the handles, literally, to cross the ocean. Note, that this unusual feature has been placed in the middle of the stone. Again, one is convinced, that the Azores are in the middle of the ocean. Below on the stone we see the 2 important islands of the West Azores, Flores and Corvo, symbolic for the gods Horus and Osiris. At the right side above are important explorations planned to the WNW in the future, and at the left side above, Flores and Corvo are repeated, for explorations planned to the WSW.

On **C2** (Fig.10), we see above the horizontal stroke, left, the easternmost Canary island, on the right the 2 islands of Madeira, and behind them, in the distance, the 3 island groups of the Azores at 38°N. Left, in the south, the endless ocean in the direction of the Tropic of Cancer. The 15 distance lines give the correct latitude: 38°-15°= 23°N.

On **C3** (Fig.11) below, the sailing route from the 2 islands of Madeira (below left) to the 2 islands of the East Azores (the 2 axeheads). The horizontal line is a distance line parallel to the coast of Africa. Above it, the sailing route from the East to the West Azores, in the endless ocean. The c.15 distance lines to the left (south) provide the latitude of the Tropic of Cancer, at 38°-15°= 23°N.

**C4** (Fig.11) appears to "fit" and be adjoined to the other end stone of the chamber, C3, but this does not help us explain it. These two end stones (C3 and C4) are earlier thought to represent the Western Azores islands of Flores and Corvo. Do you suppose the large figure on C4 is the island of Flores, with the harbor on its right side, where the town of Santa Cruz des Flores is now located? Corvo to the right looking west?

On **C5** (Fig.11), we see the sailing route from Madeira to the East Azores (below) and from the East Azores to the West Azores (above), in the endless
ocean.

**Fig.8** Four of the 23 fully engraved western menhirs of Gavrinis, c.3600 BC (Ref.4).

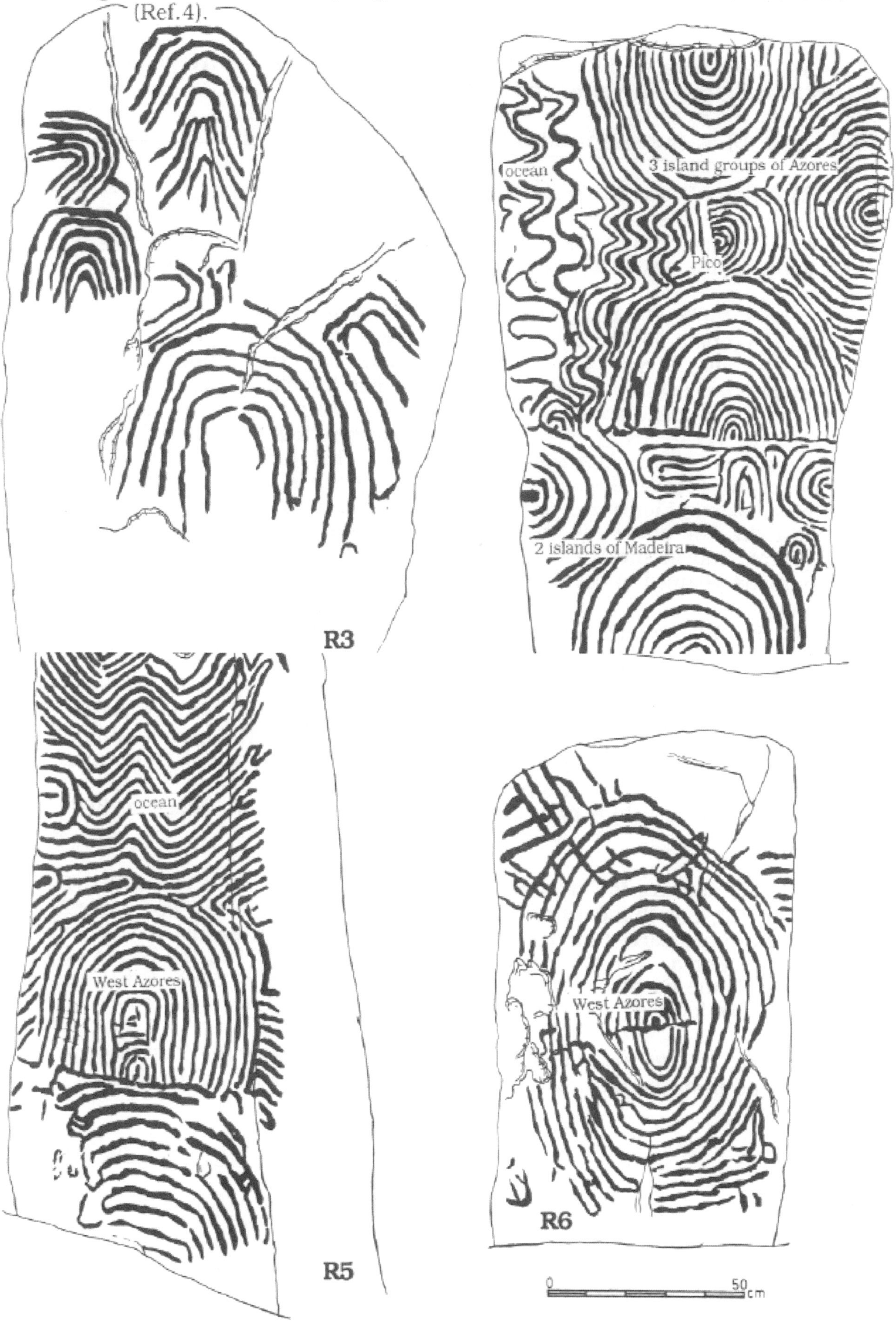

On **C6** (Fig.11), we see the large island of Sao Miguel (East Azores) at the left side below, and the West Azores at the left side above. The right side of the stone shows complicated sailing expeditions to the north, in the direction of the 50th latitude line.

Inscription **Co1 R6** (Fig.12 ) has been engraved near menhir R6, probably symbolic for the Tropic of Cancer, because R7 of quartz could not be used. It is a classical example of a Mediterranean sign. The (horizontal) handle is the (stylized) Mediterranean, the left side branch is the route from Gibraltar to the Cape Verde Islands, and the right side branch is the route around the Iberian Peninsula to Brittany. However, it is also a Jacob's Staff. This is an old instrument to measure latitude. It consists of two wooden sticks that can slide past each other. However, it is also a question. The handle means: we want to cross the ocean from the route in the Mediterranean. The side branches mean: After the Strait of Gibraltar, which way do we go, left or right? These common signs occurr in abundance in Dissignac and on the stele of Mane-er-Hroek (Brittany, Ref.2).

Petroglyph **RS 2** (Fig.12) was applied on the edge of the second coverstone. It is a height profile symbol of the ocean. The right bowl is the ocean to the Azores, and the left bowl is the ocean to the expected other side. In the middle are a number of distance lines (6dl= 666km), showing the importance of the Azores in the middle of the ocean.

The **coverstone** of the chamber of Gavrinis is the central part of what once was a huge menhir (Fig.12, top, Refs.3,7,14). (The bottom part was later used as coverstone for the Table des Marchand, Brittany.) Some petroglyphs are carved into the backside of this coverstone. Presently, these are not visible because of the mound of stones on it. At the top we see a common stylized map of the "Little Mediterranean Sea". In the east (at the right side, below) the Gulf of Sidra and the Gulf of Gabes are inscribed. The upper loop is, as usual with these symbols, the sailing route around the islands of Sardinia and Corsica. In the west (left), the unusually long foot shows the coastal routes beyond the Strait of Gibraltar, north to Iberia, and south to Cape Verde (made before the discovery of the Cape Verde Islands, c.4500 BC). Below this are two bulls, perhaps Aurochs or other prehistoric animals, with huge horns. They represent two "manifestations of the SunGod", so they had a religious meaning. Below the bulls there is another Mediterranean sign and a small Jacob's Staff. The petroglyphs from the mid-section of this menhir are older than the passage grave itself, dating from 5000-4000 BC, with the lowest the most recent.

Physical evidence found on the western coast of Corvo, in the West Azores is reported by Irwin: "In 1749 a storm ... uncovered in the sand a broken black cask. Inside the cask was a caked mass of ancient coins. The mass was pried apart and the coins studied by numismatists, who reported that these were Carthaginian coins of the fourth and third centuries BC! She also reports that "The first Portuguese who landed on Corvo reported finding on

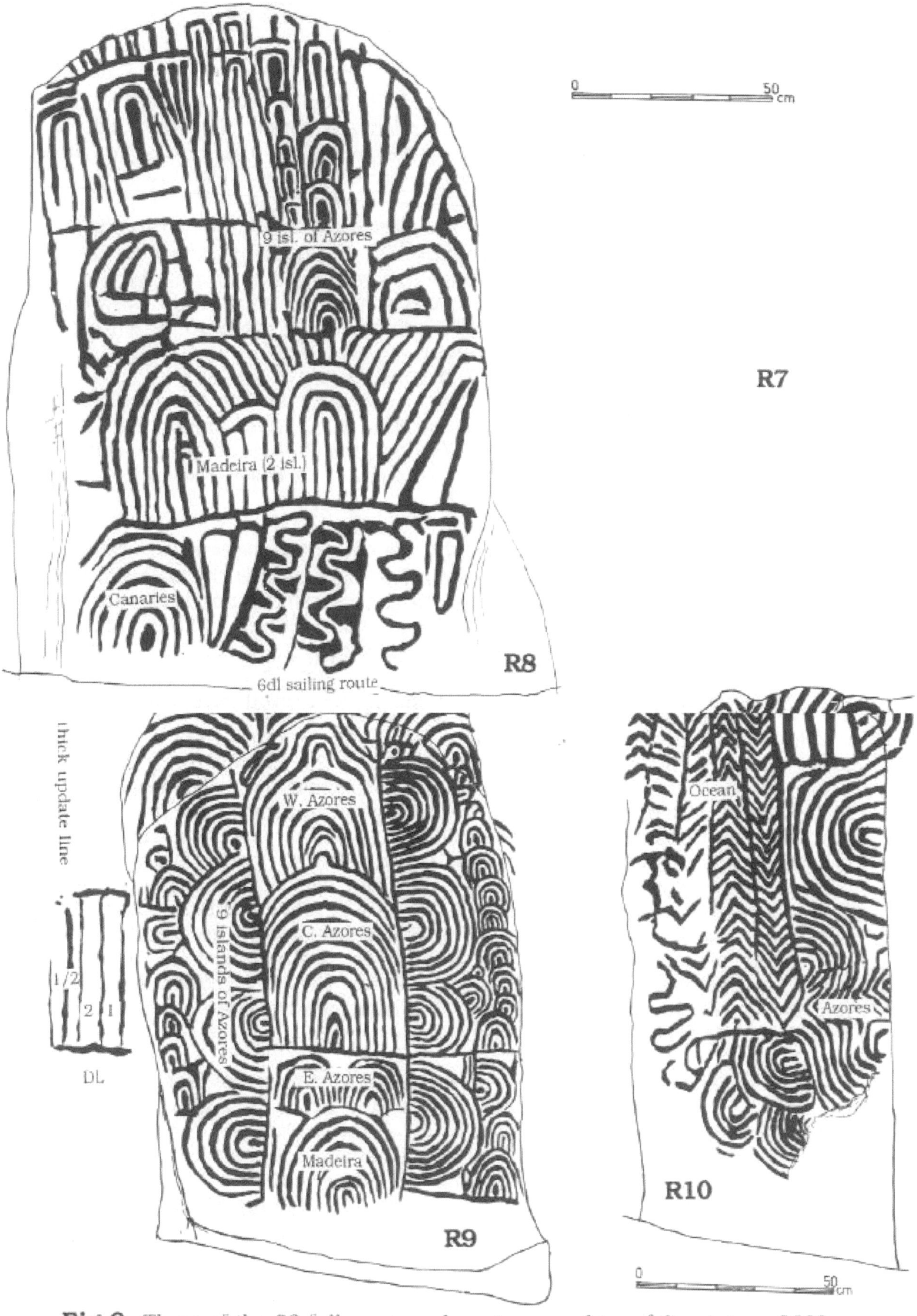


**Fig.9** Three of the 23 fully engraved western menhirs of Gavrinis, c.3600 BC (Ref.4).

a headland a statue with its right hand pointing dramatically west. Today this is heresay; the statue, if it ever existed, has long since disappeared. But for a discoverer to fabricate evidence of predecessors seems contrary to human nature" (Ref.18).

## Other Petroglyphs of the Azores

There are many megalithic petroglyphs concerning the fantastic Azores discovery. From c.3600 BC, when the Azores were visited by man for the first time, until America was discovered by the Bering Sea c.2600 BC, there was a thousand year period when the Azores were the western home of the SunGod. This was a long time to be wondering what was at the far side of the ocean, and the origin of all the myths about paradise, the land of the dead, and so on. A lot of petroglyphs about the Azores were carved during this time. All these were carved prior to the discovery of America, because none of them show features of land later found on the other side.

Figure 13 is a characteristic old petroglyph (c.3600 BC, Portugal), showing the three island groups of the Azores as circles, with the open ends pointing ESE, in the direction of Madeira. The petroglyph (Fig.14) from Cairn I at Loughcrew, eastern Ireland, shows the 5 islands of the Central Azores depicted as 5 shining suns on a clear day (c.3400 BC). Another petroglyph (Fig.17) at the same site shows emphasis on the coastal waters around the West Azores, which are explored over a distance of at least 6dl= 666km.

At Carrowmore, a huge site called a "megalithic cemetery" near Sligo in West Ireland, we found circles of kerbstones (formerly surrounding now-eroded tombs) with 38 stones (Fig.15), corresponding to the latitude of the Central Azores, at 38°N (c.3500 BC). A pretreated kerbstone face at Knowth in eastern Ireland (Fig.16), having a nice flat surface, shows the 9 islands of the Azores with distance lines around them (c.3500 BC, Refs.14,15). Figure 22 (right), also a petroglyph at Knowth, shows the extensive explorations around the Central Azores, and the West Azores at the top, producing an anthropomorphic figure that doubles as a representation of the SunGod.

After the discovery of Greenland (c.3300 BC), the Azores, having a better climate by being in the middle of the Ocean, remained important in the struggle to cross the Atlantic. They are depicted as 9 islands, or as three island groups, which is confirmed at the famous passage grave of Newgrange (Fig.18). The entrance stone and a kerbstone of this monument (Figs.18,19) show how important the Azores still were around c.3200 BC (Refs.14,15).

Newgrange is oriented on the rise of the midwinter sun, and the upper edge of the light shaft that catches the sunbeam contains 8 crosses (see Fig.19, edge of long stone at top edge of "light box"), corresponding to the 8 islands of the Azores (Flores/Corvo taken as one island). These Xs or rays are the same as those of the sill stone of Gavrinis, and have the same symbolic meaning. The letter X was named "exe", and phonetically this is "ecse", meaning "the Great Light" (Ref.16).

**Fig.10** Four of the 23 fully engraved western menhirs of Gavrinis, c.3600 BC (Ref.4).

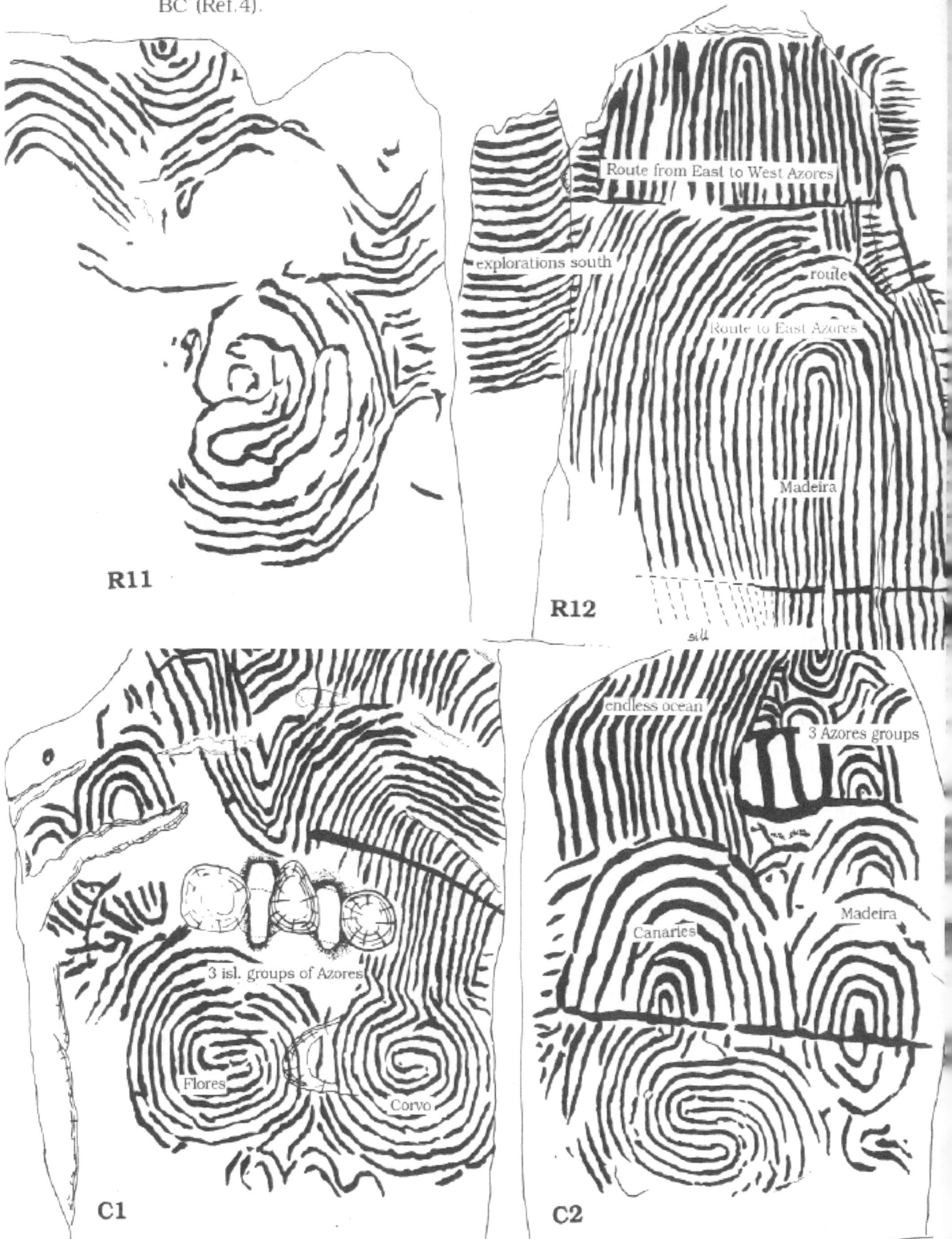

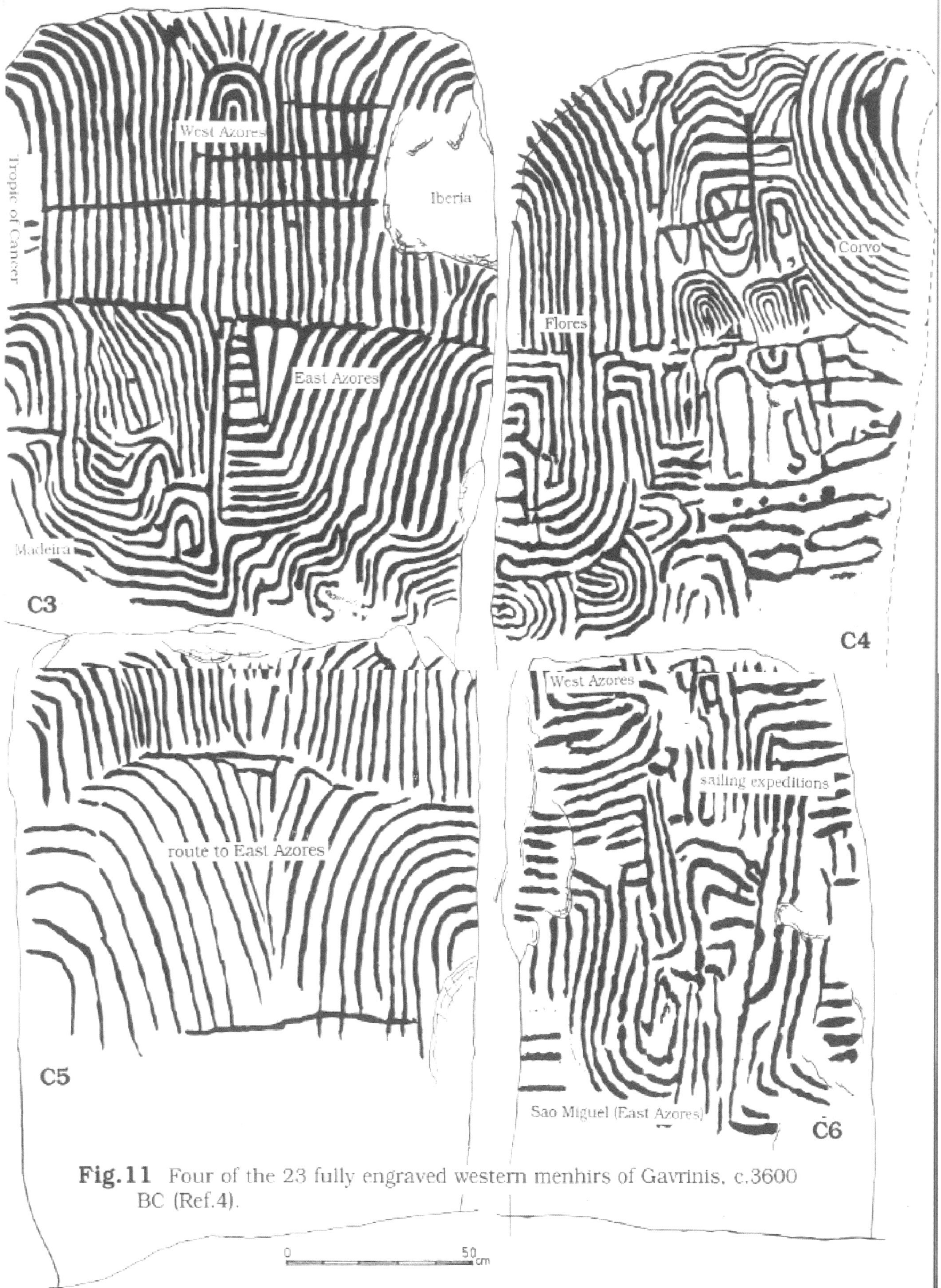


**Fig.11** Four of the 23 fully engraved western menhirs of Gavrinis, c.3600 BC (Ref.4).

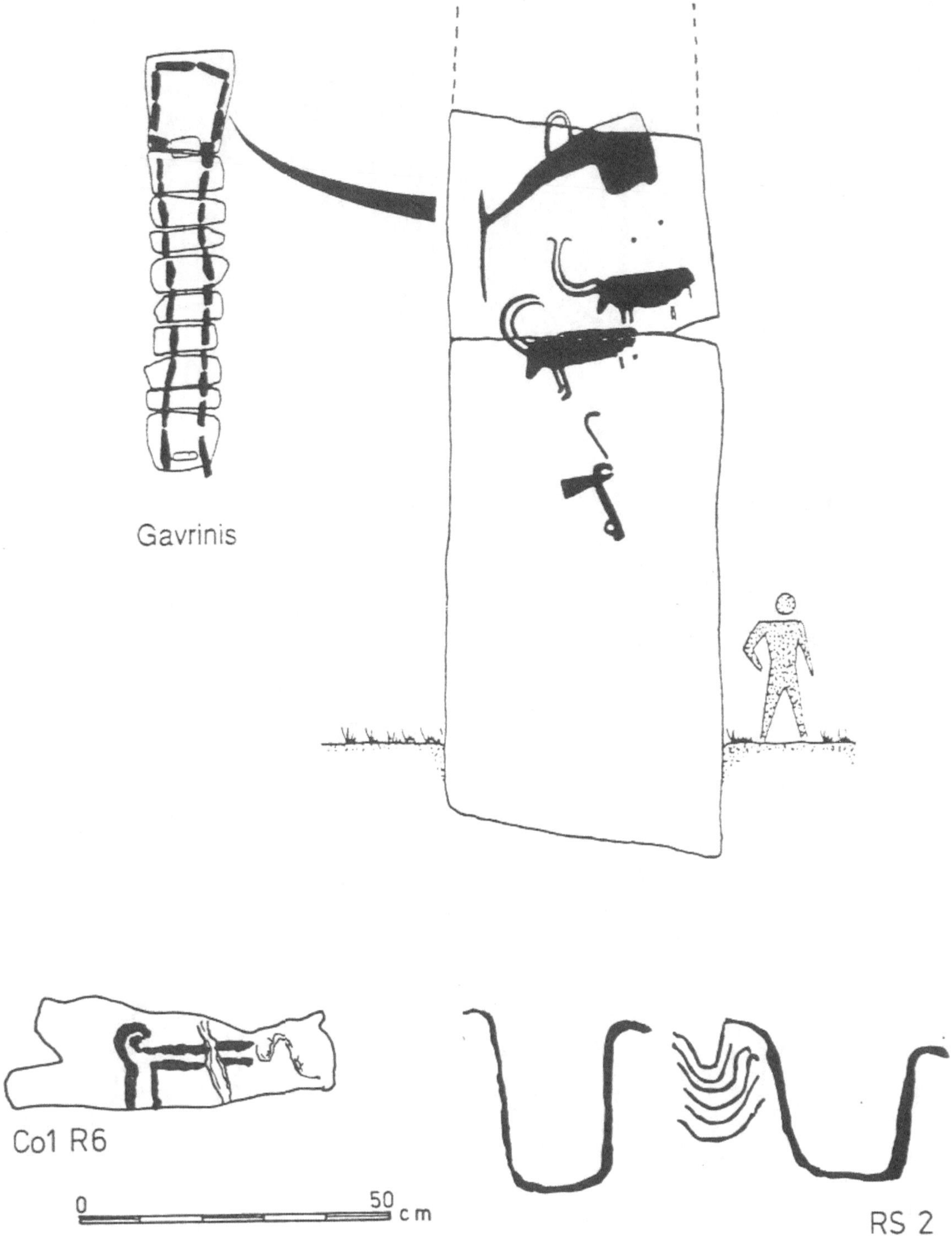


**Fig.12** **Above:** on top of the coverstone of Gavrinis is a petroglyph of the Little Mediterranean Sea, and a carving of one of two bulls (5000-4000 BC) (Ref.7). **Below:** left, a Mediterranean sign (Jacob's Staff), and right, two U-signs (crossing signs) or height profiles (c.3600 BC) (Ref.4).

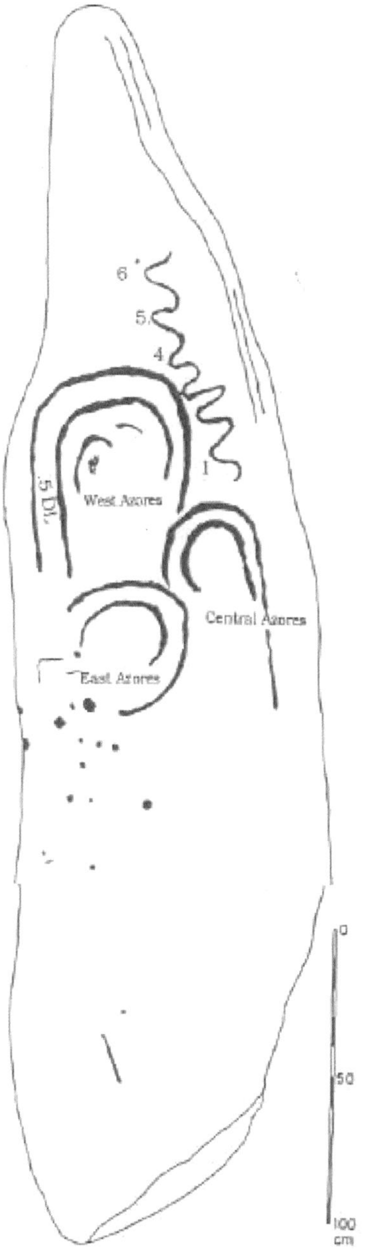


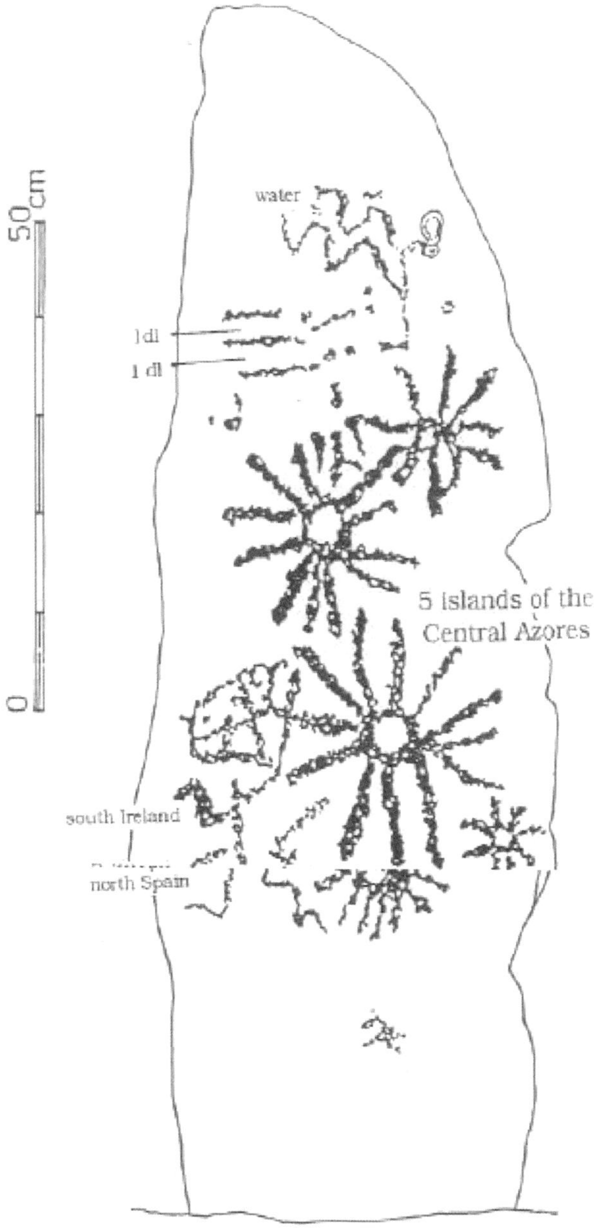


**Fig.13 Left:** The three island groups of the Azores, showing explorations of the coastal waters over 1/2 DL (=555km), to the west (the wave line) over 6dl (=666km). The open ends point ESE, so land continues in the direction of Madeira. The menhir, at the latitude of the Central Azores, 38°N, has the shape of the archipelago (Vale de Rodrigo, South Portugal, c.3600 BC, Ref.4).
**Fig. 14 Right:** The 5 suns are the 5 islands of the Central Azores. Higher, a few lines indicate the distance to the West Azores: 2dl= 222km. Behind that group of islands there is water (the two wavy lines). Left below, is the sailing route from South Ireland (above, thick) via Brittany to north Spain (below). From that coast the Azores can be reached! (Loughcrew, Cairn I, Ireland, c.3400 BC, Ref.4).

**Fig.15** We have found that kerbs of 37, 38 and 39 stones, encoding the 37°N, 38°N, and 39°N latitudes of the Azores occur in the tomb circles at the Carrowmore megalithic "cemetery". **Above**, kerbstones behind the cowshed, with Ben Bulben in background. **Below**, a circle of 38 stones, with Knockarea in background. (Carrowmore, West Ireland, c.3500 BC).

**Fig.16 Above:** The nine islands of the Azores (2 East, 5 Central, and 2 West). To the west, the ocean has been explored over 7dl=777km (Kerbstone, Knowth, Boyne Valley, East Ireland, c.3500 BC).

**Fig.17 Below:** We are looking from Ireland to the south, to the Azores. Low on the stone the East Azores (2 islands), above them the Central Azores (5 islands), and to the right behind the line, the important West Azores (Flores/Corvo). Around these western islands, the sea was explored over at least 6dl= 666km. (Loughcrew, Cairn I, c.3400 BC) (Ref.4)

**Fig.18** **Right half and deep:** the 3 island groups of the Azores; in total 3x3=9 holes in them, corresponding with the 9 islands of the Azores.
**Right half and shallow:** right, in the east, the two islands of Madeira. Left, in the west, the 9 islands of the Azores.
**Left side above:** the 2 islands of the East Azores, the Central and the West Azores. The sea (around the Central Azores) is explored over 7dl= 777km.
**Left side below:** the sea is divided in latitude lines and distance lines. The distance between the East and West Azores (shown above) is 5dl= 555km (kerbstone of Newgrange, Boyne Valley, Ireland, c.3200 BC).

**Fig.19** The nine islands of the Azores with distance lines on the entrance stone at Newgrange, with Dr. de Jonge behind it (above), and the 3 island groups of the Azores on kerbstones at Knowth and Newgrange (below) (c.3300 BC).

**Fig.20 Menhir with author's labels:** The surface is the ocean; at the right, Europe and Africa, and left, the supposed opposite side. In the middle the Cape Verde Islands (low), the Azores, and Greenland (above). Also, looking west, the East, Central and West Azores. Also the similar islets Glen East, Glen Middle, and Glen West. (Clear Island, Glen East, Ireland, c.2700 BC) (Ref.9).

**Stone ball:** A religious amulet three inches in diameter (8cm). One of well over 350 found in Scotland, showing the three island groups of the Azores, with megalithic distance lines in spirals and rings (Ref.8,17).

**Fig. 21** This symbol of the Azores, western home of the SunGod, has today become the government's symbol for the National Historic Monuments in Ireland, visible on all tourist information and admission tickets to all megalithic and Celtic monuments, as well as castles and museums in the country. (Photo by authors, passage grave of Newgrange, Boyne Valley, Ireland, c.3200 BC)

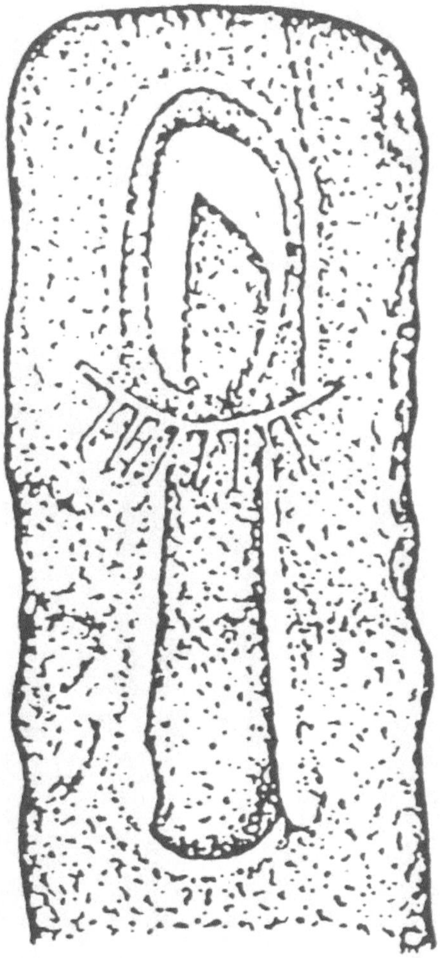

**Fig.22** **Left,** a sailing boat with 8 rowers (together 9 units, corresponding with the 9 islands of the Azores) has discovered solid rock (below) in the middle of the ocean (the stone). Also it tells that "we are sailing in the direction of the setting sun (the circle above), that projects a long shadow behind us (below)". In other words, "we do this in honor of the SunGod we have known for such a long time". This is the oldest petroglyph of a sailing boat with a sail on the ocean! Height=1.3m. (Gavrinis, Brittany, France, c.3600 BC, Ref.14).
**Right,** the Azores (below) and the West Azores (above) in particular. The sea around the Azores has been explored over 7dl=777km. We want to know what the world looks like beyond the West Azores! This is also an anthropomorphic figure that doubles as a representation of the SunGod. (Western grave, Knowth, Boyne Valley, East Ireland, c.3600 BC, Ref.14).

Figure 21 shows the most famous petroglyph of the Azores on the Newgrange site, now used as the Irish Government's National Historic Monument symbol. Figure 20 (left) shows a late carving from south Ireland, dating from before the discovery of America (c.2700 BC). At the right side is a true piece of art, a megalithic three-lobed stone ball with spirals and circles, a religious amulet which carries the magic of the Azores and the SunGod. Finally, in the passage grave of Gavrinis is a petroglyph not visible to the public, because it is on the side of a stone. This petroglyph is the oldest petroglyph in the world of a sailing ship with a sail on the Ocean (Fig.22, left) (Ref.14).

# References


1. Le Roux, C.T., Gavrinis, Ed.J.P. Gisserot, 1995
2. Briard, J., The Megaliths of Brittany, Gisserot, 1991
3. Balfour, M., Megalithic Mysteries, Dragon's World (1992) (ISBN 1-85028 -163-7)
4. Twohig, E. Shee, The Megalithic Art of Western Europe, Clarendon Press, Oxford, 1981
5. Jonge, de R.M., and IJzereef, G.F., De Stenen Spreken, Kosmos Z&K, Utrecht/Antwerpen, 1996 (ISBN 90-215-2846-0) (Dutch)
6. Jonge, de R.M., and IJzereef, G.F., Exhibition: The Megalithic Inscriptions of Western Europe, 1996
7. Bailloud, G., et.al., Carnac, Les Premieres Architectures de Pierre, CNRS Edition, 1995 (ISBN 2-85822-139-1) (French)
8. Morrison, R., et.al., Mysteries of the Ancient World, National Geographic Society, NY, 1979
9. Champion, T., et.al., Prehistoric Europe, Academic Press, London, 1984 (ISBN 0-12-167550-5)
10. McMann, J., Riddles of the Stone Age, Rock Carvings of Ancient Europe, Thames & Hudson, NY 1980 (ISBN 0-500-05033-3)
11. Wallis Budge, E.A.W., Osiris and the Egyptian Resurrection, 2 Vol., 1911, Dover Pub., N.Y. 1973 (ISBN 0-486-22780-4)
12. Wheeler, R.L., Walk Like An Egyptian, Allisone Press, 2000 (ISBN 1893774-21-X)
13. Cartographia Ltd., 1:400.000, Budapest, 1995 (ISBN 963-352-9808 CM)
14. "Art et Symboles du Megalithisme Europeen", Supplement N°8, Revue Archeologique de l'Ouest no 8, Nantes, 1995 (ISSN 0767-709-X)
15. Eogan, G., Knowth, and the Passage Tombs of Ireland, Thames and Hudson, 1986
16. Bayley, H., The Lost Language of Symbolism, Citadel Press, 1988 reprinting (ISBN 0-8065-1100-1)
17. Feats and Wisdom of the Ancients, Time-Life Books, Alexandria Va., 1990, 5th printing (ISBN 0-8094-7675-4)
18. Irwin, C., Fair Gods and Stone Faces, St. Martin's Press, New York, 1963 (pg.241)

**Fig. 1** Photo of Dr. Reinoud de Jonge explaining the "Story of Loughcrew" inscription, July 8, 1998. The flashlight points to the discovery of the Azores in the middle of the Atlantic (c. 3600 BC). This was the starting signal for the building of the monument-complex of Loughcrew, Ireland.

# The Discovery of the Faeroes, Iceland, and Greenland

## The Decipherment of Megalithic Picture Writing

(Petroglyphs of Cairn T, Loughcrew, Meath, Ireland, c.3200 BC)

Dr. R. M. de Jonge, drsrmdejonge@hotmail.com
J. S. Wakefield, jayswakefield@yahoo.com

### Introduction

Dr. de Jonge called his first interpretation of Fig.1 (seen also in Figs.4,5) "The Story of Loughcrew". This photo is of Reinoud and Stone C8, the westernmost interior stone of the megalithic "passage grave" in Cairn T, Loughcrew, Ireland, dated c.3200 BC (Figs.2,3). You are facing west toward the setting sun and the western ocean as you look at the stone inside the cairn. These cairns may not have been originally constructed as graves, but as monuments, astronomical tools ("Cairns of the Sun"), and/or mission churches. Tomkins (Ref.20) points out that "the custom of burying distinguished citizens in national monuments that were not originally designed for that purpose is common to the world, as in Westminster Abbey, the Invalides, the Pantheon, and Maes-Howe". Brennan (Ref.18) claims that "large stone bowls and burned stones [such as have been found at Knowth and Newgrange] are evidence of steambath "purification rites". There is a later Celtic tradition that having sex in the cairns will help you get pregnant through "divine coition" (Ref.11). The three hills of the complex lie near the Boyne River, west of Newgrange, in central Ireland, 60km from the east coast. This "megalithic cemetery" consists of many graves, which are situated inside cairns, on the three grassy hilltops now being grazed by cows and sheep. These tombs contain many stones with inscriptions (Ref.2), but we will primarily look at two stones in this article.

The Carbon-14 dating of megalithic tombs in Ireland has been done on teeth remaining in these tombs from low-temperature wood-fire cremations. The C-14 measurements all point to dates in the second half of the 4th millennium BC (Ref.2), which would be after the discovery of the Azores (c.3600 BC), the new western home of the SunGod (Ref.1). So, it does not surprise us, that this whole complex of monuments of Loughcrew strongly resembles the three groups of islands of the Azores. The passage graves lie on three hilltops, with an altitude of roughly 800m. The eastern hilltop possesses in fact two cairns (see Fig.2), corresponding to the two islands of the East Azores. The highest central hilltop possesses five more cairns, for seven total, corresponding to the five additional islands of the Central Azores. And the western hilltop possesses two more cairns, for nine total, corresponding to the two additional islands of the West Azores. As a consequence, the inscriptions in the passage graves should be associated with these islands. Smaller, lower cairns have been later added around these cairns, as shown in Fig.2, some of which have been shown to be

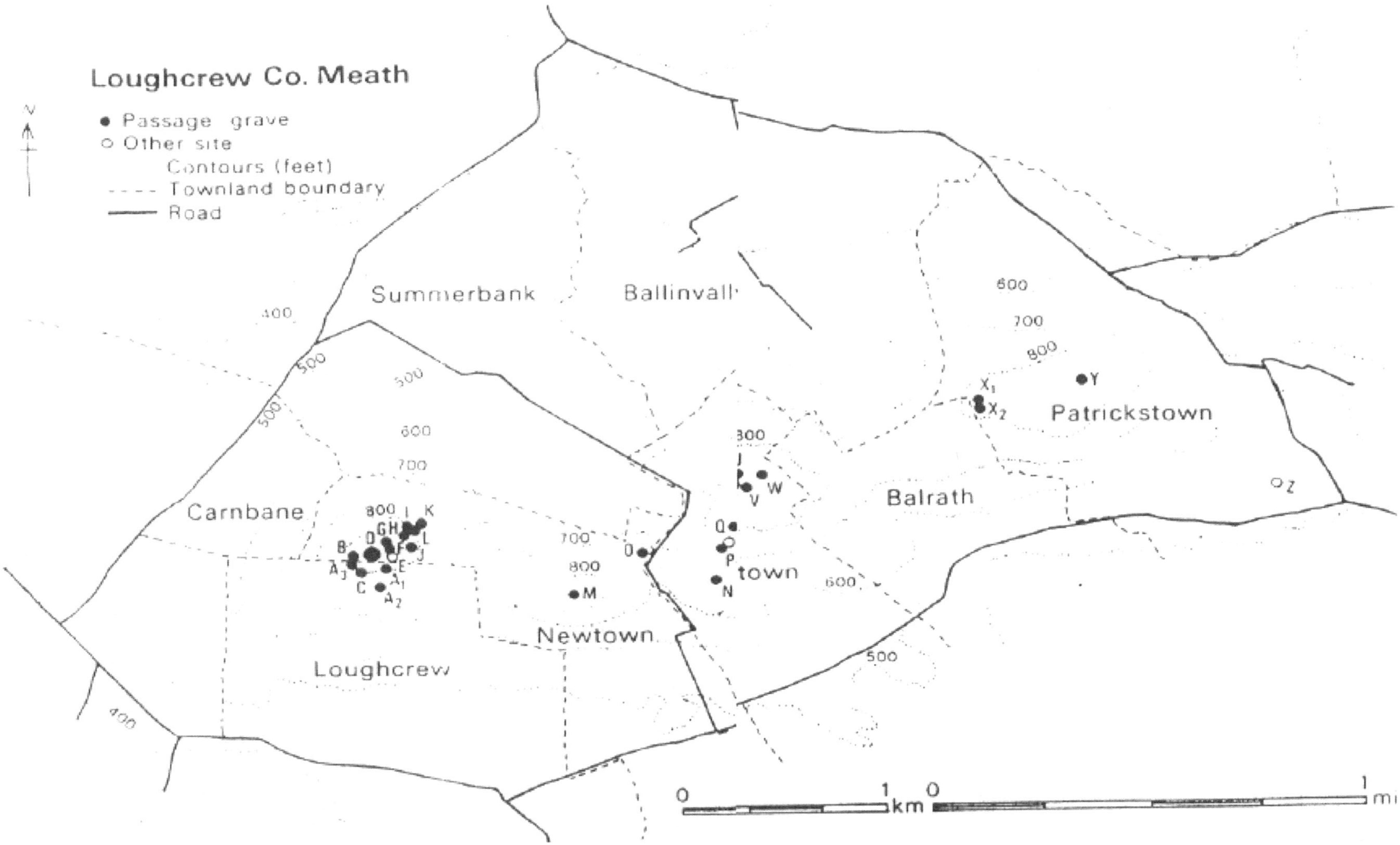


**Fig. 2.** Map of the monument-complex of icrew, Co. Meath, Ireland (3600-3100 BC) (Ref.2). The passage gravesns) lie on the tops of three hills, which resemble the East, the Ceiand the West Azores.

calendrical "warning" tools (sunlight in their passages shine on glyphs prior to those at special Holy sites (Ref.17).

The most important passage grave is in Cairn T, on top of the highest hill of Loughcrew, in Corstown (Fig.2). The cairn, including the passage grave, is oriented to the west. The westernmost chamber will be the most important one. And of course, within this chamber the westernmost stone (C8) will be of special importance. In accord with this statement, the endstone C8 has been engraved in a surprisingly beautiful fashion (Fig.4).

Stone C8 (Ref.24) is well known for its decorative "art" and "a fine array of sun-eyes" (see Mythic Ireland, by Michael Dames, Ref.11). Other authors have said this is "unsophisticated art" (O'Kelly, Ref.13). "Newgrange has much higher art, than for example, Loughcrew" (Balfour, Ref.15). These are "motifs close to patterns recorded in trances and hallucinations" (Patton, Ref.14). "A finger of light [enters the passage grave], illuminating the symbols in turn, spelling out a message in a forgotten language of symbols" (Mitchell, Ref.9). "It is highly abstract, a form of symbolic writing, many symbols seem to indicate numbers and counting connected with timekeeping" (Brennan, Ref.17). We agree that these inscriptions are artistic, and find they also contain geographic information, and convey the aspirations, experiences, and thinking of their creators.

## Stone C-8, Cairn T

This is a story in ideograms, a quantitative picture writing, actually the oldest written history on earth. It is a story of the attempts to reach the other side of the world, - that is, to cross the ocean (c.3300 BC). These images are a unique form of picture writing, and are the most historically important and most beautiful megalithic inscriptions in the British Isles. We consider them the second best in Europe, after the petroglyphs of Dissignac, Brittany. As in many other examples of cave and stone art, the natural shape of the stone has been used as part of the work, the surface of the stone being an endless sea that is now called the North Atlantic Ocean. As drawn in detail in Fig.4, note the coast of Europe on the right, Greenland in the north, and the other, unknown side of the ocean on the left. The proof of the story lies in the inscriptions themselves. Each figure represents a number. By adding the numbers, the actual and true degrees of latitude of the discovered land-points are obtained, and all in historical order!

The "Story of Loughcrew" begins with a sailing boat (see Figs.4&5) sailing west from the Strait of Gibraltar (A) to the Canary Islands (B). They expected to cross the ocean at the degree of latitude of the Tropic of Cancer, 23°N, which held religious meaning in honor of the Egyptian SunGod Ra (D), who went west every night, from the center of the Southern Egyptian Empire, at 23°N. Egyptian tombs have ships in them so the deceased pharaohs could make this trip to the realm of the dead in the west, at the other side of the waters, where the sun sets (D, behind a natural

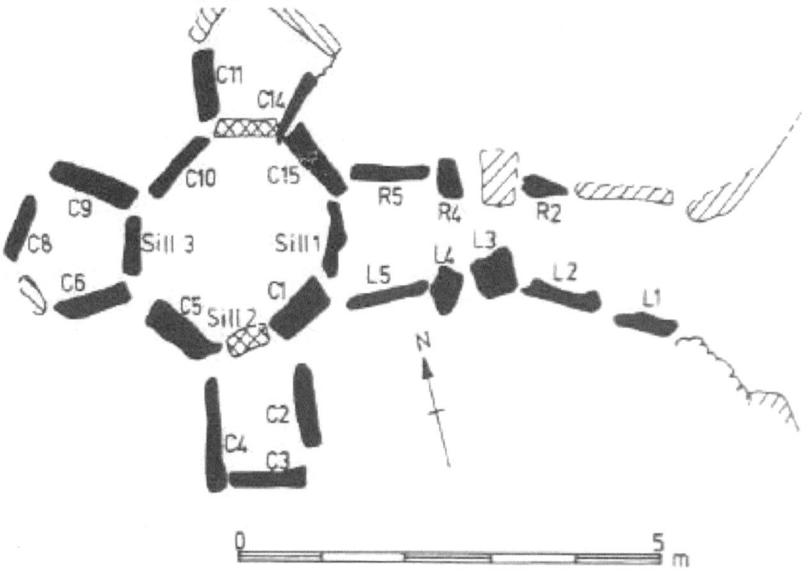


**Fig. 3** Photo by authors of Cairn T, Loughcrew, Ireland, and groundplan of passage grave in this Cairn. (c.3300 BC) (Ref.2)

relief in the stone). (The daisies, like the other important figures, mark major land points.) The "first poem written in Ireland", an archaic Druidic incantation of unknown date called "The Mystery of Amergin" asks in its 16th line: "Who teaches the place where falls the sunset?" (Ref.17). Like theorems in geometry, the proofs for Reinoud's interpretations lie within the inscriptions themselves. The ship contains six distance lines (dl) (count the spaces), three in the hull, three in the sails (follow the numbers in Fig.5). Item B has six petals, item C has three dl, and item D has eight petals, for a total of 23 (A+B+C+D= 6+6+3+8= 23). The problem with this route is that it is the tradewind route west, so the boats could not sail back against the winds and currents, and disappeared.

By sailing south and west one would sail to the Canary Islands (B) at 29°N (A+B+D+E= 6+6+8+9= 29). (Since C is Cape Verde, and further south, this item is not included in the calculation.) Along the coast of NW Africa, they sailed further southwest to Cape Verde (B to C) at 15°N (A+B+C=6+6+3= 15). (To follow the latitudes, you should use a good map, like the National Geographic's Atlantic Ocean, 1955, see Fig.8.) From there, they discovered the Cape Verde Islands (C again) at about 666km offshore (A has six dl=666km). The islands extend from east to west over 333km (C has 3dl= 333km), and they are situated at 15°N (B+D+C1= 6+8+1= 15), 16°N (B+D+C1+C2= 6+8+1+1= 6), and 17°N (B+D+C1+C2+C3= 6+8+1+1+1= 17). The petroglyph C has also the shape of the archipelago (turned 90° clockwise), and it suggests the coastal waters were explored over a distance of 6dl= 6° of latitude= 666km (right and left 3+3= 6 spaces). But they still did not succeed in crossing the ocean.

Later on, they discovered, above the Canary Islands (B), the islands of Madeira (E&F) 12°WSW (A+B= 6+6= 12) from the Strait of Gibraltar (A). These islands are situated at 33°N (A+B+C+D+E+F= A-F= 6+6+3+8+9+1 =33) at a distance of 666km (A or B=6dl=666km) from the coast of northwest Africa. They consist of the main island of Madeira (E, large inscription), and the small island of Porto Santo (F, small inscription). The discovery is modest, because the small islands (E&F are small inscriptions) are situated less west than the Cape Verde Islands, but they gave much hope (E is a little sun) that more land could be found in the west! (We hear Heyerdahl is now investigating stone pyramids in the Canaries.)

At about the same time they discovered near the British Isles (G), west of Scotland and west of the Hebrides (G'), the Islet of Rockall (G"), at 57°N (A-G,+G'+G"= 41+8+8= 57). This islet gave a little bit of hope of more land in the west, so it is carved like a star, smaller than the star of Madeira (E).

Not long afterwards, as a result of sailing explorations from Madeira (E), they discovered the three island groups of the Azores (F,H,&I). (Glyph F has double meanings.) West of the Strait of Gibraltar (A) at 36°N, the Initial Sailing Direction from Madeira is 36°NW (Ab+B+E+F+H+I= 3+6+9+1+9+8= 36), over a distance of about 9dl (E= 9, 9x111= 999km). The Terminal

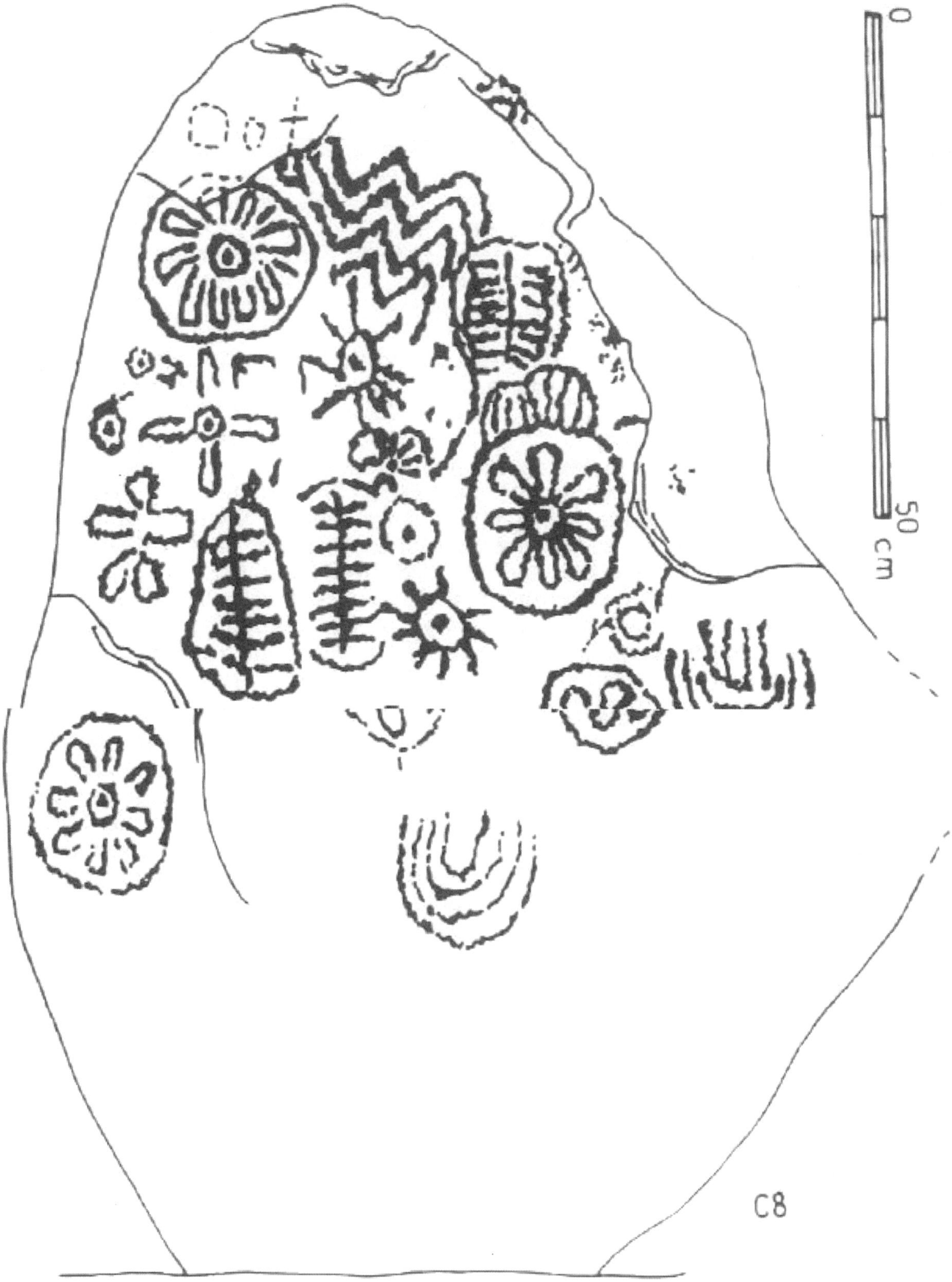


**Fig. 4** Drawing of Stone C8, Cairn T, Loughcrew, from The Megalithic Art of Western Europe (Ref.2, c.3300 BC).

Sailing Direction in the neighborhood of Santa Maria (East Azores, F) is 29°NW, corresponding to the latitude of the Canaries (B), already calculated, 29°N. The difference is due to the curvature of the Earth. The East, Central, and West Azores are situated at 37°N (Aa1+Ab+B+E+F+H+I= 1+3+6+9+1+9+8=37), 38°N(Aa1+Aa2+Ab+B+E+F+H+I= 1+1+3+6+9+1+9+8= 38), and 39°N (Aa1+Aa2+Aa3+Ab+B+E+F+H+I= 1+1+1+3+6+9+1+9+8= 39). From Santa Maria (East Azores, F), one steers a course of 27°WNW (E+F+H+I=9+1+9+8= 27) for the Central (H) and West (I) Azores. The West Azores lie approximately 1888km offshore (H+I= 8+9= 17= 17dl= 1888km). The Azores consist of nine islands (H=9), situated more to the west than the Cape Verde Islands (C), far in the ocean. To the west, the sea was explored over 8dl= 888km (I=8). H&I are small images of the ocean, with lines of latitude. They believed they were in the middle of the ocean, so they have drawn vertical lines down the middle of them. The calculated latitudes, sailing directions, and distances are correct on a modern map.Unfortunately, this discovery of the Azores did not lead to a crossing of the ocean. Note the despondent little man with his hands in the air on the west side of the Azores!

So expeditions then moved back to the British Isles (G), not to Rockall, but the Isle of Lewis in the Hebrides Islands (G') with its megalithic ruins at 58°N (A+B+C+D+E+F+G+H+G'= 6+6+3+8+9+1+8+9+8= A-H+G'= 50+8= 58). This point lies 444km (G'b=4dl=4°) north of Loughcrew at 54°N (A-H+G'a= 50+4= 54). From there, they started northward travel by going first to the Orkney Islands (J) where there are more enormous megalithic ruins, at 59°N (A-H+J= 50+9= 59). These islands (J) give them hope of crossing the ocean to paradise, just like the islands of Madeira (E), both drawn like little suns with rays.

From there they traveled a course of about 59°NW (A-H+J= 50+9= 59) for the Faeroe Islands (K), which are further north at 62°N (A-I+K= 58+4= 62). This is a distance of 444km at the most (G'a= 4dl= 444km). From the Faeroes, they sailed in a direction of approximately 30°WNW (H+I+J+K= 9+8+9+4= 30) to the southeast coast of Iceland (L) at 64°N (A-I+L= 58+6= 64). This is also a distance of 444km (G'b=4dl). It is clear that they thought that Iceland was like the Azores, on the Mid-Atlantic Ridge, or at least in the middle of the ocean (L, with central vertical line). However, Iceland is a big island, unlike the islands of the Azores (compare L with H&I, L has an edge). Then they sailed westward around Iceland to its NW peninsula (L', the edge around L, see also the map of Fig.8) at 66°N (A-I+L+L'= 58+6+2= 66). Now they have reached an area more to the west than the islet of Rockall (G", see the connection line).

Finally, they steer a course from this NW peninsula (L'), with a sailing direction of c.15°WNW (L+M− 6+9= 15), and over a distance of 555km (5dl, because the waves go up and down five times), for the southeast coast of Greenland (the edge of the depressed section of stone). They arrived at Cape Holm (M) at the latitude of the Arctic Circle (M) at 67°N (A-I+M=

# Megalithic Explorations (C8)

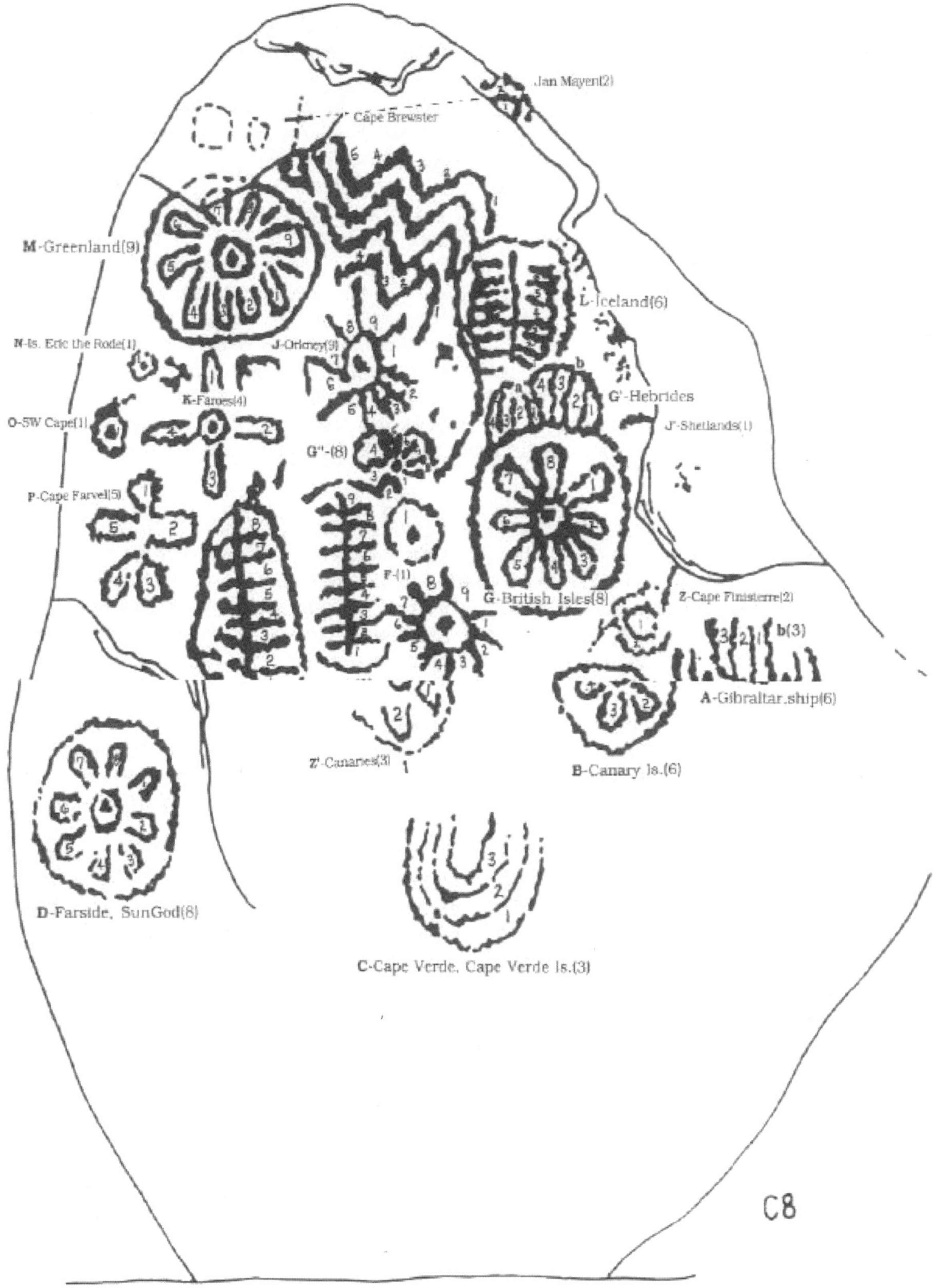


**Fig. 5** Drawing of Stone C8, with author's labels, geographic names and numbers from text.

58+9= 67). Now they have reached an area more to the west than the Azores (H&I), and they think they have crossed the Atlantic Ocean (M is almost as far west as D)! They achieved this success at the holy Arctic Circle (M), with thanks to the SunGod (both M and D)! But they encountered great difficulties sailing in the ice-cold sea of the Denmark Strait, as the western portion of the waves between Iceland and Greenland are deeply engraved!

Actually, and also of interest to the megalithic sailors, the shortest distance to Greenland is toward Cape Ravn at 444km (4 up-down wave segments at the bottom= 4dl= 444km) at 69°N (A-I+M+M'= 58+9+2= 69 where M' is the 2 marks between the petals of M). The three spaces between the waves indicate that three degrees north of the Arctic Circle is Cape Brewster, the most important cape in the North at 70°N (A-I+M+3dl= 58+9+3= 70).

At the southeast coast of Greenland they went one degree of latitude (N=1) south to Angmagssalik on the icefree Island of Eric the Rode (N) at 66°N (Arctic Circle= 67°,-1= 66°). They went further south 4° (K=4) to the icefree peninsulas of Skjoldungen (K again) at 63°N (Arctic Circle= 67,-4= 63). Then they went to 5° south (N+K=1+4= 5) along Cape Adelaer (N+K) at 62°N (Arctic Circle=67,-5= 62), the same latitude as the Faeroes (K), 62°N (A-I+K= 58+4= 62). They then went 7° (N+O+P= 1+1+5= 7) to the south at Cape Farvel (P) at 60°N (Arctic Circle= 67,-7= 60), which turned out to be the southernmost point of Greenland. The sea was explored a maximum of 5dl= 555km (P=5).

Finally, they explored 1° of latitude (O=1) to the north (P, head) to the southwest cape of Greenland (O) at 61°N, but it was tough going (O is deeply engraved). The sea could only be explored over a distance of 111km (O=1 =1dl= 111km). As a consequence, they did not reach a higher latitude on the west coast than 62°N, again the latitude of the Faeroes (K)! (This is why the Faeroes have been engraved so westerly.) So, finally they had to give up, and all explorations had to stop in dismay (P, the little man, Fig.14)!

## Details of Stone C8

The chance to cross the ocean seemed to them to be the greatest at Cape Verde Islands (C), where the known land protrudes the most to the west (Fig.8), but strong winds and sea currents are flowing from the northeast there, so to get back from the other side of the ocean, they would have to row! Glyph C is therefore a sailing boat like A, but without sails! They needed a route back that could be sailed. That seemed most likely at the Azores (H,I), where the prevailing winds and currents are flowing from the west, so the inscriptions of H and I have sails! However, if this did not turn out, the Upper North was an alternative route they could use, and so for this route one sail is shown, which is the inscription of Iceland (L).

The flower D is their symbol for the other side of the ocean (D, America), for the promised land, paradise, the land of the dead, and for the SunGod. It is a symbol with many complex and interwoven meanings, like many of

**Fig. 6** Roofstone above Stone C8, Cairn T, Loughcrew, from Megalithic Art of Western Europe, (Ref.2, c.3200 BC).

our own symbols. Inscription G is another representation of D, the SunGod, and represents Loughcrew itself. In spring, and in autumn, there is a moment at sunrise when the C8 endstone is lighted by sunshine coming through the tomb entrance lighting up inscription M, which is a flower too (Refs.9,17). That is because they were convinced that they had crossed the ocean at the holy Arctic Circle, the farthest north the sun shines on mid-winter day. Glyph B for the Canaries is a flower too, because people on the Canary Islands (B), at the latitude of the Nile Delta, have asked themselves for thousands of years: "to cross the Ocean, do we travel to the north or to the south?" Nobody knows but the SunGod!

The importance of the Azores on the route to paradise is emphasized by repeated use of 3 and 9 in megalithic art dated after the discovery of this archipellago in c.3600 BC, representing the three groups of the nine islands of the Azores. On stone C8, they made two (2) similar inscriptions, A and C, because for a way back across the ocean, they believe in Madeira and the Azores (Fig.8). They made three (3) analogous figures, H, I, and L, because they believe in the three island groups of the Azores. They made four (4) other resembling images, B, D, G, and M, because they firmly believed in the four island-groups of Madeira and the Azores. In total they made 9 (2+3+4= 9) splendid inscriptions, because they were convinced that someday the 9 islands of the Azores would play an important role in crossing the ocean. They also made two less beautiful inscriptions, K and P, because this is also true for the two islands of Madeira.

In fact, the whole complex of Loughcrew strongly resembles the three groups of islands in the Azores (see Fig.2). Cairn T, in which stone C8 is placed (see Fig.3), is oriented to the west, and has three chambers, which may be associated either with the three island groups in the Azores, or with the Cape Verde Islands to the south, the Azores to the west, and Greenland to the north. The main axis of Cairn T points 12° WNW, equal to the sailing direction from the Strait of Gibraltar to Santa Maria, the easternmost island of the Azores. The monument has 39 kerbstones (Ref.22), in agreement with the degree of latitude of the West Azores, 39° N, considered the most important stepping-stone to the other world.

The megalithic SunGod must have been the same as the Egyptian Supreme god Ra. The similar inscriptions H, I, and L represent together 23 (9+8+6= 23) units, corresponding to the latitude of the Tropic of Cancer, and the center of the Southern Egyptian Empire at 23° N. The similar inscriptions B, D, G, and M represent together 31 (6+8+8+9= 31) units, corresponding to the latitude of the important Nile Delta, the center of the Northern Egyptian Empire at 31°N. This stone C8 is in direct contact with this center, because the sum of both latitudes equal the latitude of Loughcrew at 54°N (23+31= 54). The similar inscriptions A and C (6+3= 9), and also K and P (4+5= 9) represent the nine islands of the Azores. It is not by coincidence that Madeira (E) and even the Orkneys (J) are symbolized by a sun with 9 rays. Madeira and the Azores are indicated by the inscriptions E,

# Roofstone

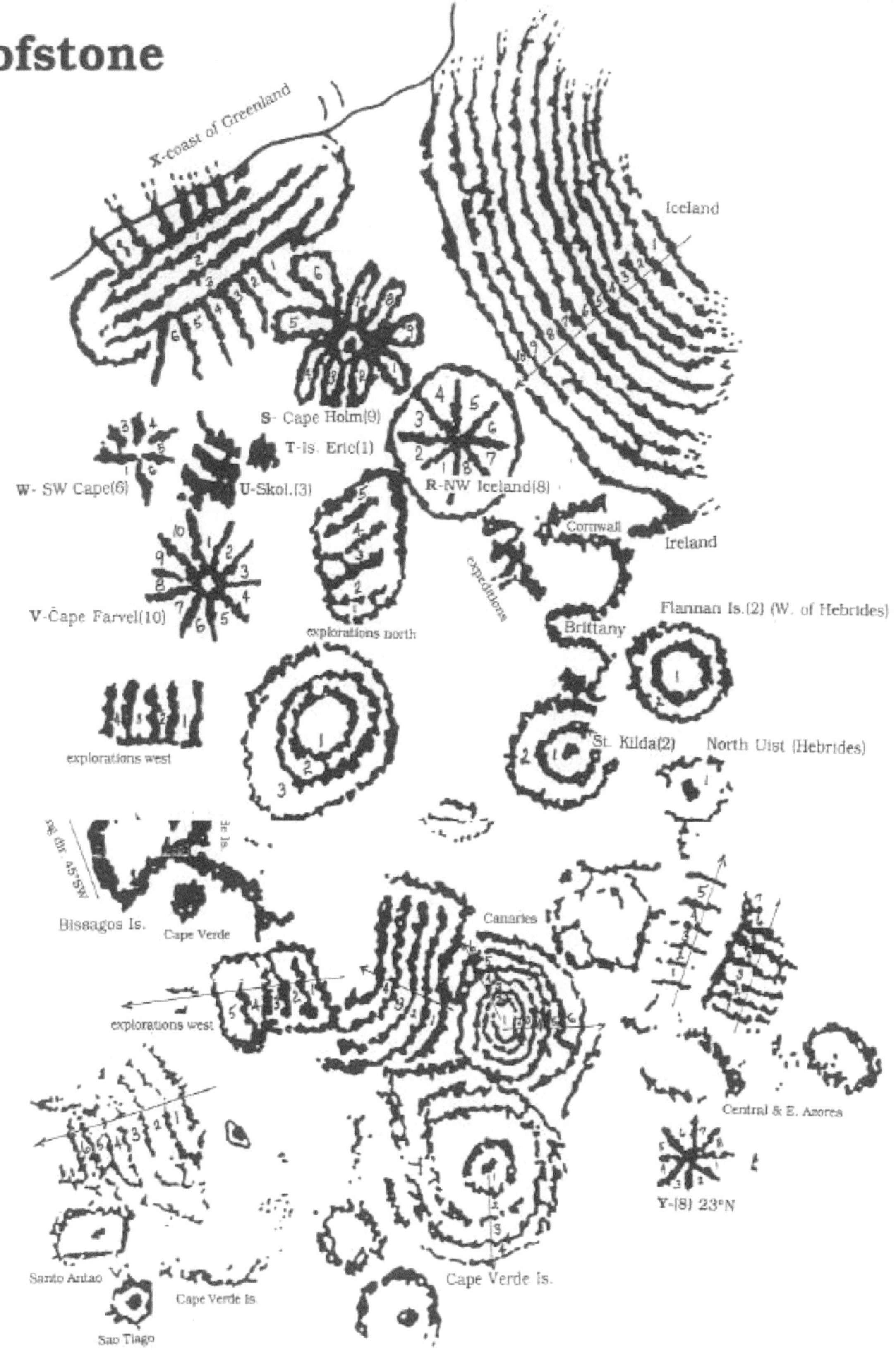


**Fig. 7** Roofstone above Stone C8, with author's labels, geographic names, and numbers from text (c.3200 BC).

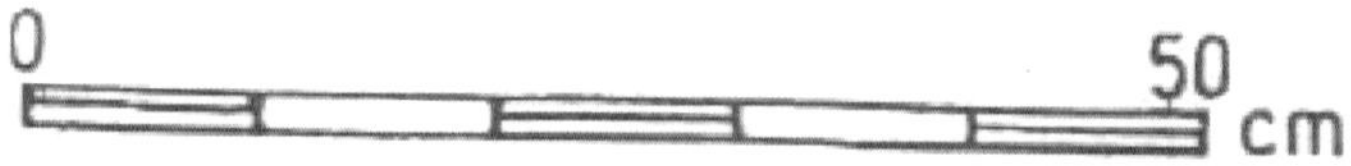

F, H, and I. They represent together 27 (9+1+9+8= 27) units, corresponding to the important latitude of the center of the United Egyptian Empire at 27°N, halfway between the Tropic of Cancer and the Nile Delta. The four "stars" E, F, G", and J contain 27 (9+1+8+9= 27) units too, confirming this latitude. (Also, the sum of all the nine major inscriptions total 63, which subtracted from a 90 degree right angle equals 27° (A+B+C+D+G+H+I+L+M= 63; 90-63= 27.) The Egyptian SunGod has told us "the realm of the dead (D) is in the west, at the other side of the waters, in the land where the sun sets. After death, you will be reunited there, with your ancestors, your family, your relatives, your friends and your acquaintances".

Other details can be deciphered on stone C8. The easterly Shetland Islands have only been indicated by the scratch of J'. They lie at 60°N (A-H+J+J'= 50+9+1= 60). The less deep carvings Z and Z' have been made at a later time, but before c.2500 BC. They do not belong to the original set of inscriptions. The double circle Z is Cape Finistere (NW Spain) between the Strait of Gibraltar (A) and the British Isles (G), at a latitude of 43°N (A-G+Z=41+2=43). The Z', south of Madeira (E), depicts the 3 westernmost Canary Islands at 28°N (A-D+Z+Z'= 23+2+3= 28). It is emphasized that both this cape and these western islands may still be of great importance. Finally, on the top of the stone, north of Iceland (L), we see a double circle, the island of Jan Mayen, discovered some centuries later (c.2900 BC). Its latitude of 71°N (69+2= 71) can be calculated from that of Cape Ravn at 69°N (A-I+M+M'= 58+9+2= 69, where M' is the two marks between the petals of M), given by the main inscriptions of the stone. The latitude is confirmed by the two circles in the Greenland shaped area on top of the rock (69+2= 71). It is further confirmed by a small piece of an extra distance line along the edge of it (Arctic Circle (M)= 67, +3+1= 71), from which a T-shaped cross arises, the horizontal bar of it pointing to Jan Mayen, the newly discovered island. This last glyph shows that it can be easily reached from Cape Brewster at 70°N (67+3= 70), with a sailing distance of 3+1= 4dl= 444km.

## The Roofstone of Cairn T

The serial "Story of Loughcrew" continues above the upright stone C8 on the roofstone in the chamber (see Figs.6&7). The western part of this roofstone (the top of the drawing) shows a continuation of the story, so we think these inscriptions are slightly later than C8. At the top right side, an inscription of the southwest coast of Iceland is shown. The sea to the SW has been explored over a distance of 10dl =1111km, which is quite a lot! To the south, it is explored until the north coast of Ireland, which is indicated with a short horizontal line. Beside this little line we see the coasts of Cornwall (above), and of Brittany (below). From the lowest peninsula a twisting line points to the northwest. So the later expeditions were undertaken from Brittany!

From the megalithic center of Brittany they sailed directly to the NW

# Megalithic Explorations, Atlantic

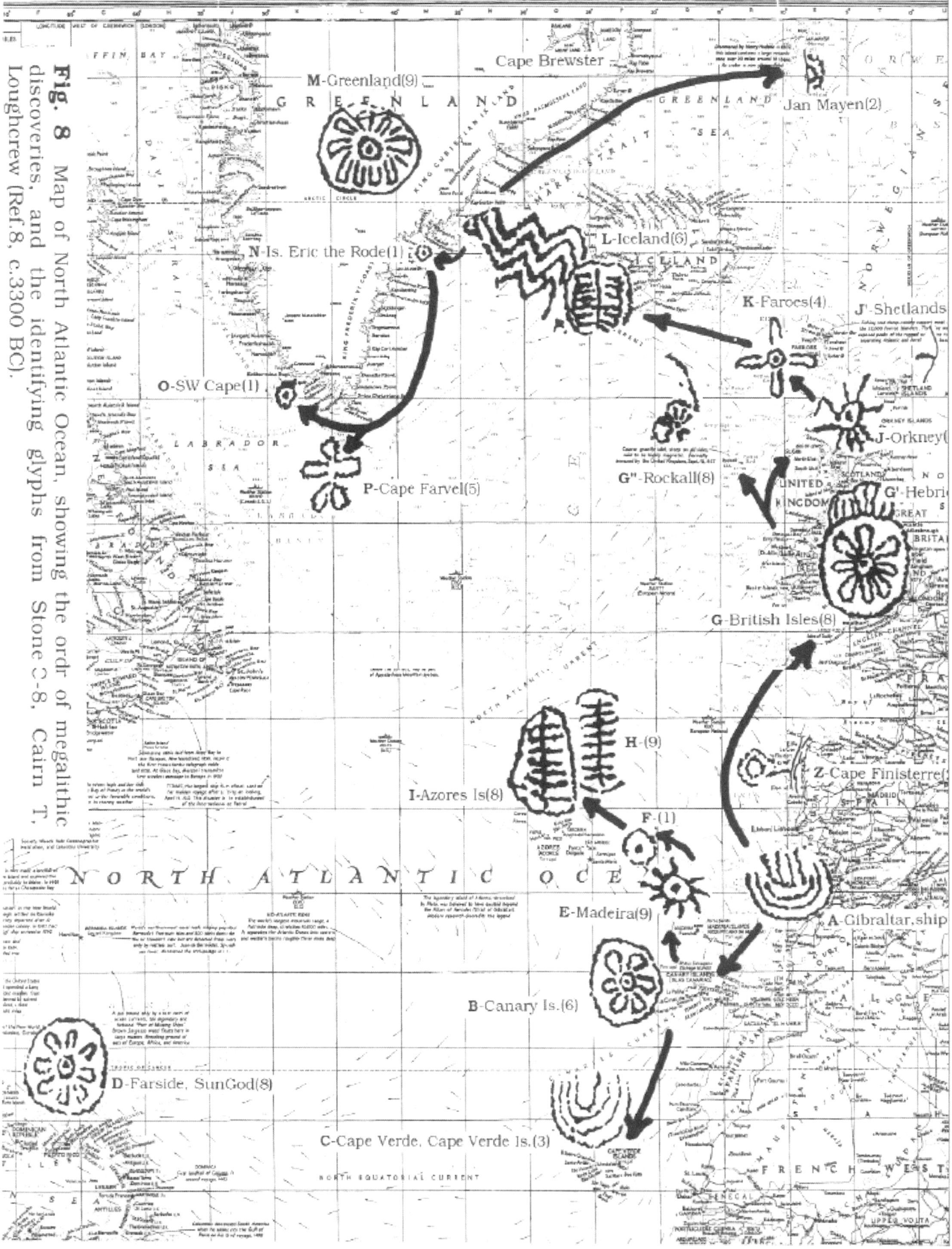


**Fig. 8** Map of North Atlantic Ocean, showing the ordr of megalithic discoveries, and the identifying glyphs from Stone C-8, Cairn T, Loughcrew (Ref.8, c.3300 BC).

peninsula of Iceland (R) (R=8, compare with L+L'= 6+2= 8 on stone C8) at 66°N (A-I+R= 58+8= 66), and next to Cape Holm (S) (S=9, compare with M=9), on Greenland at the Arctic Circle at 67°N (A-I+S= 58+9= 67). The counting of the degrees of latitude on the roofstone is closely coupled with stone C8.

From Cape Holm (S, see map of Fig.9) they sailed 1 degree of latitude (T=1) to the south to Angmagssalik on the icefree island of Eric the Rode (T) at 66°N. The inscription T (the dot) represents this island literally. Next they were another 3° (U=3) to the south on the icefree peninsulas of Skjoldungen (U) at 63°N. Again the inscription U (3 strokes) represents the three peninsulas of Skjoldungen literally. Next they reached Cape Farvel (V), the southernmost point of Greenland, at the same latitude as the Shetland Islands, 60° N (A-H+V= 50+10= 60). It should be noticed that the artist now refers to islands in the east in the same way as the artist did on stone C8, with Cape Adelaer (O+K) and the Faeroes (K), both at 62° N, and the degree of latitude is calculated in the same way as that of the Orkney's (J) on stone C8 at 59°N (A-H+J= 50+9= 59). In this manner, they are confirming that the explorations have arrived to a subsequent stage!

Finally, they arrived 6° (W=6) south of the Arctic Circle (S), at the SW Cape of Greenland at 61°N (Arctic Circle (S)-W= 61; 67-6= 61) . It should be noticed, that the two lowest beams match the shape of the south coast of Greenland. At the top left side, the edge of the stone is used to represent the SE coast of Greenland. Parallel, we see inscription X, with 6 beams at both sides (a "caterpillar"), pointing towards the south of Greenland. This means that the coastal waters from 67° to the SW Cape at 61°N (Arctic Circle-X= 67-6= 61) have been investigated completely. However, the width of the carving amounts to 3dl, so this means a distance of 333km from shore. As a consequence, America was not discovered c.3200 BC after these trips either.

At the right bottom corner on the stone, a little star is engraved having 8 beams (Y). At noon on a midsummer day at 23°N, the sun is directly overhead the Tropic of Cancer. This star marks the point where the Tropic of Cancer leaves the continent of Africa at 23°N (A+B+C+Y= 6+6+3+8= 23). From all evidence, it turns out that extreme positions of the sun are important, and also the magic number 23. The big monument of Stonehenge, England (c.3200 BC) is oriented to the rise of the midsummer sun (Refs.3,4). The total number of fully engraved stones in the famous passage grave of Gavrinis, France (c.3600 BC) (Ref.25, a monument to the discovery of the Azores) is 23. All this emphasis on 23°N shows that for more than a thousand years, people wanted to cross the Ocean at 23°N, in honor of the SunGod, but they did not succeed.

Below the "peninsula of Brittany" we see from east to west the circular petroglyphs of the Flannan Isles (2dl) with below it the west cape of NorthUist (Hebrides, Scotland), then the islet of St. Kilda (2dl), and finally

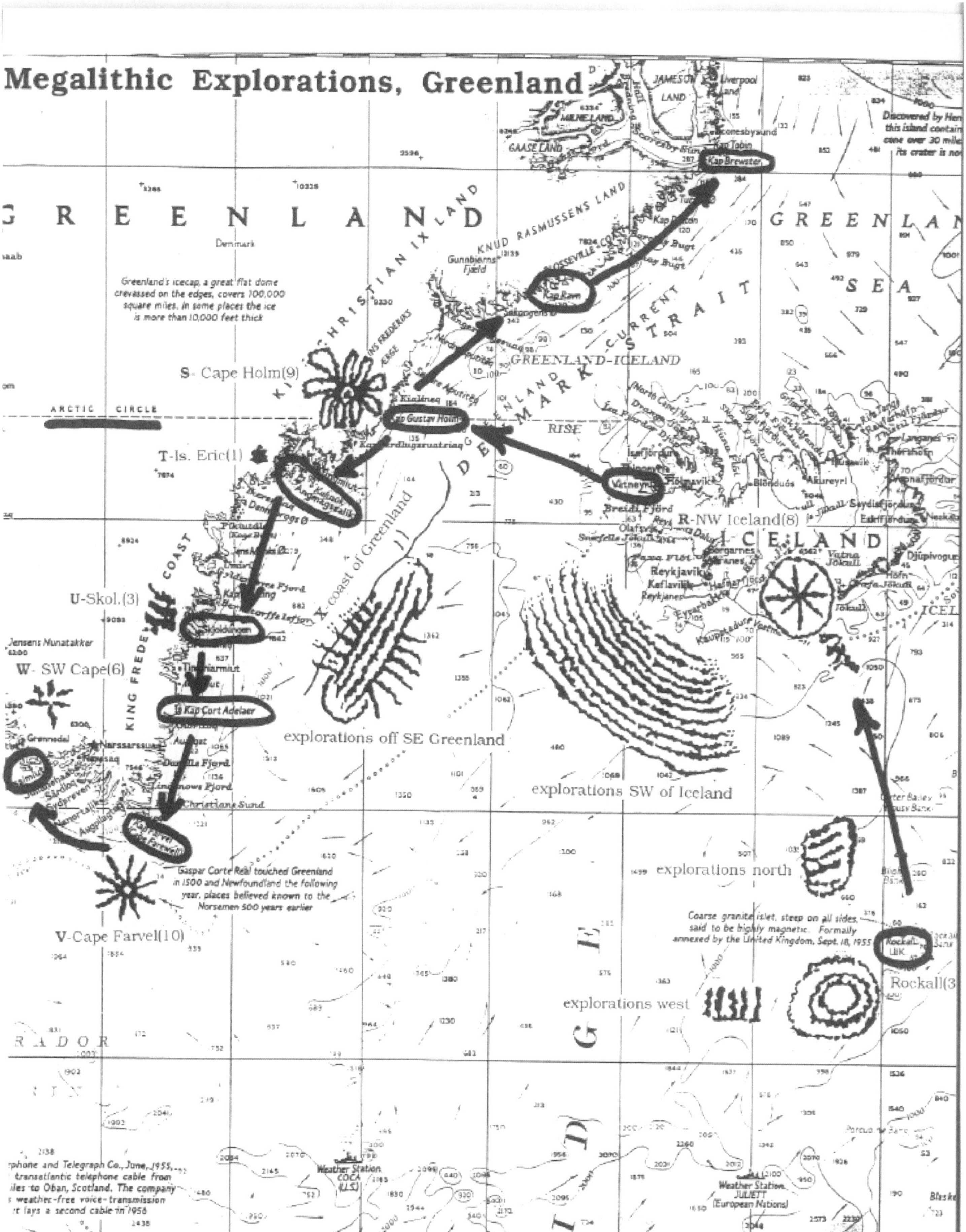


**Fig. 9** Map of Iceland and SE Greenland, showing the order of megalithic discoveries, and their identifying glyphs, from the Roofstone of Cairn T, Loughcrew (Ref.8, c.3200 BC).

the islet of Rockall (3dl) (see Fig.9). Rockall, in the west, is located at 57°N (A-H+2+2+3= 50+7= 57). Around Rockall, the sea has been explored over 333km (3dl), which is the distance to St. Kilda, and to the north over an additional distance of 555km (5dl), which is until Iceland (R) at 65°N (57+3+5= 65). To the west the ocean was explored over an extra 444km (4dl). A later visitor to the passage grave considered this round petroglyph as that of the Azores, often depicted in megalithic "art", and added a spiral with three turns, representing Madeira 3 distance lines above the 30th latitude line, at 33°N. For the Azores, the indicated explored sea areas are correct too. South of Madeira, the Canary and the Cape Verde Islands are indicated. The surroundings of the Canary Islands have been searched extensively. Below, we see right and left in both cases two Cape Verde Islands, to be precise, the main island Sao Tiago (below), and the NW island Santo Antao (above). At the left side is shown that the coastal waters north of these islands were investigated. The strong NE winds and currents evidently curtailed further explorations here.

At the extreme right side, a group of less clear carvings have been made, probably by somebody who did not understand the stone very well. He thought that the inscriptions left of the star (Y) represented the two important islands of the West Azores (which is often the case on other megalithic stones). So he carved the East and Central Azores above the star to "complete" the inscription. Then he announced that the sea north of the 3 island groups of the Azores was explored over 555km (5dl).

Finally, at the extreme left side we see a deep dark inscription -an ancient "graffiti"- dating from after the discovery of America (c.2500 BC). It really does not belong to this stone or this monument at all. Turn the drawing of the roofstone 90° anti-clockwise. The dot is Cape Verde, with left of it, the adjacent shore. The small upward scratch points to the nearby Cape Verde Islands. The long strokes from Cape Verde and from the Bissagos Islands in the south provide the global sailing direction to the NE coast of South America: 45° SW. It is the start of the Southern Crossing over the ocean, and the answer to the problem megalithic people had struggled with for so long.

## The Faeroes, Iceland and Greenland

The Story of Loughcrew is confirmed in the design of Stonehenge I in South England (c.3200 BC), but also in many European petroglyphs. Fig.10 (top) shows a simple carving from Luffang, Brittany. People sailed around South England via the Hebrides (Scotland) to the Faeroes. Next, they discovered Iceland in the northwest (c.3400 BC). This was an important discovery, because Iceland has about the size of Ireland, which is much larger than the tiny islets discovered in the past!

Encouraged by this positive result, soon Greenland was discovered, too (Refs.3,4). Fig.10 (bottom left) shows the best sailing direction from the NW peninsula of Iceland to Cape Holm, Greenland (at the edge of the stone),

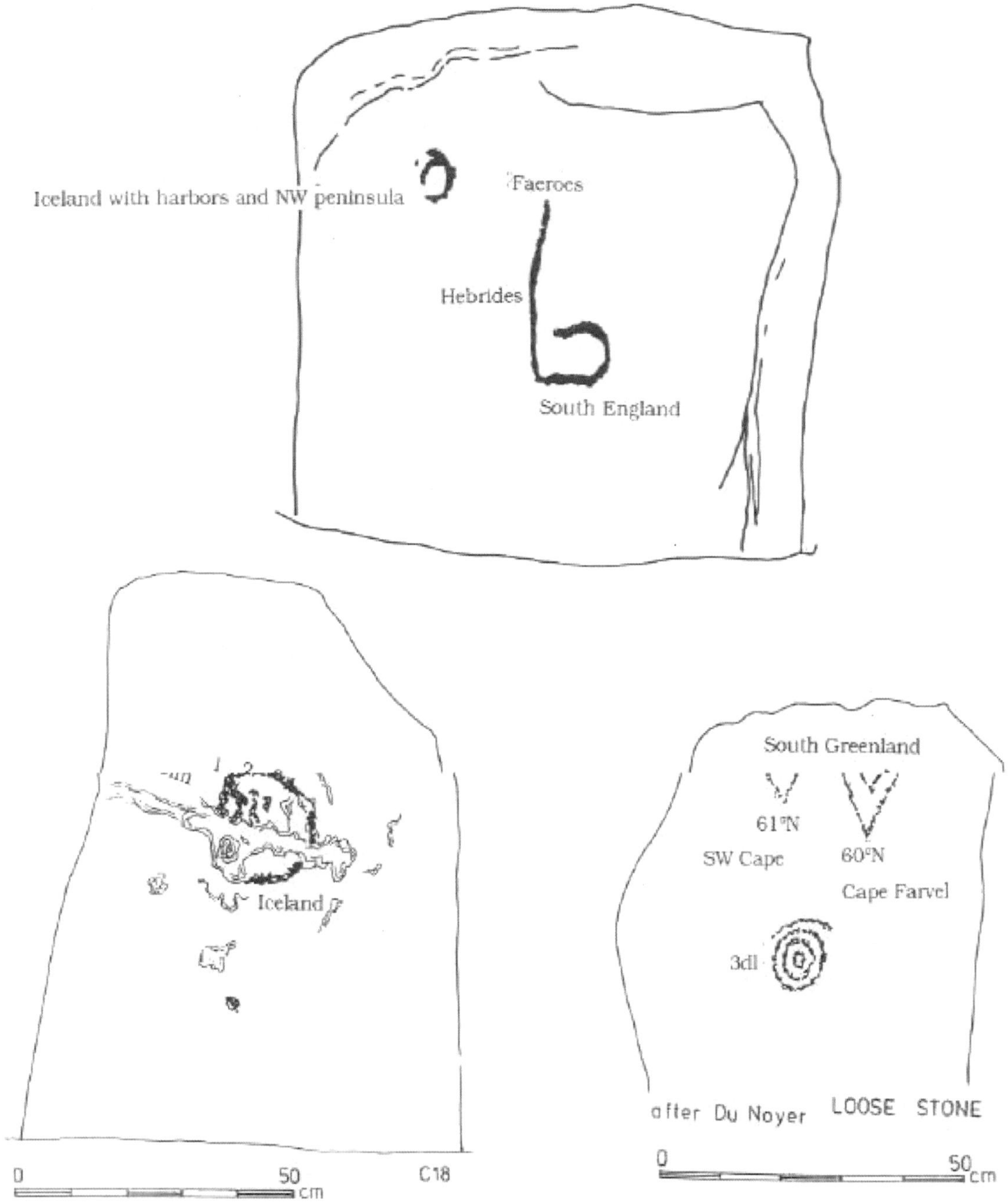


**Fig. 10** **Top:** The sailing route around south England via the Hebrides to the Faroes. Left above: the newly discovered island of Iceland (Luffang, Crac'h, Brittany, Ref.2, c.3400 BC).

**Bottom left:** Iceland, and the best sailing direction of c.15°WNW to Cape Holm, Greenland, at a distance of 5dl= 555km which is about the diameter of Iceland (Loughcrew, Cairn H, East Ireland, Ref.2, c.3300 BC).

**Bottom right:** South Greenland with, right, Cape Farvel at 60°N, and left, the SW Cape at 61°N. In the south, the sea was explored over c.6x0.5dl=3dl=333km offshore (the concentric circles are placed in the water). (Loughcrew, Cairn J, East Ireland, Ref.2, c.3200 BC)

c.15°WNW (c.3300 BC, Loughcrew). As the petroglyph shows, the sailing distance is about the diameter of Iceland: c.5dl= 555km. From Cape Holm at the Arctic Circle, people voyaged along the coast to the south, of course. Impressed by the enormous length of this coastal strip (7°=777km), they thought they had reached the other side of the Ocean. However, all the land ended at Cape Farvel at 60°N, and at the SW Cape at 61°N (Fig.10, bottom right) (c.3200 BC, Loughcrew).

Figure 11 (top), from Loughcrew too, provides us with a good overview. They sailed from the Orkneys (via the Faroes) to the SE coast of Iceland, and after a rough journey they arrived at Cape Farvel. The explorations at sea were extremely difficult, because it was bitter cold. For that reason they often switched over to units of half distance lines (hdl), corresponding to 0.5dl=55km. The Faeroes (between the spirals) are located at distances of 7x0.5dl=3.5dl=388km from the Orkneys and Iceland. Around Cape Farvel the sea was explored over only 5x0.5dl=2.5dl=277km (c.3200 BC). Next, the sea was also explored south of the SW Cape, as shown in Fig.10 (bottom right), over 7x0.5dl=3.5dl=388km (the concentric circles are placed in the water), but no land was found.

One of the Calderstones in Liverpool, England, Fig.11 (bottom), shows a complete expedition to South Greenland, c.2800 BC. As indicated by the northward foot, people sailed from the British Isles via the Faeroes (the spiral) to Iceland. The spiral has 3.5 turns, because again, the distances from the Faeroes to the Orkneys and to Iceland amount in both cases to 3.5dl=388km. From the NW peninsula of Iceland they voyaged via Cape Holm to the south point of Greenland (the natural relief). East and west of Cape Farvel the sea was explored over at least 5dl=555km, but no land was found. As emphasized by the second bow and the southward foot, they returned the same way. The foot with the frozen, broken toes shows that in Greenland, it was so cold, your toes freeze on your body. However, the southward foot near the British Isles has 6 toes, so each member of the expedition returned. As the petroglyph shows, not so long ago they discovered the island of Jan Mayen, northwest of Iceland (c.2900 BC). Note that on Iceland, the symbol of a Christian church has been engraved, showing that the meaning of this stone was still understood when Iceland was christianized, c.800 AD, which is the date of the discovery of Iceland according to official archaeology.

## The West Coast of Greenland and Beyond

Fig.12 (left), is a petroglyph from Luffang, Brittany, dated c.2900 BC. It is a stylized image of the North Atlantic Ocean with at the top Greenland, and below it the Mid-Atlantic Ridge, that divides the Ocean in two. At the upper right side is Iceland (the circle). Because of its relatively large size, they speculate about the existence of a "second Iceland" west of Greenland. They want to try to reach this second Iceland! Along the east coast of Greenland the sea was explored over c.3dl=333km, and along the west coast over c.2dl=222km. In the south of the ocean the coastal waters are indicated

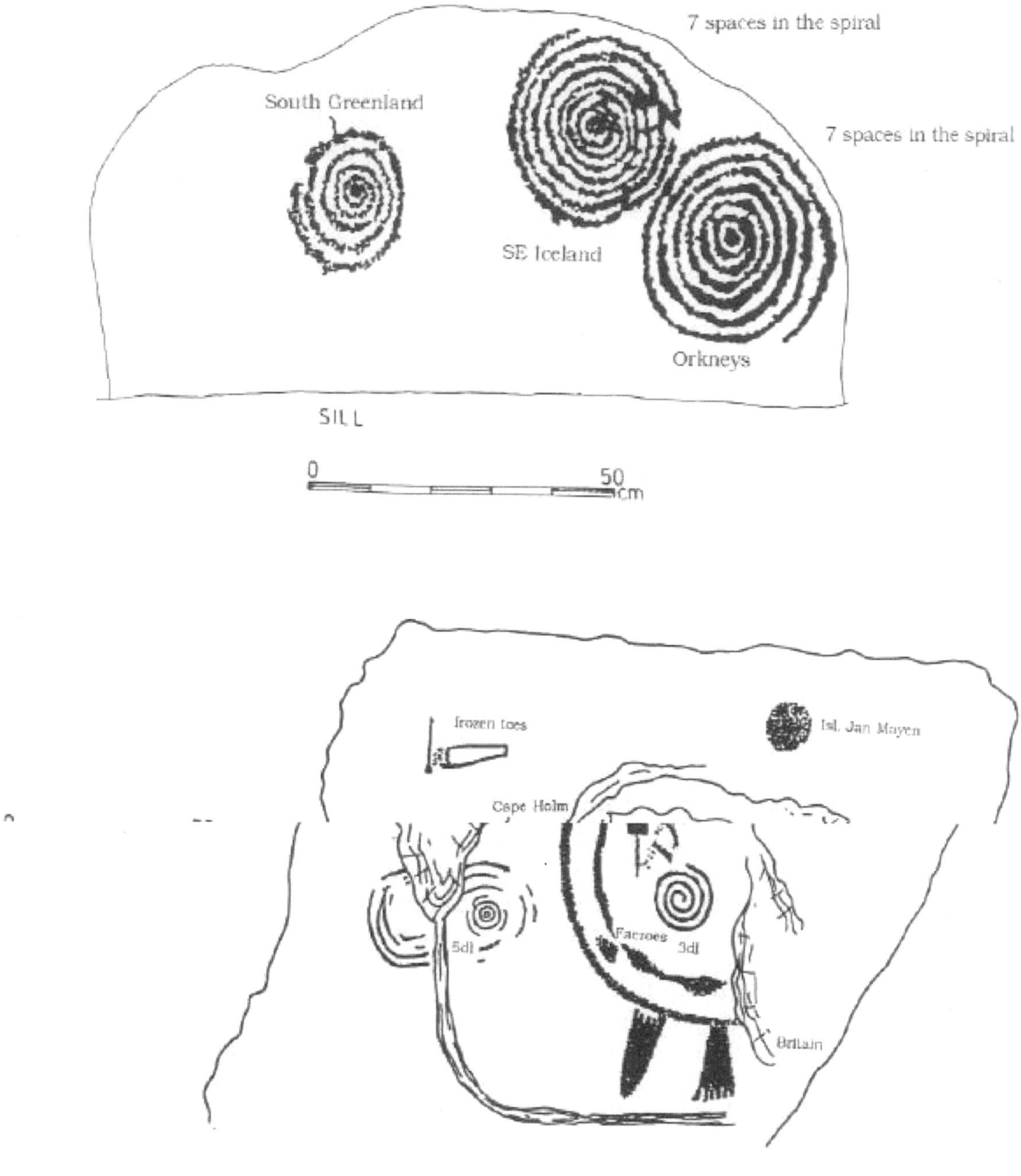


**Fig. 11 Top:** The Orkneys (right) the Faroes (between the spirals), the SE coast of Iceland (above), and Cape Farvel (South Greenland, left). The Faroes are at distances of c.7x0.5dl= 3.5dl=388km from the Orkneys and Iceland. Around Cape Farvel the sea was explored over 5x0.5dl= 2.5dl= 277km. (Loughcrew, East Ireland, Ref.2, c.3200 BC).

**Bottom:** People sailed from the British Isles (right) via the Faroes (the spiral) to Iceland (above). From there the south part of Greenland (left) was reached. East and west of Cape Farvel the sea was explored, but no land was found. It was bitter cold, but people returned along the same route without loss (the foot in the southern direction has 6 toes). (Calderstones, Liverpool, England, Ref.2, c.2800 BC)

over 2 "half big distance lines", or 2x0.5DL= 1DL=1111km. At the level of the Azores in the middle of the ocean one estimates the width of the ocean at 2x(0.5+0.5+1)DL=4DL=4444km, which turns out to be a reasonable guess. However, the petroglyph has anthropomorphic features. It is also an early representation of the mighty Ocean God, which is the same as the SunGod. It is related to the tradition of statue menhirs in the southeast region of France and on Corsica (Ref.26), which lasted about a millennium (c.3000-2000 BC).

Figure 13 (right), is again a Calderstone from Liverpool, England, with petroglyphs relating to the exploration of the sea along the west coast of Greenland (c.2700 BC). The spirals and circles on this stone are placed in the water, and again, the unit of distance in this cold sea is 1hdl=0.5dl=55km. The lowest point on the left, front side (i), is Cape Farvel at 60°N, and the heel of the lower foot is near the SW Cape at 61°N. The two feet show an area of interest along the west coast between them at 62°N. Higher on the stone is a carving of South Greenland (right), and the explored coastal waters at this latitude (62°N, left), over a distance of 7x0.5dl=3.5dl=388km offshore.

On the right, back side (ii) of Fig.13, the spiral shows the explored waters along the west coast at 63°N, again over a distance of 7x0.5dl=388km offshore. The concentric circles above show the explored waters at "the west cape", just below Disko Island (the natural relief at the top), at 67°N, again over a distance of 7x0.5dl=388km offshore. This all happened just before the discovery of America via the Bering Sea, c.2600 BC, and via the Atlantic Ocean, a century later.

## Discussion

In archaeology it is generally accepted that the idea of "degree of latitude" was used at the earliest around c.500 BC, as these words then show up for the first time in a written text (Ref.7). However, the Loughcrew inscriptions prove that the idea was already known at 3200 BC, and an older inscription about the discovery of Madeira (c.4100 BC) shows lines of latitude which are clearly indicated on a map. The five passage graves of the old eastern cairn of Barnenez (North Brittany, 49°N, c.4700 BC, Ref.5) represent the five latitudes north of that monument, from 50°N to 54°N. The "degree of latitude", in the North-South direction, is one of the oldest units of length (1°=111km). The "distance line" is nothing else but the same unit, pointing to other directions (1dl=111km).

The megalithic picture writing of Loughcrew has never been deciphered previous to this article, because people in our times have not been aware of these very old notions. If at night, one looks at the sky, the shortest angle between the polar star and the ground equals the local degree of latitude. These and similar measurements can be carried out very easily today with a cross-staff (Jacob's Staff), as they were in megalithic times (Ref.7). Megalithic inscriptions of this ancient and primitive instrument are spread

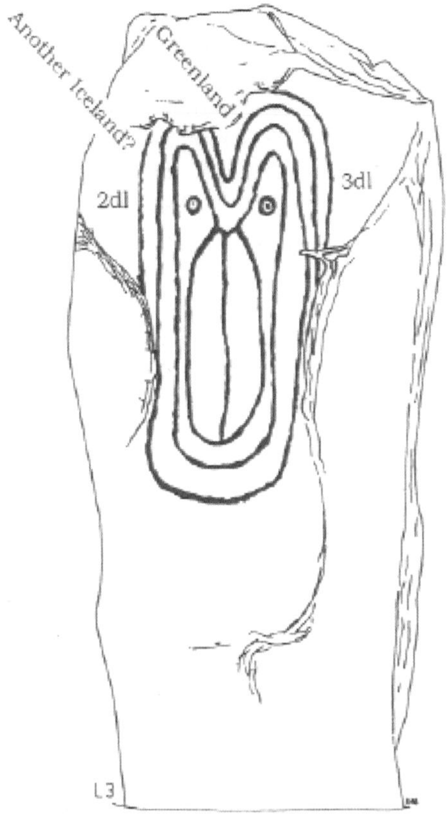


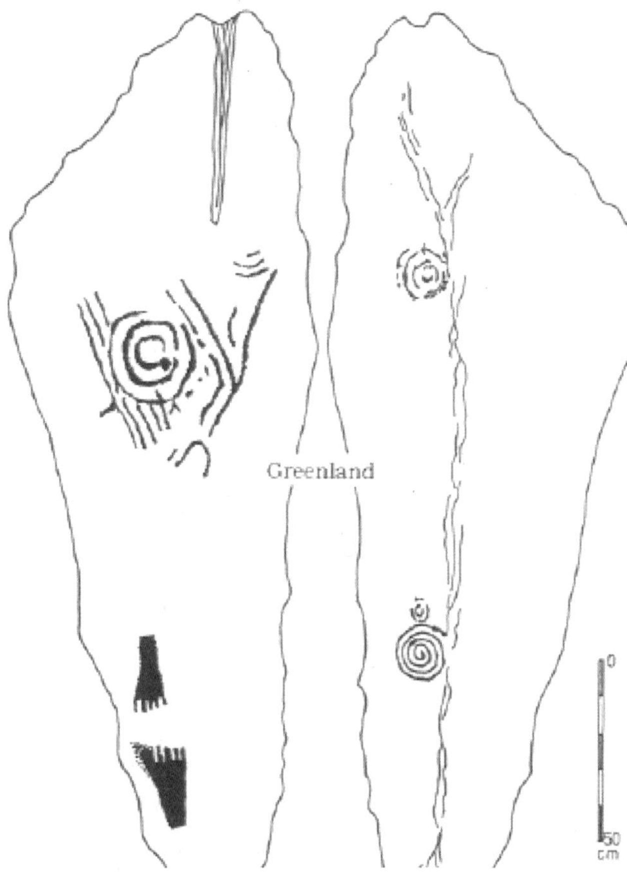


**Fig. 12** **Left:** The North Atlantic Ocean with Greenland and the Mid-Atlantic Ridge. Because Iceland is relatively large, they indulged in fancies about a second Iceland west of Greenland. Around South Greenland the coastal waters are explored over 3dl=333km (east) and 2dl=222km (west). At the level of the Azores, the width of the ocean is estimated to be 2x(0.5+0.5+1)=4DL=4444km. It is also an early representation of the God of the Ocean, which is the SunGod. (Luffang, Crac'h, Brittany, Ref.2, c.2900 BC)

**Fig.13** **Right:** Both sides of this stone are coastal maps of South Greenland.

**Center Front (i):** The lowest point is Cape Farvel at 60°N, and the heel of the lower foot is near the SW Cape at 61°N. The two feet show an area of interest between them at 62°N. Higher is a carving of South Greenland (right), and the explored coastal waters at this latitude (62°N, left), over a distance of 7x 0.5dl=, or simply 3.5dl= 388km offshore.

**Center Back (ii):** The spiral (below) shows the explored waters along the west coast at 63°N, again over a distance of 7x0.5dl. The concentric circles (above) show the explored waters at the "west cape", just below Disko Island (top, natural relief)), at 67°N, again over a distance of 7x 0.5dl. (Calderstones, Liverpool, England, 2700 BC)

widely in Western Europe (Refs.1,2). The circle angle was divided in 360°, because the cyclic year counts at 360 days. Today, we are still doing that in the same way, but it is not realized that people have "always" done that, even in the distant past.

Latitudes were in use in Egypt from the start of the Old Kingdom, but probably much earlier. The Great Pyramid in Giza (c.2700 BC) is located at exactly 30°00'N, and there is a complicated inscription on the thrones of all pharaohs since the 4th dynasty (2723-2563 BC), relating to three close values of the latitude of the Tropic of Cancer (Ref.20). In the 12th dynasty (2000-1786 BC), but probably much earlier, the point of the River Nile at exactly 23°00'N was called "Sacred Sycamore", their Tree of Life, established as the southern boundary of Egypt. In predynastic Egypt (before 3000 BC), the latitude halfway between the Tropic of Cancer and the Nile Delta was a well-known concept (Ref.20).

The circumference of the earth was very accurately known by c.2800 BC, and probably much earlier. The "moira", the distance corresponding to one degree of latitude, dates from the same time (1 moira= 1dl= 111km). A text from the Old Kingdom (2778-2263 BC) gives the value of it with an accuracy of 0.3 per thousand (Ref.20). Geographic distances were often indicated in Egypt in units of one tenth of a moira (0.1 moira=11km), called the "grand schoenia", or one sixtieth of a moira, the Egyptian mile, or a tenth of an Egyptian mile, the stadia. We have come to the conclusion that all megalithic petroglyphs and monuments were developed on the basis of one angular unit, the degree. In our time, latitudes are still divided into 60 minutes, and the minutes are each divided into 60 seconds. The corresponding distances were used in ancient Egypt, because 1°= 1 moira= 60 Egyptian miles, and 1 Egyptian mile= 60 Egyptian plethera, or the distance of a minute of a degree. Professor Alexander Thom has empirically established that all megalithic monuments are built with the aid of one unit of length, which he called the "megalithic yard", 1my=.83m (Ref.21). Possibly there is a relation between the my and the length of 1°, because, with a deviation of about 3%, 360x360=129,600, x.85733m= 111,110km, while 1°= 1dl= 111.111km.

The petroglyphs of Loughcrew (c.3200 BC) are unique, but they are so complex it is hard to understand how they could have been done without having been copied from more transient materials, which were portable, and maybe usable, outside the cairn. In the Stonehenge design in south England, the final part of the Story of Loughcrew is also told, which is another story, but there with menhirs of more than man's height (Ref.1). That monument is dated accurately at c.3200 BC (Refs. 3,4). Some details suggest that the endstone of Loughcrew predates it slightly, but Michael Dames (Ref.11) reports Cairn T has been dated at 3200 BC too.

The Loughcrew inscriptions have the same age as the oldest cuneiform of Mesopotamia, and they are also at least as old as the oldest hieroglyphics of predynastic Egypt. Gunter Dreyer of the German Archaeological Institute

**Fig.14** Reinoud de Jonge points to the last figure on the "Story of Loughcrew" inscription. It is a man telling us that the megalith builders gave up their efforts to cross the Atlantic at the SW Cape of Greenland (c.3200 BC).

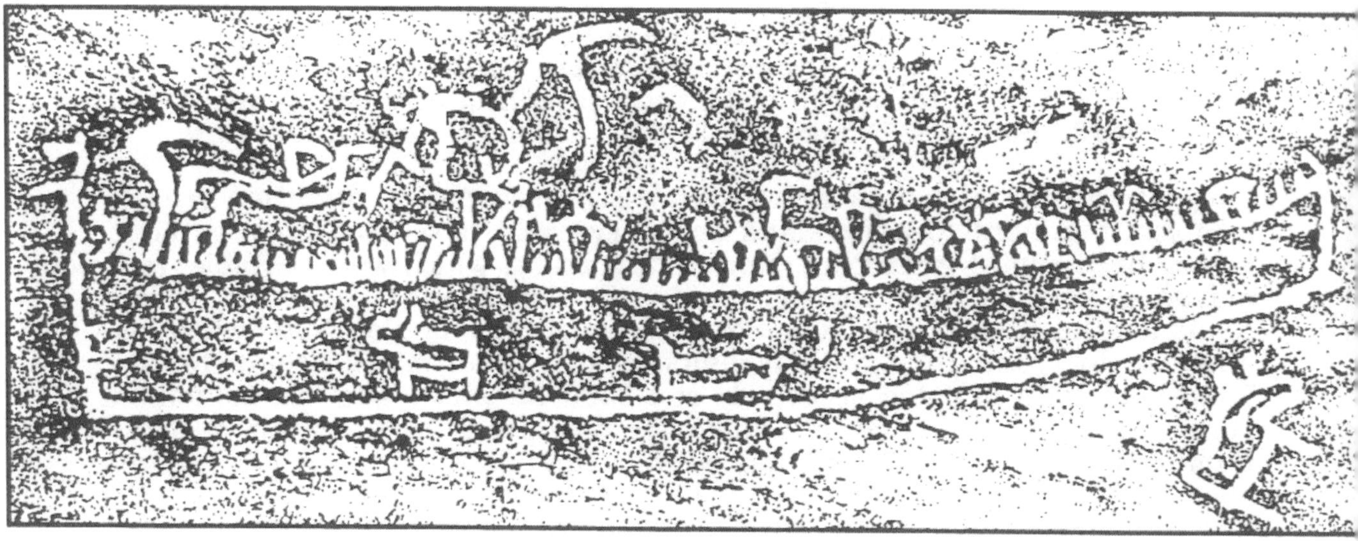

**Fig. 15** Petroglyph from Namforsen in Angermanland, near the east coast of Central Sweden. It is a huge fantasy boat (with a length of 2.4m), containing 67 rowers, showing the latitude of the holy Arctic Circle, at 67°N. This is the northernmost latitude where the sun still shines at midwinter day. At this latitude Greenland was discovered c.3300 BC. Above the crew are carvings of features of the newly discovered coast of Greenland. (Namforsen, Noton Island, Angermanland, Sweden, c.3200 BC, Ref.27)

said the Egyptian Scorpion Tomb tablets have been carbon dated with certainty to between 3300 BC and 3200 BC. According to scholars, the writings depict "the Mountains of Light, or east, and Mountains of Darkness, or west", in reference to where the sun rises and sets (Ref.19). The Loughcrew inscriptions are the oldest history in the world, ever written down. It is, as said before, the history of failed attempts to cross the western ocean.

Glyph A on stone C8 is one of the oldest images of a sailing boat ever found (Ref.6). The history starts at the discovery of the Cape Verde Islands in the south (c.4500 BC), and ends by reaching the SW Cape of Greenland in the north (c.3300 BC) and additional voyages on the roofstone (c.3200 BC). So the Loughcrew inscriptions report on a time span of exploration of c.1300 years.

Work by other investigators is showing that megalithic sites are carefully located "in a gigantic geometrical representation - a veritable prehistoric cartography" (Ref.10). Study of this "ancient science totally forgotten" has been going on since 1982 in Europe by the Association Archeologique Kergal. Their 20th publication, referenced in the bibliography, was their first in English. They note that "mathematics is not the only science used by these ancient peoples to transmit their knowledge. Symbolic expression is wholly present, not only through the architectural forms, but also through the art of engraving and the science of landscape". The "Story of Loughcrew" demonstrates all of these.

The study of prehistory is complex. The unexpected antiquity of a tomb (#4) at Carrowmore, a "megalithic cemetery" near Sligo, west Ireland, which was dated in 1998 to c.5400 BC, has led archaeologists to call it "the oldest free-standing structure in Western Europe", and is causing some re-thinking of neolithic development among the experts (Ref.12). How do the "Red Paint People", now professionally accepted as the trans-atlantic "Maritime Archaic" from 3,000 BC (Ref.23), tie in with these megalithic people? There is a need for more work on the charting of dates, to see how pieces of prehistory coming to light around the world might be fit together, or may challenge these dates.

Considering tens of other inscriptions and monuments (Ref.1), we date the discovery of the Cape Verde Islands at c.4500 BC, Madeira at c.4100 BC, the Azores at c.3600 BC, Iceland at 3400 BC, Greenland at 3300 BC, the failed attempts to cross the Davis Strait from the SW Cape of Greenland at 3200 BC, and the discovery of America via the Bering Sea at c.2600 BC. The oldest large sea-ships of Egypt are engraved near the pyramid of King Sahura (2510-2460 BC) at Abusir (Ref.16). Egyptian inscriptions tell that in the 13th year of King Sahura (2497 BC) large sea-ships of Egypt reached the far land of Punt. Thus it may be, that America was first reached via the Atlantic by this voyage in 2497 BC, about 4500 years ago.

# References


1. Jonge, R. M. de, and IJzereef, G.F., De Stenen Spreken, Kosmos Z & K, Utrecht/Antwerpen, 1996 (ISBN 90-215-2846-0) (Dutch, pg.152)
2. Twohig, E. Shee, The Megalithic Art of Western Europe, Clarendon Press, Oxford, 1981
3. Richards, J., Stonehenge, English Heritage, 1992
4. Atkinson, R.J.C., Stonehenge, 1979
5. Briard, J., The Megaliths of Brittany, 1991
6. Casson, L., Ships and Seafaring in Ancient Times, British Museum Press, 1994
7. Bailey, J., Sailing to Paradise, Simon & Schuster, 1994
8. "Atlantic Ocean", map by National Geographic Society, Dec 1955
9. Mitchell, J., Secrets of the Stones, Inner Traditions International, Rochester, USA, 1989 (ISBN: 0-89281-337-7) (pg. 95)
10. Association Archeologique Kergal, An Approach to Megalithic Geography, Booklet No20, Kergal, France 1992 (ISSN 0220-5939)
11. Dames, M., Mythic Ireland, Thames & Hudson, London, 1992 (ISBN 0-500-27872-5) (pgs.81-82,221)
12. Hearns, O., "Sligo Champion", Aug. 19, 1998
13. O'Kelly, C., Passage Grave Art in the Boyne Valley, Ireland, John English & Co., Wexford, 1975 (pg.28)
14. Patton, M., Statements in Stone, Routeledge, London & New York, 1993 (ISBN 0-415-06729-4) (pg.88)
15. Balfour, M., Megalithic Mysteries - An Illustrated Guide to Europe's Ancient Sites, Collins & Brown, 1992 (ISBN: 1-85585-3558) (pg.50)
16. Varen, Vechten en Verdienen, Scheepvaart in de Oudheid, Allard Pierson Museum, Amsterdam (1995) (Dutch, ISSN nr. 0922-159X)
17. Brennan, M., The Stars and the Stones, Ancient Art and Astronomy in Ireland, Thames and Hudson, London, 1983 (Lib. Cong. #82-50742)
18. Brennan, M., The Boyne Valley Vision, The Dolmen Press, Portlaoise, Ireland, 1980 (ISBN: 0-85105-362-9) (pg.129-130)
19. Joshi, V., "Ancient Tablets show Egyptians May Have Been First to Write", Associated Press, Eastside Journal, Feb. 1999.
20. Tompkins, P., Secrets of the Great Pyramid, Harper Colophon Books, Harper & Row, New York, 1971 (ISBN 0-06-090631-6) (pgs. 176-8, 212, 255, 293)/L.C. Stecchini (pgs.2-93, 303, 325, 345)
21. Thom, A., Megalithic Sites in Britain, Oxford, Clarendon, 1967
22. Eogan, G., Knowth, and the Passage Tombs of Ireland, Thames and Hudson, 1986.
23. Ferryn, P., "5000 Years Before Our Era: The Red Men of the North Atlantic", NEARA Journal, Vol.XXXI, No.2, pg.59 (1997)
24. NEARA Transit 9, 1997 (pg.12, Stone C8 is labeled "Stone 14")
25. Le Roux, C-T., Gavrinis, Ed. Gisserot, 1995 (French)
26. Art et Symboles du Megalithisme Europeen, Revue Archeologique de L'Ouest no8, Nantes, 1995 (French) (ISSN 0767-709-X)
27. Evers, D., Felsbilder, Botschaften der Vorzeit, 1991 (ISBN 3-332-00482-4) (German) (pg.58)

# The Discovery of America via the Bering Sea

## The Decipherment of Encoded Jacob's Staffs

(The Petroglyphs of Dissignac, Brittany, c.2600 BC)

Dr. R.M. de Jonge, drsrmdejonge@hotmail.com
J.S. Wakefield, jayswakefield@yahoo.com

### Introduction

The tumulus of Dissignac is located in the far south of Brittany, 5km west of St. Nazaire, at the mouth of the river Loire, the most important river of central France (Fig.1). The tumulus has a diameter of c.25 meters. It covers two passage graves, built side by side. Both passages of c.11 meters lead to slightly wider grave chambers (Refs.4-6). One of the passages (right, Fig.1) is oriented on the midwinter sunrise.

The monument was excavated in 1873 by Martin and Kerviler, and re-excavated and restored from 1970-73 by French archaeologist J. L'Helgouac'h (Refs.3,4). They found that the monument had two construction phases. **The first building phase dates from c.4500 BC.** At the second building phase the passages were elongated by c.4 meters, when the importance of the monument probably increased. An abundance of pottery fragments show that it was used for two thousand years, from c.4500 BC until c.2500 BC. "The ancient sub-soil near Dissignac tumulus reveals that there were few trees in the area. The considerable density of pollen traces from plantain, mugwort, and thistles reveals the extent of the clearances. Agricultural effort is also apparent from the presence of cereal pollen, as well as charred grains of corn" (Ref.23).

The smallest of the two grave chambers (left, Fig.1) is at the south side of the tumulus. At the side of the passage is a big, heavy coverstone of very hard granite. In 1968 a group of petroglyphs was discovered on this stone (Figs.2&3), covering a surface area of about a square meter (Refs.3,4). The surface seems to have undergone a substantial pretreatment. First the stone was carefully pick-dressed to produce a flat surface, then it was ground, and finally polished. The stone is of superb quality, and has such a fine grained structure, that the individual carvings are still clearly visible.

The tumulus is located close to the sea, and the passage graves with their chambers point in the northwest direction. It would appear that the people had an interest in crossing the ocean. As we shall show, the petroglyphs have geographic meanings related to this goal. Important latitudes of Egypt are found throughout the petroglyphs, indicating that this monument is related to the Egyptian SunGod and divine kingship ideas (Refs.1,2).

Before we get started studying the group of petroglyphs, we want to point out a **large geographic petroglyph** on the bottom half of the stone. Below the whole group of inscriptions are some thin lines (see Fig.2). The middle,

**Fig. 1** Photo of the tumulus of Dissignac, Brittany (France) (c.4500 BC). The left passage grave has a coverstone that bears a unique group of petroglyphs (c.6500-c.2100 BC).

curved one, represents the west coast of North Africa, (compare it with the map of Fig.8). The westernmost points of the continent are clearly indicated, Cape Verde at 15°N, and Cape Blanco at 21°N. On the left half of the stone, over the entire width of the ocean, the Tropic of Cancer is engraved. At this latitude, 23°N, megalithic people wanted to cross the ocean, in honor of the SunGod. These thin lines were carved at the same time as the carving Glyph 3, that is, at the time the tumulus was constructed (c.4500 BC), because earlier, they did not know the ocean was this big.

In the center of the group of petroglyphs we see two small, poorly inscribed glyphs, labeled "1" and "2" in Fig.4, which are **ancient maps**. Around them we see a great number of so-called **"Mediterranean signs"**, which look like "axes". These signs all have slightly different appearances, and they each have a carefully laid out position relative to the other signs in the group. Two glyphs overlap each other. At the upper left side we see a different **petroglyph, with big loops**, which dates from a late megalithic time period, as we shall explain. These figures have been drawn one after the other, with long time periods in between. The older petroglyphs are situated in the center, and the later ones near the sides. This seems like a natural progression, and fits with the record that Dissignac was in use a long time.

# Explorations of the Atlantic Ocean

## Brittany: map

In the center of the group of inscriptions (Fig.4) is a natural, rather straight crack, which, like most natural features of petroglyph stones, was used in the story being told. At the left side a rather small carving, or glyph, labeled "1" was made, attached to it (see enlarged, Fig.5). It is an image of the peninsula of Brittany (the white area in the center), with the coastal waters around it (the pecked area). The crack is the meridian along the west coast of what is now France, from the Gulf of Biscay in the south to the Cape de La Hague (Normandy) at The Channel in the north. The edge of the inner part of the figure is the coast of Brittany, from Dissignac in the south, where these petroglyphs are located, to the Gulf of St. Malo (Mont St. Michel) in the north. Some geographic details are visible, like the protruding shore near the Islet of Belle-Isle at the south coast, and the north point of Brittany near Lcs Sept Iles (Lannion). From the width of the coastal waters of glyph 1, it can be deduced that the people did not sail further than c.100km offshore. As a consequence, this central glyph is much older than the monument itself, and probably dates from c.6500 BC.

## The Discovery of the Canary Islands: map

At the left side of the previous glyph is a miniscule map labeled "2" of the "Little Mediterranean Sea" (Figs.4,5), (the western half of the Mediterranean) a common subject of early megalithic "art". It extends from Italy in the east to the Strait of Gibraltar (the neck of the figure). The Liguric Sea around the Gulf of Genoa in the north (right) is indicated correctly. Past Gibraltar, the figure splits up in a short right part, representing the Gulf of Cadiz to Cape Sao Vincente (SW corner of

**Fig.2** The petroglyphs of Dissignac according to Twohig (Ref.3) (c.6500-2100 BC).

Portugal), and a thin left part, leading to the Canary Islands. The coastal islands Lanzarote and Fuerteventura of the Canaries are clearly indicated in the proper shapes. This petroglyph was made because of the discovery of the Canary Islands. These western islands, more than 2000km south of Brittany, became the westernmost lands of the then known world. On the sailing route to the south, the coastal waters were indicated over a width of c.150km. For this reason glyph 2 is not as old as glyph 1, probably dating from c.5500 BC. Note that this glyph is looking to the west, so the orientation is 90° different from glyph 1. However, while placing this little map on the rock, the carver oriented on glyph 1, because Brittany is located due north (right) of the Strait of Gibraltar (which is at the top of glyph 2).

## The Discovery of the Cape Verde Islands: route pictogram

The right side of glyph 3 (again, Figs.4,5), has been engraved lightly. From this we can deduce that this glyph was made after glyph 2, which was also lightly executed, because most of the carvings to the left have been deeply engraved, while glyphs 6&11 on the right can be shown to be derived from deep (and later) glyph 4. **Glyph 3 is not a real map anymore, but a stylized map, derived from Mediterranean Sea maps.** The handle is now the complete Mediterranean Sea, while the head, outside Gibraltar, has a right side branch, the sailing route to Brittany, and a left side branch, the sailing route to Cape Verde. On glyph 3, the left branch was deeply engraved, and widely drawn. In other words, these people have just discovered the Cape Verde Islands! By that discovery, 3500km south of Brittany, these Islands had become the westernmost land of the then known world.

Note that the right side branch neatly runs around Iberia, and that at the end, the peninsula of Brittany is shown. However, in reality, the Mediterranean Sea is not a straight stick, and the Cape Verde Islands are not located due south of Gibraltar, as the petroglyph indicates. As a consequence, we recognize that it is a **highly stylized map, and call it a "Mediterranean sign" (Ref.1).**

**It is also a so-called "Jacob's Staff" (Ref.7).** This is an ancient instrument to measure angles and latitudes. In principle, it consists of two wooden sticks at right angles that can slide past each other. It is a forerunner of the astrolabe. When one watches the night sky, your latitude is the angle between the polar star and the ground (Fig.5). **The Mediterranean sign is used as a Jacob's Staff in petroglyphs to encode latitudes in the same way that the Jacob's Staff was used - to determine the angles of tangent lines.** This symbol (c.4500 BC) made possible the encoding of geographic data in mathematical code 1500 years prior to the start of picture writing (Egypt, c.3000 BC), and more than 3500 years prior to the development of early phonetic writing.

In glyph 3, Fig.5, we draw a line through the Mediterranean Sea; this line is the baseline. The right branch tangent makes an angle of 15° with this baseline, an encoding of the latitude of Cape Verde and the southern Cape

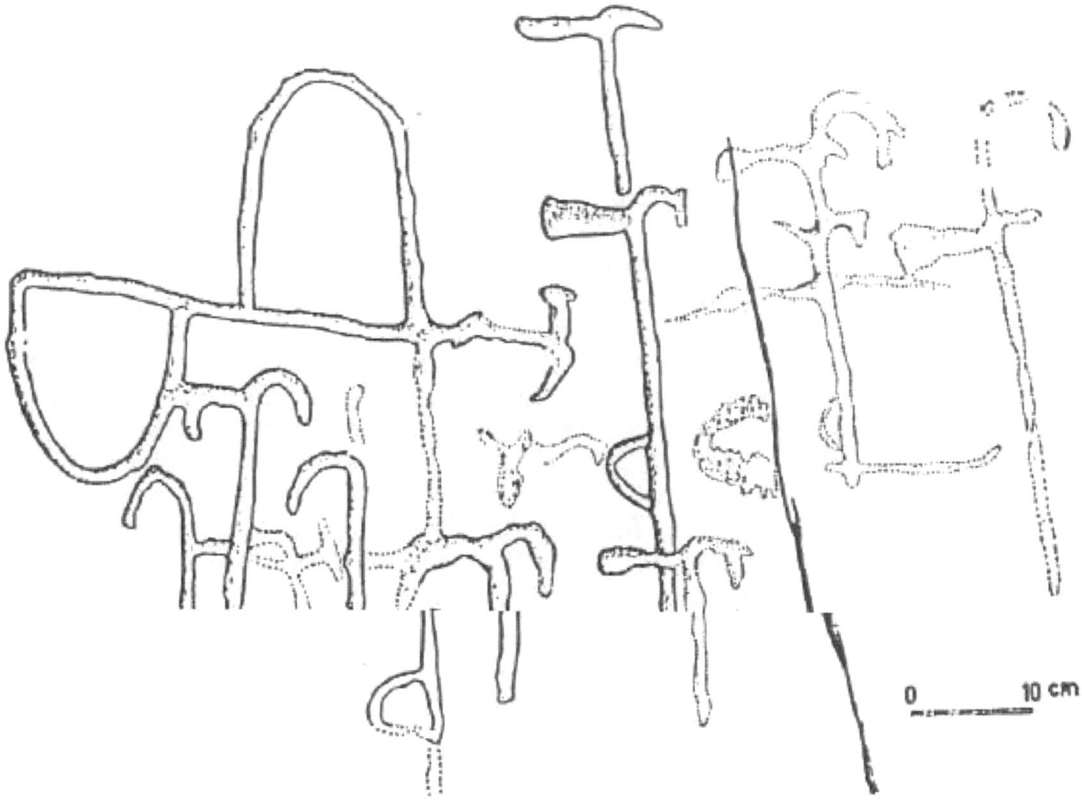


**Fig.3** The petroglyphs of Dissignac according to Briard (Ref.4) (c.6500 BC-c.2100 BC).

Verde Islands, at 15°N, the island group that was just discovered. The left branch tangent makes an angle of 21°, corresponding to the latitude of Cape Blanco, 21°N. This is the westernmost point of Africa one meets on the way to the Cape Verde Islands. The angle between both tangents is 15°+21°=36°, which encodes the latitude of the Strait of Gibraltar at 36°N, where these voyages of discovery on the ocean have started. The bottom of the handle, or far end of the Mediterranean, where the tangent lines originate, represents Egypt. In the time period this petroglyph was made, this became one of the most important civilizations on earth. From this starting point, we can draw not only tangents to the tips of glyph 3, but also those of the previous glyphs 1 and 2 (see Fig.6).

The Table 1 (on Fig.6) shows that they sailed along the coast of Africa to the south, to Cape Lopez at 1°S, and that the Cape Verde Islands were discovered. We investigated the possibility that glyph 3 also provides the sailing routes through the Mediterranean or along the coast of Western Europe, but that is not the case. Figure 6 illustrates the location code Dr. de Jonge developed for describing these angles. Compare the codes in the table, for example **"3Lb"**, with the angles of the tangents of the glyph 3 (21°) above, where **3** is the glyph number, and the tangent touches the end of its **L**eft branch (**L**), below (**b**). The lengthy table of accurate latitudes of important places demonstrates that people had developed mathematical symbolism to encode route descriptions based upon latitudes. For this reason, glyph 3 is called a **"Route Pictogram". At sea, where only latitudes could be determined, these latitude encodings were far more important than maps.** We know from study of the stone design of the Tumulus of Kercado (Brittany, c.4500 BC), that Kercado was built because of the discovery of the Cape Verde Islands (Refs.1,4,8). This center-located, early petroglyph 3, encoding the same discovery, was carved at the time the monument of Dissignac was erected, at c.4500 BC too. Note that the orientation of glyph 3 is the same as the older glyph 2 (west is at the top), that fixed upon the discovery of the Canaries. The shortest distance between Cape Verde and the Cape Verde Islands is c.600km, so by c.4500 BC people were able to cross this distance over open sea.

## The Discoveries of Madeira and Rockall

Glyph 4 (see Figs.4,7) was engraved above glyph 3, slightly to the left. From the chosen position it can be concluded that an important discovery was made WSW of the Strait of Gibraltar. So the islands of Madeira had been discovered! To further illustrate the location, the left side branch of glyph 4 is engraved widely, and the handle is directed through the thin left side branch of glyph 3 (and looking as though it was underneath it) exactly where Madeira is situated.

Glyph 4 is a Mediterranean symbol, and also a Jacob's Staff. The tangents at the important points of the petroglyph, and those of the previous figures make angles with the baseline, recording the latitudes where people have been, and sailing directions they have used. The tangent to the left side

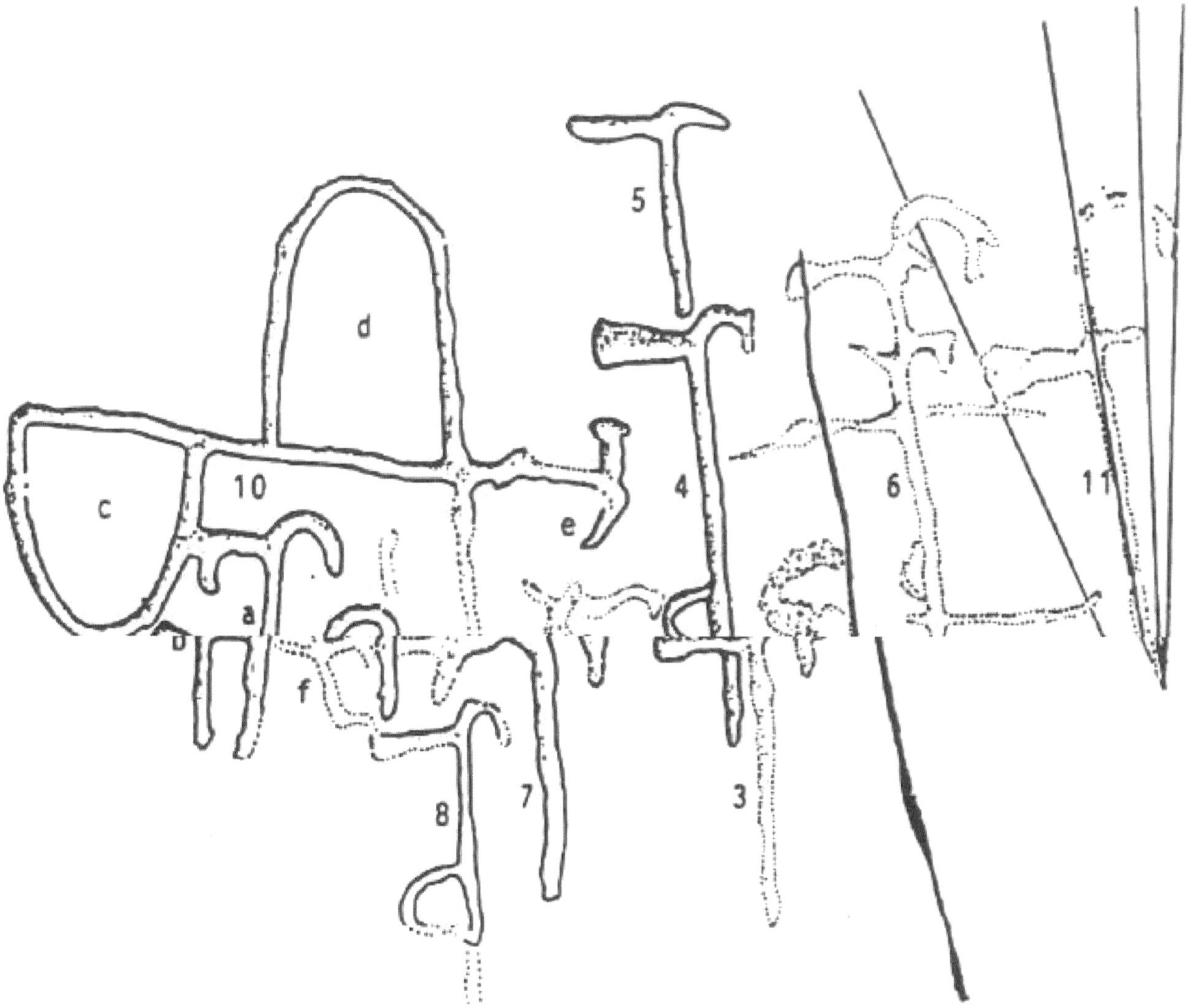


**Fig. 4** Petroglyphs of Fig.3, numbered in the order in which they have been engraved during several millenia. It is the history of the crossing of the Atlantic Ocean, and the discovery of all the land on earth (c.6500- c.2100 BC).

branch of glyph 4 makes an angle of 13° (shown on glyph 4, Fig.7), equal to the sailing direction from Gibraltar to Madeira, 13°WSW. The angle between both tangents of glyph 4 is 21°, encoding the latitude of Cape Blanco, 21°N. The newly discovered islands of Madeira are north of this cape. The right tangent of glyph 4 makes an angle of 21°-13°=8°, equal to the distance from Gibraltar to Madeira, 8dl=8° of latitude= 888km, but also showing that the coast to the south was explored until Sherbro Island, at 8°N. This is the SW point of North Africa.

The loop on the side of glyph 4 represents the peninsula of Brittany, so the handle is now the west coast of France. The ancient glyph 1 has been repeated! To confirm this, the coastal route from Iberia to Brittany has been added to the old glyph 2 (the faint horizontal leg from glyph 2 to the loop). They are trying to explain that they also had a new discovery in the north, in the neighborhood of Brittany. However, the loop of the glyph also represents the westerly coastal route around the British Isles. The faint horizontal leg now points to the islet of Rockall, in the northwest. Note that the right side branch of glyph 4 has an upward point added above its Brittany tip (Fig.7). This tip points in the direction of the Irish Sea, between England and Ireland ("Rockall" in Fig.7), confirming this discovery.

When you look at all the tangents of glyph 4, it is clear that the coast of Europe was explored to the north as far as the islet of Rockall, west of Scotland. The left tangent to the loop makes an angle of 15°, and the right tangent to the inner (left) tip of old glyph 1 makes an angle of 18° (see Fig.7). The total angle between both tangents "to the west coasts of Brittany" in glyphs 4 and 1 equals 15°+18°= 33°. This angle encodes not only Madeira at 33°N (just discovered), but also the complementary latitude of 90°-33°= 57°N, which is the latitude of the newly discovered islet of Rockall, 400km west of the Hebrides.

The stone of Paredes, NW Spain, confirms the discoveries of Madeira and Rockall at about the same time (Ref.3). The Paredes Tablet, however, has not been very well dated. We estimate that both the Madeira and Rockall discoveries occurred about c.4100 BC.

## The Discovery of the Azores

Glyph 5 has been carved above glyph 4. The higher position chosen for this glyph indicates that a new discovery has been made west of the Strait of Gibraltar. So glyph 5 must be about the Azores, later discovered far out in the ocean! Again, the figure is a Mediterranean sign. The tangents to this glyph (and from this glyph to the previous glyphs) encode important latitudes, sailing directions, distances, and number of islands, as they do for the other glyphs. Remember, these people like mathematics (Ref.35)!

Below the right side branch, a little arc has been engraved (see only in Fig.2). The right tangent of glyph 5 (running along this arc) makes an angle of 18° with the baseline of glyph 5, equal to the sailing direction from the

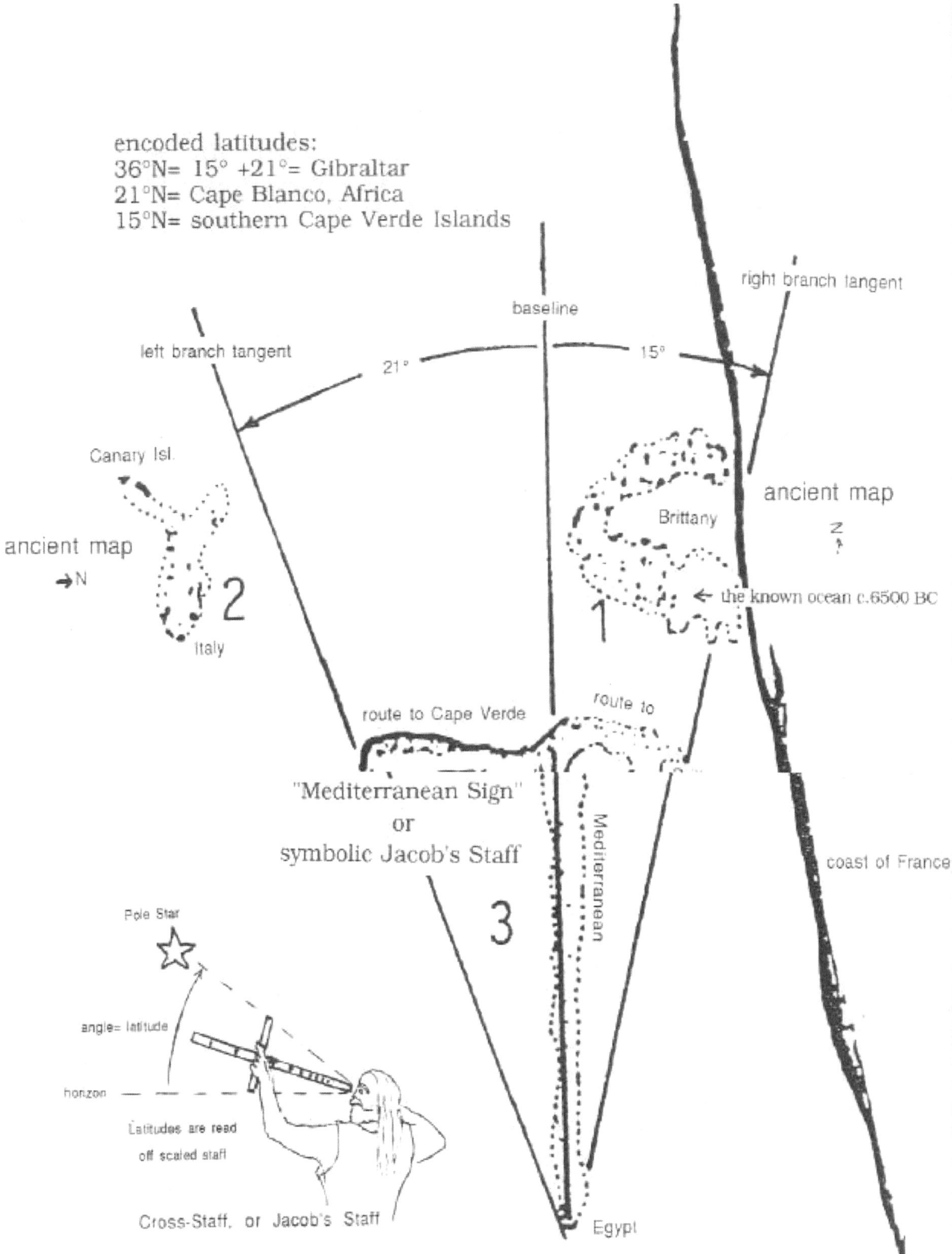


**Fig.5** **Glyph1:** Ancient map of the peninsula of Brittany, with the coastal waters over c.100km (c.6500 BC). **Glyph 2:** Ancient map of the Little Mediterranean Sea, with the sailing routes to Portugal (right) and to the discovered Canary Islands (left) (c.5500 BC). **Glyph 3:** Mediterranean sign, showing the discovery of the Cape Verde Islands (c.4500 BC), with explanation of megalithic latitude encoding system using a Jacob's Staff.

Strait of Gibraltar to the Central Azores, 18°WNW, its distance, 18dl= 18° of latitude= 2000km, and the total number of islands of the Canaries, Madeira, and the Azores, 7+2+9= 18.

Encoding the latitude of the Azores at 37°, 38°, and 39°N on glyph 5 (high, because located in the west) took some ingenuity not often seen in these petroglyphs. The left tangent makes an angle of 27°, corresponding to the latitude of Cape Bojador, 27°N, and the angle between both tangents is 27°+18°= 45°. Looking on the map, we see that the Cape Verde Islands are located 45°SW of Cape Bojador, at 27°-11= 16°N, and now they are trying to explain that the newly discovered Azores are situated 45°NW of Cape Bojador at 27°+11=38°N, which is fully correct. So, Cape Bojador at 27°N is a point of symmetry. The total number of discovered islands, including the 9 Cape Verde Islands, is now 18+9=27. The total number of islands of Madeira and the Azores is 2+9= 11, and while the Cape Verde Islands are located at 15°, 16° and 17°N, it is also true that the newly discovered Azores are situated twice 11 higher, at 37°, 38°, and 39°N. So in glyph 5, the latitudes of the East, Central, and West Azores are indirectly well indicated. The difference between the left angle and the right angle is 27°-18°= 9, equal to the 9 discovered islands of the Azores. Now, instead of the Cape Verde Islands, the Azores became the westernmost land in the known world.

You may think these complicated petroglyphs have strange features, but they do not. They describe the nautical history of the megalith builders over a time period of some 400 years. The people had complicated feelings about what happened over such a long time span. A total analysis of each glyph would require a chapter on each one, and combined, would be a book in themselves. Here, we are only showing you their methodology, by explaining a few of the glyphs, and an overview.

From study of the details of the construction of the passage grave of Gavrinis in Brittany, we know this monument was built because of the discovery of the Azores (Refs.1-3,9). Since Gavrinis is dated to c.3600 BC, we know this glyph 5 at Dissignac about the Azores was also made c.3600 BC. Santa Maria, the easternmost island of the East Azores, is located c.900km from Madeira, so by c.3600 BC people were able to sail the sea over this larger distance.

## The Discovery of the Faroes, Iceland, and Greenland

A long connection line points from glyph 6 to glyph 4 (Figs.4,7). The carver of glyph 6 was orienting on glyph 4, but placed 6 to the right, so recording new discoveries in the north. Glyph 6 is an ordinary Mediterranean symbol. The handle is the Mediterranean, the right side branch is the sailing route to Brittany, and the left side branch is the sailing route to the Cape Verde Islands. But the figure is more complicated, as it has little branches,representing important sailing routes in and around the Mediterranean.More than a thousand years later than glyph 3, the traditional Jacob's Staff (Ref.7) symbolic system was continuing to be used. Both tangents to the outside points of contact of this Jacob's Staff make an angle of 33°, (shown

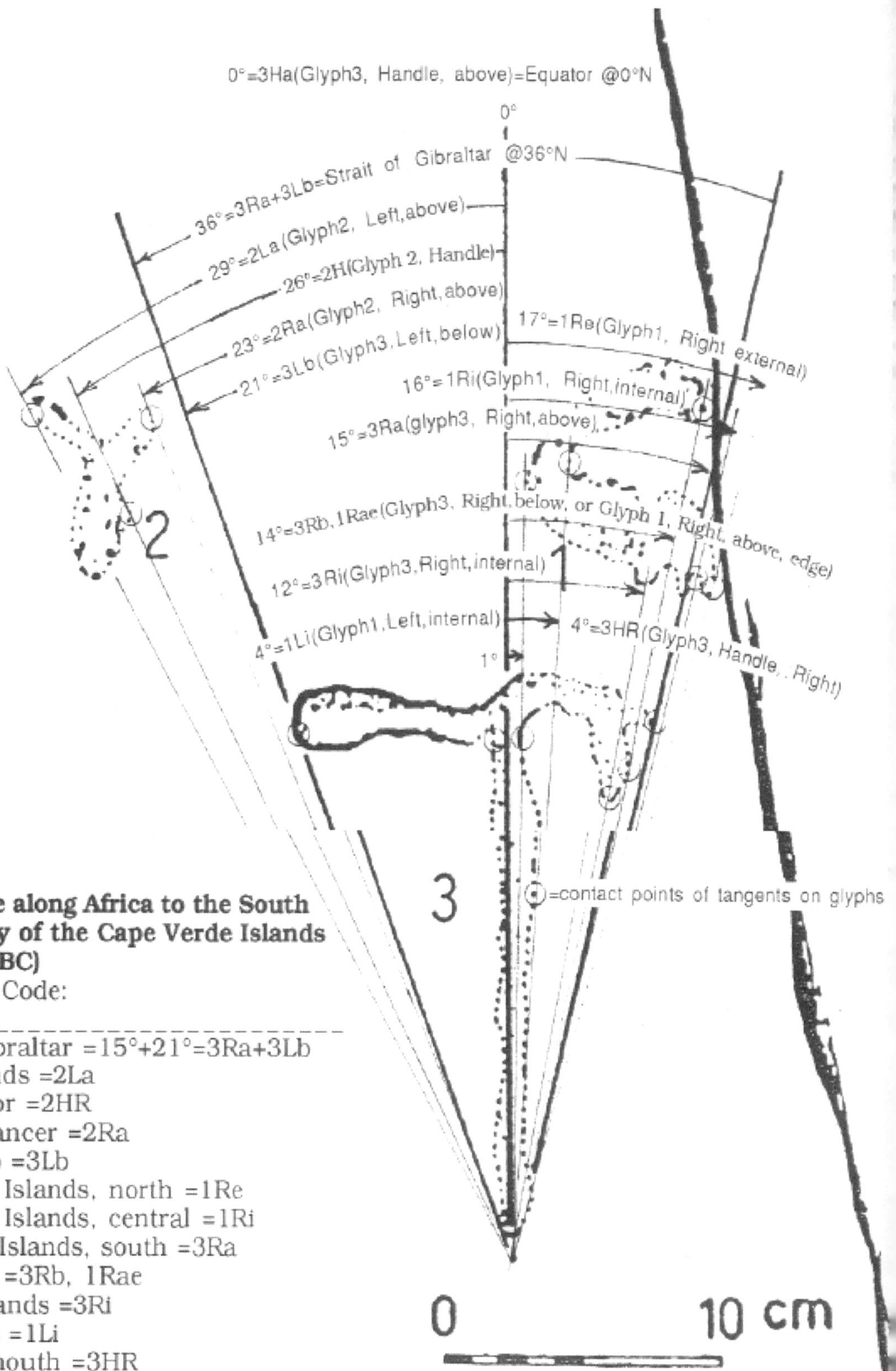


**Table 1**

**The Coastal Route along Africa to the South and the discovery of the Cape Verde Islands (Glyph 3, c.4500 BC)**

Latitude Location Code:

---

36°N Strait of Gibraltar =15°+21°=3Ra+3Lb
29°N Canary Islands =2La
26°N Cape Bojador =2HR
23°N Tropic of Cancer =2Ra
21°N Cape Blanco =3Lb
17°N Cape Verde Islands, north =1Re
16°N Cape Verde Islands, central =1Ri
15°N Cape Verde Islands, south =3Ra
14°N Cape Verde =3Rb, 1Rae
12°N Bissagos Islands =3Ri
4°N Cape Palmas =1Li
4°N Niger River mouth =3HR
0°N Equator =3Ha
1°S Cape Lopez =1Le, 3HRa/3Hla

**Fig.6** The latitude codes of glyph 3, repeated in Table 1, showing the coastal route from the Strait of Gibraltar (36°N) along Africa to the south until Cape Lopez (1°S), and the discovery of the Cape Verde Islands at 16°N (c.4500 BC).

in Fig.7), encoding the complementary latitude of Rockall and the Scottish Highlands, 90-33= 57°N. Apparently, because of explorations from the Hebrides and Orkneys, which have huge megalithic stone circles and other remains (Ref.27), a lot of new discoveries were made on the ocean. When we look at all the angles with the baseline of glyph 6, you find the latitudes of the Faeroe Islands, Iceland,and Greenland, including Cape Farvel and the SW Cape of Greenland (Ref.1). Above glyph 6 are two long thin cracks in the stone which indicate the east and southwest coasts of Greenland (Fig.2). The important sailing directions are included. Almost all the latitudes are encoded as complimentary angles. They must have been so familiar with their encoding techniques that they could use them reliably and easily at sea.

Look at the end of the glyph 6 right side branch. The notch here looks like a copy of the Brittany notch, but it is not. It is an engraving of South Greenland. This had become the westernmost land in the known world. Near the bottom of glyph 6 we see the long sailing route from the island of Crete, via the Egeic Sea and the Black Sea to the mouth of the Dnepr River, near the present city of Odessa. This route to the north is the coastal route along Norway to the North Cape. This sailing route is also indicated by the latitudes of glyph 6. This is a rather important part of the glyph, because it shows there is no land connection from Scandinavia to the west. The petroglyphs at Loughcrew and the design of Stonehenge 1 (Refs.1,3,10-12) show that the attempts to cross the sea from south Greenland to the west were given up at this point, as we have shown in other articles. Because of this dramatic decision at Greenland to give up, recorded in these monuments, we know that this Dissignac glyph 6 was inscribed c.3200 BC.

## The Discovery of Ascension and St. Helena

There is a thin, intermittent "connection line" (only seen in Fig.2), from glyph 7 to the left branch of glyph 3 on its right (Figs.4,7). This is indicating that the carver of glyph 7 is orienting on glyph 3, and that new discoveries were made in the south. For several centuries, the megalithic efforts to cross the ocean were now focused on the South Atlantic off the coast of Africa. The left side branch of glyph 7 runs from Gibraltar (the center of all Mediterranean signs) to Cape Palmas, the south point of NW Africa (see map, Fig.8). The left tangent of glyph 7 makes an angle of 29° with the baseline, equal to the sailing direction from Cape Palmas to the islet of Ascension, 29°SSW. The tangent to the right side of old glyph 2 makes an angle of 8°, encoding its latitude of 8°S. The discovery of Ascension in the Southern Ocean was important, because it became by far the westernmost land in the Southern Hemisphere.

The angle between both tangents of glyph 7 equals 45°, equal to the sailing direction from Ascension to St. Helena, 45°SE. The right tangent along this glyph makes an angle of 45°-29°= 16°, encoding its latitude of 16°S (for details, see Fig.8). So, the islet of St. Helena was discovered as well. The long hook pointed down on the right end of glyph 7 shows that a return involved coming back quite a way from the west. Figure 9 shows a

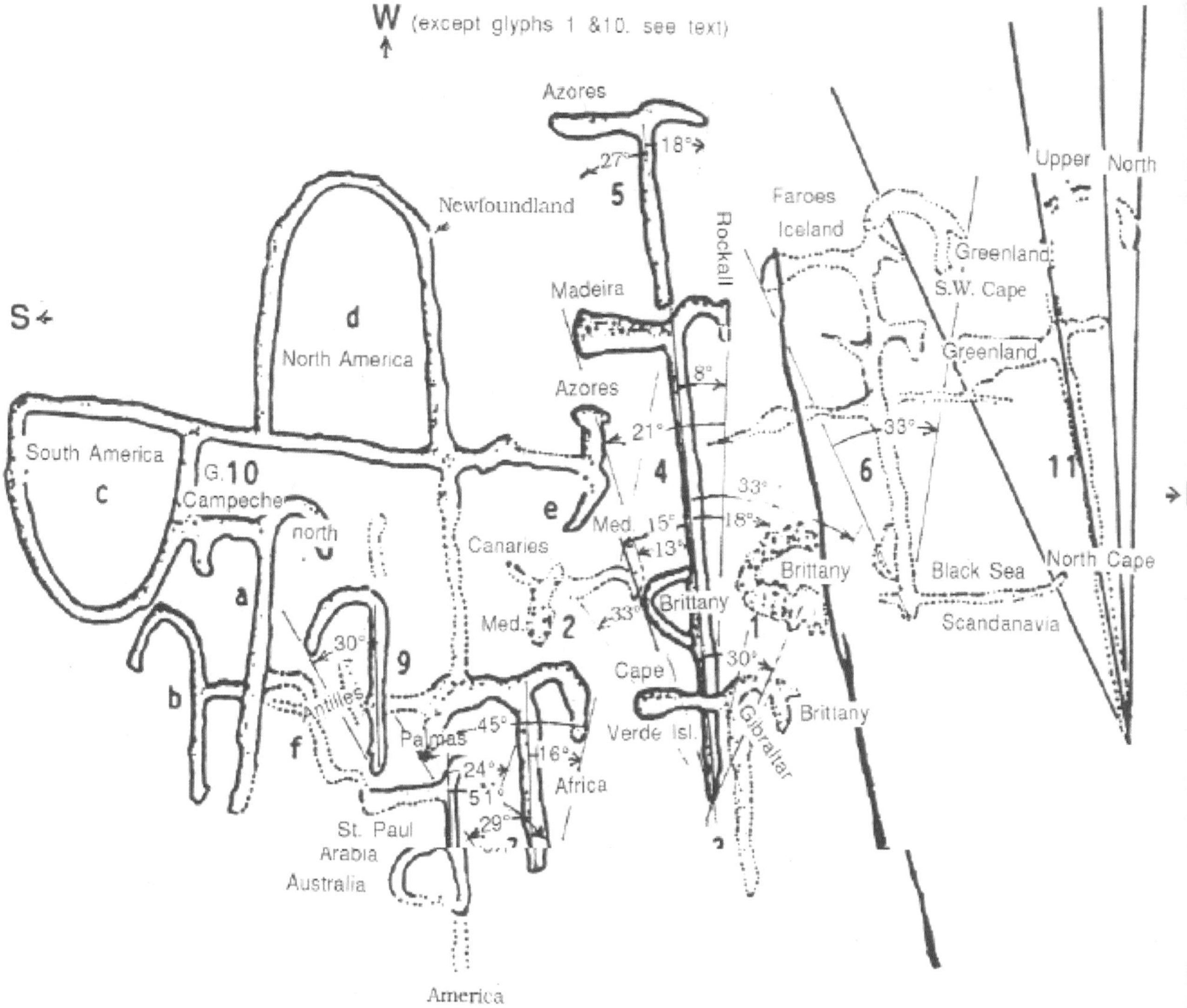


**Fig.7** Petroglyphs of Dissignac (Ref.4) with geographic labels and tangent angles. Over time, it became a megalithic symbolic map of the world using mathematics. Tangent angles of symbolic Jacob's Staffs were used to encode latitudes.

**Glyph 1:** Ancient map of the peninsula of Brittany with the coastal waters over c.100km **(c.6500 BC)**

**Glyph 2:** Ancient map of the Little Mediterranean Sea with the sailing routes to Portugal (right) and the Canary Islands (left) **(c.5500 BC)**

**Glyph 3:** The discovery of the Cape Verde Islands **(c.4500 BC)**

**Glyph 4:** The discoveries of Madeira and Rockall **(c.4100 BC)**

**Glyph 5:** The discovery of the Azores **(c.3600 BC)**

**Glyph 6:** The discovery of the Faroes, Iceland and Greenland **(c.3200 BC)**

**Glyph 7:** Discoveries of Africa, Ascension and St. Helena **(c.2900 BC)**

**Glyph 8:** The discoveries of Australia **(c.2700 BC)** and America via the Bering Sea **(c.2600 BC)**

**Glyph 9:** The sailing route to Central America **(c.2300 BC)**

**Glyph 10:** The sailing routes around the Americas **(c.2200 BC)**

**Glyph 11:** The sailing route to America via the Upper North **(c.2100 BC)**

petroglyph from Kermaillard, Brittany (Refs.2,3). Below the North Atlantic Ocean with the sea areas east and west of Greenland (where they gave up), we see a U-shaped sailing route following the Mid-Atlantic Ridge in the Southern Ocean, that confirms the discovery of Ascension and St. Helena. The petroglyph reveals a great amount of ocean sailing going on, to have found these remote islands, despite the fact that birds probably showed them the way, once they got near the islands. The sailing distance from Cape Palmas to Ascension is 1450km, from Ascension to St. Helena 1250km, and St. Helena is 1800km from Cape Frio on the coast of Africa. By now, people were able to cross long distances over the open sea.

After the discovery of Ascension and St. Helena, there were a few centuries when the megalith builders hoped they could find a way to cross the South Atlantic. The tangents of glyph 7 show that the whole western coastal route of Africa was explored over a distance of c.1500km offshore, as far as the SW cape of South Africa, the Cape of Good Hope. Because of lack of success, a new period began in the efforts of the megalith builders. They decided to start exploration of the earth to the east!

The tangents of glyph 7 show that the east coast of Africa was explored as far as Egypt at the north end of the Red Sea. At the right side of the handle of a large dot is engraved (see Fig.2), representing the large island of Madagascar, only 400km offshore, the discovery of which is confirmed by the tangent latitudes of the figure. The glyph details even show where to leave the coast for the Comoros, near Cape Delgado, at 11°S. We do not know when glyph 7 was inscribed, but from circumstantial evidence, we estimate that it probably dates from c.2900 BC.

# Explorations to the East

## The Discovery of Australia

Glyph 8 was placed close to the previous glyph 7 (Figs.4,7), so apparently it is oriented on this glyph. It was carved at the left and below, which indicates new discoveries in the southeast. The right side branch is again the sailing route around Iberia to Brittany. The left side branch of glyph 8 however, is inscribed in a very long straight manner. Latitudes reveal that the tiny islet of St. Paul was discovered, just above the equator, at 1°N (see map, Fig.8). The distance between Gibraltar and St. Paul is roughly equal to the length of the Mediterranean. Only the upper portion of the figure contains the handle that represents the Mediterranean, about the length of the left side branch. But glyph 8 extends downward substantially, illustrating that people found important discoveries east of the Mediterranean.

The circle at the bottom of glyph 8 represents the peninsula of Arabia. Further below it we see the coastal route along the Arabian Sea to the river Indus. These were the sailing routes that connected the great civilizations of Egypt, Mesopotamia, and India. These routes had been known since c.5,000 BC. The left side of the handle is the baseline of this Mediterranean symbol, and the lowest point above the circle is the starting point, because it

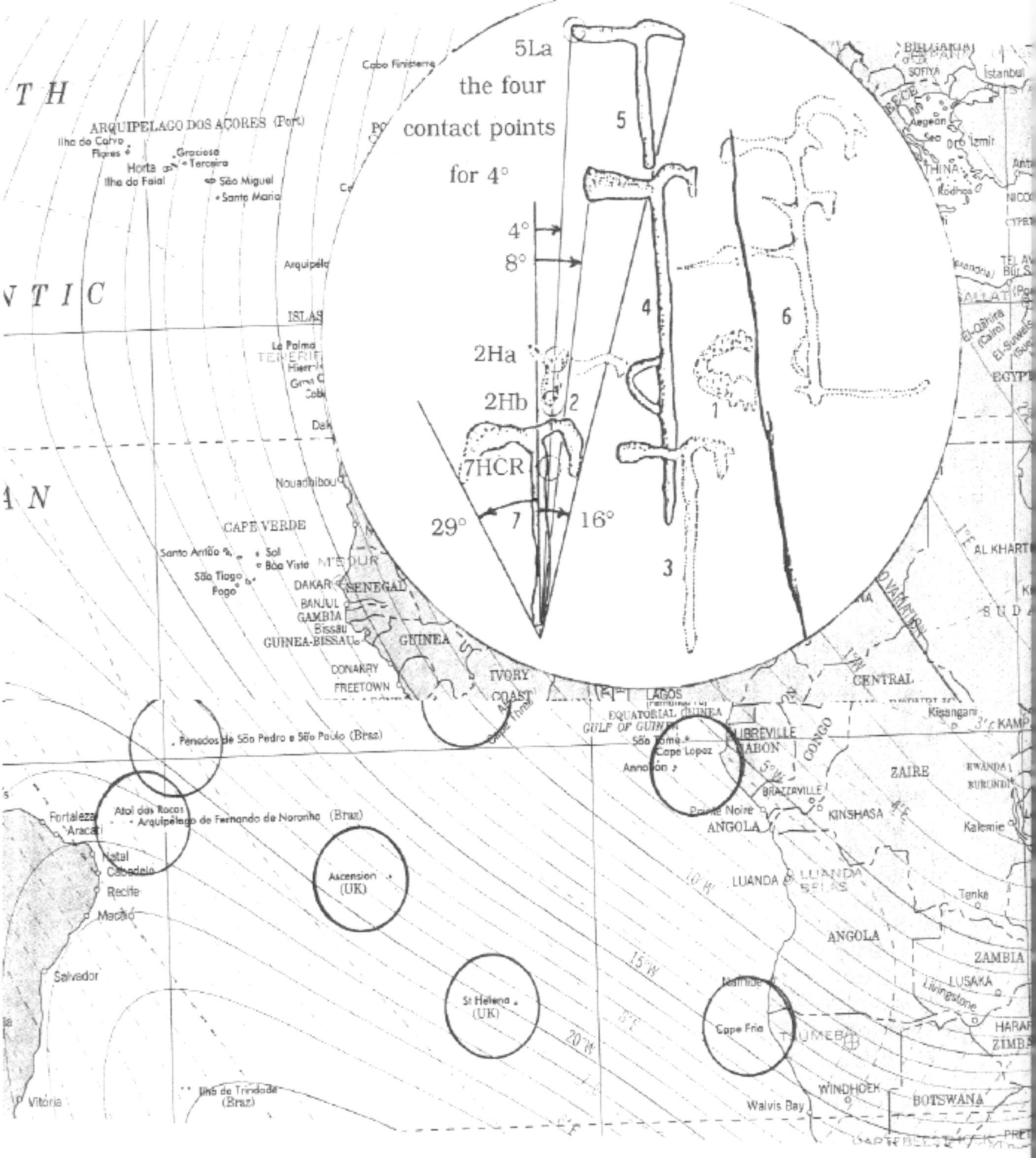


**Fig.8 Map** of the Atlantic Ocean, showing the locations of Cape Palmas, Ascension, and St. Helena, and other capes of West Africa (Ref.25).
**Above:** How the discovery of Ascension and St. Helena is encoded in the tangents of glyph 7. From Cape Palmas (7Lb) at 4°N (5La), we sailed 29°SSW (7Lb) to Ascension at 8°S (2Ra,4Lb). From there we sailed 29+16=45°SE (7Lb+7Rb) to St. Helena at 16°S (7Rb,5HRb,5Ra). (Dissignac, Brittany, c.2900 BC)

represents Egypt. Again, we can draw lines along the important points of all engraved, previous figures. The angles with the baseline represent important capes, river mouths, and islands. These tangents of glyph 8 provide the route from Egypt, along the south coast of Asia, which is along the Arabian Sea, to Sri Lanka, and then along the Bay of Bengal to Singapore. Here there is an important division of the sailing routes. One route runs to the northeast, along the continent of Asia, and the other runs to the southeast, in the direction of Australia. The petroglyph shows this fork at the top of the circle, which now is symbolic for the Pacific.

The southern route runs along the south part of Indonesia, and via the crossing of the Timor Sea from the island of Roti to Cape Bougainville, Australia (450km). This route, and the routes around Australia and Tasmania are well indicated by the latitudes given by the tangents of glyph 8. At the oldest glyph 1 (Brittany), we looked along the west coast of France to the north. When we consider glyph 8 in the same manner, the circle now represents the continent of Australia, and this part of the petroglyph closely resembles the coast map of the continent (see Fig.11). So Australia has been discovered! As just explained, this double symbolism is confirmed by the recorded latitudes, and the megalithic monuments in Australia, in the neighborhood of Kangaroo Island. We estimate this discovery of Australia by megalithic people to have been c.2700 BC. This is 40.000 years after the aboriginal people are now thought to have arrived there by boat (Ref.28).

More than 250 Egyptian hieroglyphs are found in a cave-like chamber on the east coast of Australia, 100km north of Sydney, in the Hunter Valley (New South Wales, Ref.32). Among these is a remarkable third-life size carving of the ancient Egyptian god "Anubis", the Judge of the Dead. On the sandstone walls, the story is told of the death and burial of Lord Djes-eb, one of the sons of Pharao Ra Djedef (4th Dynasty, c.2650 BC), on his expedition. Both these deciphered hieroglyphs as well as the date of them confirm our claim of the discovery of Australia, c.2700 BC.

New Zealand does not appear in glyph 8. The distance between Tasmania and New Zealand is 1500km across the famously rough Tasman Sea. Nevertheless, stone circles are being reported there (Ref.24), so at some point later it evidently was found.

## The Discovery of America via the Bering Sea

When watching the Mediterranean sign of glyph 8 to the west again (Figs.4,7), the circle at the foot of glyph 8 becomes symbolic for the Pacific. The right side of it is the route along Asia, and the tangents give the latitudes of this route clearly. Many bays like the Gulf of Siam and the Gulf of Tonkin are crossed with a sailing direction of 45°NE. After crossing the Tropic of Cancer at 23°N, the route continues along the coasts of China and Japan. The Kuril Islands are followed to Kamchatka, at 56°N, its east cape. From Kamchatka, the Aleutian Islands can be followed across the Bering Sea, with the southern-most island at 51°N. The longest crossing to North

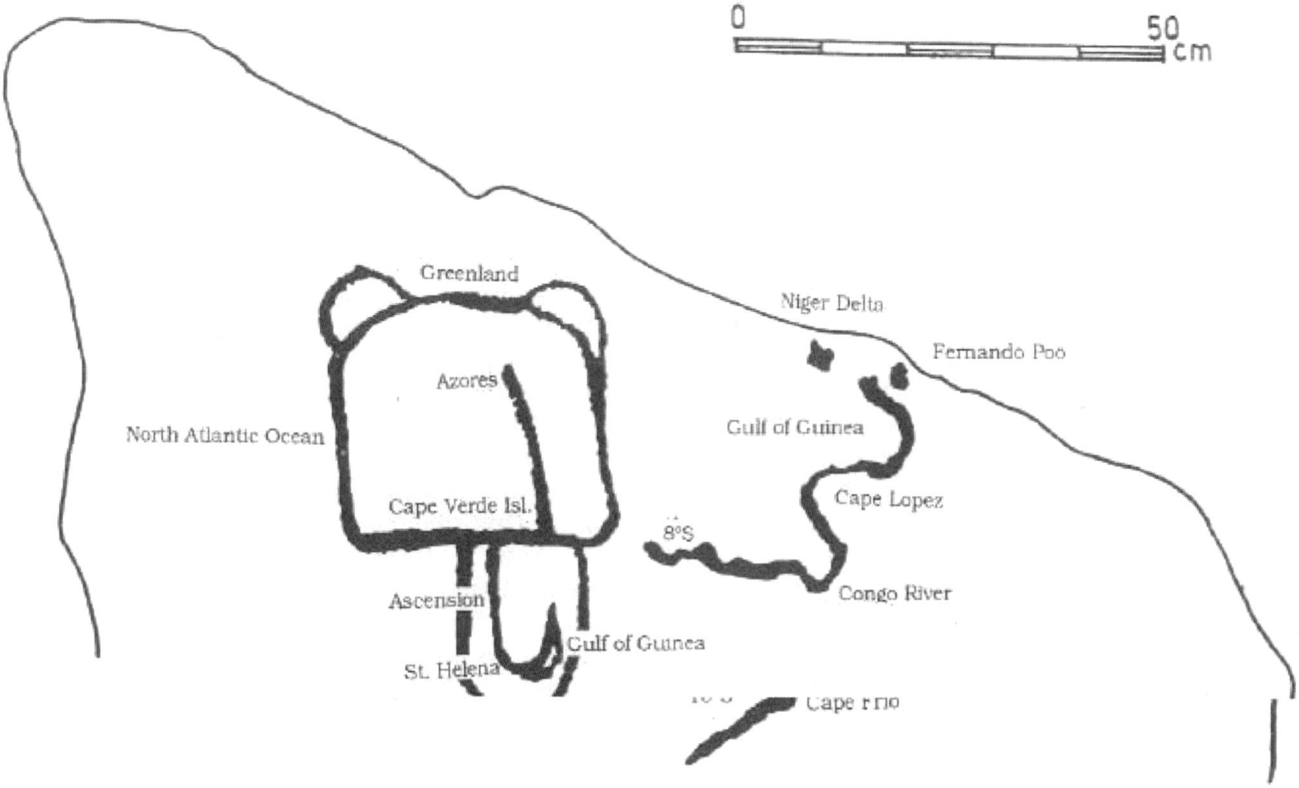


**Fig.9 Left:** Petroglyph (Ref.3) of the North Atlantic Ocean, with the sea areas east and west of Greenland, and a line across the Azores and the Cape Verde Islands in the south. Below the equator is a U-shaped sailing route in the Southern Ocean to the newly discovered islands of Ascension and St. Helena on the Mid-Atlantic Ridge, and curving back to supposedly "related" islands in the Gulf of Guinea.
**Right:** The Upper edge of the stone is the south coast of NW Africa, and the two dots are the Niger Delta (left), and the SE point of Nigeria, a high mountain of 4070m (right). The sailing route starts at Fernando Poo in the Gulf of Guinea, going widely around Cape Lopez (1°S), along the islands of Principe, Sao Tomee, and Annobon back to the coast until the Congo River (right), just north of the latitude line of Ascension (8°N). The low, straight line to the southwest shows the shallow seas SW of Cape Frio (18°S). (Kermaillard, Brittany, c.2900 BC)

America over open water is 350km. A far more northern route is to cross from Cape Providenya, at 64°N via the island of St. Lawrence with a crossing of c.200km, or further north at 66°N via the Diomedes Islands to Seward Peninsula, with a crossing of c.50km. The tangents of glyph 8 not only give the sailing routes along the west coast of North America, but along the north coast of South America. All this data is beyond the scope of this article, and needs more study by other researchers, so these tangents are omitted here.

Note though, that the tangent along the right side of glyph 8 (the point of contact is Brittany, where Dissignac is located) makes a complementary angle of 90°-24° =66°, corresponding to the latitude of the east cape of Asia at 66°N. The angle of 24° encodes the sailing direction from the East Cape to Cape Prince of Wales, the westernmost point of North America, at 24°ESE. This angle also corresponds with the latitude of the river Indus, at 24°N, the easternmost point of glyph 8. This point is symbolic for the goal of the whole expedition. The tangent line to the Dissignac site on glyph 1 (1Rib) makes an angle of 51°, shown on Fig.7), the latitude of the southern Aleutians (and Stonehenge) at 51°N). The latitudes that emerge from glyph 8 as a Jacob's Staff clearly confirm that glyph 8 is encoding the discovery of America via the Bering Sea.

The left side branch of glyph 8 is engraved to the island of St. Paul at 1°N, in the middle of the Atlantic Ocean. St. Paul is 1700km from the Cape Verde Islands, only 600km from the islands of Fernando de Noronha, and only 1000km offshore Brazil (Fig.8). At the top of the stone is a horizontal engraving (see Fig.2), showing that America has been reached. Glyph 10 to the upper left of glyph 8 was later engraved, showing the sailing routes around the Americas, and detailing them with latitudes. This large glyph 10 is attached to glyph 8, because this glyph 8 was the basis of the megalithic discovery of America. There are no other petroglyphs in Europe or anywhere else to our knowledge, that show America having been discovered via the Bering Sea. In this regard, the petroglyphs of Dissignac are unique. We do not know when glyph 8 was made, but we think it was about a century before the first crossing of the Atlantic Ocean, so c.2600 BC.

Recently (in the Spring of 2001) archaeologists announced new carbon dating (rock bag cord) of the pyramid construction Caral, in the Supe River Valley of Peru (Ref.33). The site is about a mile long, and half a mile wide.Caral features monumental architecture; it has six towering stone pyramids, large plazas, a complex irrigation system and apartment-like buildings. The city was built in a pre-ceramic era, leaving no pottery shards. The head axis of the whole complex points 31°WNW, corresponding to the latitude of the Nile Delta, at 31°N. About 500km east of Caral is one of the headwaters of the Amazon River, and a province of Brazil, both called "Acre", phonetically meaning "Great Ra" (Ref.34), which also indicates an ancient Egyptian influence. Caral is proved to be c.800 years older than the oldest previously known New World city, namely c.2600 BC, which confirms our estimated date for the discovery of America via the Bering Sea.

**Table 2**

**The way to Central America (Glyph 9, c.2300 BC)**

Latitude Location Code (sailing direction in parentheses):

----------------------------------------------------------------------

36°N Strait of Gibraltar 7La, 2LC, 4Ha/7Lib,2HRb(c)
29°N Canary Islands 8HLb/4CiL,4Cir,1ea(c)
26°N Cape Bojador 7Rib,1Le,1Lia,1Rea(c)/3Hb(c)/5HRa
21°N Cape Blanco 8bLa
17°N Cape Verde Islands, north 5La/8bLb/6bLa(c)
16°N Cape Verde Islands, central 8Lib/8CbL
15°N Brava Island 7Lb,7RCb,3La,1Reb,6bRib(c)
(60°SW) Cape Sao Roque 9La,8HLb(c)/5Ra(c)/4Cia
(60°=30°+30° SW)
South America (NE Cape) 9La+5Ra
(20°SSW) St.Paul* 4Ceb,1Rib(c)/8bRb
14°N Cape Verde* 6Hb, 6bRa(c)
13°N Gambian coast* 8RLa,3Le,3RLa,6bRb
(45°SW) ST. Paul* & Cape Sao Roque* 2Rae,6RLa/7HLb
7°N Sherbro Island* 8CLi/3RCb(c)
(20°WSW) St.Paul* 4Ceb,1Rib(c)/8bRb
1°N St. Paul* 9HLai/9aRi
(45°SW) Cape Sao Roque* 2Rae, 6RLa/7HLb
4°S Fernando de Noronha* 9aRb/4Hb(c)
5°S Cape Sao Roque 9HRa/8Ria,3Rb/1Reb-1Rib
7°S Cape Branco 8CLi/3RCb(c)
3°S Cape Camocin (Brazil) 9aRa/8CLe
2°S Bay of Sao Jose 9HLae,9aL
1°S Bay of Marajo 9HLai/9aRi
0° Equator/ mouth of Amazon 9Ha,9aLb,9aa
4°N Cape Orange (Brazil) 9aRb/4Hb(c)
9°N Orinoco River mouth (Venezuela) 7Rb,3Ra(c)/8CLi
10°N Trinidad south 8Rb(c)
11°N Trinidad, north (Antilles Route) 3Lb,3Rib(c)/9HLe
13°N Dutch Antilles/Cape Gallinas 8RLa,3Le,3RLa,6bRb(c)
10°N Panama, north 8Rb(c)
14°N Honduras, east coast 6Hb,6bRa(c)
15°N Cape Gracias a Dios 7Lb,7RCb(c),3La,1Reb,6bRib(c)
16°N Gulf of Honduras 8CbL/8Lib
17°N Bay of Chetumal, south 5La/8bLb/6bLa(c)
21°N East Yucatan (Mexico) 8bLa
22°N Cape Catoche 4Cib(c)
20°N Gulf of Campeche, north 8bRb/4Ceb,1Rib(c)
19°N Gulf of Campeche, central 8CbR
18°N Gulf of Campeche, south 8bLb/6bLeb(c)
23°N Gulf of Mexico (E)/Tropic of Cancer 8CRib/6bLea(c)

Code: H=Handle, L=Left side branch, R=Right side branch, C=Center, a=above, b=below, i=internal, e=external, (c)=complementary
*alternative crossings

According to the book "Osiris & the Egyptian Resurrection", author Wallis Budge (Ref.31) mentions that: "In the reign of King Assa (Oeser-kaf, c.2550 BC), a high official called Baur-tet brought from Punt a "tenk" (pygmy), who knew how to dance "the dance of the god", and was said to come from the "Land of the Spirits". Assa was so pleased with his officer that he bestowed great honours upon him, and we may assume that the pygmy was dispatched to Memphis (the capital) to dance before the king." These earliest contacts with Punt (Central America), c.2500 BC, also confirm our estimation of the date of the discovery of America via the Bering Sea, c.2600 BC.

In late prehistory many nomads crossed the Bering Strait, and earlier, the land bridge. Before 3000 BC, Japanese and Chinese sailors had already reached the continent, too (Refs.13-17). It is very plausible to suppose that Egyptians had heard that America existed, and how to reach it via the Bering Sea. Recently, burials of early European people have been found in China, and we have shown that megalithic age people must have explored the west coast of America, to have encoded the latitudes by the tangents of glyph 8. In the 1970's Barry Fell deciphered inscriptions of ancient Egyptian-Libyan in the "Caves of the Navigators" at McCluer Bay, New Guinea, and a similar script in a cave in Santiago, Chile, dated 231 BC. In the latter, Navigator "Maui laid claim to 4000 miles of American coastline he had traveled for the King of Egypt and his queen" (Ref.36).

## The Sailing Route to Central America

A connection line runs from glyph 9 to glyph 7 (Figs.4,7), so 9 is oriented on 7. Glyph 9 is situated at the left side above glyph 7, so there were new discoveries to the southwest. Until now, all Mediterranean symbols asked a question: if we want to cross the ocean from the Mediterrranean, after Gibraltar, do we sail north or south? The name "Dissignac" itself means literally "Place of the Two Directions"! Glyph after glyph is asking this question here, over thousands of years, at least 100 generations of sailors, astronomers, and religious/political leaders. Now glyph 9 is inscribed with only one side branch. So here the question is answered: "to cross the ocean, you have to go to the south!" Glyph 9 attaches to the left side branch of glyph 7 exactly where the Cape Verde Islands are located. The new discoveries were made in the southwest, so the new sailing route starts at Brava, the SW island. The left tangent along glyph 9 makes a complementary angle of 90-30= 60°, encoding the sailing direction across the ocean, 60°SW. It is the Southern Crossing to South America, with the tradewind and current. This crossing has a length of 2500km. In this time, people were able to cross this huge distance over open sea.

Table 2 gives the whole route by latitudes from the tangents of glyph 9, from Gibraltar to Central America. Along the north coast of South America they sailed from Cape Sao Roque, the NE cape of Brazil, via the mouth of the Amazon River and Trinidad/Tobago to Cape Gallinas, the north point of South America. Voyagers are discouraged from the Antilles route. Note that in the left side of the handle of glyph 9 a very small map of the Caribbean has

**Table 3**

**The way back from Central America (Glyph 9, c.2300 BC)**

(sailing direction in parentheses):

------------------------------------------------------------------------

22°N Cape Catoche (Mexico) 4Cib(c)
21°N Yucatan, east 8bLa
(11°ENE) Cape San Antonio (Cuba) 3Lb,3Rib(c)
(11°ENE) Cuba, west 9HLe
22°N Cape San Antonio 4Cib(c)
23°N Cuba, N/Tropic of Cancer 8CRib/6bLea(c)
(0°N) Cape Sable, Florida 9Ha,9aLb,9aa
25°N Cape Sable, Florida 5Ha/1Li,1Ria(c)
26°N Rio Grande* 5HRa/7Rib, 1Lia, 1Le, 1Le, 1Rea(c)/3Hb(c)
26°N Bimini, NE Bahamas* 5HRa/7Rib,1Lia,1Le,1Rea(c)/3Hb(c)
(32°NE) ISD Bermuda* 4CLa
(24°ENE) TSD Bermuda* 1Ria
29°N Mississippi 8HLb/4CiL,4CiR,1ea(c)
30°N Gulf of Mexico, north* 9La, 8HLb/5Ra/4Cia(c)
30°N Florida, start of peninsula 9La,8HLb/5Ra/4Cia(c)
34°N Cape Fear, North Carolina 6CRb(c)
35°N Cape Hatteras, North Carolina 2La,5Hb/2Hb(c)
(18°ESE) ISD Bermuda* 8Lib/7Lb
(23°ESE) TSD Bermuda* 8CRib/6bLea(c)
(32°N) Bermuda* 4CLa
(24° ENE) ISD West Azores* 1Ria
(5°ENE) TSD West Azores* 3Rb
41°N Long Island, New York 6La/6CLe,6CLa(c)
42°N Cape Cod 6HRia(c)
43°N America's Stonehenge 4Ra,2Rb,6Ri(c)/7HRb(c)
(0°E) 9Ha,9aLb,9aa
43°N Nova Scotia, south 4Ra,2Rb,6Ri(c)/7HRb(c)
45°N Nova Scotia, northeast 2Rae,6RLa/7HLb
46°N Cape Breton (Canada) 6Ha/2Rai(c)
47°N Cape Breton Island, north 4Ra(c)/2Rb,6Ri
47°=30°+17° idem 9Lb+5La
(45°NE) Newfoundland, southwest 2Rae,6RLa/7HLb
48°N Newfoundland, southwest 6HRia
47°N Cape Race 4Ra(c)/2Rb,6Ri
(20°ESE) ISD, West Azores 4Ceb,1Rib(c)/8bRb
(35°ESE) Terminal Sailing Direction (TSD), West Azores 2La,5Hb/2Hb(c)
39° West Azores 4Rai/6CLa(c);39=9+30°N West Azores 8CLi+5Ra
(20°ESE) Santa Maria, East Azores 4Ceb,1Rib(c)/8bRb
38°N Central Azores 4Ri/6HCa,6Rb(c)
37°N Santa Maria, East Azores 6CRi(c)
37°=20°+17° idem 8bRb+5La
(30°SE) Madeira 9La,8HLb/5Ra/4Cia(c)
33°=30°+3° Madeira 9Lb+9aRa/5La+8CLe
(20° ENE) Strait of Gibraltar 4Ceb,1Rib(c)/8bRb
36°N Strait of Gibraltar 7La,2LC,4Ha/7Lib,2HRb(c)

been engraved in mirror image. As we shall explain later, this shows the route along the north coast of Honduras, and around the Yucatan peninsula to the Gulf of Compeche, the areas which were the goals of the voyage c.2500 BC (Refs.15-17).

Just above the handle of glyph 9, the sailing route west of Gibraltar from Madeira to the Azores has been lightly engraved. This part of the petroglyph shows, how people returned from America. The whole return route of glyph 9 is shown in Table 3. From Yucatan they sailed to Cuba (c.200km), and from Cuba to Florida (again c.200km). Next, they sailed along the coast of North America with the Gulfstream all the way to Cape Race, Newfoundland. The table gives all the latitudes of the important points: Cape Hatteras, Cape Cod, America's Stonehenge, Cape Sable on Nova Scotia, Sable Island, and many more. From Cape Race (or via Bermuda) they crossed the ocean to the West Azores, with the wind and current. The Initial Sailing Direction (ISD) is 20°ESE, and the Terminal Sailing Direction (TSD) is 35°ESE. It is a distance over open sea of more than 2000km. In this manner they returned to Gibraltar and the Mediterranean.

In which century was the first crossing of the Atlantic Ocean to America? America's Stonehenge dates back to at least 2000 BC (Refs. 20,29), and Stonehenge III in England is the monument for the discovery of America, so it was before Stonehenge III was built, again c.2000 BC (Refs.11,12). Previously, Stonehenge II was partially erected (and later pulled down) for the same reason (new discoveries in the far west), so the first crossing was before c.2100 BC. During the 11th dynasty (2133-2000 BC) a nobleman called Hannu under the ruler Sankh-ka-ra conducted an expedition to Punt (tropical Central America)(Ref.13). In the text in the pyramid of Pepi I (2389-2357 BC) distinct mention is made of the "pygmy of the dances of the god (from Punt), who rejoiceth the heart of the god before his great throne" (dances for the King, Ref.31). During this 6th dynasty (2423-2263 BC) a sailor named Knemhotep visited Punt eleven times (Ref.13).

Punt is called "God's Land" 13 times in Queen Hatshepsut's temple Deir el-Bahri near Karnak (c.1500 BC). In the first relief panel, the last of 5 vessels of the departing fleet bears over its stern the pilot's command "Steer (make) to Port", an instruction to head west from the Nile (Ref.1). In a time period when Egypt traded regularly with China, celebrating "an expedition to the distant land of Punt" does not merely mean the nearby Horn of Africa. According to the reliefs "the promised land of Punt was reached long ago via land" (namely eastward via the coast of Asia), "but now more via the sea" (namely westward via the Southern Crossing) (Refs.1,4).

The oldest dated petroglyph showing features of the American coast is on the menhir of Kermorvan (see Fig.10). This menhir, situated near the westernmost tip of Brittany, was carved with an early petroglyph, half astronomical, and half geographical. It is directly related to the petroglyphs in the gallery graves of North Brittany, well dated from c.2400 BC

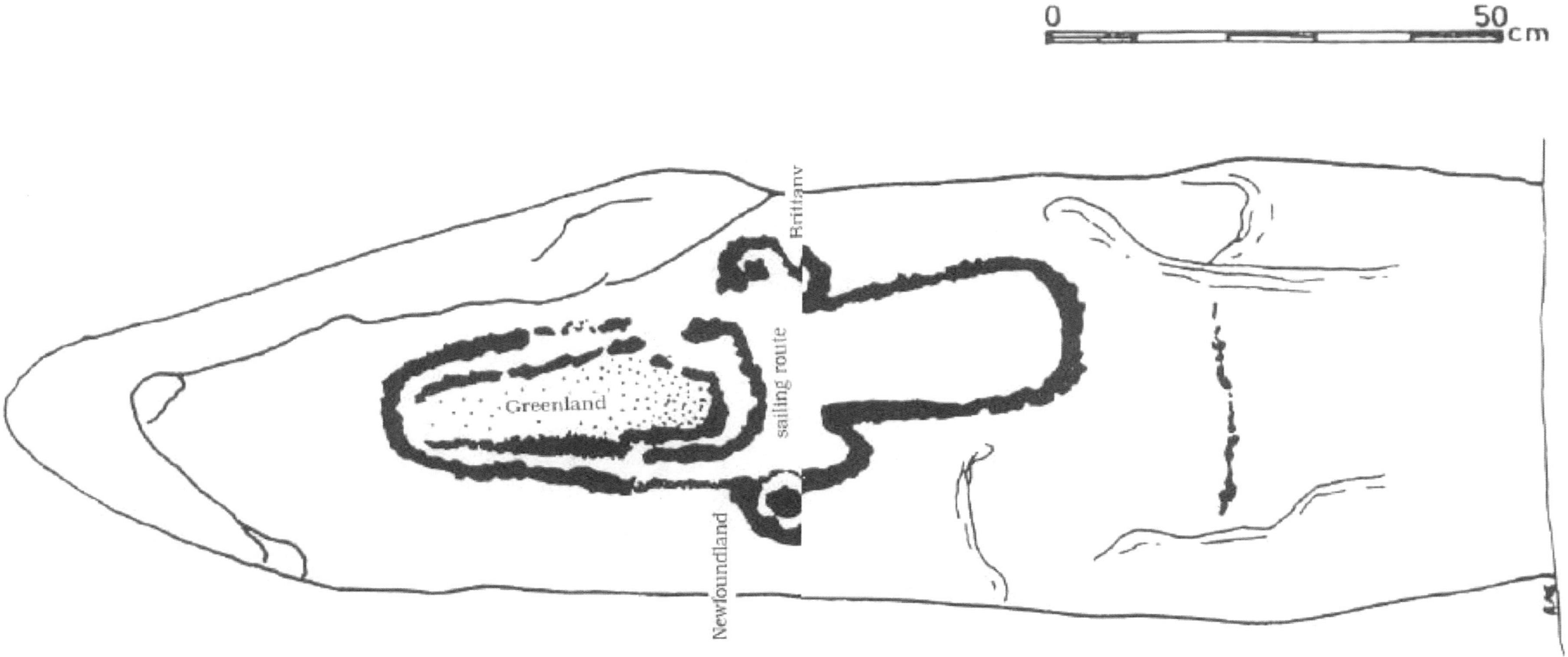


**Fig.10** Astronomic and geographioglyph of a huge comet of the size of the North Atlantic Ocean (Ref.3). Aland inside the comet is a complete map of the island of Greenland (liglicked). Just below it is the sailing route from Iceland (right) to the weist of Greenland (left), as far as the start of the shortest crossing of Datrait, at 66°N. The big dots, right and left, are Brittany and Newfoun( respectively. Lower on the stone, the latitude line between Brittany Newfoundland, at 48°N. People thought that the head of this comesisted of ice, like Greenland, which is correct. (Note that the upper 1of the menhir is steeply bevelled, resembling the top of the comet.' astronomical petroglyphs in the gallery graves of Brittany date from00 BC. (Originally, Kermorvan, Le Conquet, 48°N, Brittany, now in Preic Museum, Penmarc'h, Brittany).

(Refs.2,3,26). However, we think the first recorded crossing of the Atlantic Ocean to America was made earlier by Egyptians, in the first Punt voyage in the 13th year of the government of King Sahura, 2497 BC (see ships, Fig.12) (Refs.1,22). Glyph 9 probably dates from about 2 centuries later, c.2300 BC.

## The Sailing Routes around the Americas

Glyph 10 is by far the largest glyph of Dissignac (Figs.4,7). Connection lines run to glyphs 7 and 8, so it is oriented on those. Glyph 10 was engraved at the left side, above these glyphs, indicating huge discoveries southwest of the Mediterranean. The connection line to figure 7 runs to the spot on the left branch where the Cape Verde Islands are located. The connection line to glyph 8 leads to the spot on the tip of the left branch, where the islet of St. Paul is situated. It is clear that the Southern Crossing between Africa and South America plays a big role. This is confirmed by the glyphs 10a and 10c. Glyph 10c is the newly discovered continent of South America. It has exactly its shape (c.2200 BC).

Glyph 10a has been connected to glyph 10c. The point of connection is the Southern Crossing between the Cape Verde Islands and the west coast of South America, so here the crossing is engraved. Glyph 10d represents North America (c.2200 BC), and the horizontal line is the equator. Note that the orientation has been changed. In the case of the glyphs 10a and 10b, we look as usual to the west, and we can understand that glyph 10d, being North America, has been placed at the right side of glyph 10c, South America, with Central America between. However, when the shapes of the glyphs 10c and 10d are considered, we are looking to the north. That was also the case for the circle of glyph 8, that represented Australia (Fig.11). The reference here is again back to the oldest glyph, 1. There too, the engraver looked along the west coast of France to the north. Remember, that everything started in Brittany. About four thousand years prior to glyph 10, people sailed c.100km offshore in search of the other side of the ocean. And now, finally, this goal has been achieved, and the whole history of the discoveries along the way has been carved into this granite stone at Dissignac in a very surprising level of detail and sophistication.

The glyphs 10a and 10b are Jacob's Staffs. Glyph 10a is situated at the right side below 10d, so it encodes, through its tangents, the latitudes of all important points along the east and west coasts of North America. Glyph 10b is situated at the left below 10c, so it gives the latitudes of all important points along the coasts of South America. They are connected, so together, they also provide all necessary sailing information about Central America. Note, as shown in Fig.13, that the east coast of Central America, the principal goal of their voyages, has been engraved, as though a superimposed map (c.2200 BC).

The long vertical connection line between glyph 10e and glyph 7 is faint, but connects the equator line of glyph 10 with the Cape Verde Islands of glyph 7 (which are located just above the equator). Glyph 10e is situated to the right

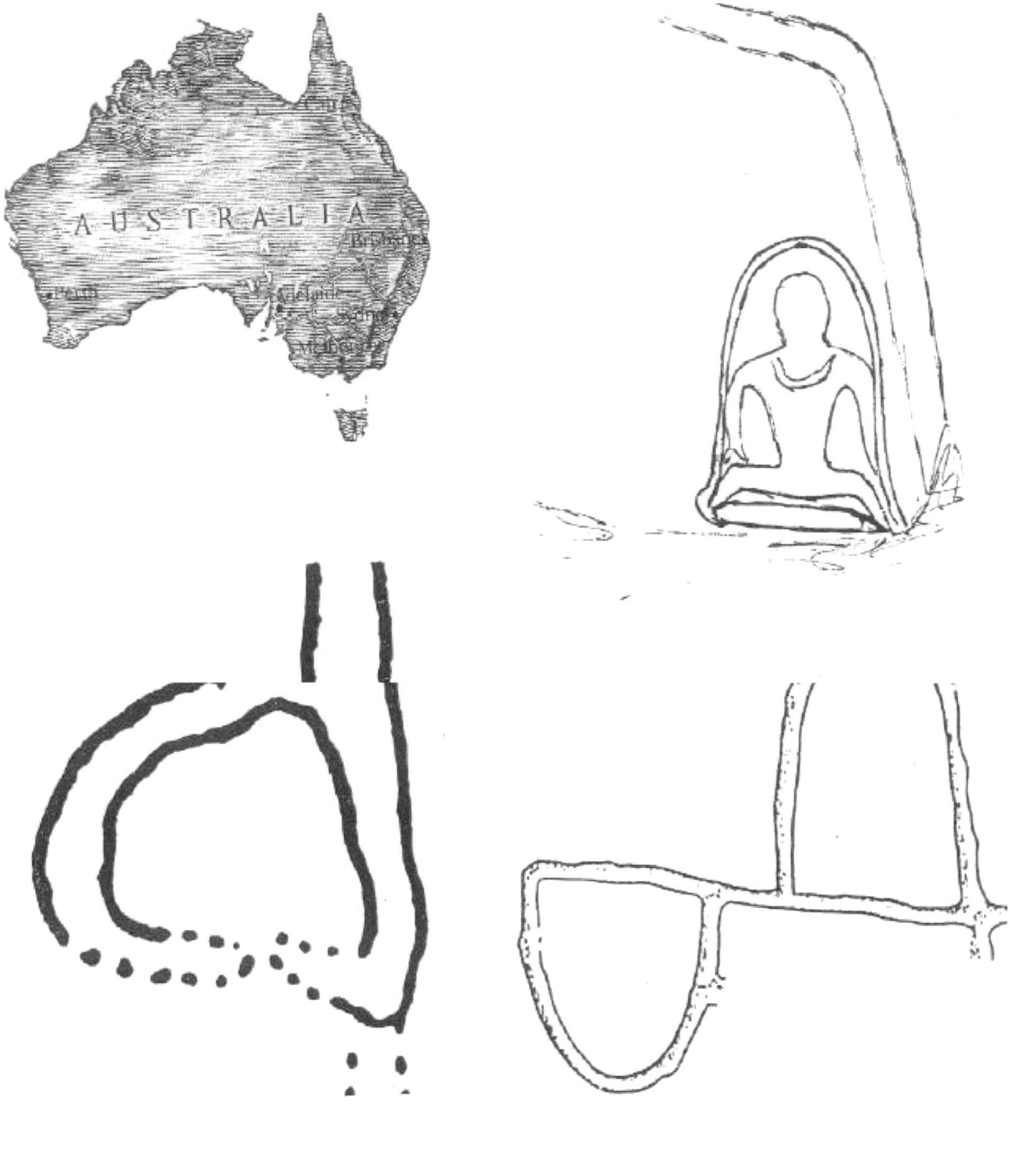


**Fig.11 Left, bottom:** Enlargement of the circle at the bottom of glyph 8, indicating the newly discovered continent of Australia, compared, **Left Top,** with shape of Australia. **Right Top :** Sketch of petroglyph found on Mineral Mountain, near Shutesbury, west/center Massachusetts, compared, **Right Bottom**, with Dissignac (France) petroglyph of North America.

of important glyph 10a. As a consequence, it is a route back, to the north, leading toward the Strait of Gibraltar. The sailing route splits with two choices, a western route ending in the Azores (it points to the left branches of glyphs 4&5, Madeira and the Azores), and an eastern route into the Little Mediterranean Sea (it even points to its map, glyph 2). Note however, the small swelling or zigzag at the start of 10e. This indicates a second meaning, with the handle running north, the west branch ending at Iceland, and the right branch ending in the North Sea. It appears that 10e means to show that these western places are dead ends of the return; that the Azores and Iceland are not places from which to cross the ocean, an important message.

Between glyphs 8 and 10a are shallow glyphs labeled 10f (Figs.4,7). The left branch of glyph 8 finishes at the islet of St. Paul, about halfway across the Southern Crossing. From St. Paul, one starts sailing to the west, after that to the south (follow glyph 10f). Next, one should sail quite a way along the north coast of South America to a cross-road. At the left side will be the big Orinoco River. It is recommended that one should continue sailing along the coast to Central America, though another option is to turn right following the Antilles route. At the end near the north coast of Cuba is a T-shaped cross-road. If one turns to the left, one sails to Central America, too. If one turns to the right, one sails to Bermuda, which may be important for the return route.

Glyph 10a has a curving right side branch. This is meant for the return route, showing that visitors to South and Central America should sail north in order to return. This side branch points to glyph 9. So it is recommended to return via the Azores, as described by this glyph for the first time (glyph 10a describes it too). However, the side branch also points just below glyph 9 to the lightly engraved route 10f from Central America to Bermuda. So, this is considered to be a good alternative.

Glyph 10 contains an enormous amount of encoded information. All this data could not be inscribed on stone without using records from complete exploration of the North and South Atlantic Oceans. Analysis of the tangents of 10 shows even Antarctica was reached (Ref.21). This glyph 10 must have been added near the end of the megalithic period, c.2200 BC, which was amazingly consistent in the continued use of this technique for the encoding of latitudes over more than two thousand years.

On Fig.11 is a sketch of a prehistoric petroglyph near Shutesbury, on Mineral Mountain (a source of white quartz) in central/western Massachusetts. The whole shape is nothing other than a copy of glyph 10d of Dissignac, meaning North America. Local authorities took this symbol from the most important petroglyphs of Europe, or at least from navigational symbols in common use at the time. Within the symbol a King was carved. The meaning of this petroglyph, which is megalithic in origin, is: "We are the people of North America, and this is our King!" The megalith builders in New England considered themselves in posession of North America (dated

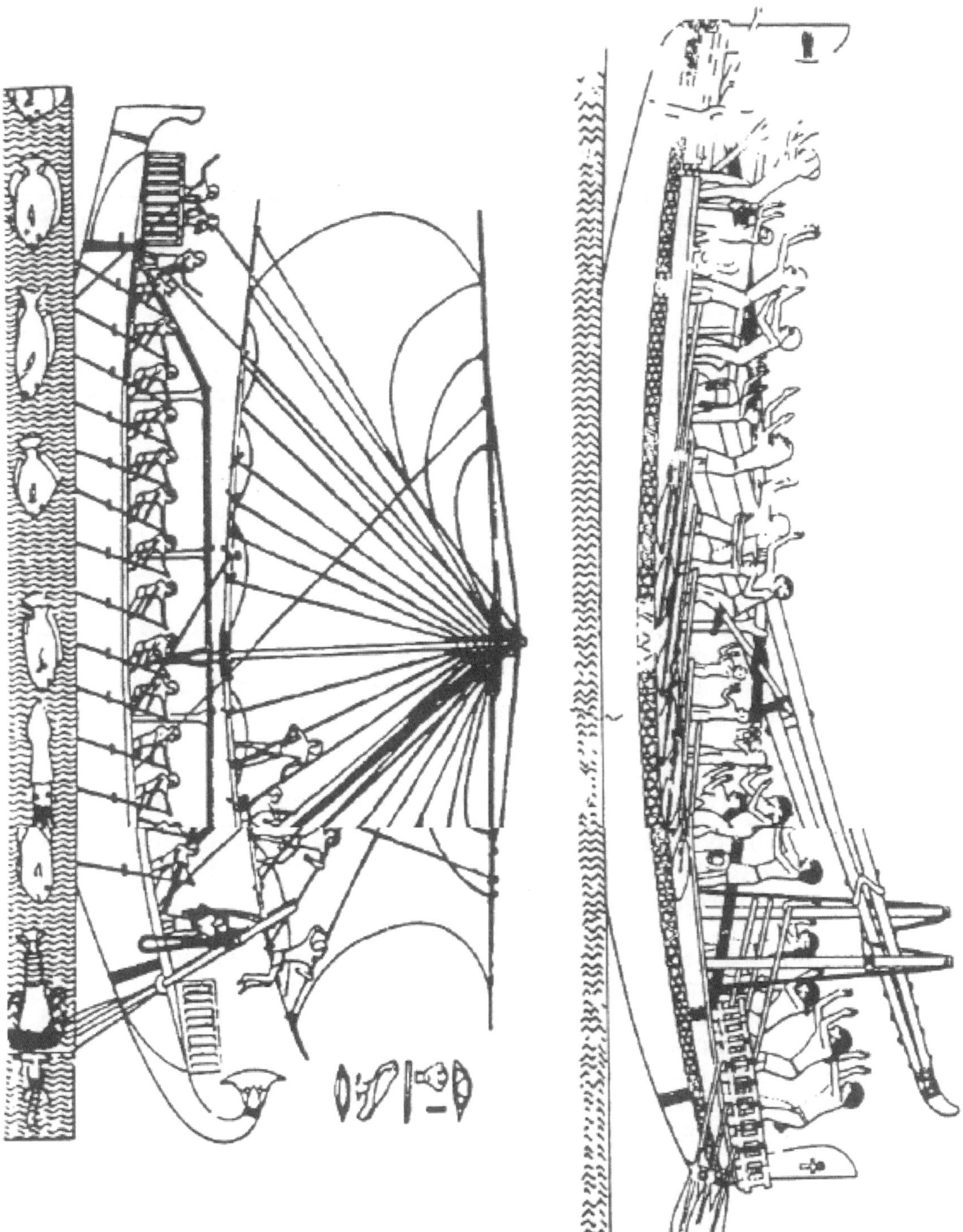

**Fig.12** **Above:** Sailing ship from the time of King Sahura (Ref.22). All passengers raise their hands and worship the discovered land of Punt (Central America). The side of the ship is decorated with equilateral triangles having angles of 60°, indicating the Southern Crossing (60°SW), and the Upper North return route (60°N) (2497 BC). **Below:** Punt ship of Queen Hatsjepsut. From the southern Cape Verde Islands (15 rowers, 15°N), they sailed to the Gulf of Campeche (at 15+3= 18°N and 18+1= 19°N), and they returned from Newfoundland (19°ESE). The 19+4= 23 passengers show the latitude of the holy Tropic of Cancer, 23°N (1470 BC).

c.2200-1700 BC).

## The Sailing Route to America via the Upper North

Glyph 11 is the last glyph, (Figs.4,7). A connection line runs from Glyph 11 to glyph 6 (Greenland), so glyph 11 is oriented on glyph 6. It has been placed at the right side of 6 because of new discoveries in the north, which must be beyond Greenland. We have seen however, that tangents of glyphs 9 and 10 provide the latitudes of the coast of North America as far as Newfoundland. Glyph 11 must contain new information located between Greenland and Newfoundland (see dot off 10d, Fig.2). This glyph was engraved because the sailing route to America via the Upper North was considered to be an alternative for the Southern Crossing. For sailors from North Europe desiring to voyage to New England, the Southern Crossing was a long way around (Ref.19). Through its tangent angles, glyph 6 has already given the sailing route via the Faroes and Iceland to Greenland. Glyph 11 repeats this, and adds the continuation of this route to Newfoundland and beyond to Central America. At the top of the stone, note the deeply inscribed horizontal line (Fig.2) confirming that America was reached via the Upper North. All the important latitudes for these areas are given by the glyph 11 Jacob's Staff. Glyph 11 was lightly executed, because this route was considered to be less important, only possible in a limited time of the year.

At the top of glyph 11 the sailing route was engraved from Greenland via Baffin Island, Labrador, Newfoundland, the Azores, and Madeira to the Strait of Gibraltar. This route is also indicated in terms of latitudes and sailing directions in the glyph 11 tangent angles. At glyph 9 the sailing route from the Azores to Gibraltar has been indicated with a shallow broken line, because it only concerned a return route. The top branch of glyph 11 is a shallow broken line for the same reason. It is only a return route.

Glyph 10d suggests that one can sail all around North America. However,glyph 11 was made to correct this. The long engraved return route to the Strait of Gibraltar says clearly this is the route home. By its absence, the implication is that a northern passage over America toward the Bering Strait is impossible. The walls at American Stonehenge in New Hampshire show that they had been exploring for such a route. We think this glyph 11 was made shortly after glyph 10, dating from the time of the construction of Stonehenge II (Refs.11,12), which is c.2100 BC.

## Tangent and Latitude Tables

Dr. de Jonge's 1996 book De Stenen Spreken (in Dutch, Ref.1) included 25 Latitude Tables derived from the tangents of these 11 glyphs. To produce a latitude location table, the positions of the baseline and the starting point have to be established in an exact manner. All possible points of contact of the tangents are used, as well as those of the previous glyphs. To indicate these points, a code was introduced. This code is valid for all glyphs of Dissignac. The first digit of the code identifies the glyph of the point of contact. Next, the code consists of some of the 4 capitals and 4 small

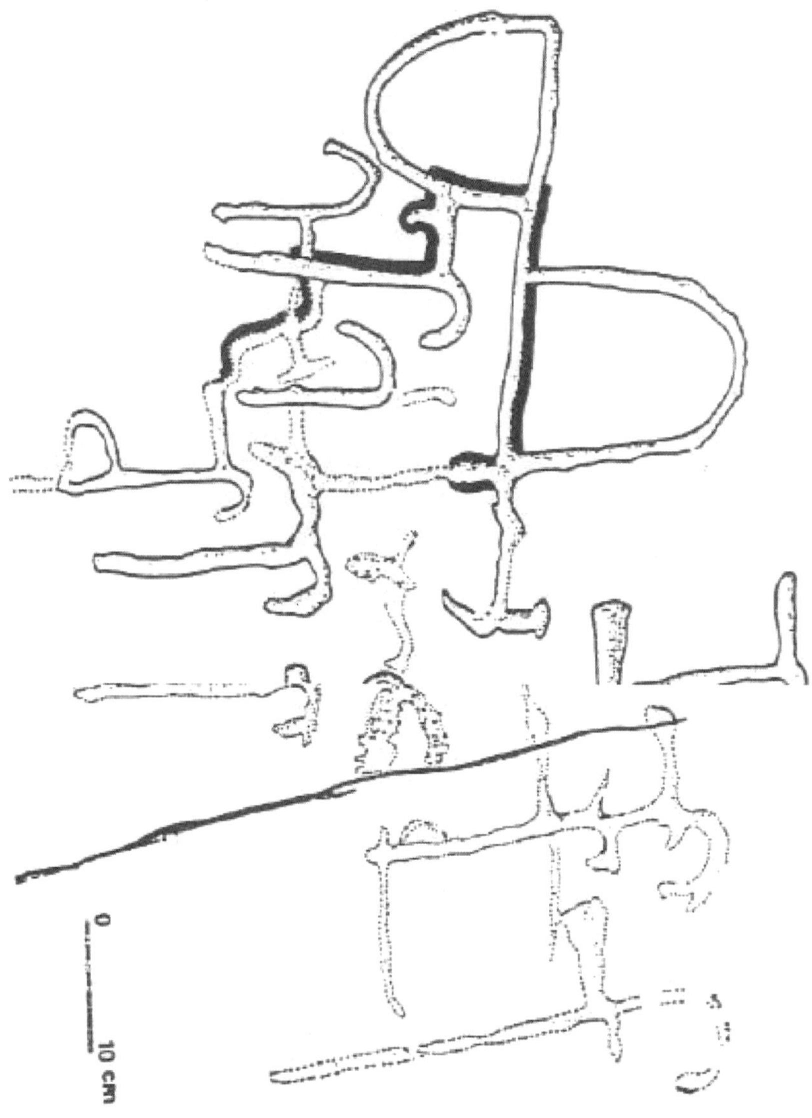


**Fig.13** Glyph 10 of Dissignac (Ref.4) contains two Jacob's Staffs (Mediterranean signs) relating to the sailing routes around the Americas. Note how a part of the entire glyph is also shaped like the east coast of Central America from Panama to Florida (outline darkened by authors). (Dissignac, Brittany, c.2200 BC)

letters: H (handle), L (Left side branch), and R (Right side branch) indicate the three parts of a Mediterranean sign. In Table 1 the use of C (Center) is not needed, because the situation is not complicated yet. The characters L (Left), R (Right), b (below), and a (above) provide further information, where the point of contact can be found. The indications i (internal) and e (external) can be very handy for the deviating glyph 1, and for Mediterranean signs of late date.

Sometimes, a latitude is indicated by several points of contact, such as Cape Lopez in Table 1 (on Fig.6). These points are mentioned in order of importance. In older glyphs and below 45°N, latitudes are given by normal angles. In glyph 6 however, and beyond, latitudes are also indicated by complementary angles (coded by c). In the northern hemisphere, and depending on the circumstances, an angle of 30° can have two meanings: 30°N, or 90-30= 60°N. In most megalithic petroglyphs and monuments we have studied, complementary angles are frequently intended, as they give clearly meaningful results. If you want to make tangent illustrations of Dissignac, you will probably notice that this is rather easy for glyphs 3 and 4, difficult for glyph 5, rather easy for glyph 6, and easy for glyphs 7-11. In glyph 10 you can combine 10a and b and treat it as one glyph, using the left tangent of b and the right tangent of a to find the new starting point. There needs to be much more study of these petroglyphs by other researchers.

## Discussion

The petroglyphs of Dissignac give the history of the crossing of the Atlantic Ocean, and the discovery of nearly all the land on the planet. It is the mission-history of the megalith builders, who believed in the SunGod. According to Diodorus Siculus (c.50 BC), the "godkings" ruled Egypt 1500 years before King Menes (c.3000 BC) (Ref.13). Below the SunGod, two other gods appeared, Horus and Osiris. It is likely that the two passage graves of Dissignac were dedicated to these two gods. If this is true, then the big solar-oriented northern passage grave is dedicated to the Sungod Horus, and the smaller, southern passage grave is dedicated to the Moongod Osiris. In that case, the petroglyphs have been engraved in the passage grave of Osiris, the god of the realm of the dead in the west.

There are many Mediterranean symbols on the stone, and always, the lowest point represents Egypt. During the whole megalithic period Egypt was the the greatest civilization on earth. Clear tangent angles of all the Jacob's Staffs show the special relationship with this area, always indicating the centers of the Southern, the Northern, and the United Egyptian Empire. The megalith builders considered the discovery of Madeira and the Azores, and then America their most important discoveries. After reaching Paradise, or the Land of the Dead on the other side of the waters, they would return via the 9 islands of the Azores and the 2 islands of Madeira. For that reason they used (9+2=) 11 glyphs of Dissignac to tell the history of their discovery of America. The petroglyphs of Dissignac are the most important

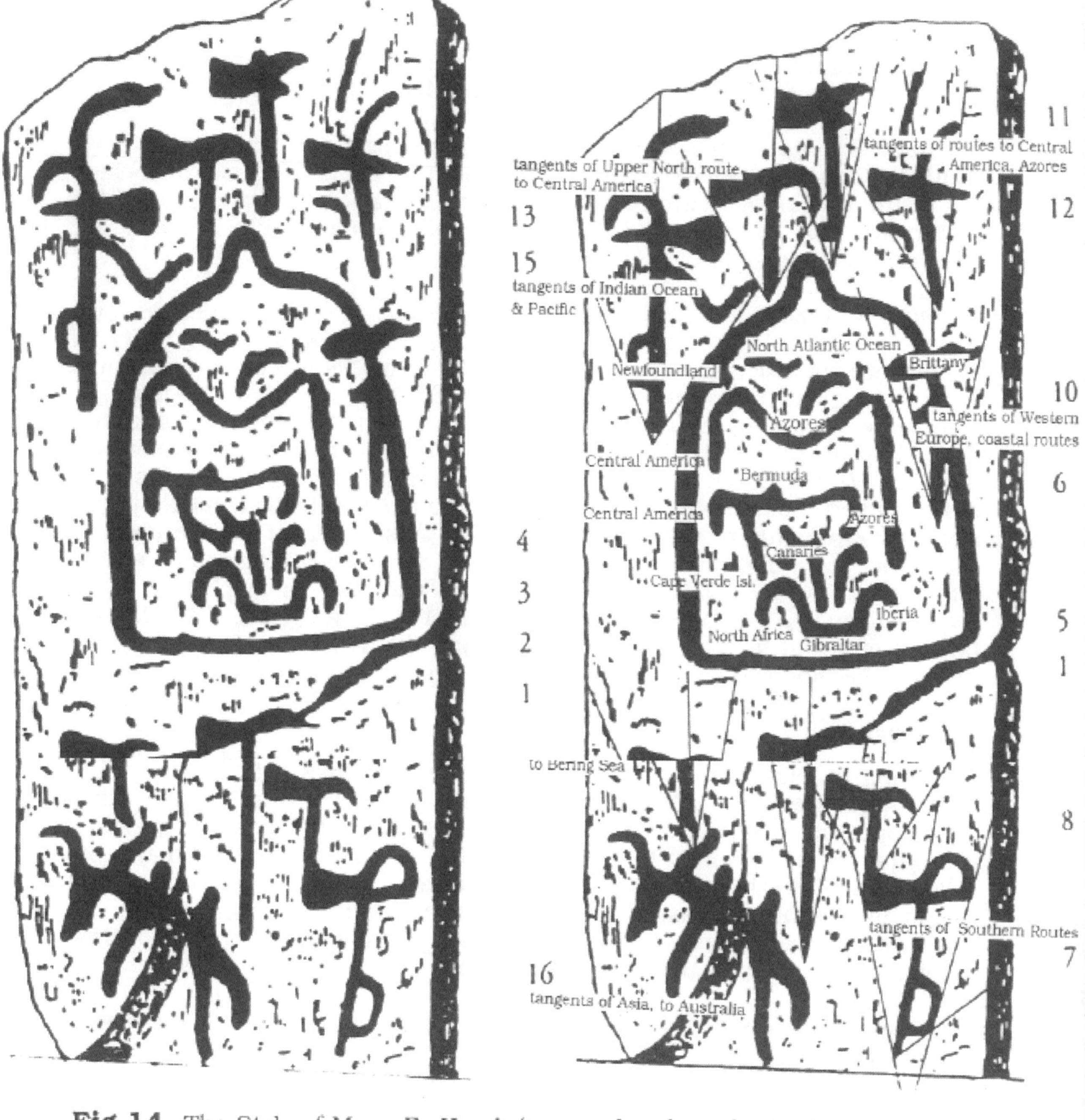


**Fig.14** The Stele of Mane-Er Hroek (repeated with author's glyph numbers, tangent angles, and labels on the right) (Refs.1,2), is the oldest world map in degrees of latitude (140x40cm). It uses the same cartogaphic system as used at Dissignac. In the center, the North Atlantic Ocean is shown (strongly formalized). On the right and left edges we see Brittany and Newfoundland, so the Stele has been made after the discovery of America. Two return routes are shown, one via Bermuda, the other via Newfoundland to the Azores. The left side of glyphs 7&8 is a map of the east coast of Central America. The Mediterranean sign glyphs above and below, when studied carefully, primarily provide encoded latitudes for the crossing of the Atlantic, though the upper and lower left glyphs encode routes in Asia and the Pacific as far as Australia. (Locmariaquer, Brittany, c.2000 BC)

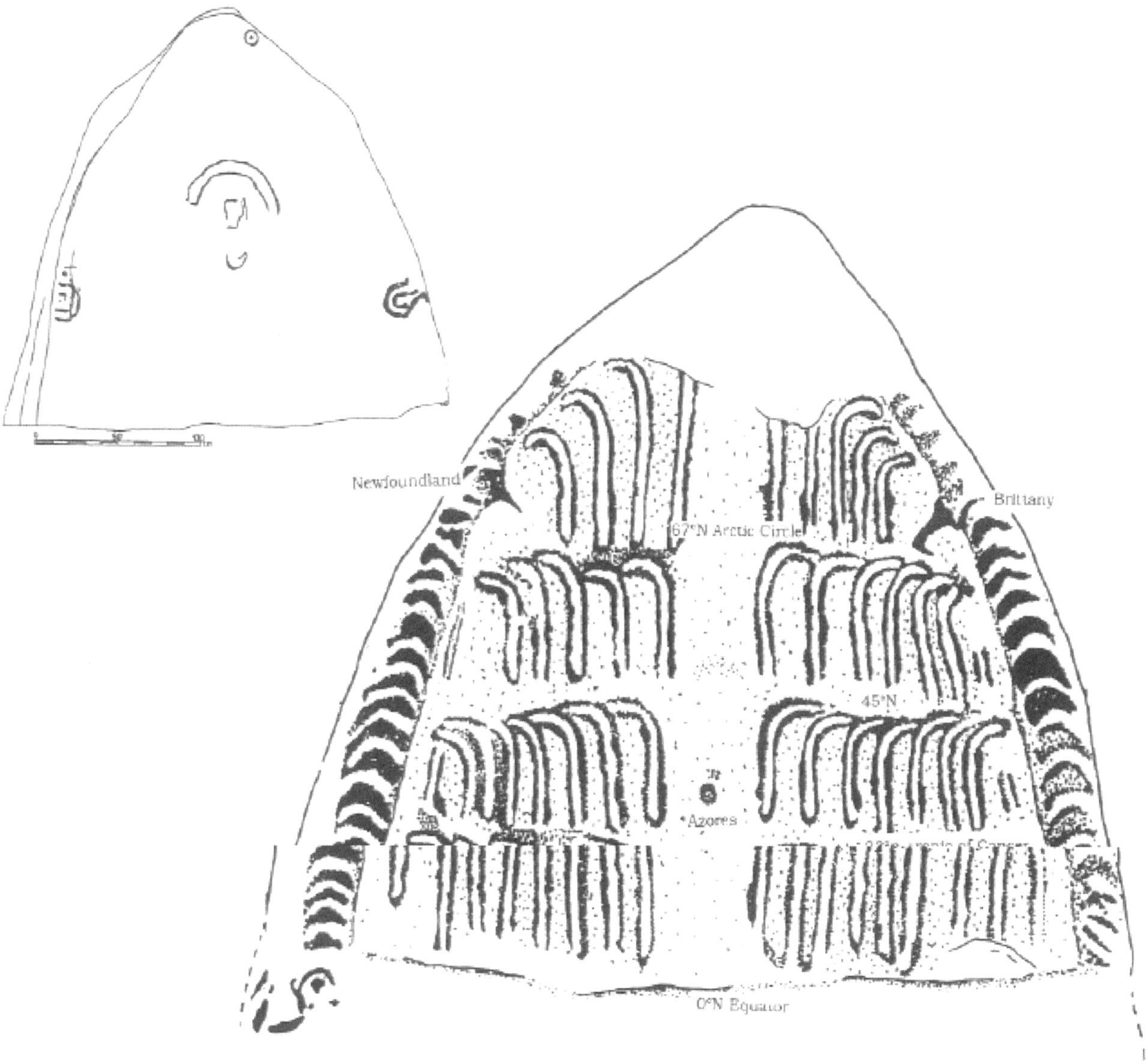


**Fig.15** Petroglyphs in haut-relief in sandstone (Ref.3). Symbolic chart of the North Atlantic Ocean, with Brittany (right, above) and Newfoundland (left, above). The ocean is divided into a well-known eastern half, and a lesser-known western half. Looking to the west, the left Mediterranean signs provide the way there via the Southern Crossing to Central America, and the right signs provide the way back via the Azores (shown as a dot in the middle of the ocean). The numbers of Jacob's Staffs form a sacred matrix. The **29 southern staffs** give the sailing direction west across the ocean from the Cape Verde Islands of 29°SSW, coming from the Nile Delta at 29°N. The **20 northern staffs** give the sailing direction east across the ocean from Newfoundland of 20°ESE , coming from the Gulf of Campeche at 20°N. The **23 western staffs** correspond to the holy Tropic of Cancer at 23°N (the SunGod), and the **26 eastern staffs** correspond to the center of the United Egyptian Empire (the Government) at 26°N, halfway between the Tropic of Cancer and the Nile Delta. **At the upper left is the backside** of the stone, showing the Azores (center), the Mediterranean (right), and the Caribbean/Gulf of Mexico (left). In the north, Greenland, represented by a circle with a dot, the Egyptian hieroglyph for the sun, or Ra, the SunGod (Refs.31,37). (Table de Marchand, Locmariaquer, Brittany, c.1800 BC)

megalithic petroglyphs in the world. They confirm what we can understand from other inscriptions, like Loughcrew in Ireland, and they confirm what we can conclude about megalithic monuments, including Stonehenge and America's Stonehenge. Loughcrew and Stonehenge I show that the megalith builders gave up their efforts to cross the Atlantic at South Greenland c.3200 BC, but Dissignac shows that America was finally discovered via the Bering Sea, 600 years later (c.2600 BC).

Megalithic petroglyphs are almost all geographic in nature, so the little maps of Brittany and the Little Mediterranean Sea (glyphs 1&2) are not unique. However, their very early date is unusual. The Mediterranean symbols of Dissignac are not completely unique either. There are more of these Mediterranean signs at Mane-er-Hroek (the oldest world map, see Fig.14, Refs.1,4), and elsewhere inside and outside of Brittany. Fourty nine Jacob's Staffs are prominent in the Table of Marchand (see Fig.15). Some images of North America (glyph 10d) are found elsewhere, but the one of South America (glyph 10c) is rare. Note though, that the enormous inner Horseshoe of Stonehenge III, the monument for the discovery of America, has the shape of 10c. Of course, what is most remarkable, is the charting of the world by latitudes that occurs in these petroglyphs. We hope someone will reverse-engineer a sailing course guide in megalithic methodology, for, say, the sailing routes around the Americas, from these petroglyphs.

# References


1. Jonge, R.M., and IJzereef, G.F., De Stenen Spreken, Kosmos Z&K, Utrecht/Antwerpen, 1996 (ISBN 90-215-2846-0) (Dutch)
2. Jonge, R.M., and IJzereef, G.F., Exhibition: The Megalithic Inscriptions of Western Europe, 1996
3. Twohig, E. shee, The Megalithic Art of Western Europe, Clarendon Press, Oxford, 1981
4. Briard, J., The Megaliths of Brittany, Gisserot,1991
5. Giot, P.R., Prehistory in Brittany, Ed JOS (ISBN 2-85543-123-9)
6. Giot, P.R., La Bretagne, des Megalithes, Ed. Ouest France, 1995 (ISBN 2-7373-1388-0) (French)
7. Fell, B., America BC, Pocket Books, Simon & Schuster, 1994
8. Batt, M., and others, Au Pays des Megalithes, Carnac-Locmariaquer, Jos, 1991 (French)
9. Le Roux, C-T., Gavrinis, Ed.J.P. Gisserot, 1995
10. O'Sullivan, M., Megalithic Art in Ireland, County House, 1993 (ISBN 0-946172-36-6)
11. Richards, J., Stonehenge, English Heritage, 1992
12. Atkinson, R.J.C., Stonehenge, Penguin Books, 1979
13. Bailey, J., Sailing to Paradise, Simon & Schuster, 1994
14. Thompson, G., American Discovery, Misty Isles Press, Seattle, 1994
15. Peterson, F.A., Ancient Mexico, 1959
16. Zapp, I. and Erikson, G., Atlantis in America, Navigators of the Ancient World, Adventures Unlimited Press, 1998 (ISBN 0-932813-52-6
17. Jairazbhoy, R.A., Ancient Egyptians and Chinese in America, Rowman & Littlefield, Totowa, N.J., 1974 (ISBN 0-87471-571-1)
18. Ferryn, P., "5000 years Before our Era: The Red Men of the North Atlantic", NEARA Journal, Vol XXXI, No.2, 1997 (pgs.5-16)
19. Coles, J., "Images of the Past", A Guide Book to the Rock Carvings of Northern Bohuslan, Bohuslans Museum, 1990 (ISBN 91-7686-110-4)
20. Tour Guide Map, America's Stonehenge, PO Box 84, North Salem NH 03073
21. Hapgood, C.H., Maps of the Ancient Sea Kings, Adventures Unlimited Press, 1996 (ISBN 0-932813-42- 9)
22. Varen, Vechten en Verdienen, Scheepvaart in de Oudheid, Allard Pierson Museum, Amsterdam, 1995 (Dutch)
23. Mohen, J-P., The World of the Megaliths, Facts on File Inc., 1990 (ISBN 0-8160-2251-8)

24. Doutre. M.. Celtic New Zealand. De Danann Publishers. Auckland. NZ. 1999 (ISBN 0-473-05367-5)
25. Map of Magnetic Variation. Epoch 1995-0. Defense Mapping Agency (US). 1996
26. "Art et Symboles du Megalithisme Europeen". Revue Archeologique de L'Ouest no 8. Nantes. 1995 (ISSN 0767-709-x)
27. Burl. A.. The Stone Circles of the British Isles. Yale University Press. 1976 (ISBN 0-300-01972-6)
28. New World and Pacific Civilizations. The Illustrated History of Humankind. Weldon Owen Pty Limited. McMahons Point. Australia (1995)
29. Lambert. J.D.. America's Stonehenge. an Interpretive Guide. Sunrise Publications. Kingston. N.H.. 1996 (ISBN 0-9652630-0-2)
30. Munck. C.P.. The Code. 1996 c/o Radio Bookstore Press. PO Box 3010. Bellevue. Wa 98009
31. Wheeler. R.L.. Walk Like An Egyptian. Allisone Press. 2000 (ISBN 1893774-21-X)
32. White. P.. "Exposure" Magazine. Vol.2. No.6. 1996. classblu@ozemail.com.au
33. "Oldest City of Americas confirmed. Peruvian Complex Contemporary with Egypt's pyramids". The Seattle Times. A4. April 27. 2001
34. Bayley. H.. The Lost Language of Symbolism. Citadel Press. 1990 reprint (ISBN 0-8065-1100-1)
35. Santillana. G. de. & Von Dechend. H. von. Hamlet's Mill. An Essay on Myth and the Frame of Time. 1999. Godine Publisher. N.H. (ISBN 0-87923-215-3)
36. Weissbach. M.M.. "A Note on the Decipherment of Maui's Inscriptions". Fidelio. Vol.VIII. No.1. Spring 1999. (ISSN 1059-9126)
37. Breasted. J.H.. Ancient Records of Egypt. Vol.2: The Eighteenth Dynasty. Histories & Mysteries of Man Ltd.. London. 1988
38. British Museum. A General Introductory Guide to the Egyptian Collections. British Museum. 1930

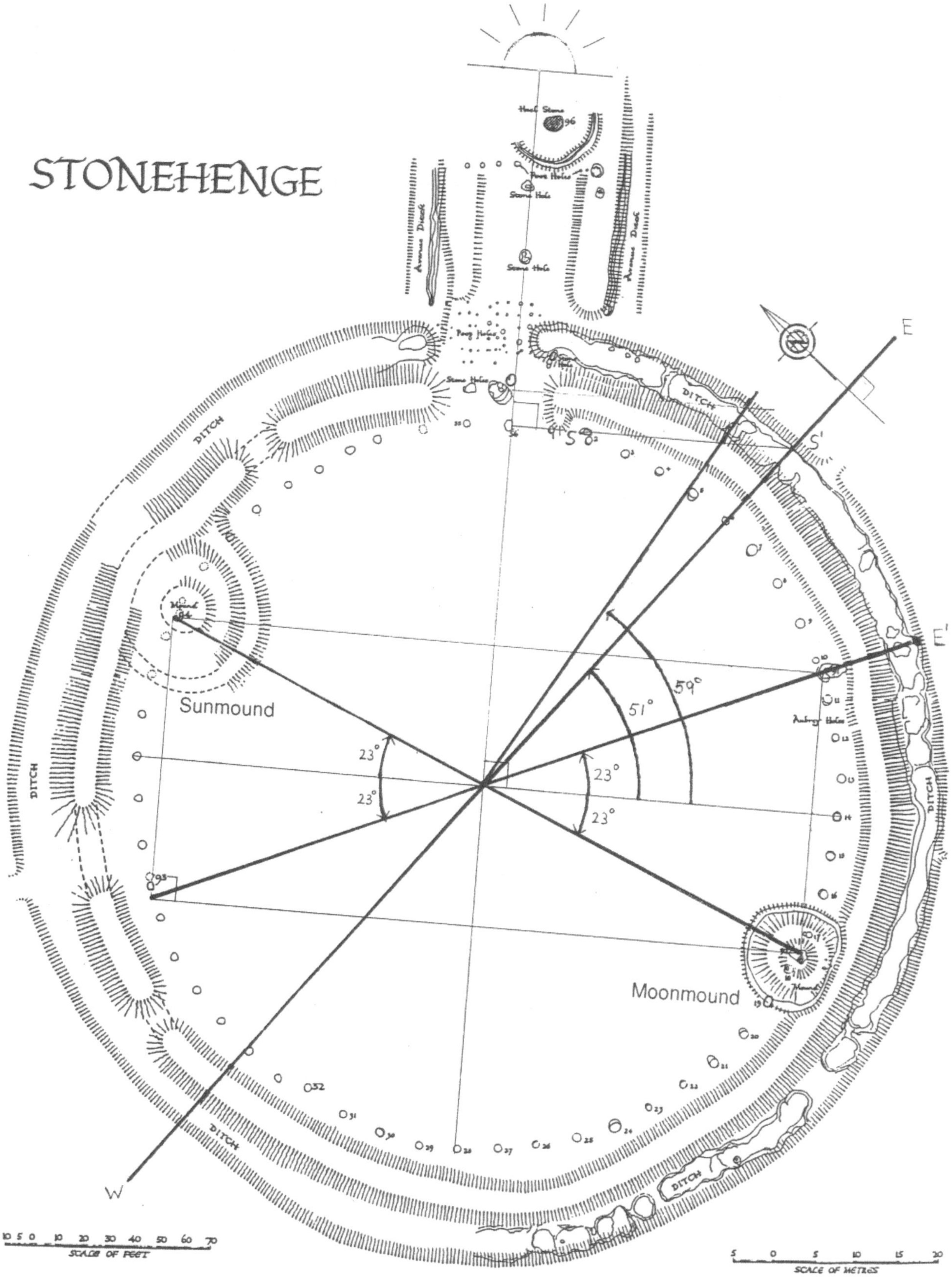


**Fig.1** Groundplan of Stonehenge I (c.3200 BC), with the earth circle where all the land (the wall) is surrounded by the sea (the ditch), the Avenue pointing to the rise of the midsummer sun (the SunGod Ra), the Sunmound (left), and the Moonmound (right). Seen from the center, the points S' and E' represent the locations of Stonehenge and south Egypt on earth.

# STONEHENGE:
## a Monument for the Discovery of America
(Salisbury Plain, Wiltshire, England, c.3200 BC and c.2000 BC)

Dr. R.M. de Jonge, drsrmdejonge@hotmail.com
J. S. Wakefield, jayswakefield@yahoo.com

## Introduction
It has been discovered that the degrees of latitude of the northeastern islands of the Atlantic Ocean are prominent in the design of Stonehenge (England), Phase I, built c.3200 BC (Refs.1-6). In the "Story of Loughcrew, Ireland", based on numeric picture-writing inscriptions, we have known that the neolithic megalith builders explored this part of the Ocean c.3300 BC. They used degrees of latitude, sailing directions, and distance lines (dl), but could not find a way to navigate across the ocean to the other side they believed was there. Further extension of the "Story" on the roofstone at Loughcrew showed that more extensive explorations in Arctic waters over the following hundred years also failed to reveal a way across the ocean. This new information from Stonehenge I, identifying known lands of the earth by correct latitudes, is contemporary with the Loughcrew inscriptions, and supportive of what has been learned at Loughcrew.

Later in this article we will show that Stonehenge III, built a thousand years later, is mathematically symbolic of the New World then discovered in the western ocean. The proof of this discovery is partially supported by several other megalithic sites in Europe, particularly the inscriptions of Dissignac, France, and by further archaeological developments in America.

## STONEHENGE I
Please examine Fig.1, a copy of the Phase I "groundplan" of Stonehenge (excepting the later-date holes and stones in the center) reproduced from the Department of the Environment's Ancient Monuments and Historic Buildings Official Handbook, by R. S. Newall (Ref.4). We have added stone hole S to the groundplan, because it is included in Refs. 3,4,5, and 6. Thus at Phase I (c.3200 BC), we have only the ring wall and ditch, the four "Stationstones", the 56 "Aubrey Holes", and the walls, ditches, stones and holes of the "Avenue" leading out of the monument in the direction of the rising midsummer sun.

The reason the 56 Aubrey Holes were placed within the circle was proposed by Gerald Hawkins in 1963 (Ref.10), that they were intended to mark the 56 years that it takes the moon to fulfil its eclipse cycle (called the Metonic Cycle), which takes place in the course of three nodal revolutions, each of 18.61 years (3x18.61=56) (Ref.18 pg.66). Julius Caesar reported that the Druids could accurately calculate eclipses of the moon long before they happened (Ref.18), showing that ancient knowledge had been passed down

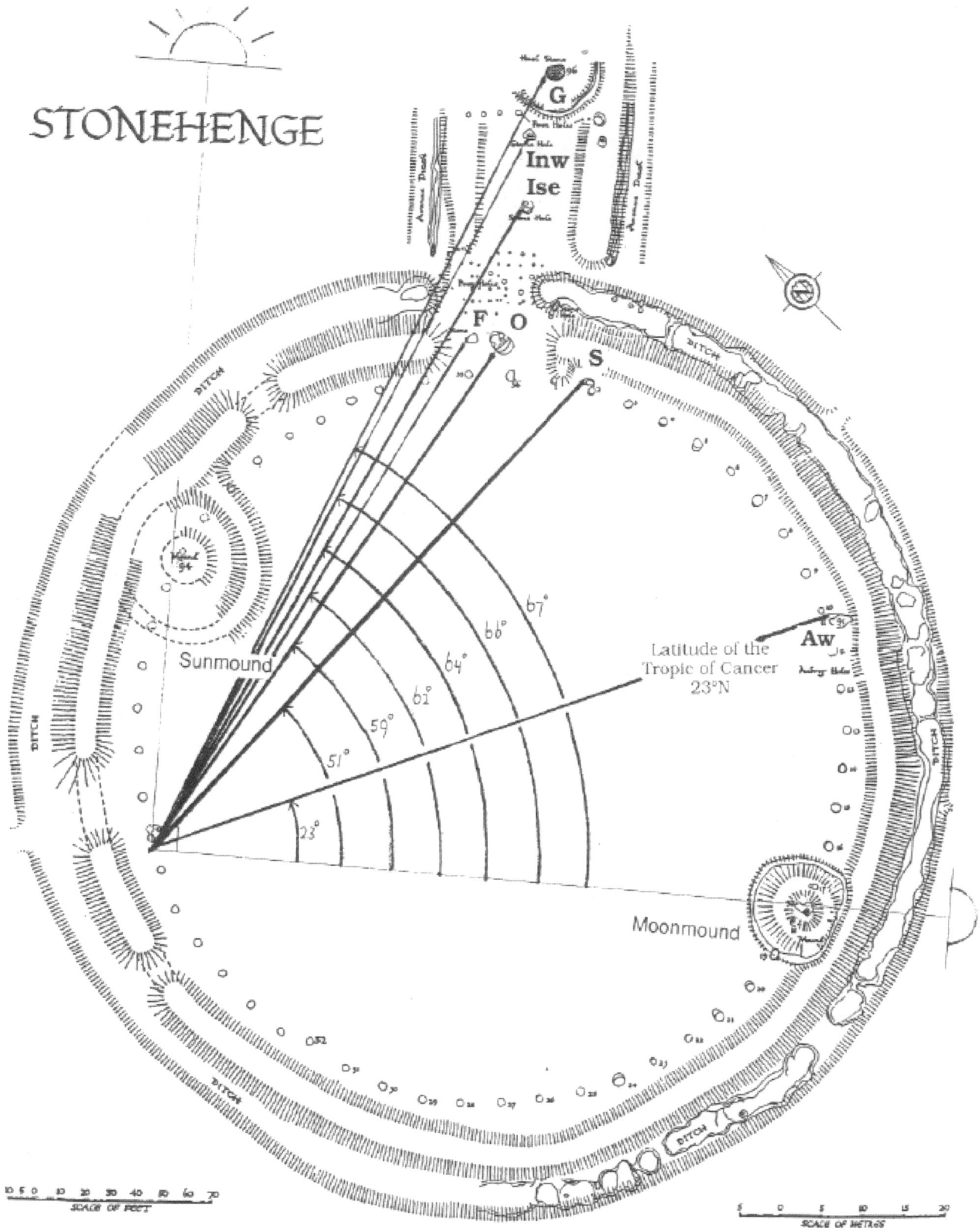


**Fig.2** Stonehenge I seen from Stationstone 93. Stationstone 91 represents the Tropic of Cancer, and the menhirs around the Avenue make angles corresponding to the latitudes of the Voyage to the Upper North. The Heelstone (G) (c.5m) represents the discovery of Greenland (c.3300 BC).

**Aw**=West Africa, at 23°N
**S**=Stonehenge, at 51°N
**O**=Orkney Islands, at 59°N
**F**=Faeroe Islands, at 62°N
**Ise**=Southeast Iceland, at 64°N
**Inw**=Northwest Iceland, at 66°N
**G**=Greenland, at Arctic Circle, at 67°N

to the Celtic Iron-Age people attacked by Caesar. Fifty-six is also the average number of days and nights (2x28=56) of the Synodical and the Siderial months (29.5 and 27.3 natural days, respectively) (Ref.1). The so-called "Stonehenge month" (28 natural days) counts exactly for 4 weeks, because 4x7=28. They correspond to the four faces of the moon: New Moon, First Quarter, Full Moon, Last Quarter. Almost 13 Stonehenge months fit into a sun year, because 13x28=364 days, so this was a pretty accurate calendar, in the Neolithic!

However, this calendrical meaning of Stonehenge is not the most interesting one, but rather its meaning geographically. The fact that the earth is a sphere which resembles the moon, was also already known more than a thousand years earlier than the start of Stonehenge. Degrees of latitude based on a 360 degree circle were already used in the stylized inscriptions of Dissignac (Brittany, starting at c.4400 BC), in the tumulus of Kercado (Brittany, c.4500 BC), and during the construction of the oldest cairn of Barnenez (Brittany, c.4700 BC)(Ref.1,2,13). This early understanding of the earth as a sphere led to an accurate knowledge of, and use of degrees of latitude, and interest in charting the earth, which supported early sailing explorations.

## Geographical Meaning of Stonehenge I

**The big ditch and bank circle of Stonehenge I, having a diameter of c.115m, represents the spherical earth**, where all the land (the wall) is surrounded by the sea (the ditch). In the north, there is a way (the Avenue) to cross the land and sea. However, even there, all land (the walls of the Avenue) is surrounded by sea (the ditches). Half of the menhirs of the Avenue are outside the main circular ditch, which is an indication that this monument deals with the problem of crossing the ocean. They did not know what was out there!

**Place a protractor on the center of the monument in Fig. 1,** and note the "crosshairs" from the center of the Avenue through the center of the monument (and through upper and lower Aubrey Holes), and its perpendicular, which is between two more of the Aubrey Holes. You find that at 51°N, the line from the center cuts Aubrey Hole #6 exactly (8/56x360=51), and creates an East-West axis of the monument at right angles to true north. Since Stonehenge is actually located at 51°N, the point at this line at the other ditch (S') represents Stonehenge itself. This point is at the same level as menhir S, which also represents Stonehenge, as we shall see later. The whole monument is directed to the rise of the midsummer sun (Refs.22,23). At noon on that day the sun is at right angles to the Tropic of Cancer. At 23°N, the degree of latitude of the Tropic of Cancer, is Stationstone 91, indicating the Southern Egyptian Empire (E') at the other side of the ditch, at 23°N, the center of the SunGod religion.

Now move the protractor to the Stationstone 93 in Fig. 2. Note that a line drawn from Stationstone 93 over Stationstone 94 on top of the mound, is

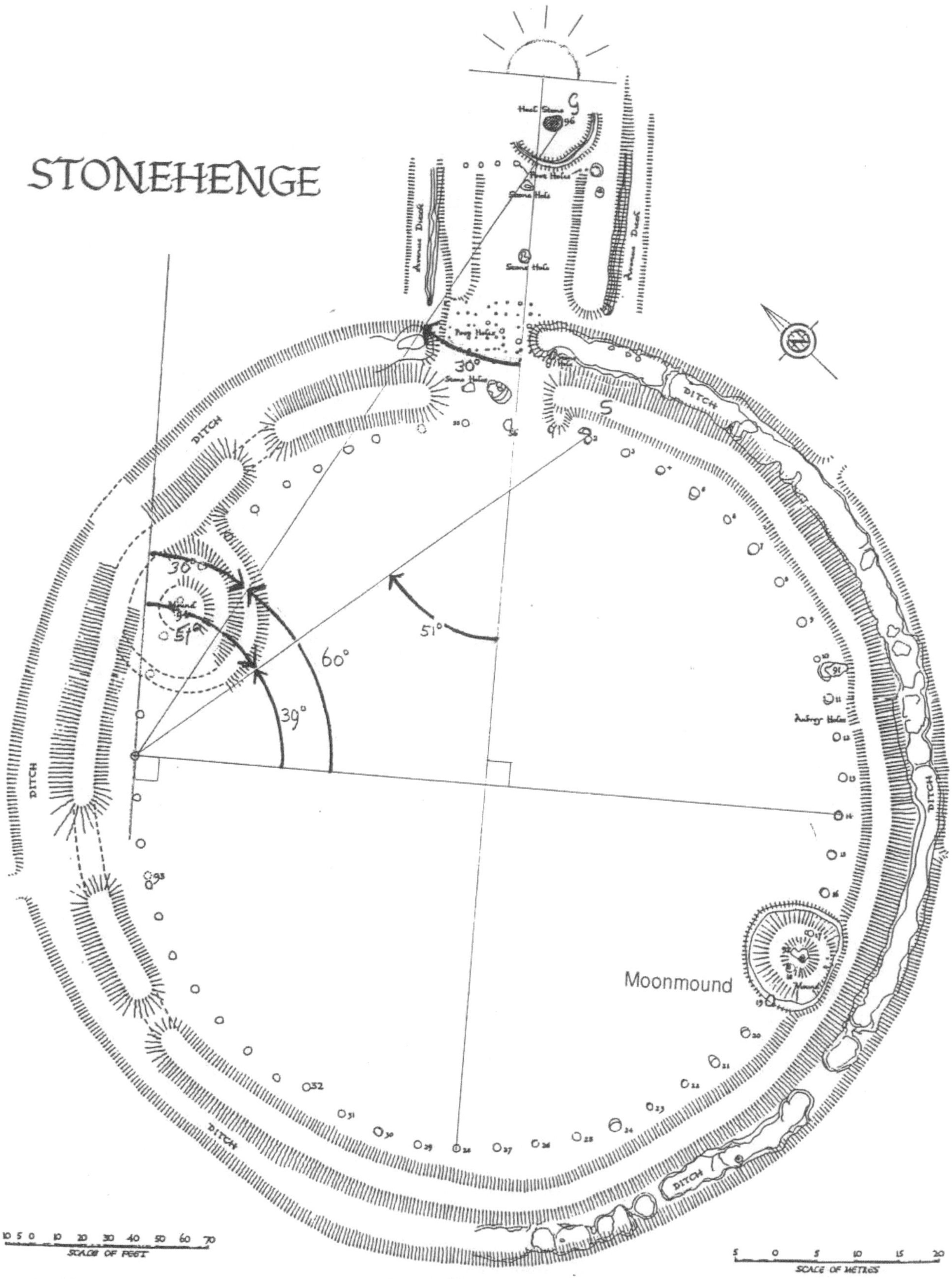


**Fig.3 Stonehenge I seen from the left-most Aubrey Hole.** The Heelstone G and menhir S represent Cape Farvel (South Greenland) at 60°N and the West Azores at 39°N, respectively. These were the westernmost lands of the then known world (c.3200 BC). Via the complementary angles, they also represent the Nile-Delta at 30°N and Stonehenge at 51°N.

parallel to the monument axis and is also in line with the rise of the midsummer sun, as are so many of the megalithic cairns (Ref.10). This is an important alignment, and the reason why this mound is the largest, and should be called the **"Sunmound"**. A line drawn from Stationstone 93 over Stationstone 92 on the other mound points to the site of the most southerly rise of the full moon, at the outside of the ditch. We shall call this mound the **"Moonmound"**. Only at 51°N, the latitude of Stonehenge, are these two sightlines at right angles to one another, and is the reason why Stonehenge is sited here (Ref.19). If one travels a day's journey to the north, or to the south, this will not be the case anymore. As you go further north, the angle gradually becomes greater. As you go to the south, the angle gradually becomes smaller. Even Neolithic man could see that only on a spherical surface would this be true, so the earth must be a sphere, like the moon appears to be.

Stonehenge is located at 51°N because it only made sense to them to draw a chart of known lands of the earth based on latitudes at a platform where they can start with a 90° angle. They could have drawn such a world-chart anywhere, but then it would not have been aligned to the sun and the moon, which was important to them, because it was the basis of their calendar and related to their SunGod (sun/moon= Horus/Osiris= Father/Son) mythology. Below the supreme God Ra (symbolized by the rise of the midsummer sun) two other gods appeared, Horus and Osirus, hence the Sunmound and Moonmound. Around the time of 3200 BC, great ideological and political changes were taking place in Egypt (Ref.23). The first Divine Kings appeared, who were substitutes for Horus and Osiris. So Stonehenge, like so many megalithic monuments, can be seen as a mission-monument for the spreading of Egyptian king theory, the divinity of the Pharaoh! The Egyptian SunGod Ra has told: "The realm of the dead is in the west, at the other side of the waters, in the land where the sun sets" (represented by Stationstone 94). "After death you will reunite there with your ancestors, your family, your relatives, your friends, your acquaintances!"

## The Discoveries in the Upper North

A line drawn at 23° from Stationstone 93 (see Fig.2) runs again to Stationstone 91, the latitude of southern Egypt, but here, seen from the west, now represents the Tropic of Cancer at 23°N where it emerges from the west coast of Africa to the other side. They thought this might be the sailing route to follow the Sungod across the ocean, from the west coast of Africa across the ocean, because on midsummer day the sun is at right angles above the Tropic of Cancer at 23° latitude. A line drawn at 51° runs to menhir S, which represents Stonehenge at 51°N latitude. Note that all other angles on Figure 2 are the correct latitudes of the islands represented by the menhirs (Refs.1-3). A line drawn at 59° runs to menhir O, the Orkney Islands, at 59°N latitude. After this, we leave the well-known mainland (the wall), and soon we cross the sea (the ditch) to more distant islands. But first, at 62°, we find a menhir F, representing the Faeroe Islands at 62°N. Then, further out to sea, at 64°, we find menhir Ise,

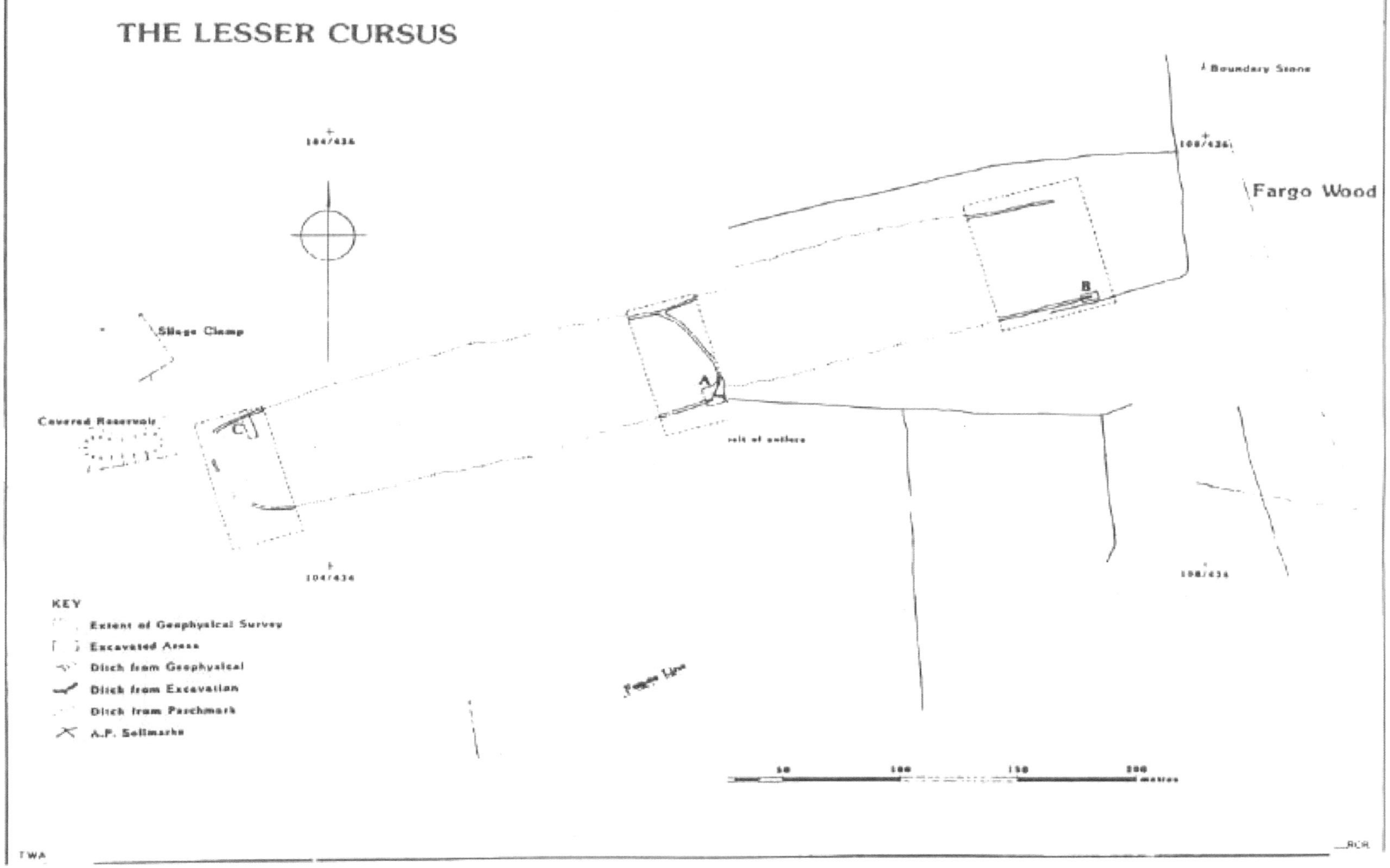


**Fig.4 The Lesser Cursus** (ground caction of a scale rod) points to the natural harbor of Padstow on the Engvest coast (15°WSW) at a distance of 231km from Stonehenge. The din at right angles (75°NW) points via the Hebrides to Iceland. The lewidth ratios applied to a unit of measure provides the accurate distato these islands. The distance to the Hebrides is 194 (length of Curs mid-section in m)/52.4 (width of Cursus in m) x 231 =855km, and thence to Iceland is (408/52.4) x231 =1799km. (Stonehenge, South Engla3400 BC)

representing the Southeast coast of Iceland, at 64°N, discovered c.3400 BC. The line running at 66° identifies the menhir Inw as the northwest cape of Iceland, located at 66°N. A line at 67° gives the large "Heelstone" (c.5m high) the identity of Greenland at the latitude of Cape Holm, on the Arctic Circle at 67°N, discovered 3300 BC. The Arctic Circle has a holy significance, because it is the furthest north line where the sun still shines at midwinter day. The sun is then at right angles above the Tropic of Capricorn, at 23°S, the latitude of Stationstone 93. Note that all the land of the coast of Greenland (the wall built around the Heelstone) is surrounded by water and ice (the ditch), making it impossible to cross Greenland.

Their most difficult discoveries can be seen in Fig. 3, when we move the protractor to the furthest-west Aubrey Hole (at the equator, perpendicular to the monument axis). This position is symbolic for the westernmost point on the earth! Note that the Heelstone makes an angle of 60°, which represents Cape Farvel at the southern tip of Greenland at 60°N. And note that the left side of the Heelstone makes an angle of 61°, corresponding to the latitude of the SW Cape of Greenland, 61°N. The 56 Aubrey Holes and the 4 Stationstones together form 56+4=60 units, or including menhir S, 61 units, confirming the important latitudes of Cape Farvel, at 60°N, and the SW Cape at 61°N. Again, all the land near the south coast of Greenland (the wall around the Heelstone) is surrounded by water (the ditch). So it was impossible to cross the sea. Probably for as long as 700 years, these maritime people were unable to cross the ocean from there. Note that, as in Loughcrew, the Faeroe Islands (menhir F) at 62°N are indicated by an extremely westerly position. This shows that they had not gone further west beyond 62°N along the west coast of Greenland at c.3200 BC.

A line drawn at an angle of 39° now reveals the West Azores (menhir S) at 39°N, with Greenland, the western-most lands in the then known world. The complementary angle of 51° (90-39=51) corresponds to the latitude of Stonehenge (menhir S again) at 51°N! The complementary angle of 30° (90-60=30) at the Heelstone corresponds to the direction to the Nile Delta of the Northern Egyptian Empire at 30°ESE, and its latitude at 30°N. The beauty of these coincidences is the determining factor in the east-west site location of Stonehenge on the 51° latitude line.

## The Cursus Monuments

About 2km northwest of Stonehenge is the so-called Lesser Cursus, 408m long and 52.4m wide (Fig.4). It is a ditch about 1m wide and 1m deep, having a very low wall at the inside of the ditch. It points 15°WSW to the natural harbor of Padstow in Cornwall, at a distance of 231km. The direction at right angles to the Lesser Cursus points 90°-15°= 75°NNW via the northernmost island of the Hebrides, Lewis, to the east coast of Iceland. The people that discovered Iceland wanted to fix down the distance, and they did so with this scale rod. The length/width ratio determines the distance. The east half has a length of 194m, and a width of 52.4m, so the distance from Stonehenge to Lewis is (194/52.4) x231=855km, which is

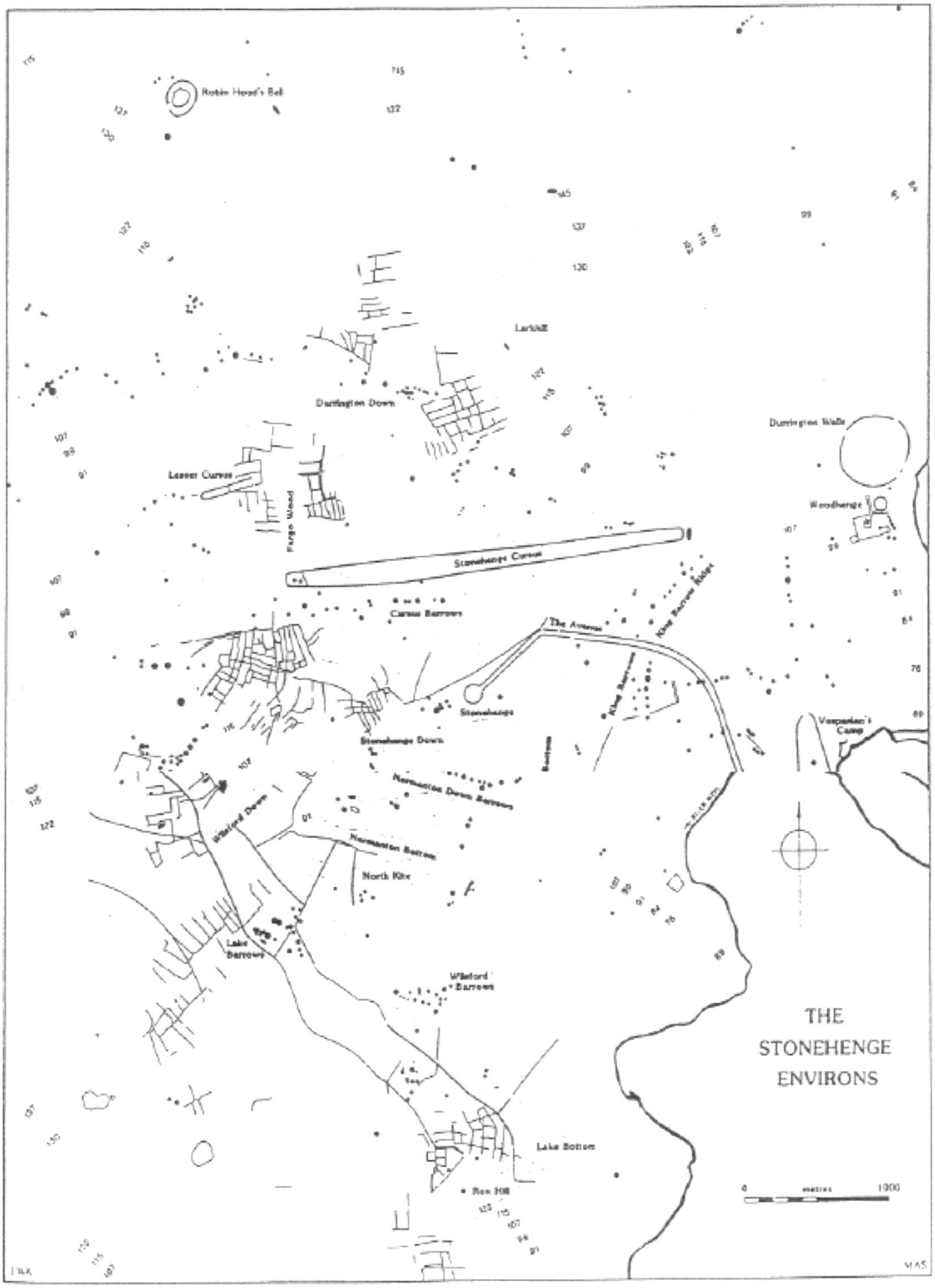


**Fig.5 The Stonehenge Cursus** (a ground construction of a scale rod) points to the mouth of the river Parrett on the English coast, a distance of 75.3km from Stonehenge. The length/width ratio of the cursus (2820/90) in meters provides the distance to Cape Holm, at the east coast of Greenland on the Arctic Circle. This distance is (2820/90) x75.3 =2360km. (South England, Ref.5, c.3300 BC)

correct. The west half has a length of 214m, with the width of 52.4m, so the distance from Lewis to the east coast of Iceland is (214/54.2) x231= 943km, which is also correct. Adding them together, the distance from Stonehenge to Iceland is 855km+943km= 1798km. So the Lesser Cursus was made after the discovery of Iceland, c.3400 BC.

About 1km north of Stonehenge is the much bigger Stonehenge Cursus, 2820m long and 90m wide (Fig.5). It points 6°WSW to the mouth of the river Parrett at the Channel of Bristol, at a distance of 75.3km. At the extreme west side is a ditch, making an angle of 67° with the axis, so this cursus provides the distance from Stonehenge to the SE coast of Greenland at the Arctic Circle, 67°N. Again, the length/width ratio provides the distance: (2820/90) x75.3= 2360km, which is correct. So the Stonehenge Cursus was made after the discovery of Greenland, c.3300 BC.

# STONEHENGE III

Stonehenge Phase III was built about 1200 years later than Phase I, in the center of the old earth circle (c.2000 BC) (Fig. 6). (Stonehenge II, dated a century earlier, had already been broken down before completion.) The new work consisted of the huge stone "Sarsencircle", an inner "Horseshoe" of even larger "sarsenstones", and the "Slaughterstone". The Sarsencircle originally consisted of 29 1/2 sarsenstones (one is of half width, #11), corresponding to the 29 1/2 days of the Synodical month (Ref.8). This is the time period between two full moons. The line through the center and along the half sarsenstone is the real North-South axis of the monument (see Fig.7). The rise of the midsummer sun makes an angle of 51° with this axis, corresponding to the latitude of Stonehenge, 51°N.

# Greenland

What meaning shall we make of the enormous "Slaughterstone" (c.6m long) shown in Fig.7? The Official Handbook (Ref.4) claims this stone was erected c.1000 years later, in Phase III, along with the large Sarsenstones: "there is reason to believe this stone once stood upright ... in the large hole..."(meant is C, overlapping O). We know from Stonehenge I and Loughcrew, that they explored Greenland to the southern point, Cape Farvel, at 60°N, but at c.3200 BC they were not able to find any land further west. Now, 1200 years later, the Slaughterstone has been later added at the time of Phase III, to emphasize the now re-discovered importance of this land at 60°N (Cape Farvel, South Greenland), because of the newly-discovered crossing from Cape Farvel to Cape Chidley at 60°N. From the most important Stationstone 93, the linc to hole C of the new Slaughterstone (Fig.7) makes an angle of 60°, corresponding to the latitude of the crossing from Cape Farvel, Greenland at 60°N, to Cape Chidley, Canada at 60°N, and by the way, also of the Shetland Islands, also at 60° N too!

The Horseshoe (Fig.6) has the shape of South Greenland, and for that reason represents this part of the land. The biggest trilithon (unit of 3 stones)

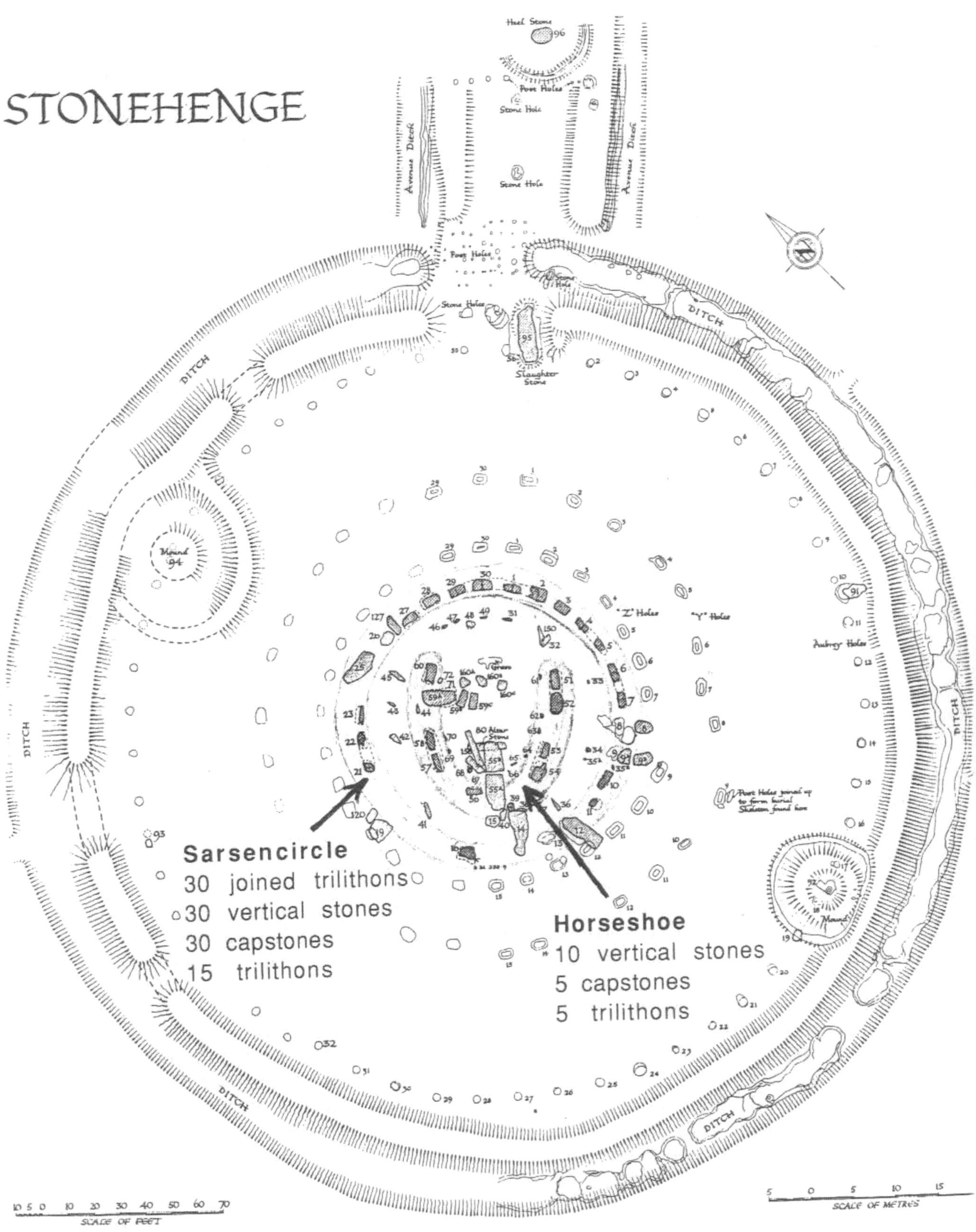


**Fig.6 Groundplan of Stonehenge III (c.2000 BC) in the center of the old monument** (see also Fig. 12). The Sarsencircle contains 30 joined trilithons, which can be seen as 15 separate trilithons, made from 30 vertical sarsenstones and 30 lintels (capstones). The Horseshoe contains 5 separate trilithons, made from 10 vertical sarsenstones and 5 lintels. The Sarsencircle represents the discovery of the "New World", with the Horseshoe representing North, and also South America.

(Figs.8&12) identified at the bottom of the Horseshoe as "T3", emphasizes by its size the importance of this crossing at Cape Farvel. The placement of many big stones indicates that much land has been found. The big encircling Sarsencircle means that a "New World" has been discovered, America! This circle contains 30 vertical sarsenstones + 30 horizontal lintels (capstones) = 60 stones, confirming the latitude of the crossing at 60°N. One simply sails to Cape Chidley, the NE point of Canada, and the "New World" has been reached (c.2000 BC) (Refs.14,15,36).

## North America

The large stones of Phase III and the way they are laid out are complex representational symbols of the New World, based upon the latitudes of the discovered continents lying in the middle of the world's ocean. Then, as now, man's symbols usually have multiple complex meanings. The sarsens and lintels are not only a monument to the New World, but are also a representational chart of it, including a focus upon the ways to reach America and their particular areas of interest. Additionally, there is religious meaning. The methodology used in this symbolism is fully consistent with what has been learned about the abilities of megalithic people from other monuments and inscriptions, such as in Loughcrew (Ireland), Dissignac (Brittany), Mystery Hill (New Hampshire), and elsewhere.

The Horseshoe (Fig.6) consists of 5 enormous trilithons, and each side contains 5 big vertical sarsenstones, both corresponding to 50 important degrees of latitude, as they knew the new land extended from 60°N to 10° N (5x10=50). It is the continent of North and Central America, and the Horseshoe has the shape of that continent, which is very important (Refs.1,2). The upper part forms an "open end", which always in megalithic petroglyphs, has the sense that "the land continues". The larger size of the central trilithon is an indication of the importance of the central area, so the land in Central America, between 20°N and 10°N appears to be the most important part of the New World to these people (c.2000 BC). After crossing from Cape Verde, this area was promptly discovered, and it included the homelands of the important people of Central America, as we shall see.

The Sarsencircle is the "New World" (later called "America" for Spanish navigator Amerigo Vespuchi, or Bristol customs agent Richard Amerike). The diameter of the Sarsencircle is about one third of the earth circle of Stonehenge I, so they thought about a third of the land on planet earth had been discovered, which is correct. The whole monument contains 30+5=35 trilithons (Fig.12), because the total surface area of America equals 35 square big distance lines (35 DL squared= 35 x 1111x1111 square kilometers= 43 million square kilometers) (1DL= 10°of latitude= 1111 km). Petroglyphs of the ocean with big distance lines (Refs.1,2,13) show that the megalith builders did use this unit of surface area in the Bronze Age. Counting the trilithons separately, we get 15+5= 20 (Figs.8&10), for a

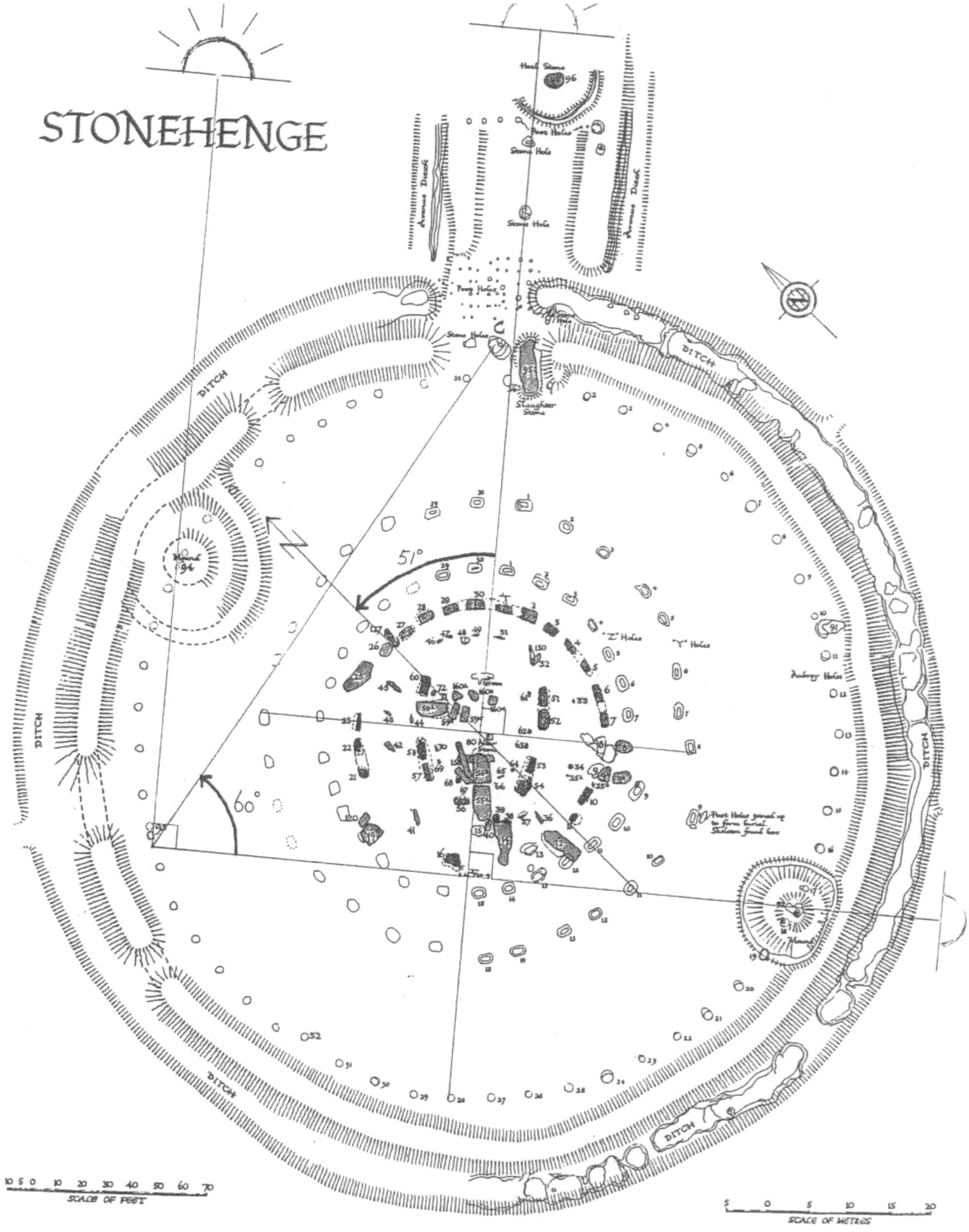


**Fig.7 Groundplan of Stonehenge III, showing the north-south axis,** which provides the latitude of both Stonehenge and the Aleutian Islands, both at 51°N, **commemmorating the original discovery of America via the Aleutian Islands.** The 60° angle from Stationstone 93 to the Slaughterstone provides the 60°N latitude of the crossing from Cape Farvel, Greenland to Cape Chidley, Canada at 60°N. Here the Horseshoe represents South Greenland, and the tallest trilithon at the base is Cape Farvel.

surface area of 20 square big distance lines for North and Central America (the Horseshoe) (20 DL squared= 20x1111x1111 km squared= 25 million km squared). Indeed, the diameter of the Horseshoe is about half that of the Sarsencircle, so there is another discovered area of about the same size, South America.

## South America

So the Horseshoe also represents the continent of South America (Fig.6). It consists of 5 enormous trilithons, and each side of the construction contains 5 big vertical sarsenstones, both corresponding to 50 important degrees of latitude (5x10= 50). The new land extends from the equator to about 50°S, and the Horseshoe has the global shape of that continent, which is very important. The upper part has an open end, which means again that the land there continues. The Sarsencircle contains 15 separate trilithons (Fig.10), because its total surface area equals 15 square big distance lines (15 DL squared= 15x1111x1111 km squared= 19 million km squared).

## The Atlantic Ocean

The Horseshoe also represents the Northern Atlantic Ocean (Fig.6). It is the old known endless sea, which is depicted in many inscriptions in megalithic monuments (Refs.1,2,13). It also represents the Southern Atlantic Ocean. The upper part of the Horseshoe forms an open end, which means that both oceans continue further to the north. It is a double symbolism, which means that they have found the northern and southern oceans to be comparable in size and shape, but also comparable in this respect with the continents of North and South America.

## The Religious Meaning

The head-axis and in particular the Horseshoe of Stonehenge III point to the rise of the midsummer sun (Fig.6, Refs.1,5-12,22,23), as did Stonehenge I, so the megalith builders were still worshipping the SunGod Ra 1200 years later, in c.2000 BC. In confirmation of this, the Sarsencircle consists of 30 joined trilithons, corresponding to the direction to and the latitude of the Nile Delta, the center of the Northern Egyptian Empire, at 30°ESE and 30°N. The trilithons of the Sarsencircle are aligned in the direction of the rise of the midsummer sun, and they even form a circle like the sun, so the whole Sarsencircle is a symbol of the Sungod. The trilithons of the Horseshoe are not aligned in the direction of the midsummer sun, but only in the direction of the most southerly rise of the full moon. In addition, the Horseshoe has the shape of a boat, and for these reasons it is a symbol of the God of both the Moon and the Sea. So, in the tradition of Stonehenge, they made two enormous new structures for their two important gods of the Sun and the Moon, Horus and Osiris.

Stonehenge III is a monument for the discovery of America, and it contains 30+5= 35 trilithons (Fig.12), each of them having the shape of a gate. The concept of there being gates or portals to the west is common in Egyptian

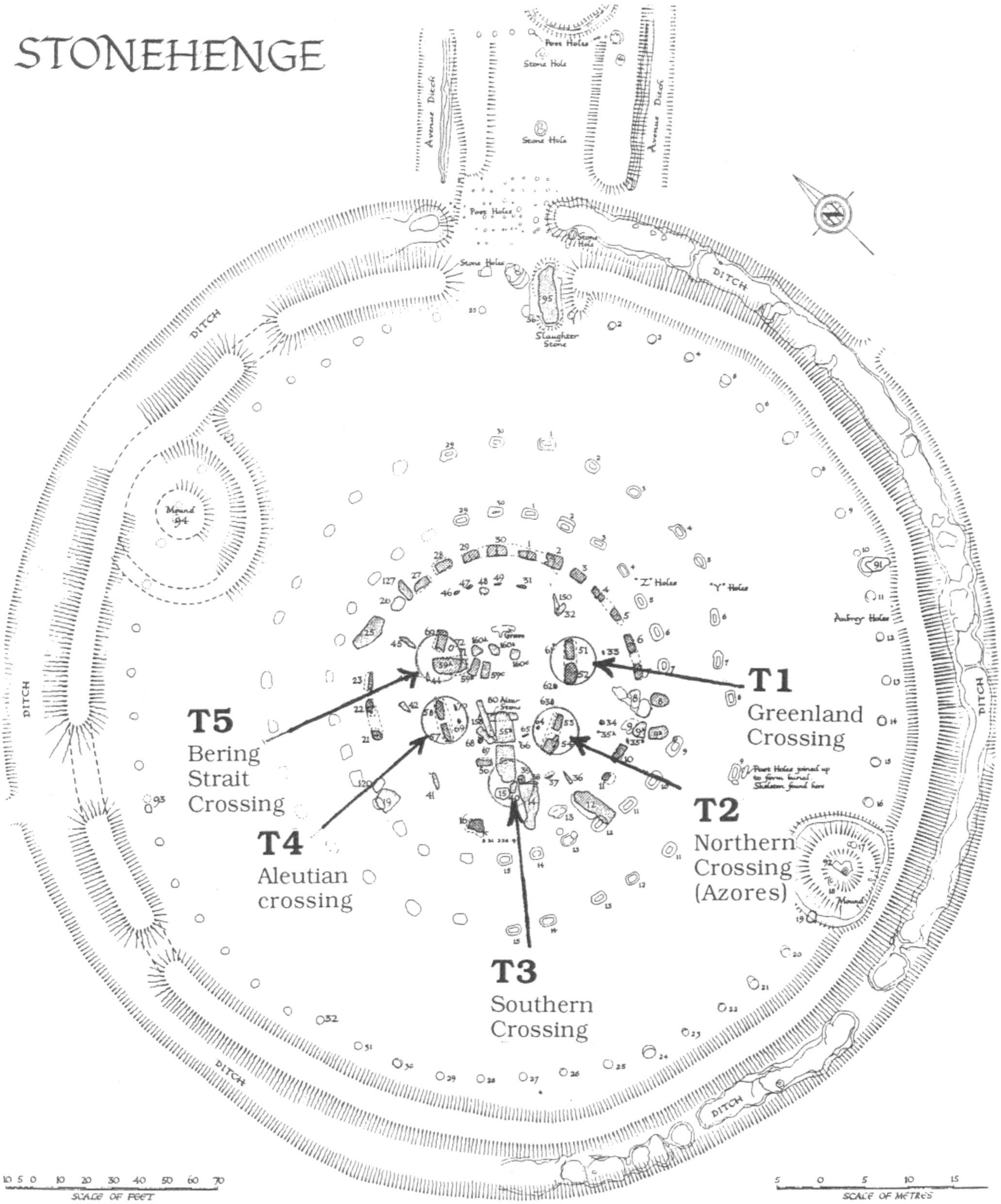


**Fig.8 The 5 trilithons of the Horseshoe, representing the 5 gates to the New World:** T1=the Greenland Crossing; T2=the Northern Crossing (the way back); T3=the Southern Crossing (the way there); T4= the Aleutian Crossing; and T5= the Bering Strait Crossing. The Horseshoe also represents the latitude of Cape Sao Roque, Brazil at 5°S, the size of the Americas, at 50° of latitude, the size of the United Central American Empire at 5° of latitude, the shape of the Gulf of Campeche 5° below the Tropic of Cancer, and the 5 islands of the Central Azores.

papyrus scrolls (funerary papyri) entombed with mummies. Written in hieroglyphics, and tucked into mummy wrappings, these have been preserved for unusually long periods of time. These texts are officially dated from 1500 to 250 BC, but many scrolls are known to be much older, at least from 2000 BC. Translated and published as The Book of the Dead (Ref.25), these "Spells" speak of "The Gates of the Underworld" (#144), and of "the Portals to the House of Osiris and the Gates of the Underworld" (#146), and speak of "the Doors of the Portals of the West" (#181). Spell #144 has a painting of the "Seven Gates", which are huge, and look just like the trilithons of Stonehenge III (Fig.9). Via the Atlantic, it is probable that America has been reached for the first time in 2497 BC, during the first "Punt" voyage of King Sahura, in the 13th year of his government (Refs.1,15,39). We think "Punt" was another name for Central America.

Note the Egyptian Jaguar skins worn in Fig.9. "From 1500 to 1200 BC the jaguar cult was formed in the Olmec world" (Ref.44.) "A painting in a nearby cave portrays an Olmec ruler wearing Jaguar skins" (Ref.45). "Blanke", or "Jaguar Sun" was one of the Mayan names of the SunGod (Ref.26). "The anthropomorphic Jaguar God ...is a dominant figure in the mythologies of most South American Peoples" (Ref.21).

## The Five Gates to America

The Horseshoe consists of five enormous trilithons, each of them having the shape of a gate. As we shall explain in the following paragraphs, these are the 5 gates, or passageways to America. Since the expectations were to find portals to the West, it should not be surprising that newly discovered routes to the west would be commemorated as stone portals. As shown on Figure 8, T1, T2, and T3 symbolize the routes across the Atlantic Ocean via Greenland, the Azores, and the Cape Verde Islands, respectively. Trilithons T4 and T5 symbolize the routes across the Pacific Ocean via the Aleutian Islands and the Bering Strait. The size of the trilithons increases toward the base of the Horseshoe, as the importance of these routes increases from north to south. We will now look at each one of these "gates" in detail.

**The Southern Crossing (T3)** is the most important sailing route to the New World, and this trilithon is the largest group of stones at Stonehenge, again showing the complex simultaneous meanings in megalithic symbolism. As shown in Fig.10, the Sarsencircle contains 15 separate trilithons, corresponding to the latitude of the southern Cape Verde Islands at 15°N. The Sarsencircle contains 30+30=60 stones, corresponding to the best sailing direction from these islands to South America, 60°SW. This is, with wind and current at your back, the easiest way there. The Horseshoe consists of 5 enormous trilithons, corresponding to the latitude of the place of arrival, Cape Sao Roque ("The Holy Rock"), Brazil at 5°S. The total number of separate trilithons amounts to 15+5= 20, corresponding to the approximate distance of the crossing: c.20dl= 2222km (1dl= 1 distance line= 1° of latitude= 111 km).

**Fig.9** In the top row (both pages), **The Seven Gates of the Underworld (portals to the West)**, each with three figures; in the bottom row, the nine Lesser Gods of Osiris (two sections of the Papyrus of Ani, (c Copyright The British Museum). These portals were the conceptual archetype for the trilithons of Stonehenge III, particularly the gates of the Horseshoe (c.2000 BC).

10470 11.

Figure 11 shows a pictograph of this important T3 crossing on a menhir in a megalithic passage grave in North Portugal, dating from c.2300 BC. "This is one of a group of four monuments in the Viega de Maos de Sallas, on the right bank of the river, on the Orense side of the cemetery" (Refs.2,13). The curved line at the right side is the stylized west coast of Africa, from the Canary Islands in the north to Cape Frio in the south. The short piece of line at the left side is the northeast coast of South America (Brazil), from the Holy of Holiest Bay in the south to the Amazon River in the north. The line in between is the literal crossing. The vertical stroke is the imaginary line half-way, which points to the magnetic north pole, because they already used a compass in those days. This magnetic north pole was located behind Greenland in the Queen Elizabeth Islands, at 77°N. At the east side the crossing takes place in known waters (thin line), but at the west side the route is new (thick line). Note also the strong emphasis on a northern direction after the crossing. The proportions of the whole inscription suggest that their main goal is Central America!

The wiggly line was a common megalithic symbol for water. In the Egyptian hieroglyphic language, the wiggly line meant water, as it still does, in a symbolic way, to us today. Again, at the east side the crossing takes place in known waters (small waves), but at the west side the route is new (big waves). They want to emphasize that they know the water all the way to the land at the other side (a very big wave ending at the left edge of the stone)! The wiggly line goes due west up and down ten times, which means that they have calculated that Cape Sao Roque, the place of arrival, is only 10 distance lines (= 10°of latitude = 1111 km) west of the Cape Verde Islands, which is correct.

**The Northern Crossing (T2)** is the most important route back (Fig.8). The Horseshoe contains 5 enormous trilithons, corresponding here to the latitude of the place of departure for return to Europe, Newfoundland (new-found-land), at 50°N (5x10°/DL=50°). It is almost the latitude of Stonehenge itself, a pleasing symmetry! The total number of separate trilithons amounts to 15+5=20, corresponding to the initial sailing direction to the West Azores, 20°ESE, and to the approximate distance of this crossing: c.20dl= 2222 km, with the wind and current again at your back. The total number of trilithons equal 30+5= 35, corresponding to the terminal sailing direction in the neighborhood of the West Azores, 35°ESE. The difference is due to the curvature of the Earth. They form, together with the four Stationstones, 35+4= 39 units, corresponding to the latitude of the West Azores, at 39° N. There are nine islands in the mid-ocean Azores. The two Stationstones at the left side (93 & 94) represent the two islands of the West Azores, the five trilithons of the Horseshoe represent the five islands of the Central Azores, and the two Stationstones at the right side (91 & 92) represent the two islands of the East Azores. The Sarsencircle consisted of 30 joined trilithons, corresponding to the initial sailing direction from Santa Maria (East Azores) to Madeira, 30°ESE. The total number of trilithons is 30+5= 35, corresponding to the terminal sailing direction in the neighborhood of

Madeira, 35°ESE. The 30 trilithons of the Sarsencircle together with the first 3 trilithons of the Horseshoe (finishing at the biggest one), form 30+3= 33 units, corresponding to the latitude of Madeira, at 33°N. All this sailing knowledge may have been carried about through some sort of memory system, like an early version of the knot-string quipus of the Chinese and the Incas. Monuments and inscriptions elsewhere show that this crossing (T2) was a popular sailing route in megalithic times, as well as later (Refs.1,2,13,32,38).

**The crossing via the Upper North (T1)**, along the islands of the Faeroes, Iceland, and Greenland (and sometimes the other way around) has already been discussed. This route was known to be difficult (Refs.2,13,14, 24,30,32), and was often replaced by the easier southern routes via the Cape Verde Islands (T3) and the Azores (T2) for exploration, trading, settlement and political consultations over the next four thousand years.

**America had been discovered via the islands of the Aleutians (T4)** at 51°N, as shown by the inscriptions of Dissignac, Brittany (Ref.1) c.2600 BC, and not via Greenland. The north-south axis of Stonehenge III makes an angle of 51° with the head-axis of the monument, which confirms this latitude (Fig.7). No wonder these big stones were built here in the then-ancient monument at 51°N: the discovery of the New Land was at this very latitude! This original discovery of America via the Bering Sea is confirmed in Mystery Hill (New Hampshire), and is indicated indirectly by the center of Stonehenge III, which is slightly above the center of the old monument (showing the new construction is not a logical consequence of Stonehenge I). This T4 route (Fig.8) starts at the Kamchatka Peninsula (Asia) at 56°N, and finishes at the Alaska Peninsula (America) at 56°N, both represented by the latitude of the southern Aleutians, and Stonehenge, at 51°N, combined with the five trilithons of the Horseshoe (51+5= 56). The size of the trilithons, and the importance of these islands increases from north to south, the southernmost being at the latitude of Stonehenge at 51°N. In addition, the sailing route along the southern Aleutians has the global shape of a Horseshoe.

**The crossing of the Bering Strait (T5)** is the shortest, but also the most northern route, from Siberia to the Seward Peninsula of Alaska. The Sarsencircle contains 15 separate trilithons, because this crossing is 15° north of Stonehenge, and north of the southern Aleutians at a latitude of 66°N (51 +15= 66).

## Central America

Central America has a great importance, because the Sarsencircle contains 15 separate trilithons, corresponding to the latitude of the culture along the north coast of Honduras, at 15°N. The Sarsencircle and the Horseshoe together contain 15+5= 20 separate trilithons (Figs.8&10), corresponding to the latitude of the civilization around the Gulf of Campeche at 20°N. The Horseshoe even has the shape of the Gulf of Campeche! The Horseshoe is

# STONEHENGE

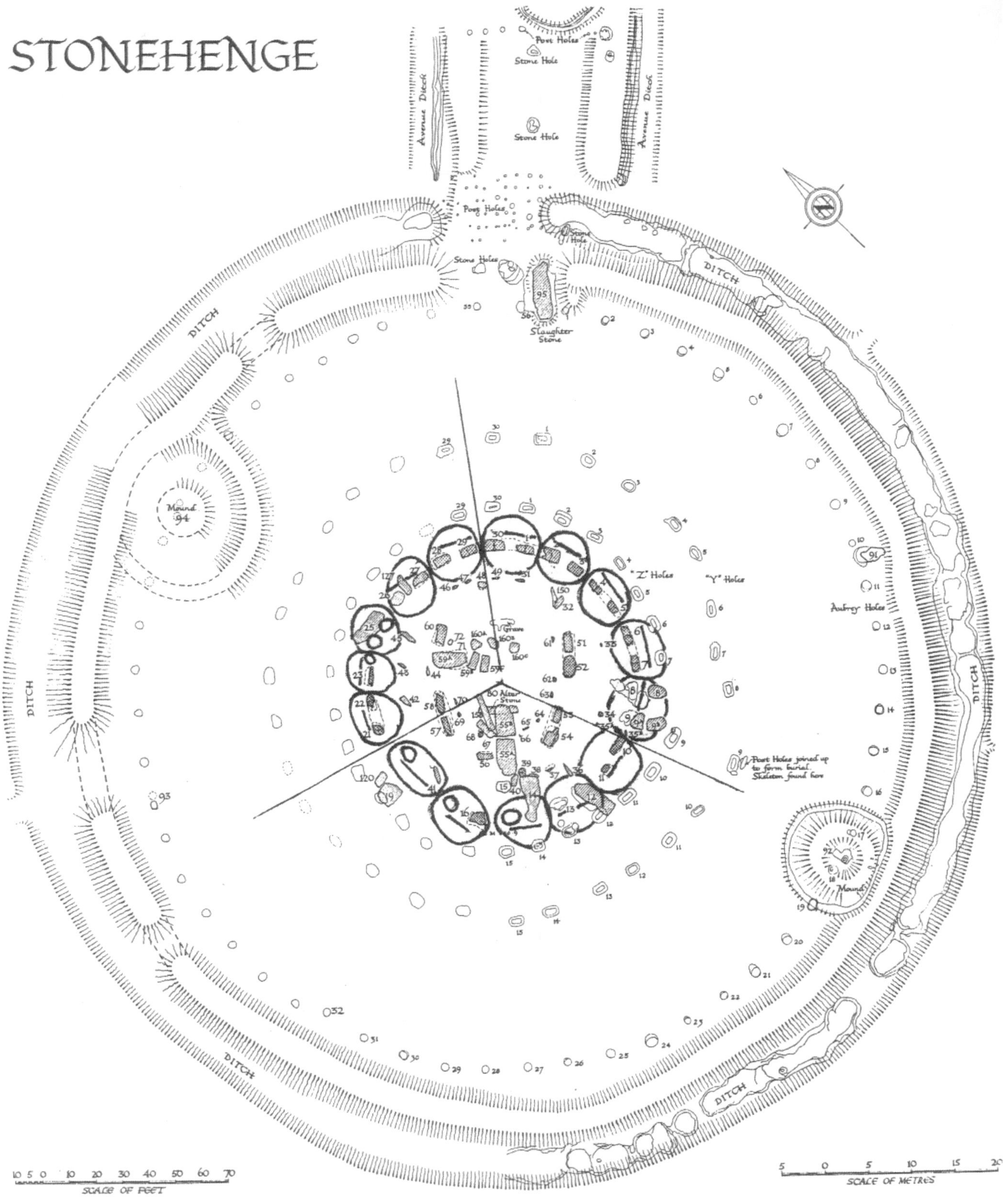


**Fig. 10 The 15 separate trilithons of the Sarsencircle** represent the southern Cape Verde Islands at 15°N, the surface area of South America at 15DL squared, the culture along the north coast of Honduras at 15°N, and the Bering Strait, which is 15° north of Stonehenge. When viewed as 30 joined trilithons, they represent the Mississippi Delta and the Nile Delta, both at 30°N. When seen as 60 stones, they represent the sailing direction from the Cape Verde Islands to Cape Sao Roque, Brazil, 60°SW, and the crossing from Cape Farvel, Greenland to Cape Chidley, Canada at 60°N.

constructed of bigger trilithons than the Sarsencircle, so this area must have been considered to be of great importance. The most southern trilithon in the Horseshoe is the tallest, so there lies the center of the New World. The number of trilithons of the Sarsencircle (15) and the first three trilithons of the Horseshoe (finishing at the biggest one) equal 15+3= 18, representing the latitude of this important place (where the Olmec originated) at 18°N.

Stonehenge III contains two big structures, the Sarsencircle representing the culture along the north coast of Honduras, and the Horseshoe representing the civilization around the Gulf of Campeche. So, Stonehenge III as a whole symbolizes the United Central American Empire! The Horseshoe contains five standing trilithons, and at both sides five standing sarsenstones, corresponding to 5 important degrees of latitude. At 18+5=23°N, the Tropic of Cancer is located (the line between Stationstones 91 and 94). So it is clear that the Gulf of Campeche is the bay below the Tropic of Cancer. It follows, consequently, that it was likely that in Central America there was introduced the worship of the Egyptian SunGod Ra, and indeed there are a lot of astronomically-oriented sun and moon temples found there, as well as a host of other transplanted goods and ideas (Refs.14,15, and others).

## The Mississippi Delta

As its primary representation, the Horseshoe is a symbol of the continent of North America (Fig.6). The trilithons T2 and T4 are bigger than T1 and T5, but smaller than that of T3. So the land between 40°N and 20°N is more important than the north, but less important than the southernmost area. The Sarsencircle contains 30 joined trilithons, corresponding to the latitude of the people of the Mississippi Delta, at 30°N, the same 30° latitude as the Nile Delta, the location of the greatest civilization then on earth.

Other megalithic monuments and inscriptions from after the discovery of America in c.2500 BC also show that people felt great interest in the area around the Gulf of Campeche and along the north coast of Honduras. It is here, that the cultural tradition of the Olmec started, around 2000 BC (Refs.20,21,26,37,40). There are a small number of indications that they were interested in the area around the mouth of the Mississippi, Poverty Point, or the river itself. Items have been found upriver in North America which document at least explorations starting there (Refs.14,15). On the British Isles, studies of pollen have shown that "thc impact of agriculture decreased (2500-2000 BC) when it was thought there was a widespread and extended phase of woodland regeneration" (Ref.27). These were the years of exploration and colonization of the New World, and the construction of Stonehenge III.

## Egypt and Central America

The Sarsencircle consists of 30 joined trilithons, corresponding to the direction to, and the latitude of, the Nile Delta at 30°ESE and 30°N,

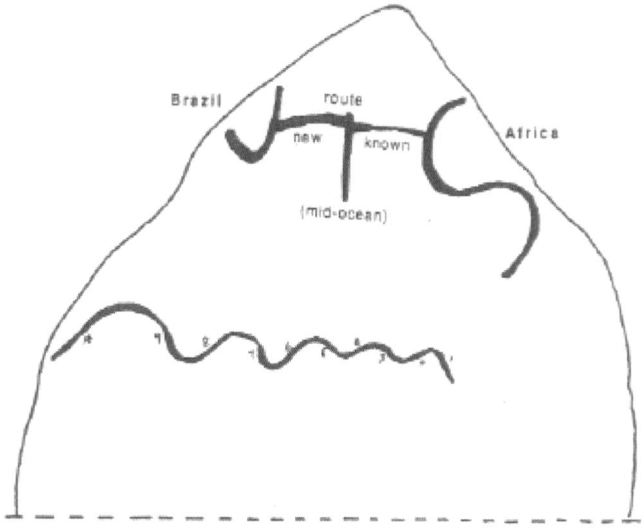


**Fig.11 Pictograph from Sallas, Orense, North Portugal, in red ochre ( Ref.13, c.2300 BC).** "This is one of a group of four monuments in the Viega de Maos de Sallas, on the right bank of the river, on the Orense side of the cemetery." At the pointed top of the stone, the Southern Crossing of the Atlantic Ocean between the stylized west coast of Africa (Cape Verde Islands, right) and the northeast point of South America (Cape Sao Roque, Brazil, left). The vertical stroke is the imaginary line halfway, which points to the magnetic north pole. The wiggly line shows that now the water is known all the way to the other side of the ocean (the left side of the stone), which is only 10dl (ten wiggle legs) to the west.

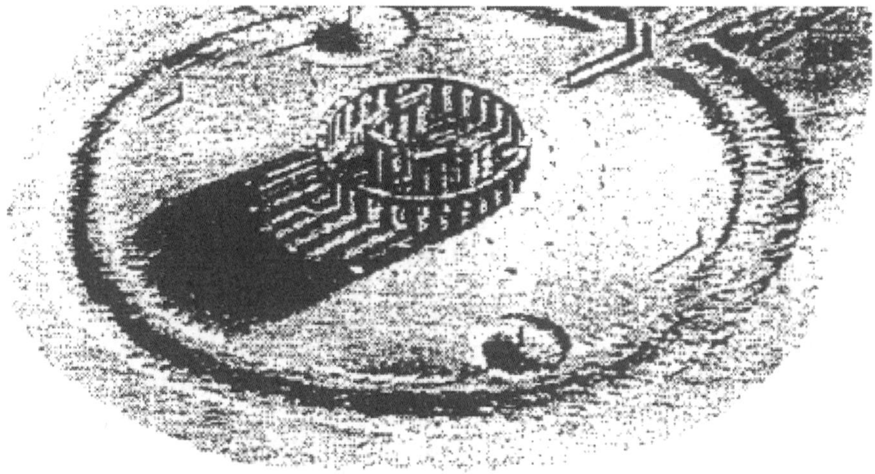

**Fig12 Drawing of the completed monument of Stonehenge**, as seen from south and above (Ref.5). The 15+5= 20 separate trilithons (5-7m high) represent the civilization around the Gulf of Campeche, at 20°N, the surface area of North America at 20 DL squared, the two longest sailing distances across the Atlantic of 20 dl, and the initial sailing direction from Newfoundland to the Azores, 20°ESE. The 30+5=35 trilithons represent the surface area of America, 35 DL squared, the latitudes of Cape Hatteras and the Strait of Gibraltar at 35°N, and the terminal sailing directions near the Azores and Madeira, both 35°ESE. This seems pretty fancy for "illiterate" folks, but with no phonetic writing, it had to be! That is why the stories we have deciphered from the Neolithic and the Bronze Age are told in numbers!

respectively. From the center of Stonehenge, the four Stationstones make angles of 23°, all four corresponding to the latitude of the Southern Egyptian Empire, 23°N. Stonehenge III possesses 15+5=20 separate trilithons; together with the 4 Stationstones, the Slaughterstone and the Heelstone, the center of Stonehenge forms 20+4+1+1= 26 units, encoding the latitude of the United Egyptian Empire at 26°N. Each trilithon contains 2 Sarsenstones. The one is the Southern, and the other is the Northern Egyptian Empire. It has also a lintel. So, each whole trilithon symbolizes the United Egyptian Empire. The Sarsencircle possesses 30 such joined trilithons, which confirms this 30 times, so the Sarsencircle as a whole symbolizes the United Egyptian Empire.

Each trilithon contains 2 sarsenstones. One is North America, the other is South America. It has also a lintel, Central America. So each whole trilithon symbolizes the double-continent of America. But also one sarsenstone of each pair is also the civilization of the Gulf of Campeche, and the other is also the culture along the north coast of Honduras. So each whole trilithon also symbolizes the United Central American Empire! The Horseshoe possesses 5 of these trilithons, which confirms this five times, and it has the shape of the Gulf of Campeche. The empire extends over 5 degrees of latitude, and the center is located 5° below the Tropic of Cancer, so the Horseshoe as a whole represents the United Central American Empire.

Stonehenge III consists of the Sarsencircle with 30 trilithons, and the Horseshoe with 5 trilithons, together 35, which correspond to the degrees of latitude of the Strait of Gibraltar, and Cape Hatteras, both at 35°N. This is a connection-line between the Old and New Worlds, across the Ocean. This is the shortest link between the Mediterranean and North America, between Egypt and Central America. The biggest trilithon is made of two gigantic Sarsenstones, the one is Egypt, the other is Central America. The huge lintel capstone is their literal connection-line, their indissoluble friendship, their everlasting alliance, both homes of the SunGod. Now over the entire earth, the light of the SunGod will shine forever, not only in the Old World, but also the New World, because if the sun shines in Egypt, the moon shines in Central America, and vice versa, because the light of the moon is descended from the sun! The gigantic trilithon is the union of the two Countries. Stonehenge contains 35 such trilithons (Fig.12), which reconfirms this symbolism 35 times!

According to the complicated Sun-religion, Egypt is the land of Horus (the Sarsencircle) where the Pharaoh is the living god (Horus) on earth, and Central America is the land of Osiris (the Horseshoe), the "Ruler of the West" (Ref.23). The realm of the dead in the west, at the other side of the waters, in the land where the sun sets (America) has finally been found (Refs.1,20,23,36,37,40). Stonehenge III, built within the Stationstone connection lines at 23° (see Fig.1, so built in the Tropics), celebrates that the Home of the SunGod in the West has been found, and this New World has become a part of the Great Egyptian World Empire of 2000 BC.

## Discussion

These findings would not be a big surprise to Thomas Jefferson. The American statesman has been called the "first scientific digger" after his sectional examination of the strata in a Virginia mound in c.1780. He states: "Great question has arisen from whence came those aboriginal inhabitants of America? Discoveries, long ago made, were sufficient to shew that a passage from Europe to America was always practicable, even to the imperfect navigation of ancient times. In going from Norway to Iceland, from Iceland to Groenland, from Groenland to Labrador, the first traject is the widest; and this having been practiced from the earliest times of which we have any account of that part of the earth, it is not difficult to suppose that the subsequent trajects may have been sometimes passed" (Ref.31).

It should be mentioned, that nowadays a lot of stones of Stonehenge have tumbled down, and that some of them are even completely missing. However, due to extensive archaeological research, the positions of these stones are accurately known, which made it possible for us to use these data without any hesitation. It is hard to know at this point which of the many possible simultaneous symbolisms described above were intended, or most important, to the creators. From our extensive study of megalithic inscriptions, monuments, and religion, we think many simultaneous symbolisms were intended. It is generally agreed by professional archaeologists that a lot of thought went into the design of Stonehenge, as a number of partially completed features, such as the different locations of the Bluestones, and the Y and Z holes, appear as part of the construction history.

The stories of these Bronze Age people are told in numbers. Very few researchers have recognized this. One of these, Ralph Ellis, suggests that Avebury Henge is a map of the planet, with its 23° tilt from true north, and Stonehenge and the South Sandwich Islands encoded by stone count. He and John Mitchell claim Avebury is on a ley line through hilltop shrines of southern England. We note that on a globe, this line runs to Central America at 16°N! Stonehenge is on a ley-line running to the Azores at 38°N. This is in accord with the work of Carl Munck and that of Zapp and Erickson, whose studies demonstrate worldwide ley lines. Because of the complexity of the subject, and because they have been unknown to each other, and unaccepted in professional literature, they have drawn both parallel and different conclusions, which have not yet been sorted out, and painted into a coherent picture (Refs. 26-30,32,33). Much work remains to be done in this promising and exciting area of research.

The comparison of latitudes to the stone construction that we have done in this article is similar to the approach we have taken to all the sites we have worked on, including American Stonehenge. Now we know that latitudes were an important part of the geometric, astronomic, and navigational skills of the people at this time. These insights are solving a number of problems in archaeology, illuminating an interesting period of late prehistory, and creating a new respect for the abilities and achievements of early man.

# References


1. Jonge. R.M. de. and IJzereef. G.F.. de Stenen Spreken. Kosmos Z&K. Utrecht/Antwerpen. 1996 (ISBN 90-215-2846-0)(Dutch)
2. Jonge. R.M.. de. and IJzereef. G.F.. Exhibition: "The Megalithic Inscriptions of Western Europe". 1996 . (Dutch)
3. Jonge. R.M. de. "Stonehenge als Zeekaart". BRES 158. 1993 (Dutch)
4. Newall. R.S.. Stonehenge. Dept of the Environment. Official Handbook. Her Majesty's Stationery Office. London. 1977 (ISBN 0-11-670068-8)
5. Richards. J.. Stonehenge. English Heritage. 1992 (pg.127)
6. Atkinson. R.J.C.. Stonehenge. Penguin Books. 1979
7. Atkinson. R.J.C.. Stonehenge and Neighboring Monuments. English Heritage. London. 1993 (ISBN 1-85074-172-7)
8. Postins. M.W.. Stonehenge: Sun. Moon. Wandering Stars. Malthouse Lane. Kenilworth. Warwickshire. 1982
9. Chippindale. C.. Stonehenge Complete. Thames and Hudson. 1994 (ISBN 0-500-27750-8)
10. Hawkins. G. S.. Stonehenge Decoded. Barnes & Noble. 1993 (ISBN 0-88029-147-8)
11. Gibson. A.. Stonehenge and Timber Circles. Tempus Press
12. Castleden. R.. The Stonehenge People. Routledge Press
13. Twohig. E. Shee. The Megalithic Art of Western Europe. Clarendon Press. Oxford. 1981 (pg.158)
14. Fell. B.. America BC. Pocket Books. Simon & Schuster. 1994
15. Bailey. J.. Sailing to Paradise. Simon & Schuster. 1994
16. Mallery. A. and Harrison. M.R.. The Rediscovery of Lost America. the Story of the Pre-Columbian Iron Age in America. E.P. Dutton. New York. 1951 (ISBN 0-525-47545-1)
17. Peterson. F.A.. Ancient Mexico. .1959
18. Mitchell. J.. Secrets of the Stones. Inner Traditions International. Vermont. 1989 (ISBN 0-899281-337-7) (pg.66)
19. North. J.. Stonehenge - A New Interpretation of Prehistoric Man and the Cosmos. The Free Press. New York (1996). (ISBN 0-684-844512-1) (pg.496)
20. Stuart. G.E.. "New Lignt on the Olmec". National Geographic. Nov. 1993
21. Campbell. J.. Historical Atlas of World Mythology. Part III. The Middle and Southern Americas. Harper & Row. New York. 1989 (ISBN 0-06-055159-3)
22. Bleeker. C.J.. Egyptian Festivals. 1967 (pg.6-17)
23. Strelocke. H.. Egypte: Geschiedenis. Kunst en Cultuur in het Nijldal. Cantecleer. de Bilt. 1981 (Dutch)
24. Ferryn. P.. "5000 years Before our Era: The Red Men of the North Atlantic". NEARA Journal. Vol XXXI. No. 2. 1997
25. Faulkner. R.O.. transl.. The Ancient Egyptian Book of the Dead. University of Texas Press. Austin. 1997 (ISBN 0-292-70425-9) (pg.135)
26. Bradley. M.. The Black Discovery of America . Personal Library. Toronto. 1981 (ISBN 0-920510-36-1)
27. Roberts. N.. The Holocene. An Environmental History. Blackwell. R.R Bowker 1998 (ISBN 0-631-18638-7). from Whittle. A.W.R.. 1998. pg. 146 (from Whittle. A. W.R.. "Resources and population in the British Neolithic". Antiquity. 52 (pgs. 34-52.146).
28. Vecht. C.F. Ph.D. van der. De Steenen Spreken. Den Haag. 1949 (Dutch)
29. An Approach to Megalithic Geography. Ass. Arch. Kergal. transl. from French. Booklet No.20. 1992 (ISBN 2902727-20-8)
30. Munck. C. P. The Code and Whispers from Time. 1996 and 1997. P.O. Box 147. Greenfield Center. NY 12833 or The Radio Bookstore. PO. Box 3010. Bellevue. Wa 98009-3010 USA
31. Butler. A.. Bronze Age Computer Disc. Quantum. London. 1999 (ISBN 0-572-02217-4)
32. Lawlor. R.. Sacred Geometry. Thames and Hudson. London. 1998 (ISBN 0-500-81030-3)
33. Jefferson. T.. Notes on the State of Virginia. Ch.11. "Aborigines". Univ. of North Carolina Press. (ISBN 0-8708-4588-4) (pg 101)
34. Ellis. R.. "The Henge of the World". Atlantis Rising. #27. April 2001. from his book Toth. Architect of the Universe. 1997.
35. Zapp. I.. and Erikson. G.. Atlantis in America. Navigators of the Ancient World. Adventures Unlimited Press. 1998 (ISBN 0-932813-52-6)

36. Thompson, G., American Discovery, Misty Isles Press, Seattle, 1994
37. Jairazbhoy, R.A., Ancient Egyptians and Chinese in America, Rowman & Littlefield, Totowa, N.J., 1974 (ISBN 0-87471-571-1)
38. Lambert, J.D., America's Stonehenge, an Interpretive Guide, Sunrise Publications, Kingston, N.H., 1996 (ISBN 0-9652630-0-2)
39. Wallis Budge, E.A., Osiris and the Egyptian Resurrection, 2 Vol., 1911, Dover Pub., N.Y., 1973 (ISBN 0-486-2278-0-4) (Vol.1, pgs.232,233)
40. Evers, D., Die Ware Entdekkung Americas, Weissbach, 2000 (ISBN 3- 930036-45-2) (German)
41. Silva Ramos, B.A., da, Inscripcoes 'e Tradicoes da America Prehistorica Especialmente Do Brasil, 1937 (Portuguese)
42. Feats and Wisdom of the Ancients, Time-Life Books, Alexandria, Va., 1990 (ISBN 0-8094-7675-4)
43. Severin, T., The Brendan Voyage, Hutchinson, London, 1978
44. Bernal, I., The Olmec World, University of California Press, Berkely, 1969 (ISBN 0-520-02891-0)
45. Stuart, G.E., "New Light on the Olmec", National Geographic, Vol.184, N0.5, Nov.1993
46. Brandao, A., A Escripta Prehistorica do Brasil, (Ensaio deInterpretacao), Civilizacao Brasileira S.A., Rio de Janeiro, 1937 (Portuguese)
47. Wilkins, H., Mysteries of Ancient South America, Citadel Press, Secaucus, N.J., 1956 (original, 1947)
48. Wilkins, H., Secret Cities of Old South America, Adventures Unlimited Press, Kempton, Ill., 1952 (original, 1919

# The Oldest Chart of the Atlantic Ocean

(The Petroglyphs of Kercado, Brittany, c.2200 BC)

Dr. R.M. de Jonge, drsrmdejonge@hotmail.com
J.S. Wakefield, jayswakefield@yahoo.com

## Introduction

On stone C1 of the chamber in the passage grave of Kercado, Brittany, dated c.4500 BC, is a petroglyph of the North Atlantic Ocean (see Fig.1, left stone C1) (Refs.1-3). The more or less horizontal strokes cut into the stone are big latitude lines every 10°, as can be observed nowadays on every globe. The vertical-looking strokes are not meridians, which would all converge on the North Pole, but big distance lines (DL). In reality, these lines run parallel, and they are in all aspects similar to big latitude lines. Each unit of distance, vertical or horizontal, can be calculated from the circumference of the earth, and the circle angle, based on a 360° circle. The unit of these Distance Lines is 1 DL, which is equal to the 40,000km circumference of the Earth divided by the 360° in a circle, times 10, which equals 10° of latitude (1111km), or 10 moira (since 1 moira= 1°). The DL also equals 600 Nautical Miles (since 60 NM= 1°). These big latitude and distance lines often appear on megalithic petroglyphs of the ocean, where these features of orientation and distance are so important (Refs.6,19).

The slanting edge on the left side of the stone is the east coast of North America, from Greenland in the north to Central America in the south (see Fig.2). At the level of Central America the stone and also the ocean have maximum width. Further downwards the left edge of the stone runs inwards just like the long NE coast of South America; the ocean narrows again. The less important right edge of the stone follows the coast line of Europe and Africa, from Norway in the north to the equator in the south, with Africa protruding to the west. Because of the layout of the distance lines, we can see people are extremely interested in the width of the northern part of the North Atlantic Ocean, so therefore this petroglyph dates from after the discovery of America in c.2500 BC (established by the inscriptions at Dissignac, Brittany, and other sites, Ref.2).

## Africa and Europe

Close inspection of the petroglyph leads to the conclusion that the tall vertical column was made first (Fig.2). Then they started to count the degrees of latitude from below, from the equator, at 0°N. At the right side we see a thick dot at a latitude of 15 to 16°N. This dot is Cape Verde at 15°N, and the Cape Verde Islands at 16°N. Cape Verde is the westernmost point of all the mainland, and between c.4500 and c.3600 BC the Cape Verde Islands were the westernmost archipelago of the then known world (Ref.2). Slightly higher and at the left side of the vertical column we observe a long broken horizontal line, running in the direction of the left edge of the stone. That is the Tropic of Cancer, at 23°N, where people originally

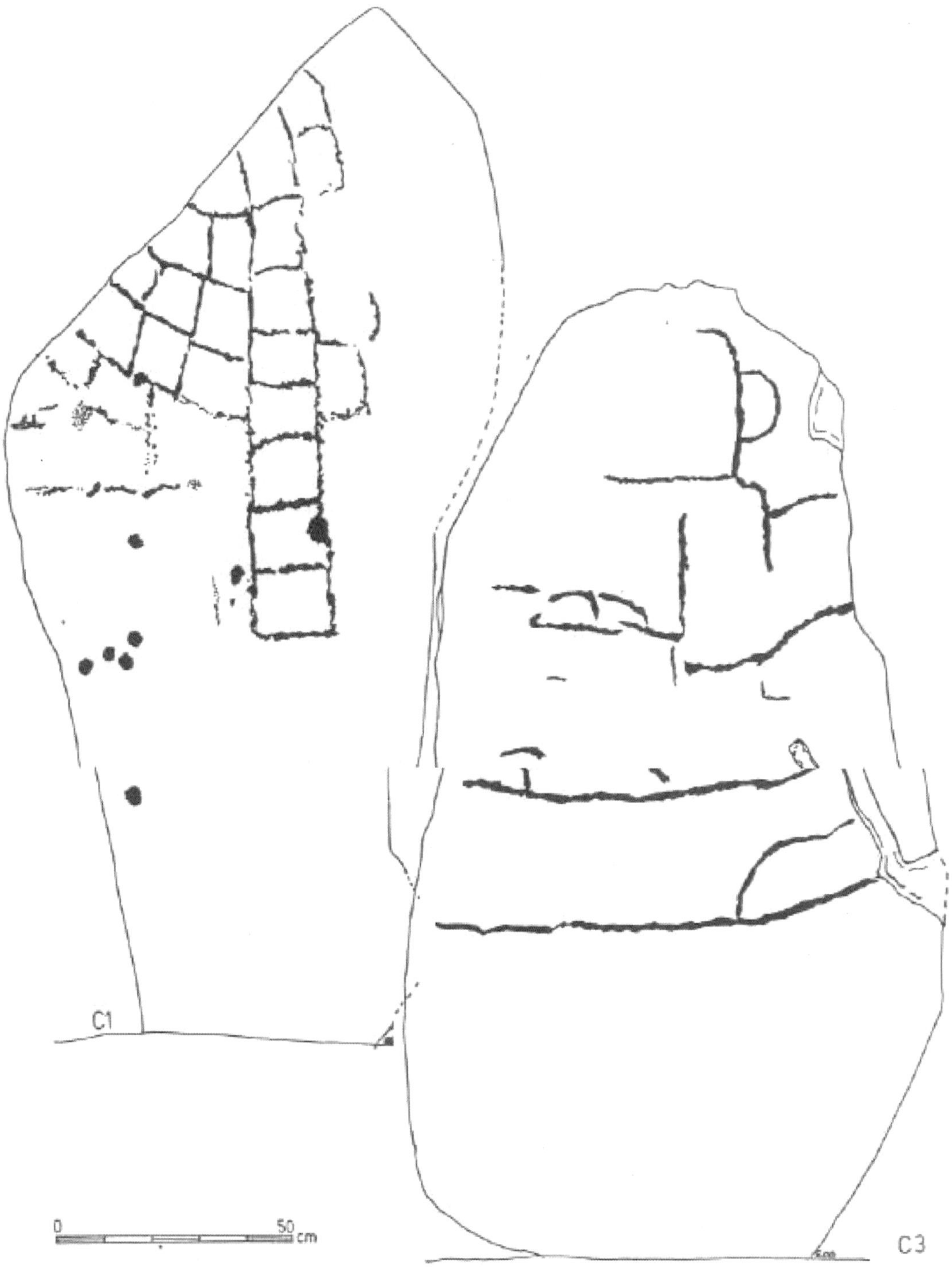


Fig. 93 Kercado, Carnac. C1, C3.

**Fig. 1 Left:** Petroglyph of the North Atlantic Ocean (c.2200 BC) on stone C1 of the chamber of the passage grave of Kercado, Brittany (Ref. 1).
**Right:** Petroglyph of the crossings of the ocean on stone C3 (c.2200 BC) (Ref. 1).

wanted to cross the ocean, following the path of the Sungod Ra.

Still higher, and at the right side of the column, we see a compartment that represents the sea area west of Iberia. The dot at the left side below represents the islands of Madeira at 33°N. These islands were discovered c.4100 BC (Ref.2). The horizontal line to the right leads to the Strait of Gibraltar (the dot at the right side below), at 36°N. The line directly upwards is the west coast of Iberia, and the horizontal line at the end to the left provides the latitude of Kercado in the megalithic center of Brittany at 47°N, where this petroglyph is located (Ref.3). Opposite this compartment and at the left side of the column, at the end of the stroke, are the West Azores, at 39°N. (The exact position of these islands is difficult to recognize.) The Azores were discovered from Madeira c.3600 BC (Refs.2,3).

The bow at the right side, above the isolated compartment mentioned before, is the Irish Sea between England and Ireland, of some importance for the sailing route to Iceland. Just below the uppermost horizontal stroke of the column at 70°N, at the right side a small bend is seen followed by an interruption of the vertical line (Fig.2). Although hard to see, it is the west coast of Iceland, at 65°N. Iceland was discovered from Scotland, c.3400 BC (Refs.2-5). This west coast of Iceland is located due north of the east side of the Cape Verde Islands (the big dot), at 16°N, so the right side of the tall column is the north-south line (the meridian), which connects these two areas.

Above the west coast of Iceland is Cape Brewster, the east cape of Greenland, at 70°N. The thickening just below it at the left side of the column is Cape Holm at the Arctic Circle, at 67°N. According to the Stories of Loughcrew and Stonehenge, Greenland was discovered here, c.3300 BC. At the right side above the column we observe still higher latitude lines, because of the North Cape of Norway at c.70°N, Spitzbergen at c.80°N, and the theoretical presence of the North Pole at 90°N. This is the highest point of the petroglyph.

## The Southern Crossing

The official point of departure for the Southern Crossing of the ocean is Brava, the southwestern Cape Verde Island. Brava is indicated with the thick dot on the column at 15°N (Fig.2). The official point of arrival is Cape Sao Roque (the Holy Rock), which is the NE cape of Brazil. This cape is indicated by the leftmost dot near the edge of the stone, just below the column, so south of the equator, at 5°S. This is accomplished with the tradewind and current at your back all the way. Here no latitude or distance lines are present, because this is considered to be an easy crossing. In general, when heading for Central America, it is even recommended to cut the corner, and to land somewhere along the coast north of the Amazon River. However, if one runs into trouble, one can go ashore prematurely on the island of St. Paul (the first dot, 1°N), or on one of the two tropical islands of Fernando de Noronha (the next two dots, at 4°S) (Ref.2).

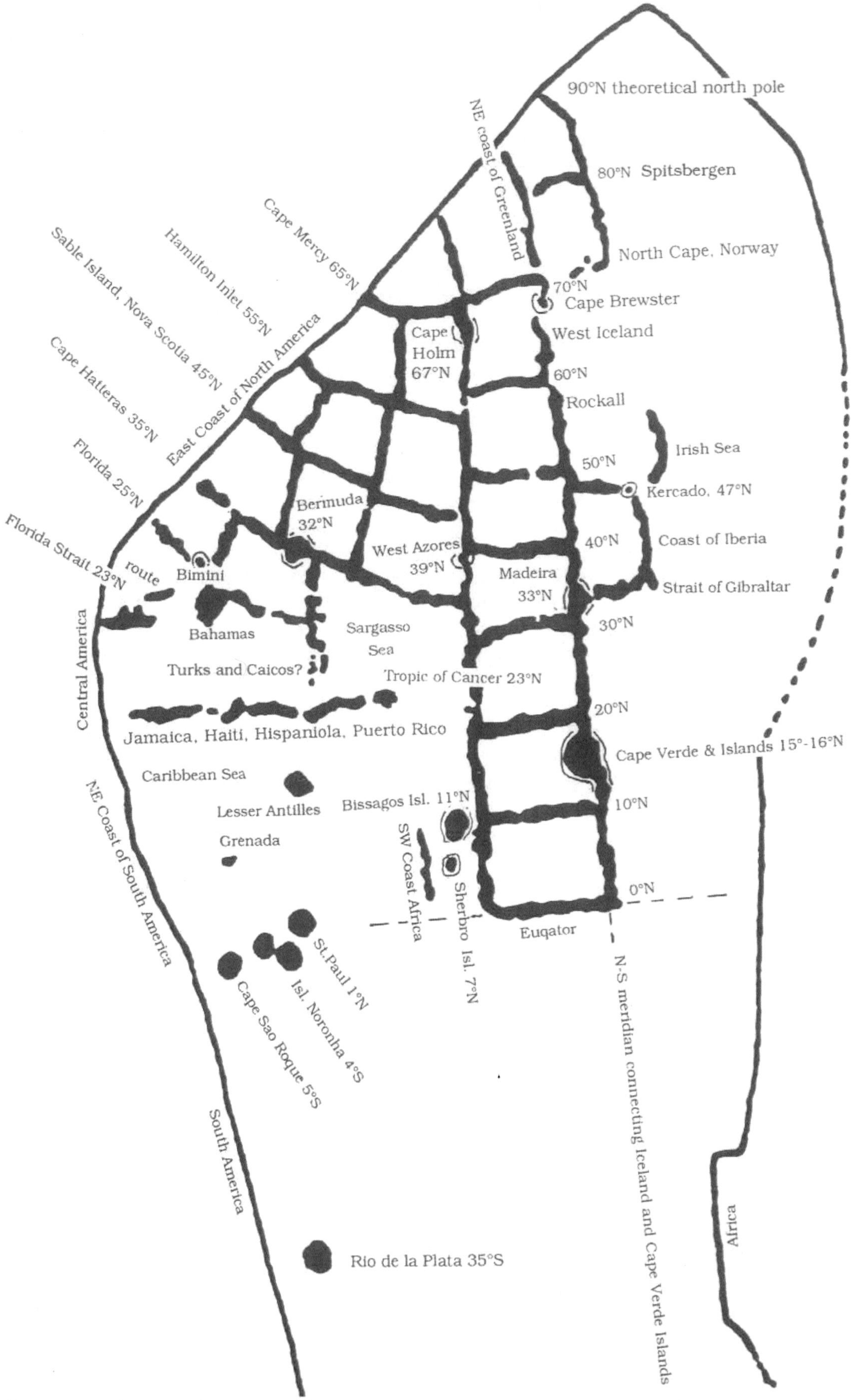


**Fig.2** Petroglyph of the North Atlantic Ocean on stone C1 (Fig.1, c.2200 BC), with geographic meanings. The "horizontal lines" are big latitude lines, the "vertical lines" are big distance lines, both spaced every 10°.

If one crosses the ocean for the first time and if one is afraid of the long distance over open sea, one can follow the African coast south a while, then depart further to the south from the Bissagos Islands (the dot directly left of the column, at 11°N), or from Sherbro Island (the little dot below it, at 7°N), or in general from the SW coast of Africa (the small vertical line). This part of the inscription has been placed at the left side of the column to indicate that in reality, because of the usual nice weather and the steady mild trade winds, this crossing (of just over 2.5DL= 2,777km, or 1,501 nautical miles) is not as long as it seems to be on the stone. One can sail intentionally toward St. Paul (the first dot, 1°N), or to one of the two islands of Fernando de Noronha (the next two dots, 4°S.), to decrease the length of the crossing. After that one lands on the NE coast of Brazil, near Cape Sao Roque (the fourth dot, 5°S). This route is recommended too, when one is heading south (downwards from the four dots). The lowest dot on the stone is the Rio de la Plata River at c.35°S, the second largest river of South America, which is reported to be navigable a long way (toward the ore-rich Andes). The correct latitude of the River at 35°S can be verified by using the vertical column as a measuring stick.

## North America and the Northern Crossings

In general, most folks sail along the NE coast of South America toward the Caribbean and Central America (upward from the four dots) (Refs.2,6-14). Because of the wind and its current, the Southern Crossing cannot be accomplished in the reverse direction. Instead one starts sailing via Cuba to the north, along the east coast of North America (along the slanting edge of the stone upwards). From there one is able to return to the Old World with wind and current astern. To sail the North Atlantic they needed to know the distances across the northern part of the Ocean. That is what this inscription is about.

It is clear that the compartments at the left side on the stone have been carved some time after the vertical column, because the junctions with the column are rather forced. When one looks west from the coast of Brittany, where this petroglyph is located, all latitude lines turn off to the north as a result of the curvature of the earth. That is shown in this part of the inscription in the right way. If we take our bearings on the vertical column, we see latitude lines at 25°, 35°, 45°, 55°, and 65°N. However, the first line at 25°N has not been extended all the way to the right, and the junction of especially the last line with the column has not been carved in the right manner. The places along the American east coast which are of importance for these lines are: Bimini, Bahamas at 26°N, Cape Hatteras at 35°N, Sable Island, Nova Scotia at 44°N, Hamilton Inlet at 54°N, and Cape Mercy on Baffin Island at 65°N. The places along the European west coast at the end of these lines, including the column, are: SW Portugal at 37°N, Western Brittany at 48°N, Northern Ireland at 54°N, and West Iceland at 65°N. From the lines of latitude of the petroglyph and the compartments above them, including the column, the distances across the ocean can be estimated, as shown below in Table 1:

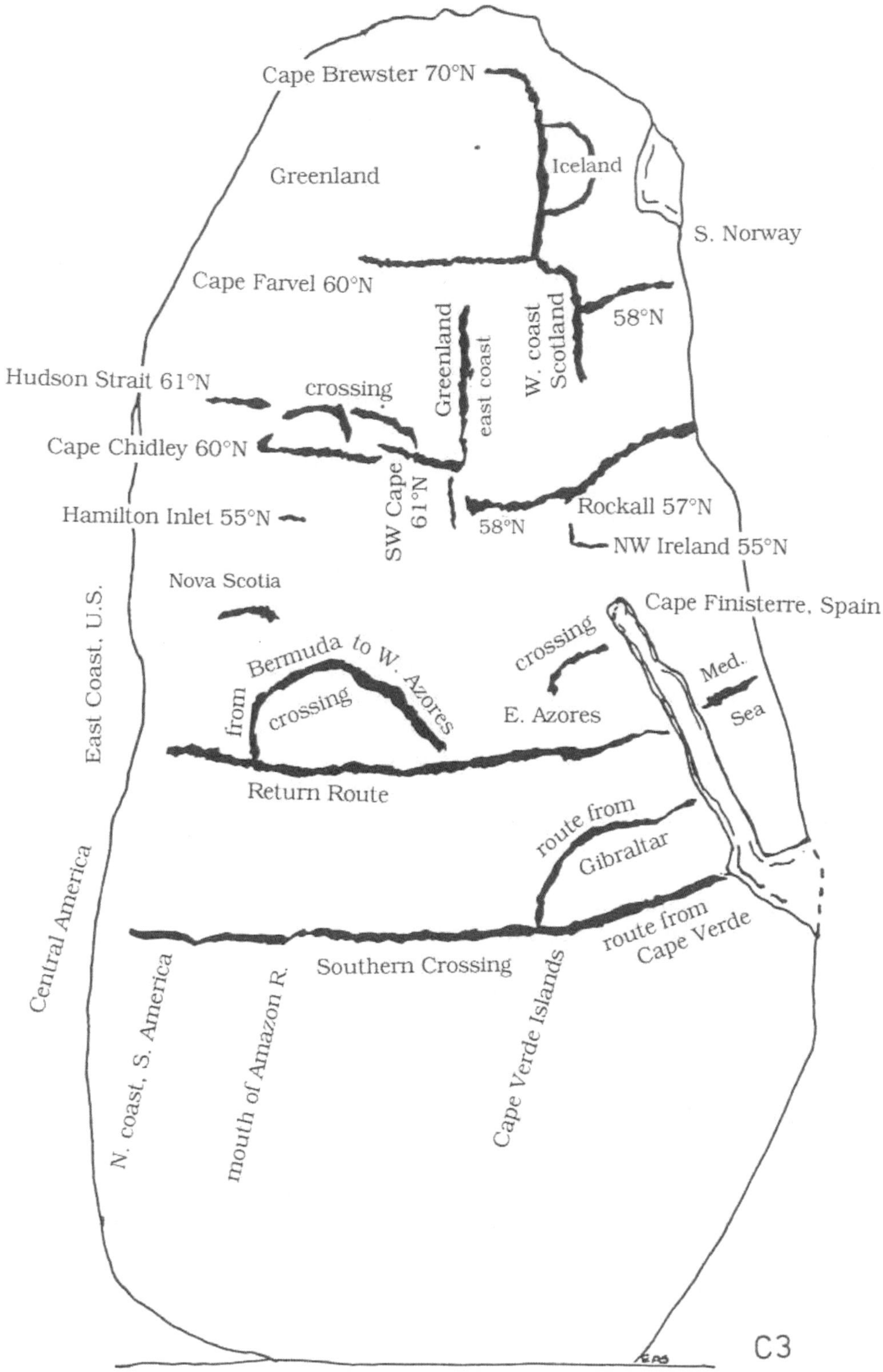


**Fig.3** Petroglyph on stone C3 (Fig.1, c.2200 BC), with geographic meanings. The Southern Crossing, the Northern Crossing, and the Crossing via the Upper North (shown twice) of the North Atlantic Ocean.

Table 1

| | | | | |
|---|---|---|---|---|
| Bimini to SW Portugal | (c.30°N) = 5.5DL= | 6111km= | 3,300 nautical miles |
| C. Hatteras to SW Portugal | (c.35°N) = 5DL= | 5555km= | 3,000 nautical miles |
| C. Sable to W. Brittany | (c.45°N) = 4DL= | 4444km= | 2,400 nautical miles |
| Hamilton Inlet to N. Ireland | (c.55°N) = 3DL= | 3333km= | 1,800 nautical miles |
| Cape Mercy to W. Iceland | (c.65°N) = 2DL= | 2222km= | 1,200 nautical miles |

As the crow flies, these distances match the real distances! In all cases, the errors are less than one third of a big distance line (1/3DL=370km, or 200nm), which is negligible in view of the scale and size of the petroglyph (c.2200 BC)!

At the start of the slanting edge of the stone is Bimini, off Florida, on the edge of the Bahama Bank at 26°N. The nearest vertical stroke upwards is half a big distance line offshore, to show that the total crossing of the ocean from this island is c.555km longer than from Cape Hatteras at 35°N. This has been shown in the table above. When we follow the latitude line from Cape Hatteras to the east, Bermuda is carved just over one DL offshore (the dot at 32°N). Bermuda was discovered from the American Coast c.2200 BC.

At the extreme left side of the stone the Strait of Florida is indicated above Cuba, with left and right the coastal waters around the south cape of Florida. The little thin line to the west is the crossing to Bimini. According to the next compartment, Bermuda is located about one DL (big distance line) north and east of this island, which is correct. Below Bermuda, a vertical line runs to the south in the direction of the Tropic of Cancer, at 23°N. The inscribed part of this line passes Puerto Rico, an easternmost island of the Greater Antilles at 18°N. This last island has only been indicated to clarify the exact position of Bermuda, because of its importance in the crossing of the ocean. Great Inagua and the Caicos lead to the Bahamas. If the sea were lower, the Caicos Banks and Silver Banks would have been islands too. Around the Tropic of Cancer no latitude or distance lines are indicated, simply because this part of the ocean is unsuitable for sailing. There are no islands here; this is the Sargasso Sea, the famous "Bermuda Triangle", where seaweed and hulks of dead ships float in a dead windless circle created by encircling winds and currents.

## Discussion

In the Bronze Age, the most important sailing routes from the mentioned places along the American coast, in the direction of the Old World, were (Ref.6):

**Bimini - Bermuda - Azores**
**Cape Hatteras - Bermuda - Azores**
**Sable Island - Cape Race - Azores**
each followed by:
**Azores - Madeira - Canary Islands - NW Africa.**

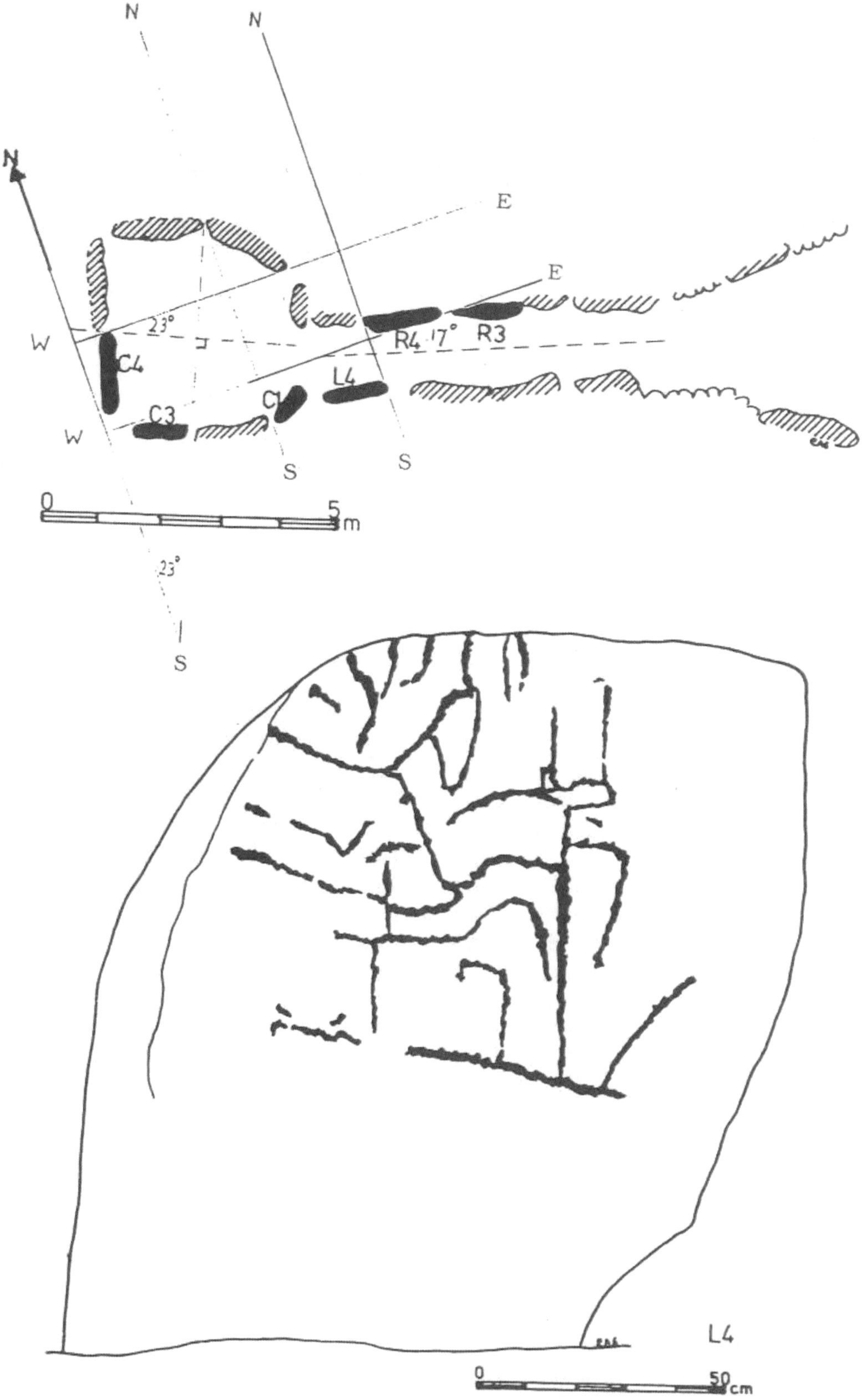


**Fig.4 Top:** Groundplan of the passage grave of Kercado (Ref.1), made because of the discovery of the Cape Verde Islands (c.4500 BC). Between the menhirs are NS and EW axes. The black menhirs contain petroglyphs on their inside surfaces. **Bottom:** Petroglyph of explored sea area on stone L4 (c.4500-3500 BC).

More northern routes were also used, although not often, because of the unfavorable climate:

**Hamilton Inlet - SW Cape of Greenland - Iceland**
**Cape Chidley - SW Cape of Greenland - Iceland**
each followed by:
**Iceland - Faeroes - Scotland**
Surely other crossings, which are less obvious, have been used in the remote past also:
**Cape Hatteras - Azores (directly)**
**Sable Island - Cape Race - Ireland (a daring crossing)**
**Cape Mercy - Greenland at 65°N (a very old route).**
In nearly all cases the sailing distances can be estimated accurately using this Kercado petroglyph.

This petroglyph on stone C1 is the oldest complete chart of the North Atlantic Ocean in the world. It has never been described previously to this article, except in Dutch in Ref.2. Probably, it was carved in three phases, finishing with the area around Bermuda in c.2200 BC. (We have given the glyph the late date of c.2200 BC because the glyph includes Bermuda, which was discovered at that date.) It is a proof that in the Bronze Age people visited America by sailing boat, and that they returned via the northern part of the ocean to the Old World. The nautical center of American Stonehenge in New Hampshire (c.2200 BC, Refs.6,20), at the latitude of Cape Sable, the south Cape of Nova Scotia (43°N), confirms this. The location of this petroglyph in Brittany proves that the sailors did not merely return to the Mediterranean Sea, but often visited or inhabited the European west coast.

## Other Petroglyphs at Kercado

The surface of stone C3 (Figs.1,3) is again the North Atlantic Ocean, dating from c.2200 BC. The lowest line is the Southern Crossing from the Cape Verde Islands to the NE coast of South America, followed by the coastal route to Central America at c.10°N. The second line above it is the way back from the East Coast of the US, via Bermuda and the Azores, to Iberia and the Mediterranean Sea at 35°N. (Above Bermuda, the south coast of Nova Scotia is indicated.) The third line is the 58°N latitude line from Norway to the east coast of South Greenland (the vertical line), followed by the crossing from the SW Cape of Greenland (at 61°N), to Cape Chidley in Canada at 60°N. The fourth line is again the 58th latitude line from Norway, followed by the 60°N latitude line beyond the west coast (the vertical line) of Iceland (the semicircle), to Cape Farvel (South Greenland) at 60°N. The highest point of the carving finishes at Cape Brewster, the east cape of Greenland, at 70°N.

The Tumulus of Kercado was built to commemorate and document the discovery of the Cape Verde Islands, c.4500 BC. The corridor of the passage grave (Fig.4, top) contains 4 southern and 5 northern menhirs, corresponding to the 4 southern and 5 northern Cape Verde Islands. As a consequence, menhir L4 symbolizes the SW island of Brava, and the petroglyph on this menhir L4 (Fig.4, bottom, and Fig.5) deals with the

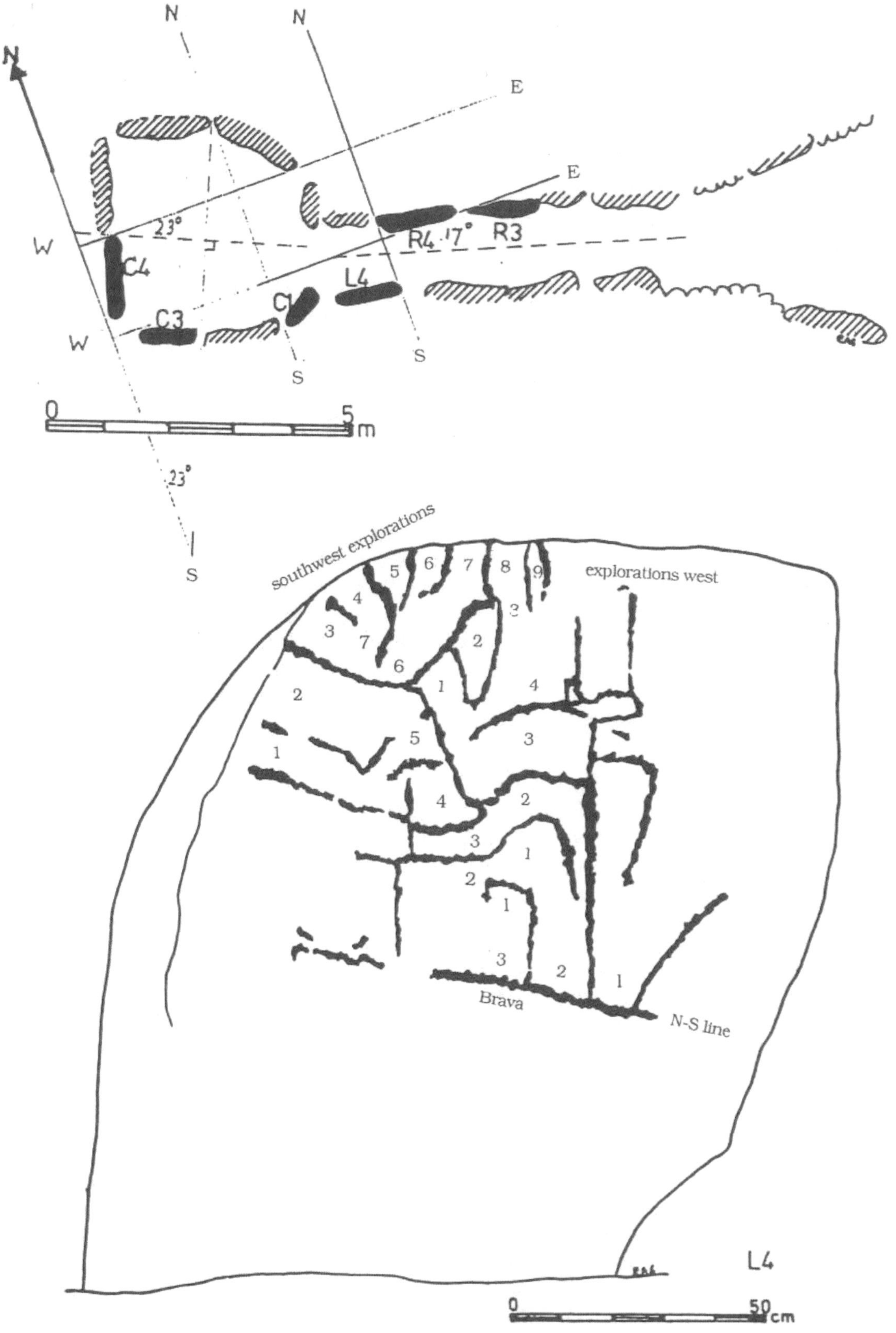


**Fig.5** Petroglyph on stone L4, with geographic meanings (4500-3500 BC). The sea was explored southwest of the island of Brava, over a distance of 7dl =777km.

exploration of the waters around and beyond this island which has a diameter of only c.10km. The distance lines in this old glyph each correspond to 111km, or 1° of latitude. When we compare the petroglyph with a map of the archipelago, the most probable scenario is, that explorations were focused in the southwestern direction, shown in the upper left corner on the stone. As a consequence, the long vertical line at the right points to the west, and Brava is located at the start of that line, where it meets the more or less horizontal NS line. Ocean explorations are shown in a variety of directions, looking for unknown lands. However, in the SW direction, we count a total of 7 spacings between the lines, which means that they explored the ocean in that direction over a distance of 7dl= 7° of latitude= 777km. The lower half of the inscription, corresponding to 4dl= 444km, may date from the construction of the monument, c.4500 BC. But the upper left part is probably contemporary with the stone semicircle around the tumulus of Kercado, c.3500 BC (Ref.3).

The west side of the burial chamber of Kercado (Fig.4, top), is made of two menhirs. The SW menhir C4 symbolizes the SW island of Brava, and the NW menhir C5 symbolizes the NW island of Santo Antao (Ref.1). In the upper right corner of menhir C4 (Fig.6, top, and Fig.7), is an old petroglyph. Both the menhir and the semicircle of the carving, exaggerated in size, relate to the island of Brava. At the right side below Brava, the T-shaped carving shows the east side of the archipelago, and the start of the sailing route back to the mainland. The right edge of the stone is the 16th latitude line, pointing to the west, between Brava at 15°N and Santo Antao at 17°N. The distances between the lines correspond again to 1dl= 1° of latitude= 111km. The petroglyph shows that the ocean was explored to the west over a distance of only 4dl= 4°of latitude, 444km, and south 4° to 16°N-4°= 12°N (c.4500 BC). Because of wind and current, the exploration in the southwesterly direction was more difficult.

The other inscriptions on this stone C4 date from c.2200 BC, after the discovery of America, because in the center of the stone we observe a complete description of the North Atlantic Ocean. The ocean has been divided into a well-known eastern part, and a largely unknown western part. The width of the eastern part is 2DL= 2 big distance lines= 20° latitude= 2222km, comparable with the width of the western part. The latitude lines are at levels of 10, 20, 30, and 40°N. The bottom line right is the Southern Crossing of the ocean, the left line is the NE coast of South America. At the left end of the 40°N latitude line is a junction representing the West Azores at 39°N.

In the upper north (top of middle glyph of Fig.7) we see at the right side a stylized crossing from Norway around the north coast of Iceland to East Greenland, at 65°N. At the left side we see the stylized crossing from Greenland, (north with the West Greenland Current) to Baffin Island and down again toward the Labrador coast. In the center is the south part of Greenland, with below it the crossing from Cape Farwell to Cape Chidley

**Fig.6 Top**: An old petroglyph in the upper right corner (c.4500 BC), and a few others elsewhere (c.2200 BC) on stone C4 (Ref.1). **Bottom**: Petroglyph on underside of capstone (6000-5000 BC), (Ref.1).

(Canada), at 60°N. This upper part of the glyph has been added later.

The thick vertical line glyph in the upper left corner is the north-south line along the west coast of Iceland. West of it is the south part of Greenland, with the crossing from the SW Cape of Greenland to the coast of Labrador, at the sailing direction shown of 45°SW. South of Labrador is meant to be shown that the best route is into Belle Isle Strait, rather than the outside route around Newfoundland. In the lower left corner are two parallel lines, which are the coasts of Florida. For the crossing to Cuba, it is shown that one is advised to follow the east coast of Florida until the north coast of Cuba, which has been indicated as a dot and a horizontal stroke. The bottom line of the petroglyph is the Tropic of Cancer at 23°N, which runs all the way back to the Old World. East of Florida, the east coast of this peninsula is repeated, and then islands of the Bahamas are shown, of importance for the way back via Bermuda to the Old World.

On the capstone of the burial chamber of Kercado is a beautiful stylistic petroglyph (repeated in other megalithic sites, Ref.2), of the so-called Small Mediterranean Sea, from Italy to Gibraltar, of c.4500 BC. Note that this example of the inscription is drawn so carefuly (see Fig.6, bottom, and Fig.8) that it includes the Gulf of Gabes on the Tunisian Coast. The bottom line is the north coast of NW Africa, from the Strait of Gibraltar in the west, to Cape Bon at Tunis; then the figure runs south and east past Gabes to Tripoli. The right side runs up from Tripoli via Malta to the east coast of Sicily, and the Strait of Messina. The horizontal upper line runs west from Messina, along the north coast of Sicily, the south coast of Sardinia, and south of the Baleares to the south coast of Spain in the west. The large ox-like bow to the north is the sailing route around Corsica and Sardinia. On the left end is an upward bump, representing the Gulf of Cadiz, beyond the Strait of Gibraltar. The lower part of this glyph is far more accurate than the northern part, indicating that in this time period, the southern side of the Small Mediterranean was far more important than the northern side, interesting in light of recent archaeological findings in the Sahara.

## Other Ocean Petroglyphs in Brittany

Figure 9 Above is a petroglyph of the Ocean with parallels of latitude and distance lines. It dates from after the discovery of America, probably c.2200 BC. The distance from Brittany (right upper corner, 47°N) to South Greenland (60°N) amounts to 3DL= 3333km, to Canada (at 55°N) 4DL= 4444km. The distance from America's Stonehenge (New Hampshire, 43°N, (Refs.6,20) to the Azores (39°N) amounts to 3DL= 3333km; from Cape Hatteras (35°N) to the Azores (39°N) 3.5DL= 3888km. Finally, the distance from Florida (Bimini, 26°N) (via Bermuda and the Azores) to the Canary Islands (29°N) amounts to 6DL= 6666km. These last three distances all concern return routes.

Figure 9, middle, is a petroglyph of the southern part of the ocean. The distance from the Cape Verde Islands (lower right corner, 15°N) to the

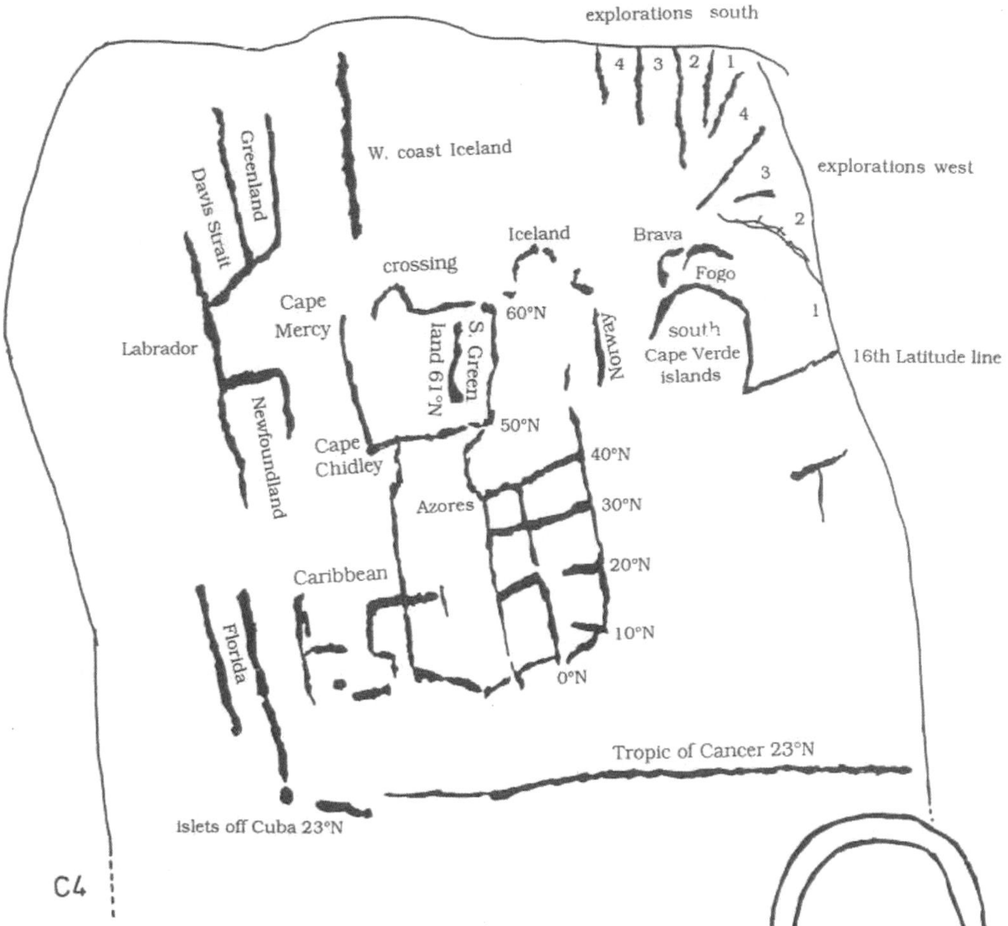


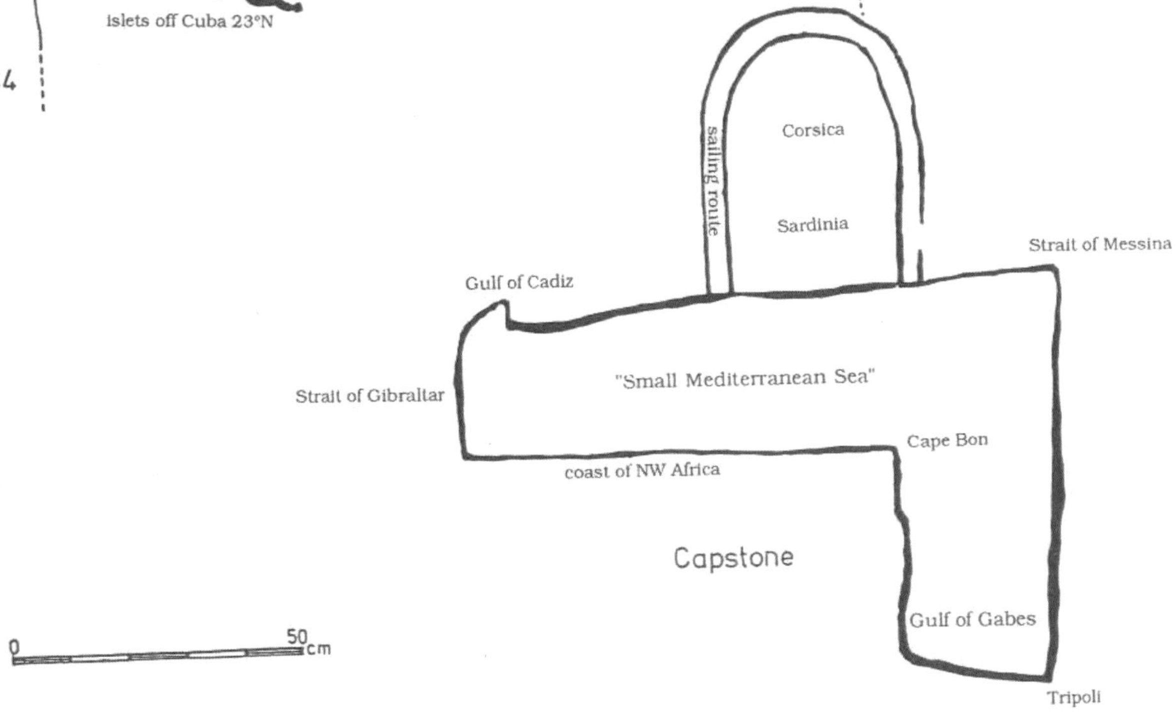


**Fig.7 Above:** Petroglyphs on stone C4, with geographic meanings. Exploration of the sea west of Brava (top right) over a distance of 4dl=444km (c.4500 BC), and inscriptions of other important crossings of the ocean, and sailing routes along Central and North America (c.2200 BC).

**Fig.8** Below: Petroglyph on capstone of the Small Mediterranean Sea, with geographic meanings (6000-5000 BC).

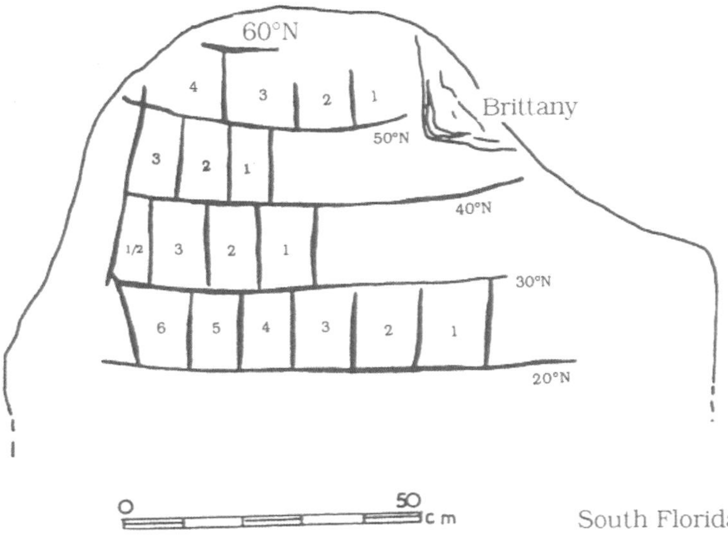


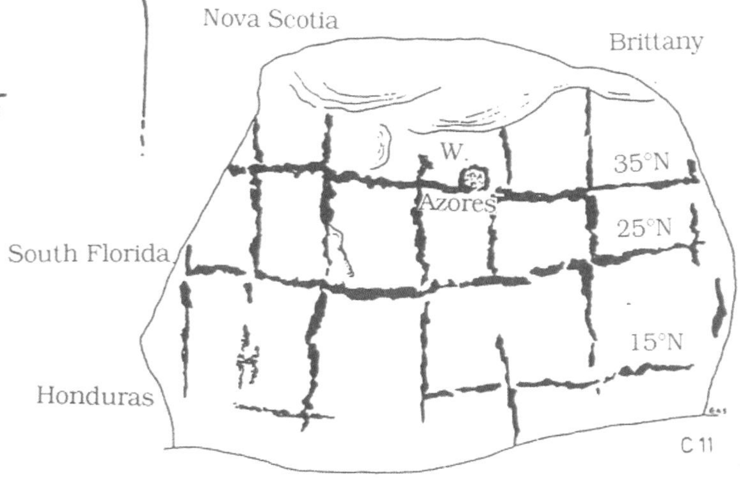


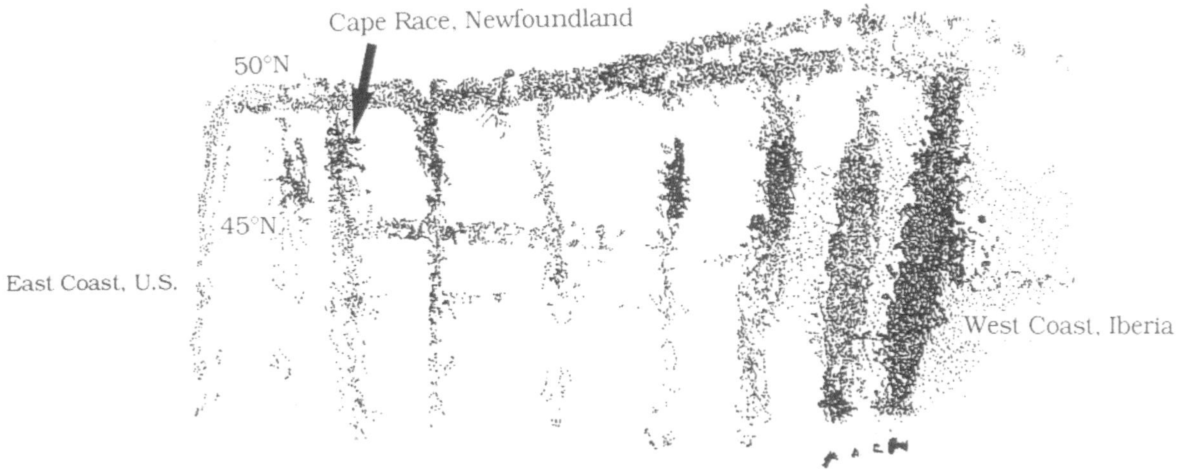


**Fig.9 Above:** The North Atlantic Ocean with parallels of latitude and distance lines. At the top, the distance from Brittany to Greenland (3DL), and to Canada (4DL). In the middle, the distance from the US to the Azores (3 to 3.5DL), and at the bottom, the distance from Florida to Africa (6DL). (Butten-er-Hah, Ile de Groix, Brittany, Ref.1, c.2200 BC)
**Middle:** The broad, lower part of the Ocean with parallels of latitude (15°, 25°, and 35°N) and distance lines. The upper corners are Brittany and Nova Scotia. In the middle the Azores. The bottom of the stone is the 10°N latitude line. (Poulguen, Penmarc'h, Brittany, Ref.1, c.2100 BC)
**Below:** A charcoal drawing in the Pileta Cave near Malaga in South Spain shows the distances of Sable Island, Cape Race, West Azores, East Azores, and Madeira to Iberia as 7,6,3,2, and 1 hDL respectively (1hDL= 5°of latitude= 555km). These distances are correct. This drawing was dated to 2394-1975 BC, or c.2200 BC by the C-14 method ("arrow shows the sampling zone"). (Ref.21)

north coast of South America is 2DL= 2222km, to the Antilles is 4DL= 4444km, and to Nicaragua is 6DL= 6666km. Between 25° and 35°N, the Ocean is broadest, 6DL= 6666km. The Azores in the middle of the Ocean are clearly indicated. From Florida (Bimini, 26°N) to the Azores (38°N) is 4DL= 4444km, and from America's Stonehenge (New Hampshire, 43°N, Refs.6,20) to the Azores (38°N) is 3DL= 3333km. From the Azores to the Strait of Gibraltar (36°N) is 2DL= 2222km.

# References


1. Twohig, E. Shee, The Megalithic Art of Western Europe, Clarendon Press, Oxford, 1981
2. Jonge, R. M. de, and IJzereef, G.F., De Stenen Spreken, Kosmos Z&K, Utrecht/Antwerpen, 1996 (ISBN 90-215-2846-0) (Dutch)
3. Briard, J., The Megaliths of Brittany, Gisserot, 1991
4. Richards, J., Stonehenge, 1979
5. Atkinson, R.J.C., Stonehenge, 1979
6. Tour Guide Map, America's Stonehenge, PO Box 84, No. Salem, NH 03073
7. Ferryn, P., NEARA Journal, Vol. XXXI, No. 2, p. 59 (1997)
8. Lenik, E.J., and Gibbs, N.L.,"The Frost Valley Petroglyph, A Catskill Mountains Enigma", NEARA Journal, Summer1999 (ISSN: 0149-25-51)
9. Fell, B., America BC, Pocket Books, Simon & Schuster, 1994
10. Bailey, J., Sailing to Paradise, Simon & Schuster, 1994
11. Thompson, G., American Discovery, Our Multicultural Heritage, Misty Isles Press, Seattle, 1994 (ISBN 0-9621990-9-5)
12. Mallery, A., The Rediscovery of Lost America, The Story of the Pre-Columbian Iron Age in America, E.P. Dutton, New York, 1951 (ISBN 0-525-47545-1)
13: Peterson, F.A., Ancient Mexico (1959)
14. Stuart, G.E., "New Light on the Olmec", National Geographic, Nov. 1993
15. Hall, G.P.D., Ocean Passages for the World, Hydrographer of the Navy, Ministry of Defence, Taunton, Somerset, 1973
16. Thom, A., "Megalithic Geometry in Standing Stones", New Scientist, Mar. 12, 1964.
17. Butler, A., Bronze Age Computer Disc, Quantum, London, 1999 (ISBN 0-572-02217-4)
18. Munck, C., The Code (publication of author, @PO Box 147, Greenfield Center, NY, 12833).
19. Thompkins, P., Secrets of the Great Pyramid, Harper & Row, NY, 1971 (ISBN 0-06-090631-6)
20. Lambert, J.D., America's Stonehenge, an Interpretive Guide, Sunrise Publications, Kingston, N.H., 1996 (ISBN 0-9652630-0-2)
21. Sanchidrian Torti, J.L., and others, "International Newsletter on Rock Art", No. 29, 2001 (pg.17, Fig.1)

# The Southern Crossing

## The Sailing Route from Portugal to Central America

(Petroglyphs of Chao Redondo, North Portugal, c.2200 BC)

Dr. R.M. de Jonge, drsrmdejonge@hotmail.com
J.S. Wakefield, jayswakefield@yahoo.com

### Introduction

The tumulus of Chao Redondo is located in the neighborhood of the little town of Viseu between the rivers Douro and Mondego in Portugal (Fig.2). On a loose slab, probably the former coverstone of the passage grave, is a petroglyph of remarkable size (170x60cm, Figs.1,3,4) (Refs.1-3). The slab was excavated in 1960, and "interpreted unconvincingly as a zoomorph". At first glance, the glyph looks like an elk, with broken big horns, prominent ribs, a big butt and a curly tail. However, like many other megalithic inscriptions, it is a geographic chart, and conforms to the usual megalithic cartographic methodology, such as use of distance lines and dual symbolisms.

The most important part of the petroglyph is situated at the far left side of the petroglyph. This is the "unknown area", or new area of primary interest. Here the notches are deep and the figure is complex. The tall vertical lines across the petroglyph are so-called "big distance lines" (Refs.2,3). These conceptual lines run exactly parallel with one another. They are comparable with big latitude and longitude lines which are seen on every globe today, except that the longitude lines in the figure do not come closer together toward the poles. The longitude lines are more like latitude lines. The distances between the lines have similar units: 1 big distance line= 1DL= 1 big latitude line= 10°of latitude= 1111km. This distance is equal to 10 (Egyptian) moira, a very ancient unit of measure (Ref.17).

### The Route to Central America

At the right side there is a circular area (Fig.3). It has the shape of the Iberian Peninsula, where the petroglyph is located. In the northeast of the circle, the line is a little rough, because of the Pyrenees Mountains on the border with France, and in the south the line is thin, because of the narrow Strait of Gibraltar. So this circular area represents Iberia.

To the left side of this area, a dark line (#7) runs downwards. This line starts at the latitude of North Portugal, where the dolmen of Chao Redondo is located. This is the stylised and then well-known route to the south along the west coasts of Portugal and North Africa down to a right turn at the left. This is the crossing from Cape Verde, the western-most point of all the mainland, to Brava, the southwest Cape Verde island, both at a latitude of 15°N (Refs.2,3). At the end of this line the first vertical distance line (#6) appears, showing that we are at a distance of 1DL= 1111km west of Portugal, and this is very close to being accurate!

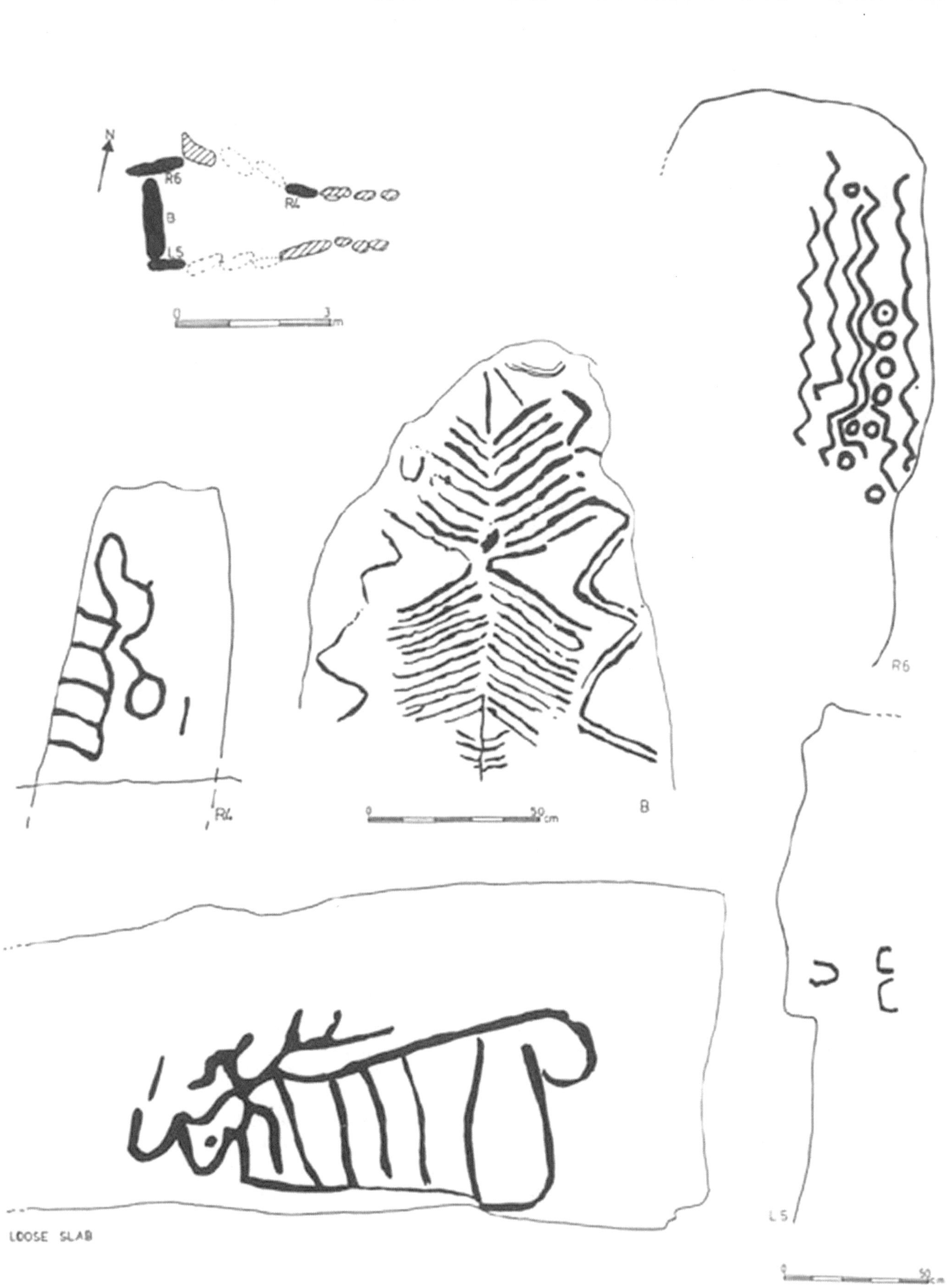


**Fig. 1** The petroglyphs of Chao Redondo (Ref. 1), which are on the insides of the black stones in the groundplan of the passage grave (upper left). The monument has 17 upright menhirs, corresponding with cultural centers in Central America at 17°N (Chao Redondo, Talhadas, Sever do Vouga, North Portugal, c.2200 BC)

Along the lower edge of the stone a very thin line runs to the left over three distance lines. This is a scary mid-ocean (therefore thin line) Southern Crossing of the Atlantic with the tradewinds and currents over a distance of 3DL= 3333km. To emphasize that it is a crossing over water, the two thin central distance lines (#4&5) have not been inscribed all the way to the bottom line. On the contrary, the distance lines at the beginning and at the end of the crossing (#6&3) have been engraved deeply up to the very ends, because at both ends there is land. The fact that the thin, horizontal line runs so close along the lower edge of the stone, confirms that we are dealing here with the Southern Crossing.

From the length of the crossing shown on the stone (3 big distance lines), and the westerly sailing direction, it can be concluded that sailors did not take the shortest crossing to Cape Sao Roque, the NE point of Brazil at 5°S. This cape is at a distance of a little more than 2.5DL= 2777km from Brava in the Cape Verde Islands, and is 60°SW of Brava. The inscription recommends cutting the corner, to disembark at the shown distance in French Guyana (bottom of distance line #3), at c.5°N. That is a distance of 3DL= 3333km from Brava, and c.20°WSW of Brava. Looking at a modern chart shows that this was good advice. As the inscription indicates, we now are at a distance of approximately 4DL= 4444km west of Portugal.

After the crossing of the ocean, the petroglyph shows a thick horizontal line, which is the sailing route along the north coast of South America. Heading west, we first arrive at a distance line (#2) which is in a slanting position. Distance lines are parallel lines on the surface of the earth, in a grid pattern. When many lines are placed behind the other, as shown in this petroglyph, especially the outermost lines curve, because of the curvature of the earth. It is like putting a piece of fishnetting over a big ball: you get distortions of the netting around the edges. Distance line #2 shows this effect markedly, as should be expected. Another illustrative example of curving DL is the petroglyph of the North Atlantic Ocean at Kercado (Refs.1,21).

When we continue heading west, we arrive at a thick vertical line (#1), that has not been extended all the way to the bottom line, which is quite remarkable. Primarily, this vertical line (#1) is an ordinary distance line, but it also symbolizes the Antilles islands sailing route. The "vertical" part is the Lesser Antilles, and the "horizontal part" is the Greater Antilles. The line is so thick, because it is a coastal sailing route as well as a distance line. It is not connected to the bottom line, because when heading for Central America, one is dissuaded from taking this route. The islands are beautiful, and are easy to sail on a reach, with the easterly trade wind between them. But the route along the north coast was direct to Central America, and being coastal, was felt to be more safe.

## Central America

Most sailors continued to travel to the last vertical line, which represents the east coast of Nicaragua. This east coast runs due north, as indicated by

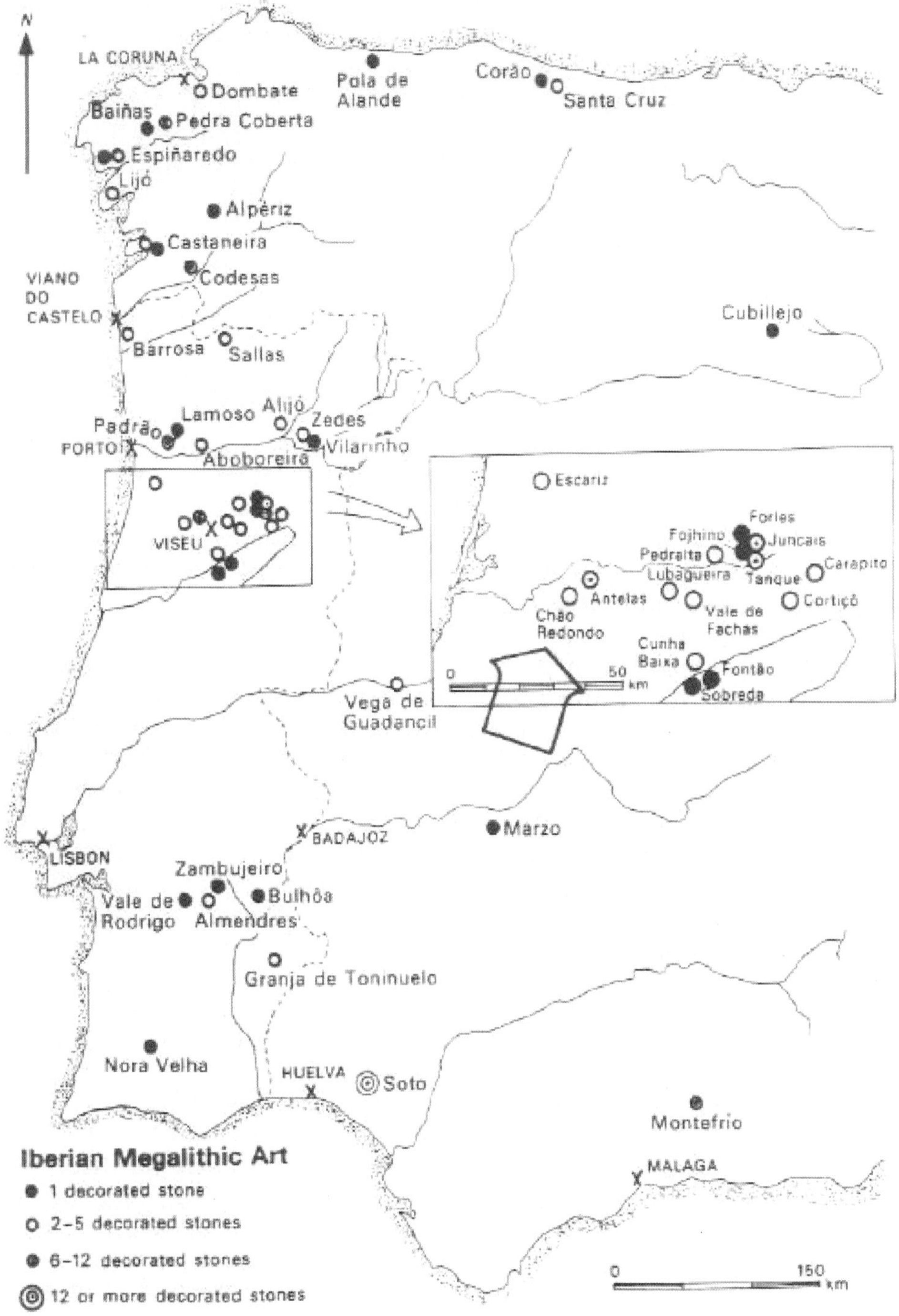


**Fig.2** Map of central and west Iberia, with geographic distribution of megalithic "art", showing the location of the Chao Redondo petroglyphs near Viseu in North Portugal (From Towhig, Ref.1).

the petroglyph. But this vertical line is also the last distance line. According to the inscription we are 7 distance lines= 7DL= 7777km west of Portugal, which is correct! The error is less than 1/4 DL= 277km (Ref.2). The last vertical line is followed upwards to the junction, but in reality we range the east coast of Nicaragua up to Cape Gracias a Dios, the NE cape on the border of Honduras. It may be noticed that from this point a continuing line runs at the upper side of the petroglyph toward the Iberian Peninsula in the northeast. Below this line 7 distance lines are present, confirming the distance of 7DL= 7777km, which is correct.

The short piece of line to the southwest, from distance line #1 toward the junction is the sailing route from the Windward Passage between Hispaniola and Cuba (Greater Antilles) to the island of Jamaica, followed by the crossing, usually with the wind astern, along some of the tiny Mid-Caribbean Islands to Cape Gracias a Dios, the NE cape of Honduras (c.650km, Ref.27). The inscription indicates that the distance between Windward Passage and Cape Gracias a Dios is 1 big distance line= 1DL= 1111km, which is correct. The line #1, that represents the island-hopping Antilles route has been indicated thickly, so we must conclude that sometimes people used this route.

At the junction, we now follow the line to the left, but in reality we skirt the north coast of Honduras to the west. Here, the maker of the petroglyph makes an error! The north coast of Honduras runs due west, while a bay towards the north has been engraved. In his enthusiasm, he mistakes it for the Gulf of Campeche, which will be visited afterwards. He calls this gulf the "first bay". Here is the old culture along the north coast of Honduras, Guatemala and Belize (Refs.2,7-12). This is the first goal of these voyages. The dot is the Bahia Islands, situated along this coast. Such coastal islands were important for trading ships, because fully packed boats could safely disembark there without too much danger for theft of cargo. When we follow the line in the inscription upwards to the second junction, we are actually sailing due north along Belize and the peninsula of Yucatan to Cape Catoche, the NE cape of Yucatan.

At the junction, again we choose the line to the left. But in reality we sail to the west, around the peninsula of Yucatan to the old civilization around the Gulf of Campeche, which is engraved in the right way (Refs.2,7-14). This is the second goal of these voyages. The furthest left line of the glyph has not been extended further north than the level of the north coast of Yucatan. This means that this civilization did not extend further north than the latitude of c.22°N, which is the mouth of the Panuco River, at today's city of Tampico, Mexico. According to Juan de Torquemada, a late sixteenth-century Spanish chronicler, legends tell that it was here that the first SunKing, Quetzalcoatl, arrived in America (Ref.26). At the upper left of the inscription, a later visitor to the dolmen of Chao Redondo engraved a thin vertical line with a different tool, showing the coastal area to the north up to Corpus Christi, Texas. (This line was probably engraved 2000-1500 BC.)

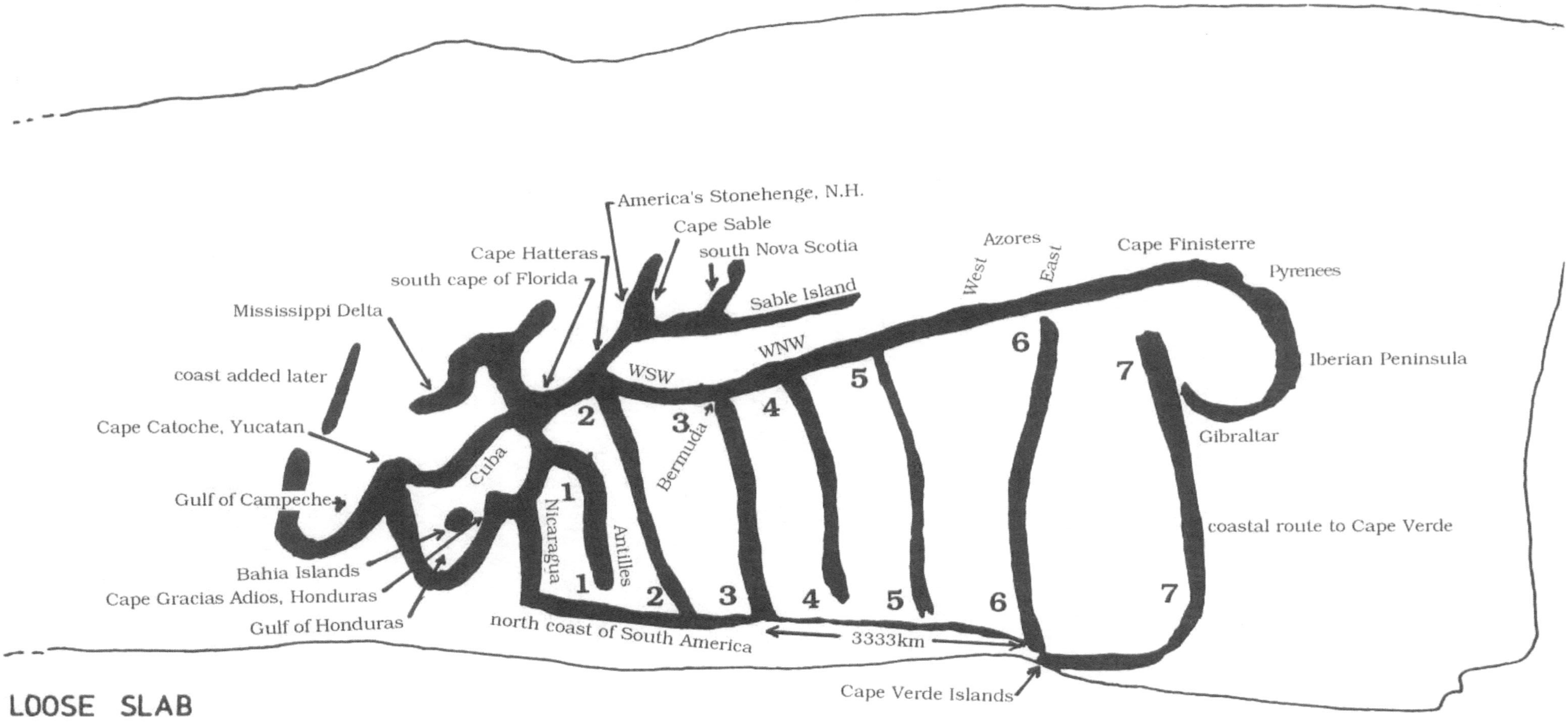


**Fig.3** Petroglyph (with geographic meanings) of the sailing route across the North Atlantic Ocean from the Iberian Peninsula (right) to Central America (left). The vertical lines are "big distance lines", each corresponding to a distance of 10° of latitude, or 1111km. It is the route from Portugal in the east via the Southern Crossing (3333km) to the Gulf of Campeche in the west, and the way back from Cape Hatteras (US) in the west, via Bermuda and the Azores, to the Iberian Peninsula in the east. (Chao Redondo, Talhadas, Sever do Vouga, North Portugal, c.2200 BC)

## Departure from Central America to the North

From the last junction, we now follow the line to the right. This is the small crossing (c.200km) from Cape Catoche, the NE cape of Yucatan, to the SW cape of Cuba, followed by the sailing route along the north coast of Cuba as far as the Tropic of Cancer (also shown at America's Stonehenge, New Hampshire, c.2300 BC, Refs.4,20). A joining line from the right, from Cape Gracias a Dios, along the tiny Mid-Caribbean Islands, indicates the less important route via Windward Passage, also along the north coast of Cuba, to the same point at 23°N. However, it is not called Windward Passage for nothing. Although it seems that the petroglyph shows it as a return route, we assume it was not very popular, because the long crossing from Honduras to Jamaica (650km) was against the winds, especially during winter months. The current is always flowing against the wind in the Windward Passage, which makes the water there famously rough. (The tradewinds of the Southern Crossing are forcing a lot of water into the Caribbean further south, and some of it is flowing out here.) Of course, sailing at the proper seasons is very important, and their knowledge of such information, whatever it may have been, is not shown on this inscription. In any case, the inscribing of this crossing to Jamaica is proof that these folks had pretty fully explored alternative routes.

The two crossing routes described above come together in one point, which is deeply inscribed. This is the north coast of Cuba, at the latitude of the Tropic of Cancer, 23°N. When the deeply inscribed notch in the petroglyph is followed upward, and to the left, we cross the Florida Straits from Cuba to Florida (c.200km), and we skirt the coast along the Gulf of Mexico to the west. The exaggerated zigzag at the end of this line is the sailing route around the Mississippi Delta. (The Three Rivers Petroglyph in New York (Ref.5), c.1500 BC, illustrates the river trade from New York all the way to the Mississippi Delta.)

For a moment, focus at the point of the inscription representing the south cape of Florida (Fig.3). From here a continuing line runs to the right, toward the circular area symbolizing Iberia. Below this line are 6 distance lines. This indicates that the direct distance (as the crow flies) between the south cape of Florida and the west coast of Portugal equals 6DL= 6666km, which again is correct. The error is less than 1/4 DL= 277km.

Following the dark line to the right over merely one distance line, we sail from the south cape of Florida at 25°N, along the American east coast to Cape Hatteras at 35°N. The real distance is 35°-25°= 10°of latitude= 1DL = 1111km! The remaining 5 distance lines to the right again provide the direct distance to Portugal, this time between Cape Hatteras and the west coast of the Iberian Peninsula, 5DL= 5555km. Again, the error is less than 1/4 DL= 277km.

## The Way Back from Cape Hatteras

Starting at Cape Hatteras, first the continuing line to the right runs over one

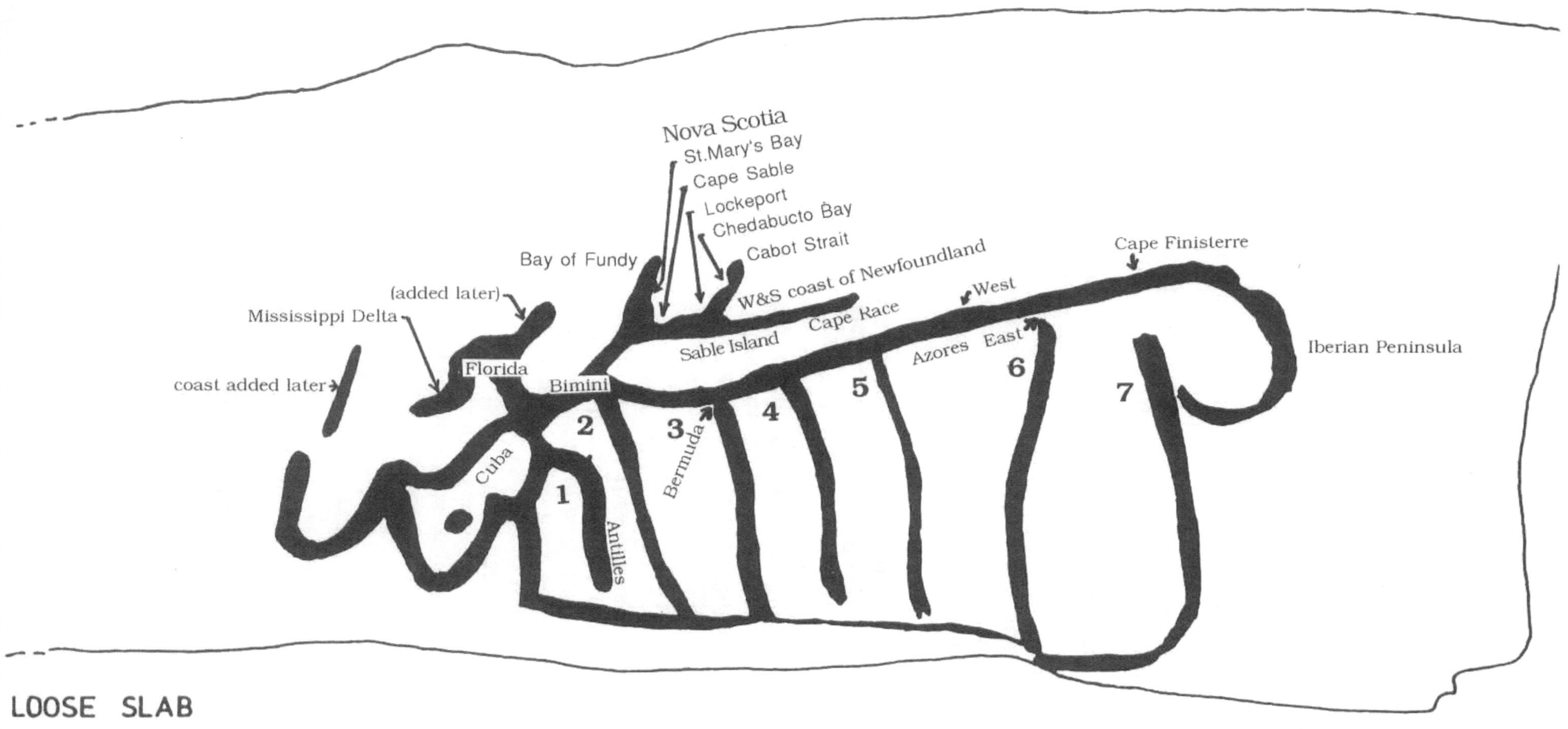


**Fig.4** Same petroglyph as Fig.3, with geographic meanings. The alternative return routes from Nova Scotia via Sable Island and the Azores, and later from the Bahamas via Bermuda and the Azores, to the Iberian Peninsula. (Chao Redondo, Talhadas, Sever do Vouga, North Portugal, c.2200 BC)

distance line WSW, and next over the other distance lines WNW. This encodes the sailing route from Cape Hatteras at 35°N in the direction of Bermuda at 32°N, and the next sailing leg via the Azores at 38°N in the direction of Cape Finisterre, NW Spain at 43°N (see also Fig.8). The distance from Cape Hatteras to Bermuda is 1DL= 1111km, which is correct. This part of the inscription shows, by the way, that the petroglyph was made after the discovery of Bermuda, c.2200 BC (see America's Stonehenge, N.H., a site greatly affected by the discovery of Bermuda, Refs.4,20).

In practice, Bermuda may have not been visited very often, or only in emergencies. The West Azores in the middle of the ocean are much more important. If we look at distance lines #5 and #6, it appears that they diverge considerably, because the maker of the carving has allowed for the West Azores, which are located halfway between the two lines. According to the inscription, the sailing distance from Cape Hatteras to the West Azores amounts to 3.5DL= 3888km, and that is correct. So, at the time of this petroglyph, this route was the most important return route across the ocean to the Old World.

The remaining "half line" toward line #6 is the sailing distance from the West Azores to the East Azores, half a DL, or 1/2 DL= 555km, which is correct. So distance line #6 points to the East Azores! This line #6 curves at the upper side to the right. This means that the last important sailing distance is less than one big distance line. And indeed, the sailing distance from Santa Maria (East Azores) to Madeira is less than 1DL (1111km), namely c.900km. However, this distance line #7 has not been extended all the way to the upper line. As a consequence, sailors were to be disuaded from sailing to Madeira, which was a long way out of the way on the return trip. Instead, one is recommended to travel directly to the Iberian Peninsula, as indicated. This sailing distance is more than one distance line, namely (1 1/3 DL)= c.1400km, as shown correctly at the upper right side of
the glyph. Note that a blank is left between the last vertical line #7 and the upper line, so it extends to the additional circular area (the Iberian Peninsula), so the total distance becomes one and a third. (Compare the heights of the enclosed areas, and see the one-third ratio.)

## The Alternative Way Back

The upper left or "Elk horn" part of this petroglyph has a second symbolism, or dual meaning. Let us return to the point in the petroglyph that represents Cape Hatteras, at the north end of distance line #2 in Fig.3. If we follow the slanting line upwards, we can choose an alternative way back across the ocean to Portugal. In this case, we start sailing north along the American coast. According to the inscription, we arrive at a junction, which is correct. We have to sail around the peninsula of Nova Scotia. The details of the carving clearly show the geographic characteristics along the Bay of Fundy, and the east coast (see Fig.4). The south point of Nova Scotia, Cape Sable, is at the same latitude, and very close to the most important megalithic monument of North America, America's Stonehenge, New

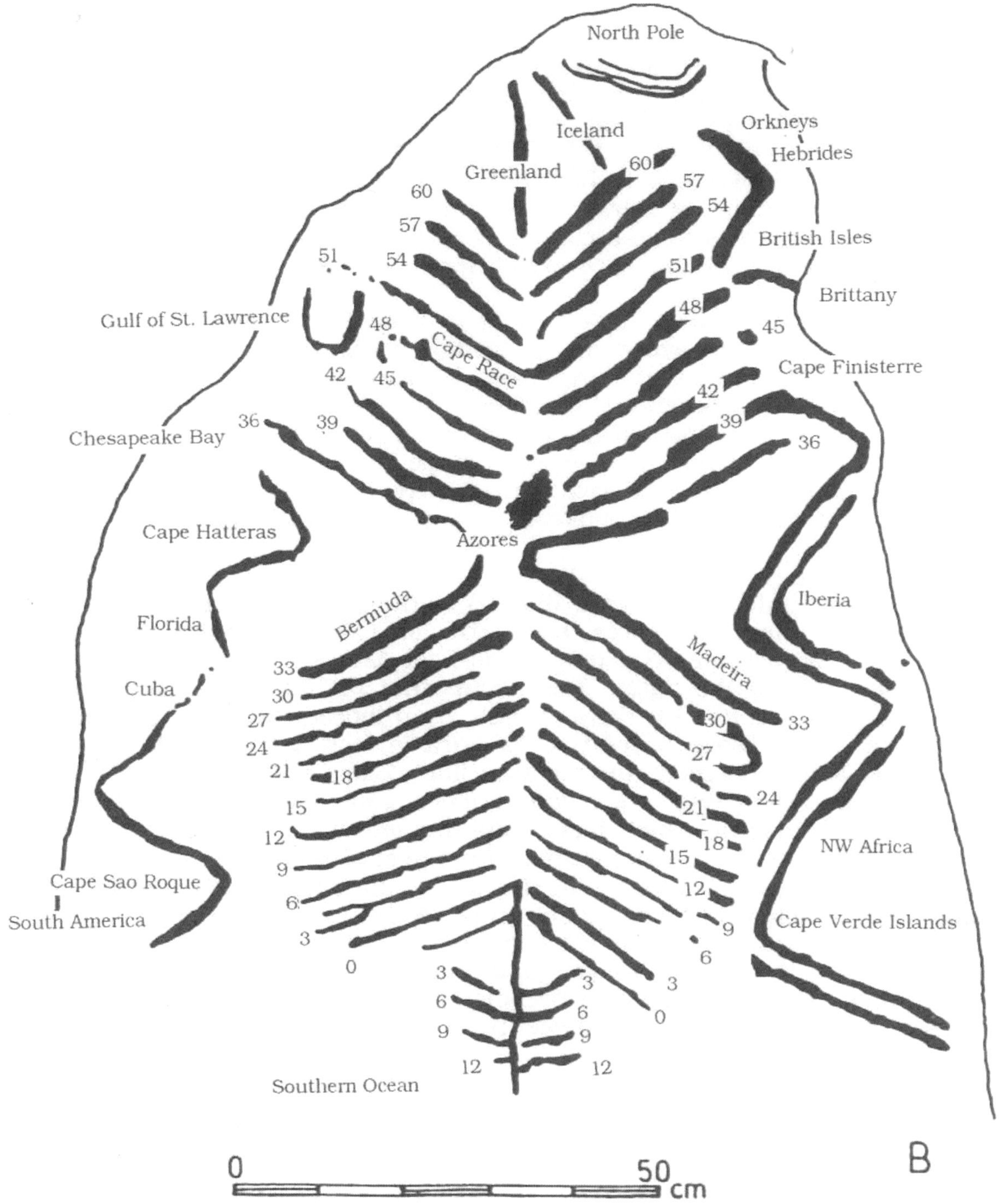


**Fig.5** Petroglyph (with geographic meanings) of the North Atlantic Ocean with in the middle the Azores. Right, the Old World with the Iberian Peninsula and NW Africa, left, the New World. Iceland and Greenland are above. Latitude lines are shown each 3°. The New World can be reached from West Africa to South America or via Iceland and Greenland. The return routes are via Bermuda, the Azores and Madeira, or from the Gulf of St. Lawrence (Cape Race), via the Azores to Cape Finisterre (Iberia). (Chao Redondo, Talhadas, Sever do Vouga, North Portugal, c.2200 BC)

Hampshire, which deals with the problem of crossing the ocean to the Old World (Refs.4,6,20).

To the right (Fig.3), the route reaches a second junction, but the lowest line continues to run almost horizontally. So here, one is advised to sail from the south coast of Nova Scotia at 43°N to the east, to Sable Island (c.300km) at 44°N and beyond. The first "latitude line" started near the south point of Florida at c.25°N, the second horizontal line started at Cape Hatteras at 35°N, and the the third horizontal distance line starts at Nova Scotia at 45°N. The third vertical distance line (#3) points via Bermuda to the north, to Nova Scotia, as inscribed in the Devil's Head petroglyph in Maine (Ref.22, c.2200 BC). The distance from Cape Hatteras to the West Azores amounted earlier to 3.5DL. According to the petroglyph, the distance from Nova Scotia to the West Azores now amounts to only 3.5-1= 2.5DL= 2777km, which is correct. The upper (distance) lines of the petroglyph confirm this value (Fig.4): from the Bay of Fundy to the West Azores is 2.5DL= 2777km.

While sailing from Nova Scotia to the east, to Sable Island and beyond, people oriented on Newfoundland. According to the petroglyph (see Fig.4), they did so over a distance of almost 2DL= 2222km (from line #3, pointing at Nova Scotia, you cross two line segments to line #5, opposite the tip of Cape Race). This is because Newfoundland is the easternmost land area of North America, and Cape Race, the east cape at 47°N, is only 2° north of the 45th latitude line. In cases of emergency, ships were able to turn in to Newfoundland. So the second meaning of this part of the petroglyph is a representation of Nova Scotia and Newfoundland, and the tip represents Cape Race, the easternmost point of North America. Stone L5 in Fig.1 (also) shows the crossing from Newfoundland to the 2 islands of the West Azores, Flores and Corvo (c.2200 BC).

## The Mississippi River and the Return Route from the Bahamas

The left "elk horn" portion of the petroglyph has a dual meaning. Look again at the line which represents the sailing route along the southeast coast of the United States, to the big black route junction labeled "south cape of Florida" in Fig.3. The "T" shaped carving rising to the upper left probably does not belong to the original petroglyph. It was added later, after more explorations had taken place. The right side of the "T" is difficult to identify, but in the first instance it is the Mississippi River at a distance of 1DL= 1111km from the East Coast, which is correct. (Compare the "west and east coasts of Nova Scotia" on Fig.4, which turn out to be distance lines as well.) This part of the glyph extends beyond 45°N. Actually, the whole top of the "T" is a serpent having 4 zigzags or sections, corresponding to a waterway having a length of 4 half DL= 2DL= 2222km (equal to the width between the distance lines #2 and #4). Now, the first two sections in the south are the Mississippi River, the third section, emphasized in the glyph, is the Illinois, and the fourth slightly different section is Lake Michigan. Thus this later-added "elk horn" part of the petroglyph clearly refers to the ancient copper trade from Upper Michigan to the south. (Compare this to

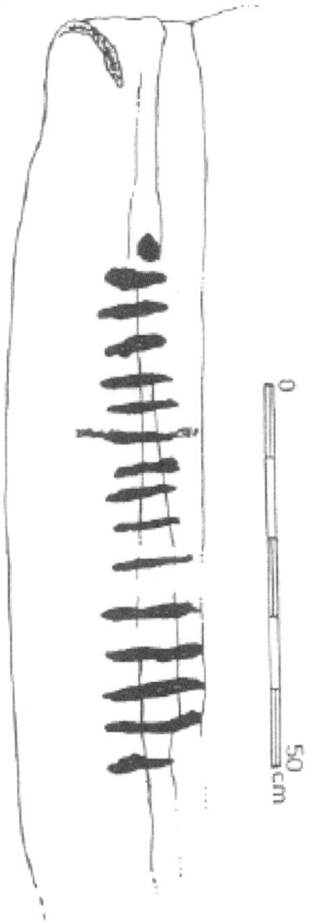


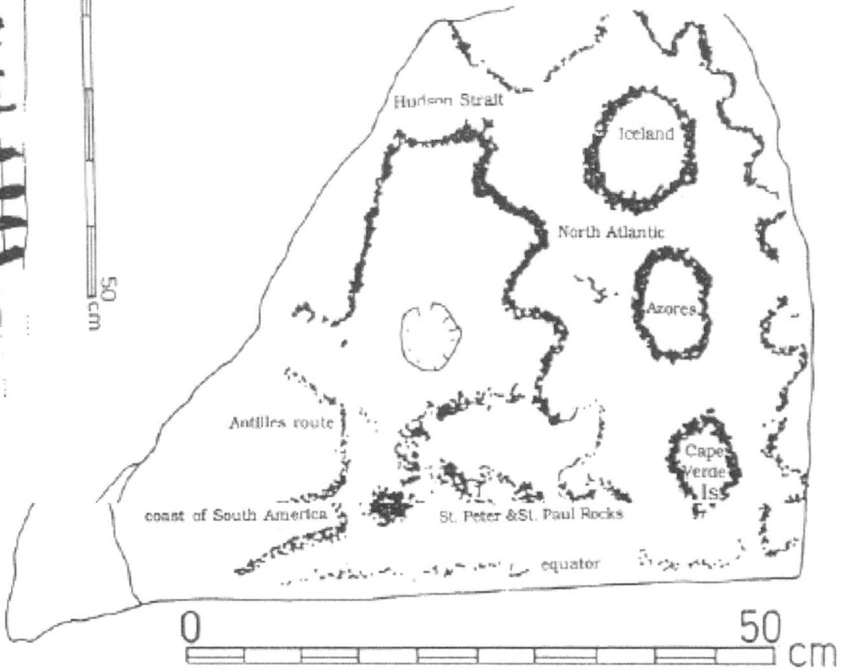


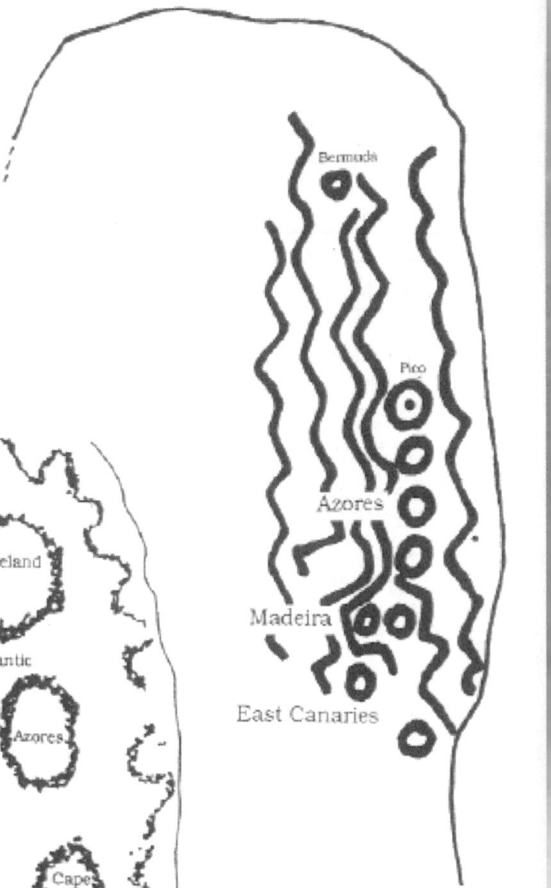


**Fig.6 Left:** Brava (the dot at the top), the southwest Cape Verde island, at 15°N (the 15 latitude lines). The Southern Crossing starts at this island. (Cunha Baixa, North Portugal, Ref.1, c.2200 BC)

**Fig.7 Middle:** At the right side the North Atlantic Ocean from the equator to c.60°N, with the Cape Verde Islands, the Azores, and Iceland shown as three circles. At the upper left, Hudson Strait into Hudson Bay. Below, a later inscription, showing the coastal islands of St. Paul and Fernando de Noronha, important in the Southern Crossing. Left, the coast of South America and the Antilles route. (Drumreagh, NE Ireland, Ref.1, c.1900 BC)

**Fig.8 Right:** First meaning: The 9 islands of the Azores, with the main island of Pico (dot), in the middle of the ocean. Second meaning: Bermuda (top), the Azores (most important, with dot), the two islands of Madeira, and the two eastern Canaries (bottom). The sailing distance of this route from Central America (left) to the Old World (right) amounts to 5.5DL= 6111km (the 5.5 wave lines). (Chao Redondo, Talhadas, Sever do Vouga, North Portugal, Ref.1, c.2200 BC)

the dragon petroglyph in Embden, Maine, Ref.23, c.1500 BC, dealing with the copper trade on the East Coast.)

The space between the right branch of the "T" and the "East Coast" now closely resembles the peninsula of Florida, and the point that previously represented Cape Hatteras (at the north end of distance line #2, Fig.3), now represents the NE Bahamas (Bimini or Great Abaco or both), see Fig.4, situated just off the east coast of Florida. There are strong indications, that slightly later in time the northern Bahamas had become the most important point of departure for the crossing of the ocean. Probably this had happened in the first half, but surely in the second half of the 2nd millennium BC.

## North Atlantic Ocean Sailing Routes

Chao Redondo has several other geographic petroglyphs, which further explain trans-atlantic sailing routes. The glyph on stone R4 (shown in Fig.1) we call the "Sailing Route of The Upper North", which we have written about in detail. This route follows the coasts of Ireland and Scotland, via Iceland, to the south point of Greenland, and then crosses the Davis Strait behind Greenland to Baffin Island (Canada, c.2200 BC).

The petroglyph on the backstone "B" (150cm wide, 220cm high) is shown in Fig.5. It represents the North Atlantic Ocean, with the Old World in the east, and the New World in the west (c.2200 BC). The most important part of the ocean is between the equator (at 0°N) and Cape Farvel, the south point of Greenland (at 60°N). For that reason, 20 latitude lines are engraved every 3° (20x3°=60°) to show this portion of the ocean in detail.

When the upright stones of the dolmen of Chao Redondo are counted in the direction of the movement of the sun (see Fig.1), as the SunGod wishes it, the backstone B is stone #9, corresponding to the 9 islands of the Azores. The Azores are represented by the big prominent dot in the middle of stone B (Fig.5). The return routes via the Azores at 38°N are the most important return routes, and these are the focus of the petroglyph. The makers of this carving clearly distinguish between a southern and a northern part of the ocean, because the Upper North route is so different from the important Southern Crossing, **and because they have learned that there are two (northern and southern) return routes which both use prevailing winds and currents to the Azores** that are opposite the winds and currents of the Southern Crossing.

The **southern** return route is from Florida, or Cape Hatteras, via Bermuda (at 32°N) and the Azores to Madeira (at 33°N) or the SW point of Iberia. This petroglyph of Chao Redondo contains one of the best carvings of this return route (see also Fig.8). The **northern** return route is from the coast above Cape Hatteras, via Cape Race (at 47°N) and the Azores, to Cape Finisterre (NW Spain, at 43°N). Other destinations are Brittany and SW Ireland (also shown). The glyph dramatically illustrates that they distinguished between southern and northern return route portions of the ocean.

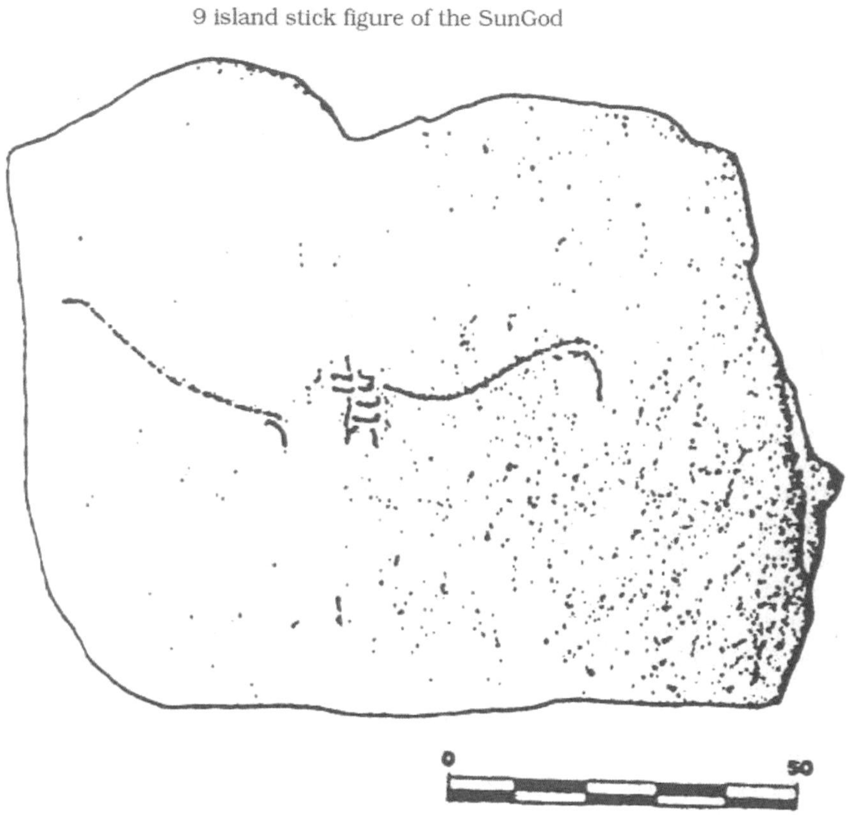


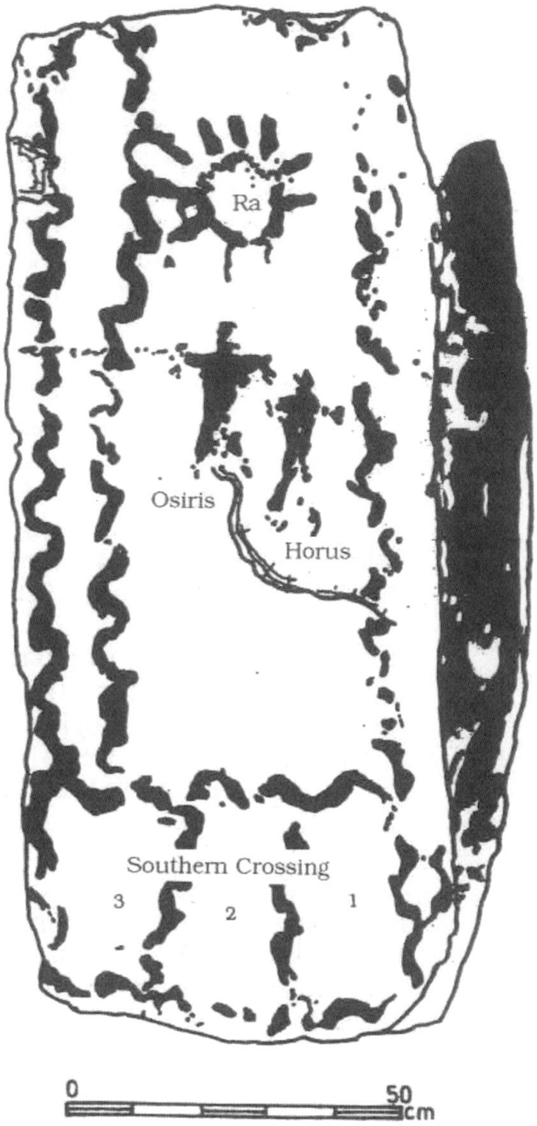


**Fig.9 Left:** The SunGod Ra (in the middle) involved in his difficult "Royal Crossing" of the Atlantic Ocean. Right, the crossing from NW Spain to the Azores (middle, 9 lines, 9 islands), left, the even longer crossing to Newfoundland, the east point of North America. (Portillo de las Cortes, Guadalajara, West Spain, c.2000 BC) (M.Devignes, Ref.16)

**Fig.10 Right:** The Atlantic Ocean, with the Southern Crossing of 3DL= 3333km. Below SunGod Ra, the Egyptian gods Horus and Osiris crossing their difficult windward route, via the Azores and Newfoundland, to Tuat, the realm of the dead in the west, at the other side of the waters, in the land where the sun goes down. (in red ochre, Porto University, Portugal, Ref.1, c.1900 BC)

The most important route to the west is the Southern Crossing between the Cape Verde Islands and Cape Sao Roque, the NE point of South America (indicated). At the strart of this route people sail along the coasts of Iberia and North Africa to the south. For this reason the coastal waters are here engraved over 3dl= 333km. (The starting point is the island of Brava, the SW Cape Verde Island at 15°N, as shown in Fig.6.) Some latitude lines in the Southern Ocean are shown as well, because Cape Sao Roque at 5°S is situated on the Southern Hemisphere. The dolmen of Chao Redondo has 17 upright stones (see Fig.1), corresponding to the center of the United Central American Empire at 17°N, halfway between the north coast of Honduras at 16°N, and the south point of the Gulf of Campeche at 18°N. This area is the goal of these voyages. The alternative route to America is the Crossing via the Upper North, from the British Isles via Iceland and Greenland, also included in this petroglyph.

It may be noted, that west of Cape Race the Gulf of St. Lawrence has been engraved. Possibly this glyph is related to the copper trade from Lake Superior, that originally took place via this inland sea. The Embden dragon petroglyph of Maine also deals with the copper trade in this region (Ref.23). All petroglyphs of Chao Redondo date from just after the discovery of America via the Atlantic (c.2500 BC), but also after the discovery of Bermuda, c.2200 BC, shown in America's Stonehenge, and the Devil's Head petroglyphs in Maine (Refs.4,20,22). A charcoal drawing in the Pileta Cave near Malaga in South Spain (Fig.9, Chapter7, Ref.24) shows the distances of Sable Island, Cape Race, West Azores, East Azores, and Madeira to Iberia as 7, 6, 3, 2, and 1hDL (1hDL= 5°of latitude= 555km). These distances are correct. Using the C-14 method, this drawing is dated to c.2200 BC, too.

Figure 7 is a petroglyph of the North Atlantic Ocean from Drumreagh in northeast Ireland. At the right side is the Ocean, with the Cape Verde Islands, the Azores, and Iceland represented by three big circles. Left and right of them are the continental coasts, symbolized by wave lines from the equator (indicated) to the 60th latitude line (every 5°), in the north changing into distance lines (3DL= 3333km up to South Greenland). In the northwest we see Hudson Strait into Hudson Bay to the mouth of the Albany River, possibly also of importance for the copper trade from Lake Superior (Refs.7,8,15).

At the left lower side of Fig.7 is a later petroglyph with details about the southwest part of the Ocean. The two circles are the islets of St. Paul and Fernando de Noronha, so important for the Southern Crossing. The connection line with the western coastal waters of the original petroglyph confirms that these are American coastal islands. West of them, the characteristic coastline of South America has been carved, including the Antilles Route. Northwest of this route is a possibly natural circle, probably representing Bermuda, discovered from the American coast (c.2200 BC), important for the return route to the Old World.

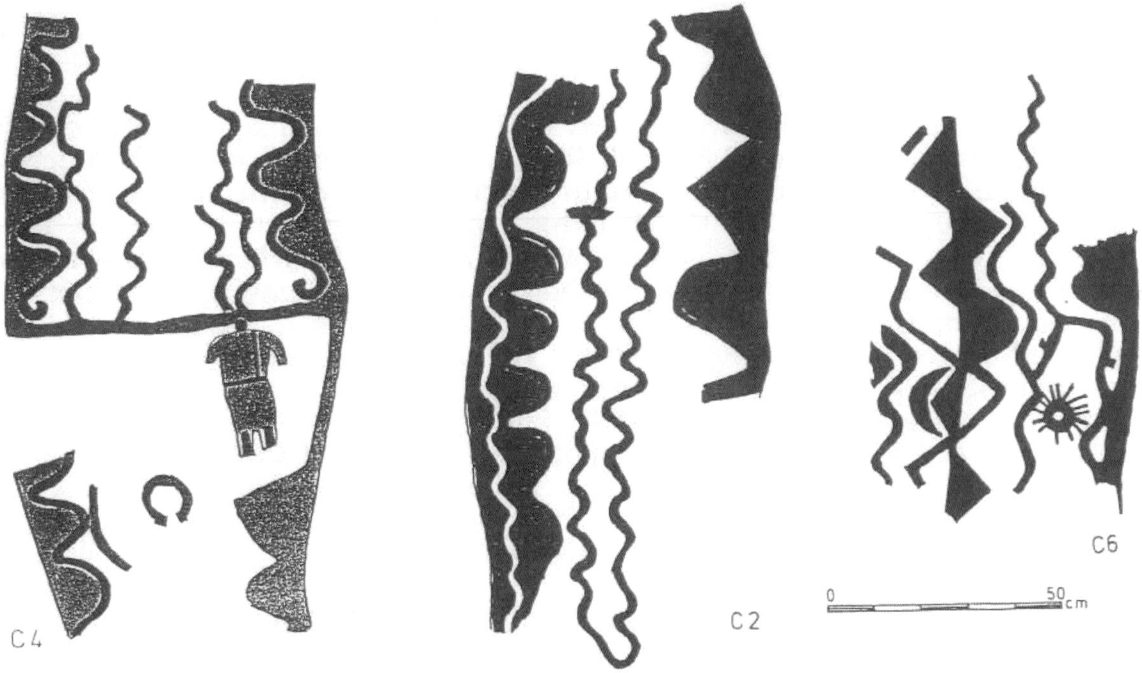


**Fig.11** Three paintings of the Ocean, with literal, but above all religious meaning: The King (C4) leaves the land of the living (right) of the Egyptian god Horus (the sun, C6). After the crossing of the waters (the ocean) he reaches the land of the dead (left) of the Egyptian god Osiris (the moon, C6). But he returns with a boat from the realm of the dead (center, C2), for he has eternal life (C4). Red ochre with some black lines. (Antelas, Viseu, North Portugal, Ref.1, c.1800 BC)

**Fig.12** The nine gods of the Tuat (the "Underworld" in the west) explain why it was considered encouraging that the Cape Verde Islands and the Azores were each found to be composed of nine islands. The number of the Great Goddess Isis was also nine "for the four cardinal points of heaven and earth joined by the Tree of Life, the Sacred Sycamore of Hawthor" (Ref.18). (Louvre, Paris, France, c.2400 BC)

## The Religious Meaning of the Crossing of the Ocean

The passage graves along the coast of Western Europe always point with their important chamber to the west, to the other side of the ocean. For this reason it should not surprise us that crossing of the Ocean has a religious meaning. The chambers are mission churches of Egyptian King theory. In Fig.9 we see the "Royal Crossing", against the governing current and wind direction, from Cape Finisterre (NW Spain), where this passage grave is located, via the Azores to Newfoundland. There is no need to explain that nobody was able to carry out this crossing except the SunGod, and only in a symbolic manner. The SunGod Ra is here represented by an enthusiastic stick figure King, drawn with 9 lines, corresponding to the 9 islands of the Azores, in the middle of the Ocean.

Basically, in Fig.10 the same thing is shown in a painting. Below the glyph of the SunGod Ra we see the Egyptian gods Horus and Osiris (left and right), busy with their "Royal Crossing" from Iberia (where this monument is located), via the Azores, where they are standing, to Newfoundland. According to the Egyptian Religion the kings and pharaohs were the substitutes of these gods. The sun has 9 rays, corresponding to the 9 islands of the Azores. In the west the waters along the coast of North America are shown over a distance of 0.5DL= 555km. Below we see the usual Southern Crossing (from Africa to South America) over 3DL= 3333km, used by ordinary people to cross the Ocean to the west.

In the dolmen of Antelas near Viseu in North Portugal (see map, Fig.2) are the most beautiful megalithic paintings of Western Europe, shown in Figure 11. On the stones C2, C4, and C6 the North Atlantic Ocean is shown to about 60°N (6 to 7 vertical waves), having a width of c.4DL= 4444km (5 vertical lines). In the east is the Old World, and in the west is the New World. On stone C6 we see in the east a sun with 15 rays, representing the SW Cape Verde Island of Brava, at 15°N. It is the starting point of the Southern Crossing. In the west we see the coastline of South America, from the Falkland Islands in the deep south via Cape Sao Roque, the NE point of Brazil, to the Antilles Route in the north. The paintings primarily have a religious meaning: the King (on stone C4) leaves the Land of the Living of the Sungod Horus in the east (the sun on stone C6). After the crossing of the waters (the Ocean), he reaches the Land of the Dead of the Moongod Osiris in the west (the moon on stone C6). But he returns from the west, from the realm of the dead, with a boat (visible in the middle of stone C2), for he has eternal life (the King on stone C4 lives forever).

Figure 12 shows a religious painting on an Egyptian sarcophagus from the time "subsequent to the period of the pyramids at Sakkara" (Memphis, Refs.18,19), which is just after the 5th dynasty, or after the discovery of America, c.2500 BC. We look over the invisible North Atlantic Ocean to the north. Osiris (left) is the great God of the Tuat, the Underworld in the west ("Punt", or Central America). The earthgod Maat (before him) has decided, that from this land of the dead he should return in a boat over the waters

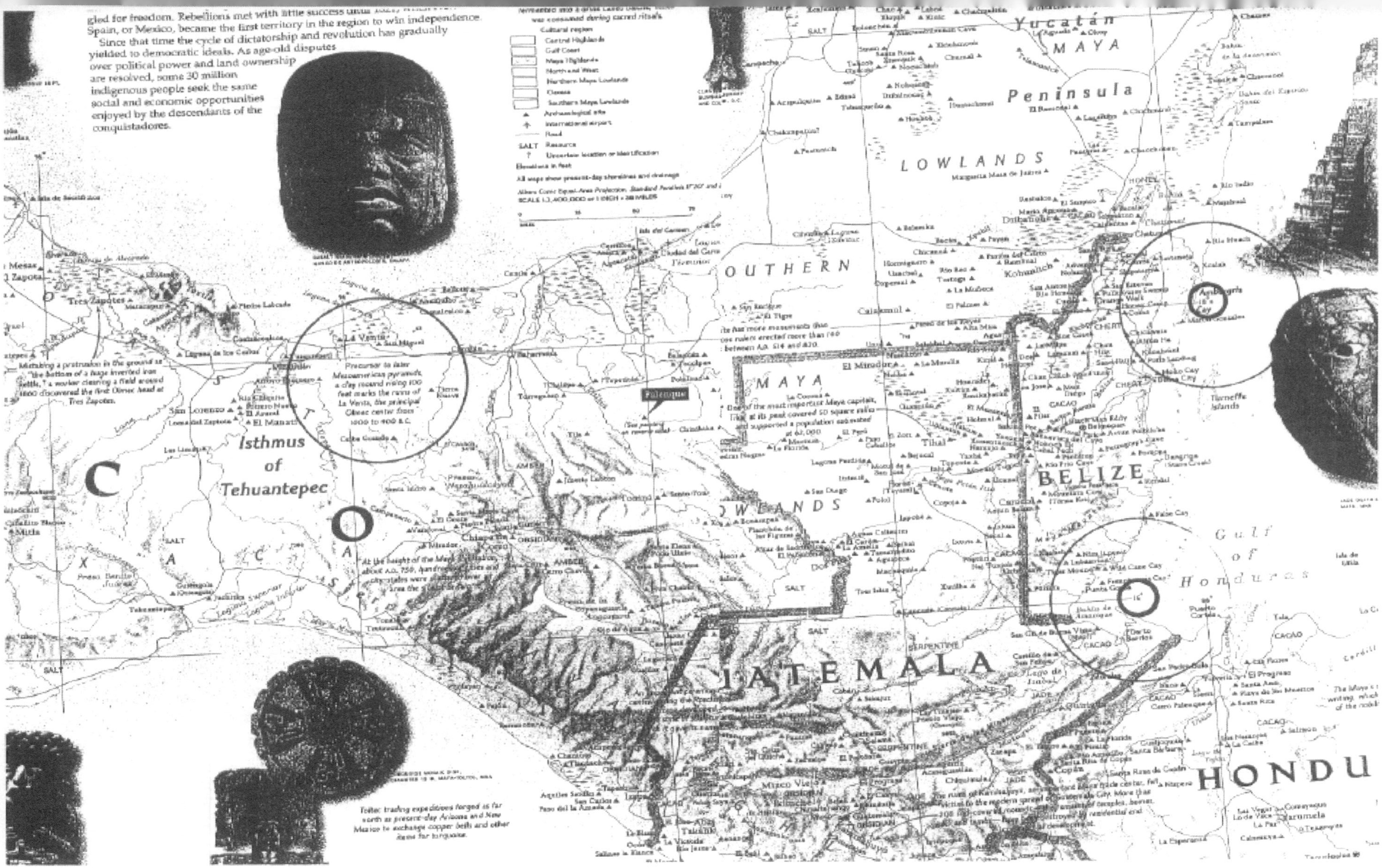


**Fig.13** The ancient cultural c of Central America, where the Olmec Civilization started, between thth coast of Honduras at 16°N (right), and the Gulf of Campeche at 1&ft). (National Geographic Map of Meso America, 1997).

(the ocean) to the land of the living (Egypt), which is in the east (upper right corner). During this voyage he will change into other, less important gods, as indicated. He will be helped by 4 birds (the 4 wind directions, or the four "gospels"), but above all by his 9 fellow gods (the 9 islands of the Azores) in the middle of the waters (the ocean), who are characterized (above) by 38 bundles of Nile reed (for the Azores latitude of 38°N). An almost identical painting on another side of the sarcophagus (Ref.18) has 18 bundles of Nile reed along the upper edge, corresponding to the latitude of the south point of the Gulf of Campeche, the center of Tuat, at 18°N (Fig.13). The painting of Fig.12 illustrates beautifully how important Central America was in the early Egyptian religion, but it also shows the religious meaning of the Azores in the return route to the Old World.

## Discussion

Unfortunately little remains of these megalithic people, and their four thousand year effort to explore the other side of the world, other than their petroglyphs and stone monuments. It is difficult for many to accept history that is not written in a phonetic language, which is after all the definition of history: written records. Even worse, one expert who ought to know better claims that these were "people who were only capable of crude inscriptions", only because all that we have from them is stone inscriptions we have failed to understand! And further, that "I would not expect to find latitude information in a monument, as it could be memorized by a pilot, or recorded on rawhide or parchment. This is 'mundane' information that is not likely to show up on a religious monument". We should not prejudge the matter without actually counting the rocks, or belittle ideas that seem mundane today, but appear to have been cutting-edge technology four and five thousand years ago. What we are finding is actually a fairly sophisticated encoding methodology involving mathematics, symbols, and petroglyphic maps. These are making it possible to understand some of man's accomplishments in prehistory, and in particular, his discovery and exploration of the backside of the earth!

**Fig.14** Two Mediterranean symbols. The literal meaning of each: To the left to the Cape Verde Islands, to the right around the Small Mediterranean Sea, and upwards to the Azores. The real meaning of the two: To the south is the way to the inland seas of Central America, and in the west the return route via the Azores. The religious meaning: Left, the Sungod Horus, right, the Moongod Osiris. (Sournan, Brittany, Ref.1, c.2000 BC)

# References


1. Twohig, E. Shee, The Megalithic Art of Western Europe, Clarendon Press, Oxford, 1981
2. Jonge, R. M. de, and IJzereef, G.F., De Stenen Spreken, Kosmos Z&K, Utrecht/Antwerpen, 1996 (ISBN 90-215-2846-0)(Dutch)
3. Jonge, R.M. de, and IJzereef, G.F., Exhibition: The Megalithic Inscriptions of Western Europe (1996)
4. Tour Guide Map, America's Stonehenge, PO Box 84, No. Salem, N.H. 03073
5. Lenik, E. J. and Gibbs, N.L., "The Frost Valley Petroglyph, a Catskill Valley Enigma", NEARA Journal, Summer 1999 (ISSN 0149-25-51)
6. Ferryn, P., NEARA Journal, Vol. XXXI, No. 2, pg 59 (1997)
7. Fell, B., America BC, Pocket Books, Simon & Schuster, 1994
8. Bailey, J. Sailing to Paradise, Simon & Schuster, 1994
9. Thompson, G., American Discovery, Misty Isles Press, Seattle, 1994
10. Peterson, F. A., Ancient Mexico, 1959
11. Stuart, G. E., "New Light on the Olmec", National Geographic, Nov. 1993
12. Bernal, I. , The Olmec World, University of California Press, London, 1969 (ISBN 0-520-02891-0)
13. Allen, J.M., Atlantis, The Andes Solution, Windrush Press, Gloucestershire, 1998 (ISBN 1-900624-19-2)
14. Hapgood, C. H., Maps of the Ancient Sea Kings, Adventures Unlimited Press, Illinois, 1996 (ISBN 0-932813-42-9)
15. Joseph, F., Atlantis in Wisconsin, Glade Press Inc., Lakeville, Mn. 1998 (ISBN 1-880090-12-0)
16. Art et Symboles du Megalithisme Europeen, Revue Archeologique de L'Ouest no. 8, Nantes, 1995 (ISSN 0767-709-X)
17. Tompkins, P., Secrets of the Great Pyramid, Harper Colophon Books, Harper & Row, New York, 1971 (ISBN 0-06-09-0631-6)
18. Wallis Budge, E.A., Osiris and the Egyptian Resurrection, 2 Vol., 1911, Dover Pub., N.Y., 1973 (ISBN 0-486-22780-4)
19. Wheeler, R.L., Walk Like An Egyptian, Allisone Press, 2000 (ISBN 18937747-21-X)
20. Lambert, J.D., America's Stonehenge, an Interpretive Guide, Sunrise Publications, Kingston, N.H., 1996 (ISBN 0-9652630-0-2)
21. Briard, J., The Megaliths of Brittany, Gisserot, 1991
22. Carlson, S., www.neara.org/CARLSON/
23. Strong, R., "Did Glooskap Kill the Dragon on the Kennebec?", NEARA Journal, Vol.XXXII, no.1 (pg.38)
24. Sanchidrian Torti, J.L., and others, "International Newsletter on Rock Art", No 29, 2001 (pg.17, Fig.1)
25. Map of Mesoamerica, National Geographic, Dec. 1997
26. Irwin, C., Fair Gods and Stone Faces, St. Martin's Press, N.Y., 1963 (pgs.35-40)
27. Atlas of Pilot Charts of Central American Waters and South Atlantic Ocean, Second Edition 1955, reprinted 1969, U. S. Naval Oceanographic Office
28. Cruxent, J.M. and Rouse, Early Man in the West Indies, pg 43-45

# The Sailing Route of the Upper North, Using a Compass

(Petroglyph of Chao Redondo, North Portugal c.2200 BC)

Dr. R.M. de Jonge, drsrmdejonge@hotmail.com
J.S. Wakefield, jayswakefield@yahoo.com

## Introduction

The dolmen of Chao Redondo is situated in the vicinity of the town of Viseu, between the rivers Douro and Mondego in North Portugal. On menhir R4 at the north side of the passage of the dolmen is a large petroglyph (Fig.1)(Refs.1-3). Like almost all megalithic inscriptions it has a geographic meaning, and as so often, the surface of the stone represents the North Atlantic Ocean, probably chosen by the carver for its similar shape.

## From Iberia to Greenland

Near the center of the stone is carved a circle. This represents the Iberian Peninsula, where the passage grave of Chao Redondo is located. Seen on a globe, and as known by sailors, the peninsula is close to this shape.

At the right side a vertical stroke has been carved. Note on a map that east of Iberia, across the "Little Mediterranean Sea" often depicted in megalithic "art", are the large islands of Corsica and Sardinia. Together, they look like this vertical line. This ties the rest of the petroglyph to sailors familiar with the Mediterranean, and helps scale the long distances indicated on this petroglyph.

A small circle is also a megalithic symbol for a starting point, in this case Iberia. The petroglyph shows a sailing route starting at or near Santander on the north coast of the Iberian peninsula, that runs along the shores of France, England, and Ireland to the northwest (Fig.2). Ireland is just west of Iberia, as clearly shown in the petroglyph. Next, the sailing route bends around Ireland and the west coasts of Scotland, to the Faeroe Islands via Iceland, all the way to Cape Farvel, at the south tip of Greenland (at 60°N). The small side route on the right leads along the west coast of Iceland to the north. It is the start of the alternative route via the shortest crossing of the Denmark Strait to Cape Holm, which might be chosen for safety reasons.

Here, however, we face a curious problem. The petroglyph has been carefully made, and we know that boats used in the Bronze Age were small and simple, making ocean crossings perilous. But the spacings on the petroglyph between the indicated places (Fig.2), do not appear to be represented in the correct proportions, as far as their true north latitudes are concerned. The correct latitude for Ireland is 53°N, ten degrees above the north coast of Iberia, which is correct, but Ireland is only seven degrees south of Cape Farvel, Greenland, which is shown a much larger distance north. What is going on here? In a few moments, we will show that the upper horizontal

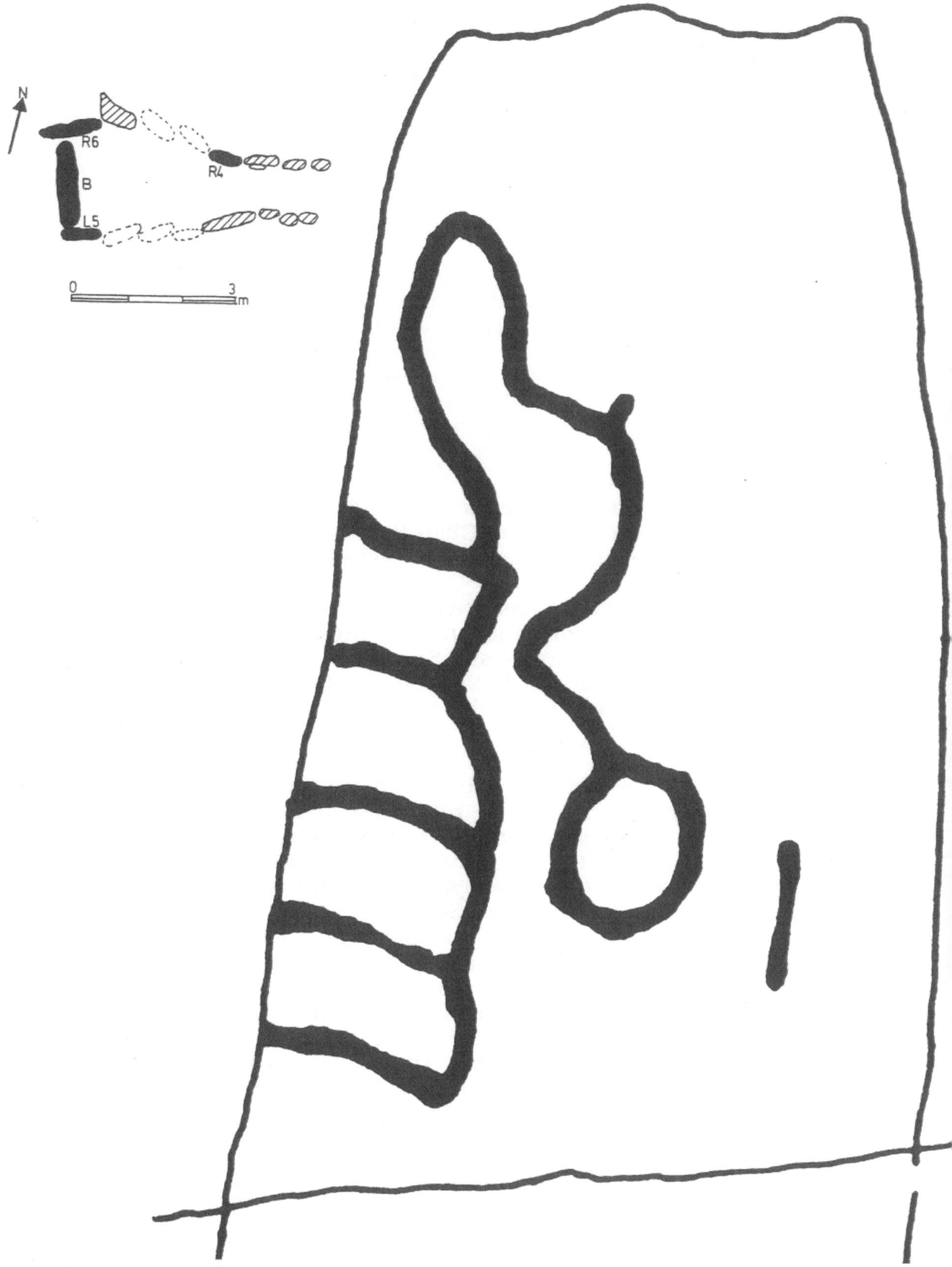


**Fig.1** Petroglyph of the sailing route from the Iberian Peninsula (the circle) via Iceland and Greenland, and via the Arctic circle (top) to North America (left). The horizontal strokes are big latitude lines between 10°N and 50°N (Ref.1). (Menhir R4, Chao Redondo, North Portugal, c.2200 BC)

line at the left side is the 50°N latitude line. How can this be drawn above Ireland, located at 53°N? Also, the north coast of Corsica lies exactly at the same level as the north coast of Iberia (43°N), yet, the upper part of the vertical stroke has been clearly carved below the upper part of the circle, where the sailing route starts. The whole line representing these Mediterranean islands seems placed too low. Most of the differences between the true latitudes of the land features on the route (shown on right side of Fig.2) are not sensible.

## The Compass

An explanation for these conflicts of spacing and latitude, is that while sailing this route, these people were not orienting themselves to the geographic north pole (by the stars), but were orienting on the magnetic north pole with a compass!! Today, this magnetic pole is located behind Greenland in the Canadian Queen Elizabeth Islands at 77°N. It is known that the location has changed greatly over geologic time, and pole reversals are used in dating. However, the Bronze Age is only yesterday in geologic time. If in the Bronze Age, the magnetic pole was where it is today, the petroglyph represents the sailing route, with way-point latitudes established using a compass, with much more correct proportions (see Fig.3)!

When we look from Iberia toward the magnetic north pole, with global latitude lines drawn around the magnetic north pole (as in Figs.3&4), it is logical that the 50°N latitude line on Fig.2 in the west appears above Ireland, because it now appears to be at 55°! Note how the Mediterranean islands below the north coast of Iberia, now fall on the latitude of 39°, not 43°N. Look at Cape Chidley in the Canadian Arctic, in Figure 4. Its True North latitude of 60°N reads as close to 70° when based on magnetic north. The differences are most dramatic in the upper north portion of the route, where the distance between the two poles is relatively greater compared to the distance of the route from the poles (see Fig.5). The whole explanation has far-reaching consequences, because it appears that this petroglyph is proving that people were using the compass in c.2200 BC! It would then follow, that all later crossings were probably aided with this technology.

The painting of the Southern Crossing in Sallas, Orense, in North Portugal (Fig.11, Chapter 6, Ref.1), of almost the same date, confirms the use of a compass in the same way. The mid-ocean line in this glyph, at right angles to the carved sailing route from Africa to South America at 45°SW, points to the same Magnetic North Pole, which is west of the top of the rock, representing the Geographic North Pole.

In the 6th century before Christ, the Greeks knew the mineral magnetite, and called it "the stone of Heracles" (Ref.8). This type of stone has oriented mineral grains, so it reacts in a magnetic field, including the Earth's magnetic field, and for that reason was found useful for making a compass. A Greek legend attributes the compass to a 12th-century BC hero named Hercules (Ref.9). The Olmecs knew magnetite too, as in c.1550 BC they

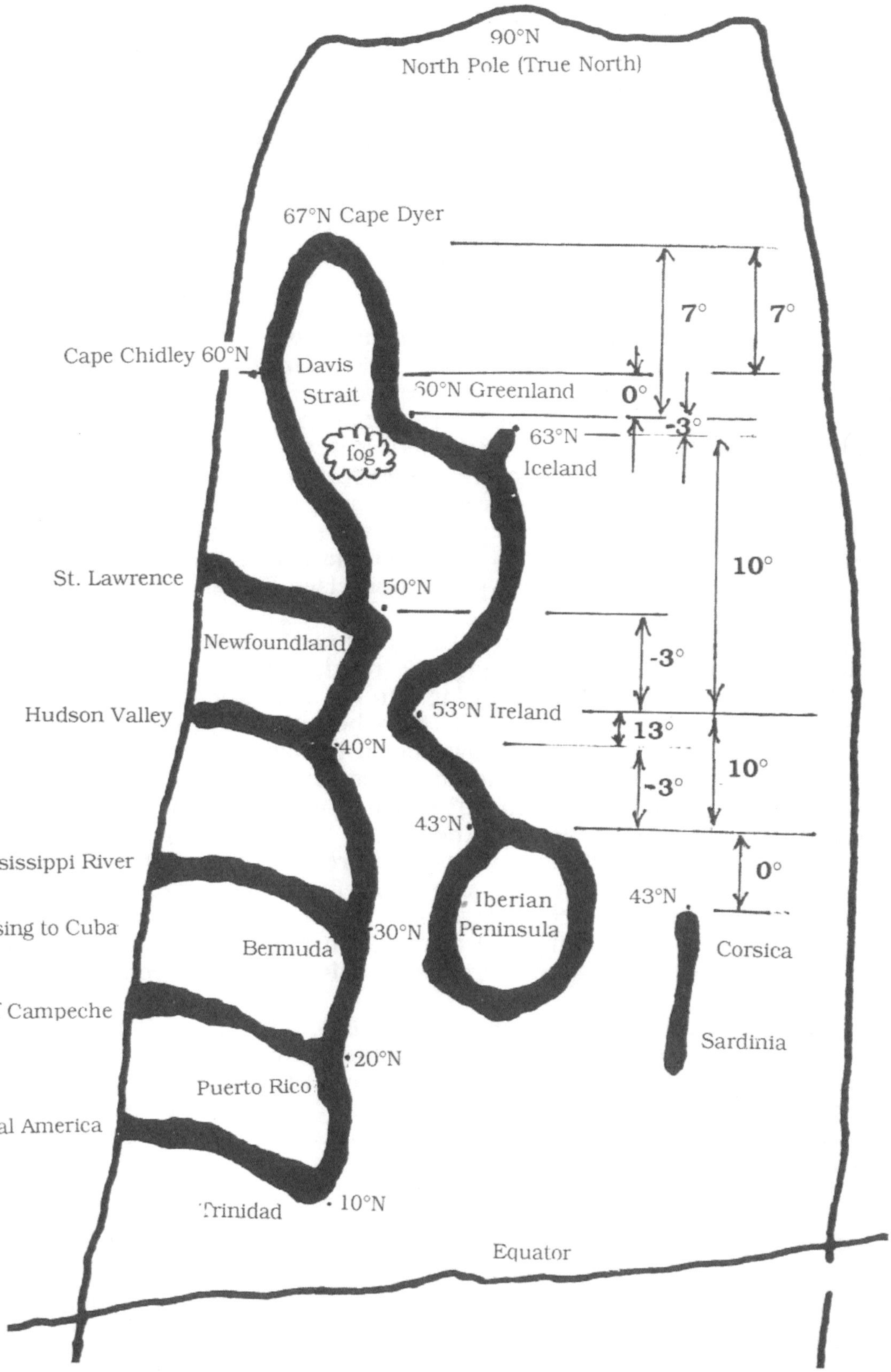


**Fig.2** Fig.1 petroglyph with geographic locations and latitudes **based on true north.**. At the right side the curious differences in latitude, showing that, during the first half of the voyage, people did not orient on the true north. (Menhir R4, Chao Redondo, North Portugal, c.2200 BC)

erected earth and stone buildings oriented on the magnetic pole (Refs.8,13), and their human statuary in basalt boulders has been found to have magnetic right temples or navels, while in Guatemala a statue of two men siting on a bench both have magnetic poles where their arms cross, and a likeness of a jaguar has magnetic paws (Ref.26). In the Gilgamesh epic, dating from c.2800 BC (600 years older than this petroglyph), a passage is written in which the "sailing stone" plays an important role (Ref.8).

## Route from Greenland to Central America

According to the Story of Loughcrew and the design of Stonehenge I (Ref.2), the megalith builders discovered Cape Farvel, the south point of Greenland,c.3300 BC. They did not succeed in crossing the Davis Strait, and they gave up their efforts to reach the other side of the earth. According to Stonehenge III, inscriptions at Dissignac, and the Egyptian tradition, America was discovered by these people seven hundred years later, via the Bering Sea (c.2600 BC). Shortly afterwards, it became also possible to sail from Greenland to North America.

On the petroglyph (Figs.1&2), the route is seen from Cape Farvel along the west coast of Greenland, sailing north. The shortest crossing is chosen, to Cape Dyer on Baffin Island, Canada, at 67°N, the holy Arctic Circle. This crossing, with a moderate favorable current, has a length of only c.350km. Next, they sailed south to Cape Chidley, the NE Cape of the mainland of North America at 60°N, which is indicated by a faint kink in the route, as it is on a global map. Note that this kink has been placed higher than Cape Farvel (60°N too), still showing their orientation on the magnetic north pole (see Fig.3). Thus in this Chao Redondo petroglyph the crossing of the Atlantic Ocean in the Upper North is shown. The glyph itself is carved on menhir R4 (see top of Fig.1) at the center of the north side of the passage, another confirmation that this glyph is about a northern route (Ref.1).

South of Cape Chidley the petroglyph follows the characteristic curve of the east coast of Labrador to the north coast of Newfoundland (new-found-land) at the 50°N latitude line, indicated by the nearly horizontal stroke (Refs. 6,7). Note that an alternative route from South Greenland to the east coast of Labrador (800 to 1000km) has not been indicated (Fig.2), though that would have meant a considerable saving of traveling distance. This area, where the Labrador Current running south meets the Gulfstream running north, is famous for icebergs and fog. They obviously gave a high priority to safety here, by going further north.

In this time period, much longer routes were being discovered and sailed, like the Southern Crossing, which was 2500km over open sea, yet became preferred. The Royal Navy guide for sailing ships, first published in 1895, describes the Newfoundland Banks, including these comments:

"...Between the S limit of the Labrador Current and the Tail of the Bank, the

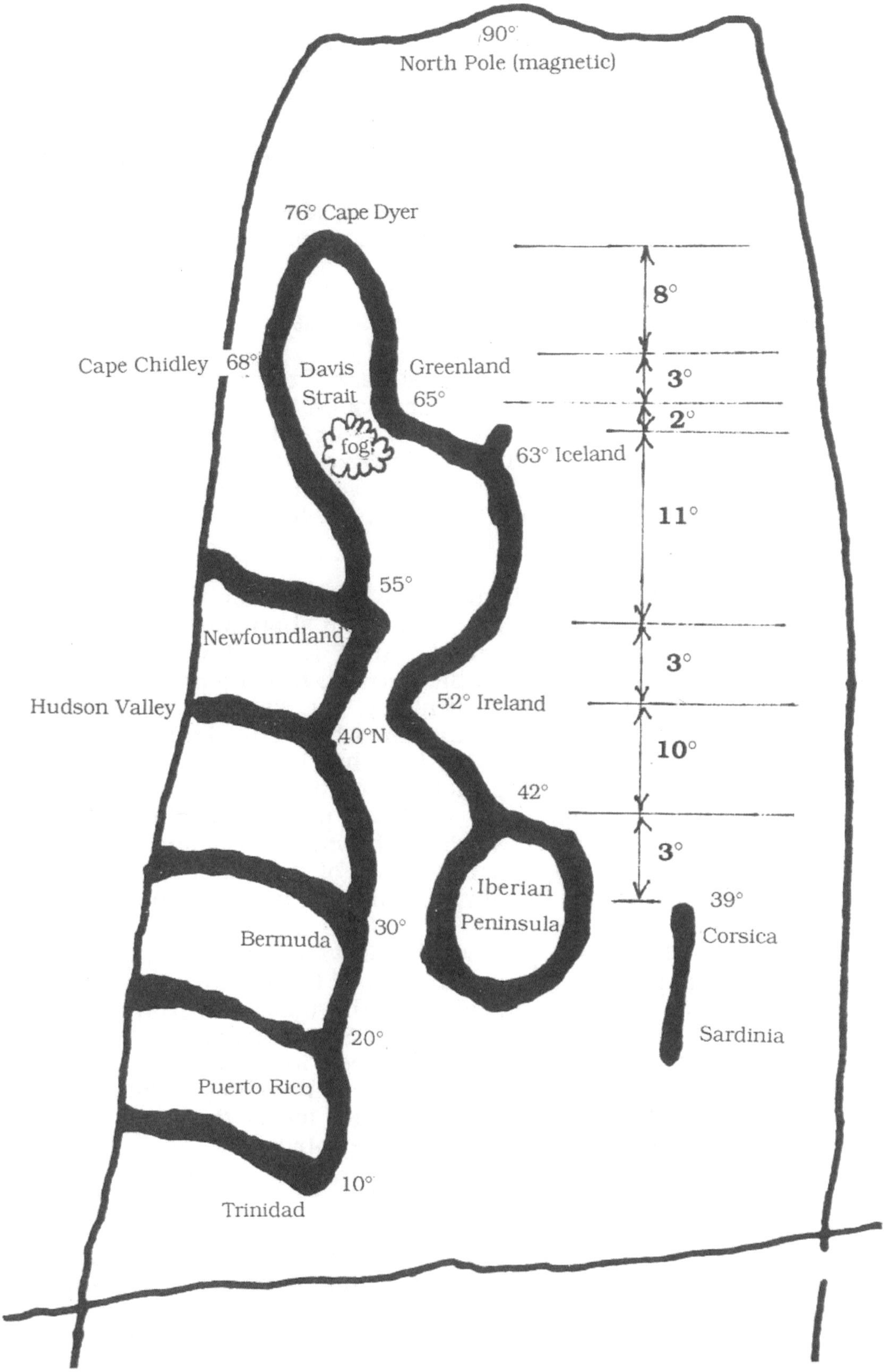


**Fig.3** Fig.1 petroglyph with geographic locations and latitudes **based on the magnetic north pole**. At the right side the differences in these latitude values show that, during the first half of the voyage, people oriented on the magnetic north pole. (Menhir R4, Chao Redondo, North Portugal, c.2200 BC)

warm and cold waters converge on a line which is known as the Cold Wall. The E end of the Cold Wall presents the greatest hydrographic contrasts to be found in the world, the water changing from the olive or bottle green of the Arctic side to the indigo blue of the Gulf Stream, with temperature changes of 11° or more over short distances.....Almost all the icebergs which menace the North Atlantic routes originate in the glaciers of the W coast of Greenland where they are calved at a rate of several thousand a year. They are carried S by the Greenland, Baffin Land, and Labrador currents, and when they finally reach the shipping routes they may be several years old. ...In the region of the Grand Bank, the worst season for icebergs is between March and July, with May as the month of greatest frequency. ... Denmark Strait is normally free of ice on its E side throughout the year, but on rare occasions, as in the spring of 1968, the ice spreads across from Greenland to close the strait. ... As fog is exceedingly prevalent off the S coast of Newfoundland, especially in summer, vessels should guard against the set of the current ..." (Ref.15, pgs. 39,40).

The island of Newfoundland (Fig.2) is indicated with a protruding point in the route to the right, because here is the east cape of North America, Cape Race. (This became an important departure point for the return to the Old World.) The 50°N latitude line is indicated to the left, coinciding with the Belle Isle Strait, and the south coast of Labrador. At the far left a small bow indicates the entrance of the St. Lawrence River.

The more-or-less horizontal strokes at the left side of the stone are big latitude lines at 50, 40, 30, 20, and 10°N. Using this scale as a measuring stick, note that the top of the stone, the geographic north pole, is exactly at 90°N (proving the use of degrees of latitude). The lines are at right angles to the left side of the stone, which is in good approximation the meridian (the NS line) over Florida. This peninsula has the westernmost ocean shore of North America (important near the Gulf of Mexico), and from Florida one has to cross due south to Cuba in order to reach Central America. Seen from the Iberian Peninsula, the latitude lines in the far west turn upwards, which is shown in the right manner, as we have seen in other petroglyphs.

Following the roughly vertical lines of the petroglyph (Fig.2), the route is shown from Cape Race (Newfoundland,47°N), via Nova Scotia ("New Scotland") and Sable Island (44°N), to Cape Cod (42°N), and Nantucket Island (41°N) in the south. All these areas are important departure points for the return route to the Iberian Peninsula, as indicated in American Stonehenge (New Hampshire, 43°N) (Refs.4,23). This is the most important megalithic monument in North America. The east coast of America bends westward here, as indicated in the petroglyph. The 40°N line has been engraved, which encodes the important waterway of the Hudson River. It also points eastward to the West Azores at 39°N, which are important on the return route. It points also to Chao Redondo (41°N), where this petroglyph is located.

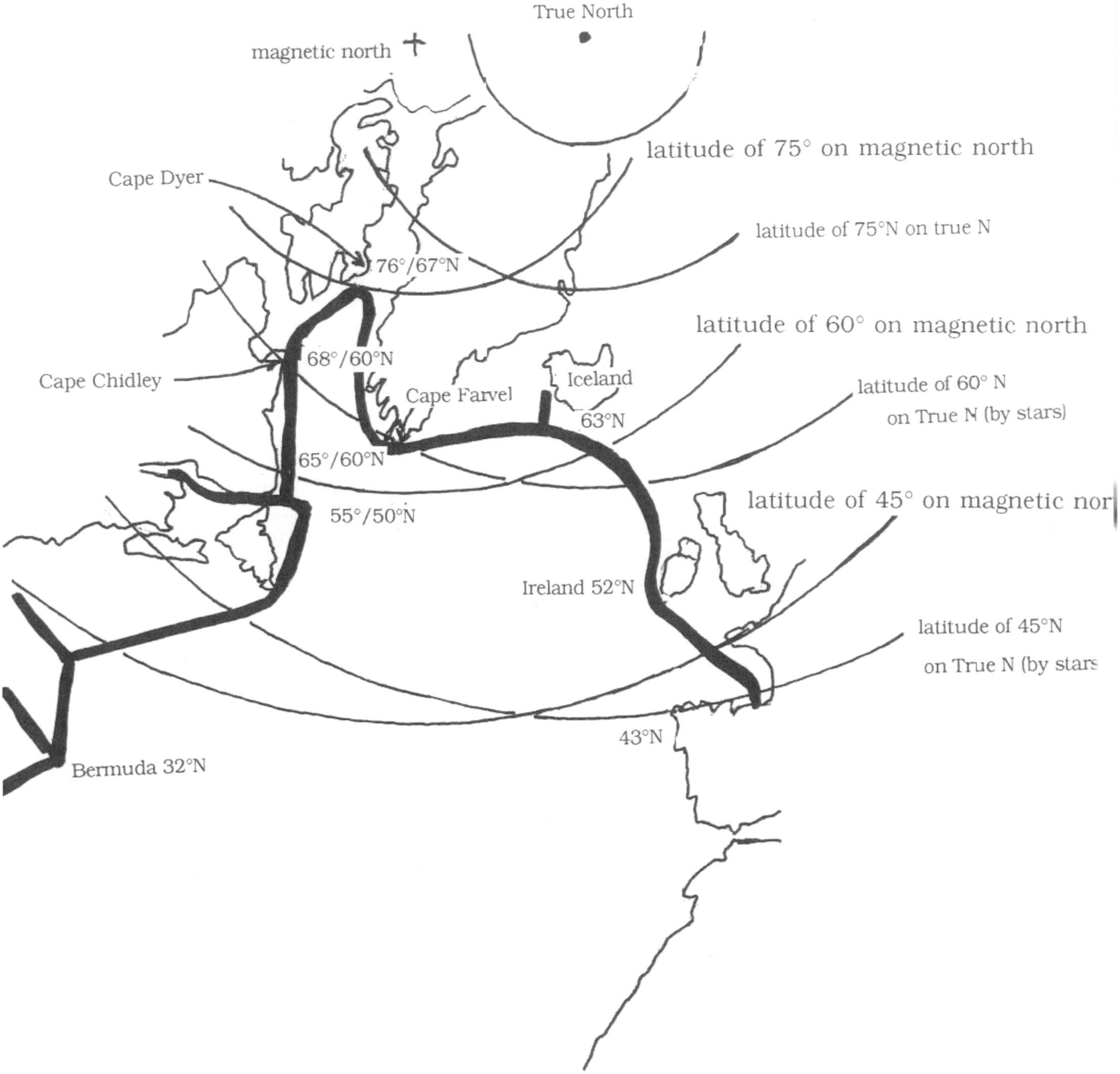


**Fig.4** Tracing from one foot globe, showing Upper North Crossing Route, and latitude lines based both on true and magnetic north. Note how the Cape Dyer latitude can be seen as 76° or 67°N depending upon whether you are using a compass (76°) or the stars (67°N) to determine north. Perhaps the summertime fogs of Davis Strait led to dependence upon the compass for navigation here.

Further south an eastward bending line is seen, which connects Cape Cod (42°N) to Bermuda (32°N). Bermuda formed an important waystop for returning voyagers from Central America, who cross from Bimini or Great Abaco (26°N), or Cape Hatteras (35°N), via Bermuda to the Azores. This route was good for sailing, with prevailing winds and currents from the stern. The inclination of this line on the petroglyph shows that this inscription was made after the discovery of Bermuda, or after c.2200 BC.

The 30° latitude line has been carved, because this line points from Bermuda via the south coast of the U.S. to the mouth of the Mississippi. This huge river was of great importance for the trade with the interior (Refs. 5,8,13). To the east the 30° latitude line points to Madeira (33°N), the eastern Canaries (29°N), and the Nile Delta (30°N), at that time the most important center of civilization on earth (Refs.7-9).

Further to the south a line is seen, which connects Bermuda (32°N) to Puerto Rico (18°N). Puerto Rico is the easternmost island in the Greater Antilles. The 20° latitude line has been carved because this line runs along all the important islands of the Greater Antilles, and because it forms the north side of the civilization around the Gulf of Campeche. This was the location of the greatest civilization in America at the time this inscription was made (Refs.2,7-13). Note that the dolmen of Chao Redondo (which has this petroglyph in it) has 17 upright menhirs (see Fig.1), corresponding to the latitude of the center of the United Central American Empire, 17°N. This is halfway between the north coast of Honduras (at 16°N), and the south point of the Gulf of Campeche (at 18°N). The edge of the stone between 30° and 20° coincides with the peninsula of Florida, and the important crossing of the Florida Straits to Cuba (c.200km).

When we continue heading south, a slightly eastwards bending line is observed, coinciding with the Lesser Antilles, from Puerto Rico (18°N) in the north to Trinidad/Tobago (10°N) in the south. The 10th latitude line has been carved, because this line coincides closely with the north coast of South America, pointing westwards to Panama and Costa Rica, important for crossing over land to the Pacific (Refs.2,13). A major part of the edge of the stone coincides here with the east coasts of Costa Rica, Nicaragua and Honduras, from 10°N to 16°N. The bottom of the stone corresponds with the equator, the southern border of the Atlantic Ocean.

When counting the upright menhirs of the dolmen of Chao Redondo in the direction of the movement of the sun (see Fig.1), as the SunGod wishes it, this petroglyph is applied on stone #14, corresponding to the latitude of Cape Verde, at 14°N. This has been done, because Cape Verde is the westernmost point of all the mainland of the Old World, and because the Upper North Crossing Route is an alternative for the Southern Crossing (also shown in this dolmen), which starts at this Cape.

The petroglyphs of Dissignac in Brittany (Refs.2,3) show that directly after

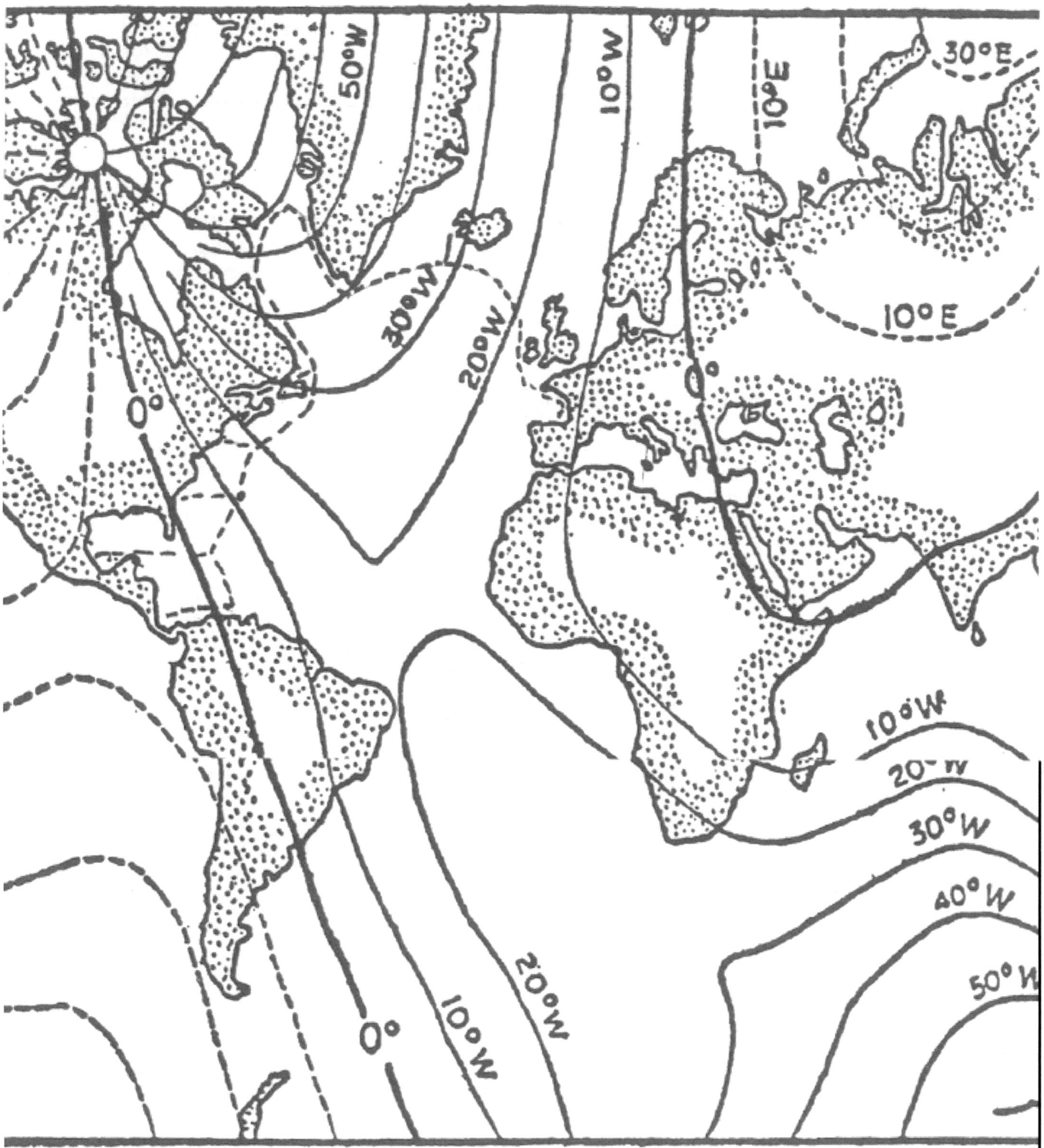


**Fig.5** Upper North Crossing Route, on map of Mixter (Ref.14). Lines of Equal Variation run to the Magnetic North Pole (in the upper left corner). When sailing from Iberia to the Arctic Circle at Cape Dyer, the variation changes from 10°W to 60°W.

the discovery of America (c.2500 BC), the megalith builders were very interested in finding a northern passage from the Atlantic to the Bering Strait. It was the last geographic problem they were dealing with. America's Stonehenge in New Hampshire (c.2200 BC) confirms this interest in the Upper North very convincingly, by extensive stone wall constructions of the Upper North. Figure 6 shows four archaeological sites around the Baffin Bay area (Ref.21), dating from c.2500 BC, and one from c.2000 BC. Archaeological sites in the Upper North can be identified and excavated much more easily than elsewhere, because of the lack of population, trees, brush, grass, and sometimes even dirt. Three of the five sites are clearly related with finding a northern passage to the west. All five sites may be megalithic in origin, although most archaeologists are not even aware of the fact that this may be a option.

## Other Upper North Sailing Routes

Figure 7 (top) is one of the Calderstones from Liverpool, England. It shows the crossing of the Denmark Strait from Iceland (right) to Greenland (left). The double circle is the NW peninsula of Iceland at 66°N, 2° higher than the place of arrival at the SE coast (at 64°N). The southward foot shows that they were trying a more southern crossing to Greenland, as indicated. Left of it is the first depiction of a compass card in the world, c.2700 BC, which is before the discovery of America. The feet say: "we have been here, and we are heading west!" The upper left foot points to Cape Holm at the Arctic Circle, 67°N. Thanks to the SunGod, Greenland had been discovered there, c.3300 BC (Refs.2,3).

Figure 8 (bottom) is a petroglyph applied on the endstone of the passage grave of Mane Lud in Brittany, France. It was made shortly after the first Upper North Crossing to North America. After a lot of attempts, over seven centuries (c.3200 BC - c.2500 BC), they finally discovered the secrets of how to reach North America. In that mood, they made this huge stylized petroglyph. The double lines indicate the important sailing route, from the NW peninsula (or west coast) of Iceland across Denmark Strait to Cape Farvel at 60°N. The voyage continues along the west coast of Greenland to Cape Mercy on Baffin Island at 65°N. From there, a coastal route leads south to Cape Chidley at 60°N, the same latitude as Cape Farvel. This important sailing route is indicated by a double line. It is also possible, though not recommended, to use shortcuts at 60°N, as indicated by single lines. The one between Cape Farvel and Cape Chidley is a half double, half single line, because in the future this crossing may become important, if boats improve! Lower on the stone is the return route from Newfoundland to South Ireland along the 50°N latitude line, followed by the coastal route around Ireland to Rockall, at 57°N. Because of its size (1.30m), this is definitely an important petroglyph. However, it is a rude "seaman carving", totally different from the carvings at Dissignac, done by "academic people".

The upper line in Fig.10 shows the Upper North Crossing Route from Ireland via Iceland to Greenland, and then around South Greenland to its

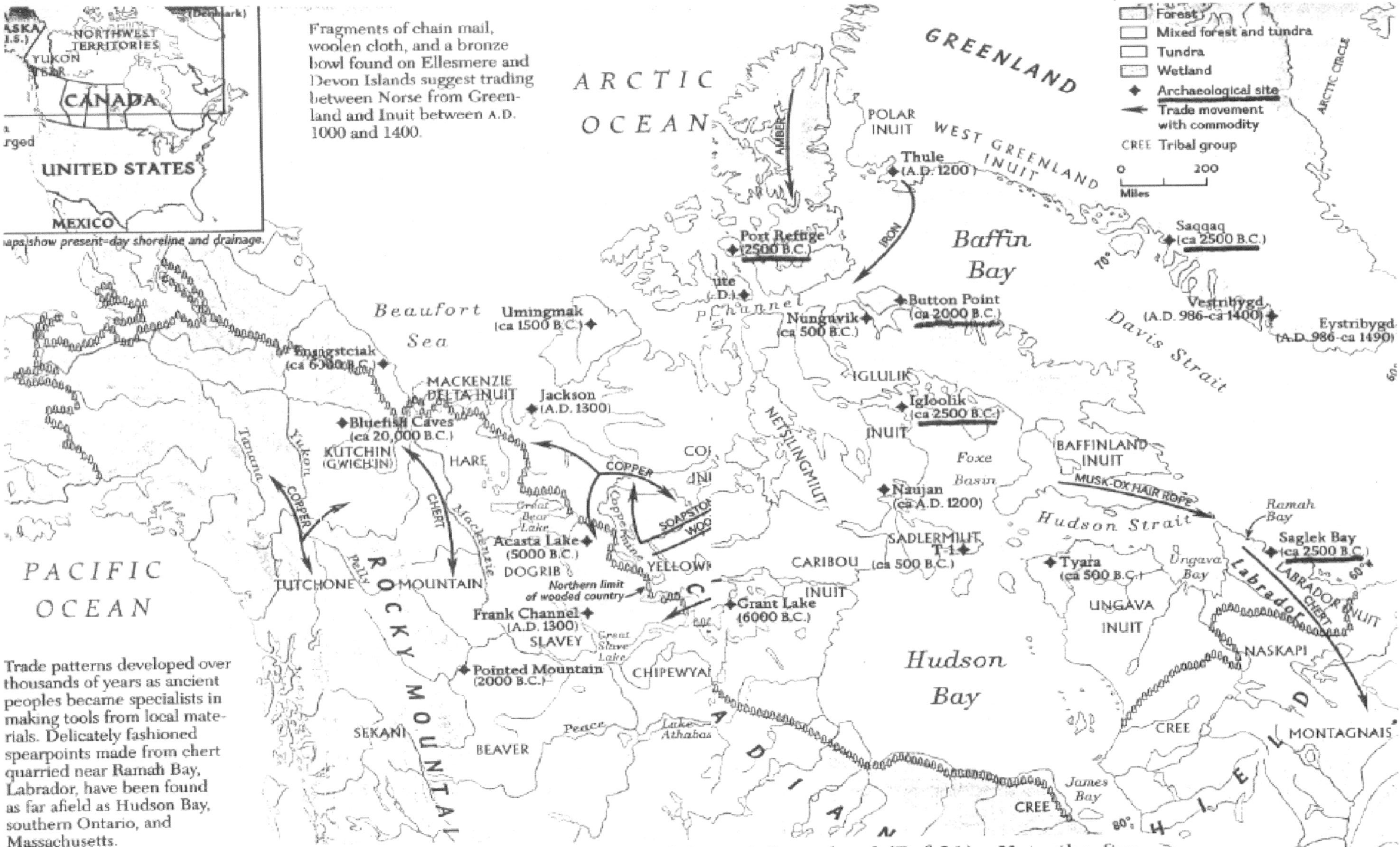


**Fig.6** Archaeological Map of Caand Greenland (Ref.21). Note the five sites dated at c.2500 BC and c.2C in the Baffin Bay area. All five sites may be megalithic in origin.

west coast, followed by the crossing to Cape Mercy on Baffin Island. Cape Chidley, at the same latitude as Cape Farvel, 60°N, is on the other side of Hudson Strait. The lower line of the glyph is added for emphasis, to show the necessity to cross Davis Strait behind Greenland at Cape Mercy, 65°N, and not at the level of Cape Farvel (60°N). South of Newfoundland, the Gulf of St. Lawrence has to be crossed to reach the important latitudes of the east coast of the U.S.: Sable Island at 44°N, Cape Hatteras at 35°N, and the south point of Florida at 25°N. These are also the most important departure places for the return route to the Azores, indicated by a height profile or crossing sign in the middle of the Ocean (Brittany, c.2300 BC).

Figure 11 is a menhir in Anglesey, Wales, showing the North Atlantic Ocean, with right, the Old World, and left, the New World. Coastal waters and latitudes in the north are shown over 1/4 DL= 2.5°= 277km. In the south are indications of latitude lines at 10°N, 20°N, 30°N, and 40°N. At the left side below an extra line was added for the coast of Central America behind the Antilles. The sailing distances from the Faroes to Iceland, and from Iceland to Greenland (top) are indicated as 5dl= 555km, which are not the shortest distances (4dl= 444km), but the most economical ones. Slightly lower, the ocean is shown over a distance of 5DL= 5555km (numbered), which is the west coast of Hudson Bay. The petroglyph shows that people from Anglesey sailed via the Upper North Crossing to the northern part of North America (sometimes also to Central America), and they returned via the 2 islands of the West (and the East) Azores. The northern part of the Ocean, north of the Azores, is "their territory", probably important for fishing and trade. The decorative style of this petroglyph and the way of using big distance lines is typically late megalithic (c.1900 BC).

Figure 12 shows the Upper North route used by the Vikings (c.1000 AD, Ref.22), first from Iceland to Greenland by the Denmark Strait, then from West Greenland to Baffin Island, crossing the Davis Strait. Because of the fog and the icebergs between South Greenland and Labrador, the last crossing was performed at the latitude of the Arctic Circle, 67°N, the same route as the megalithic people three thousand years before them. Recently half of a Viking compass was excavated at the coast of Greenland (Refs.2,19). Originally, it was a wooden disk with a hole in the center, having a diameter of c.7 cm. Along the outer edge it had 32 notches, corresponding with the 4x4x2= 32 sailing directions. With a lodestone of magnetite attached, it could float on water, and it also might have been used as some sort of a sundial.

## Lodestone Compass found

Surprisingly, a prehistoric compass was found in Southern Mexico in 1967, and reported in Science, 1975 (Ref.17, pg.446): "the artifact, henceforth designated M-160 (Michigan sample) was found in situ at the Early Formative Olmec site in San Lorenzo, Veracruz. It was excavated by P. Kroster of the Yale University excavaton project headed by archaeologist M.D. Coe. ...dated by radiocarbon methods at 1400 to 1000 BC ... in general

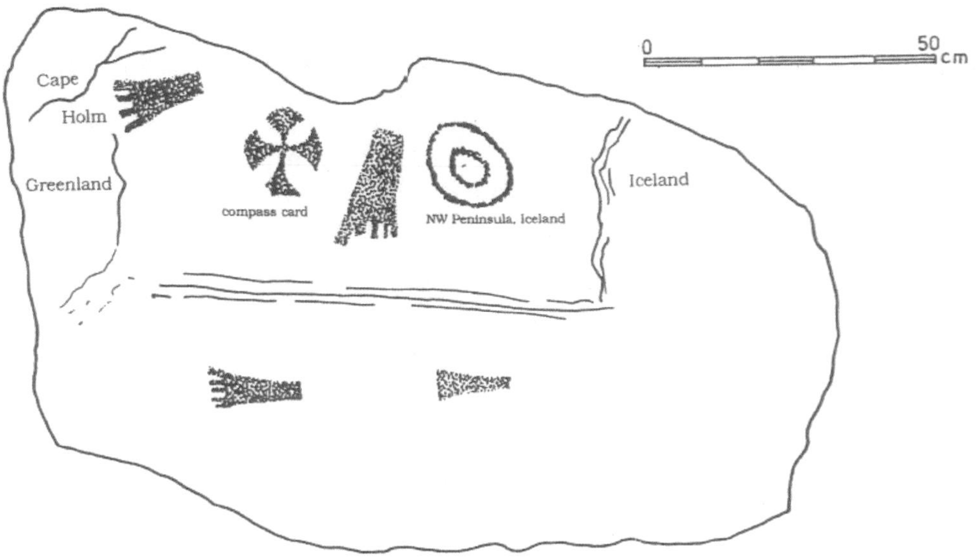


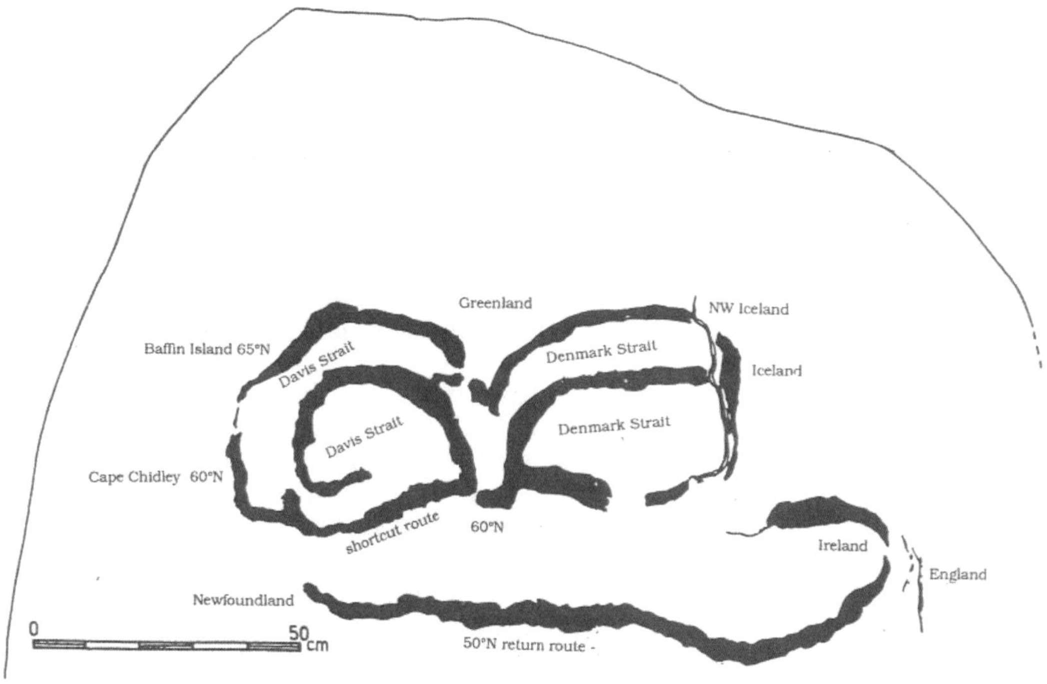


**Fig.7 (Top):** The crossing of the Denmark Strait from Iceland (right) to Greenland (left). The circle is the NW-peninsula of Iceland at 64°+2 circles= 66°N, with left of it, the first depiction of a compass card in the world. The feet denote: "we have been here!" The left foot points to Cape Holm at the Arctic Circle (at 67°N). (Calderstone, Liverpool, England, before the discovery of America, Ref.1, c.2700 BC)

**Fig.8 (Bottom):** The sailing routes for the Crossing of the Upper North. From Iceland across Denmark Strait to the coast of Greenland and to Cape Farvel at 60°N. From there, behind the west coast of Greenland, across Davis Strait to Cape Mercy on Baffin Island at 65°N, followed by the coastal route to Cape Chidley, again at 60°N. In the south is the return route from Newfoundland to south Ireland at 50°N. (Endstone, Mane Lud, Brittany, Ref.1, c.2400 BC)

appearance, the artifact ... immediately suggested to Coe that it might be part of a compass. To test the possibility, he cut a piece from a cork mat,placed the object on it, and floated them in a plastic bowl full of water. It consistently oriented itself to the same direction, which was slightly west of magnetic north ... could be used to align the bar to within half a degree ... small, carefully shaped, highly polished rectangular bar of hematite ... except for the broken end, all sides are flat and highly polished. Great care and purpose are exhibited in the production of M-160. The mineral is hard and brittle and its finishing and polishing must have required great skill and much time. To my knowledge, M-160 is unique in morphology among all known examples of worked Mesoamerican iron ore. ... original bar probably no longer than 10cm. ... The artifact and several iron ore mirrors were examined under a microscope with magnification up to x400. The high degree of polish and optical flatness is quite amazing. The techniques used to shape and polish the artifact are unknown. ... no function for the object other than that of a compass pointer has been suggested by anyone who has examined it critically. ... The observation of Olmec site alignments 8°W of north is a curiosity in its own right, ... predates the Chinese discovery of the geomagnetic lodestone compass by more than a millennium. ... At present, M-160 is a unique artifact and San Lorenzo a unique site: the first civilized center of Mesoamerica and probably of the New World".

## Discussion and Dating

This typical megalithic petroglyph of Chao Redondo dates from after the discovery of America via the Bering Sea in c.2600 BC (Refs.1-13). The fact that this petroglyph illustrates the crossing of the Davis Strait at the Arctic Circle points to an early date, not long after c.2600 BC. For safety reasons they preferred sailing this route rather than a faster route straight from Greenland to Labrador. Current climatic research based on study of the salinity of deep Greenland ice cores is showing that following the difficult discovery of America, the weather became more favorable in the following centuries. "The Northern Hemisphere was experiencing extreme summer -like conditions around 2200 BC ... this was a very big event, a remarkable event that must have covered a large area" Ref.27).

As concluded earlier in this article, the glyph itself indicates it dates from just after the discovery of Bermuda, c.2200 BC. A C-14 analysis from a charcoal drawing in the Pileta Cave in Spain (Fig.9, Chapter 7, Ref.25), showing the important distances across the Ocean, confirms this date. The glyph date cannot be much later, because in Southern Europe (including Chao Redondo) the megalithic culture ended c.2000 BC. America's Stonehenge in New Hampshire is related, in its second building phase, to Bermuda too, so megalithic sites on both sides of the ocean bring contributing pieces of information to the story of Bronze Age exploration of the New World. The crossing of the ocean via the Upper North was an alternative to the Southern Crossing from Africa to South America. Due to the cold climate it was of less importance. This is also true of the return route, which became focused on the Azores.

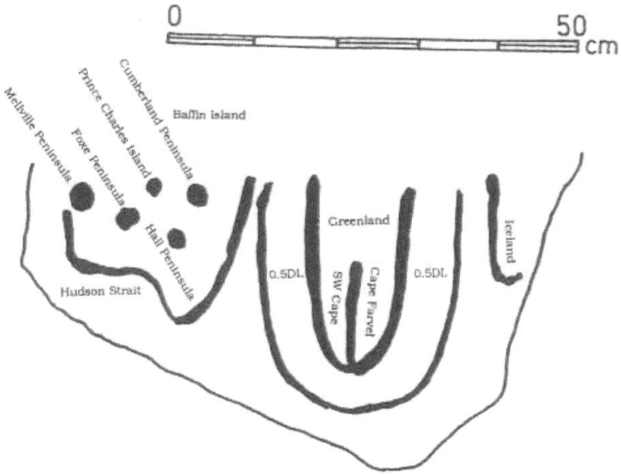


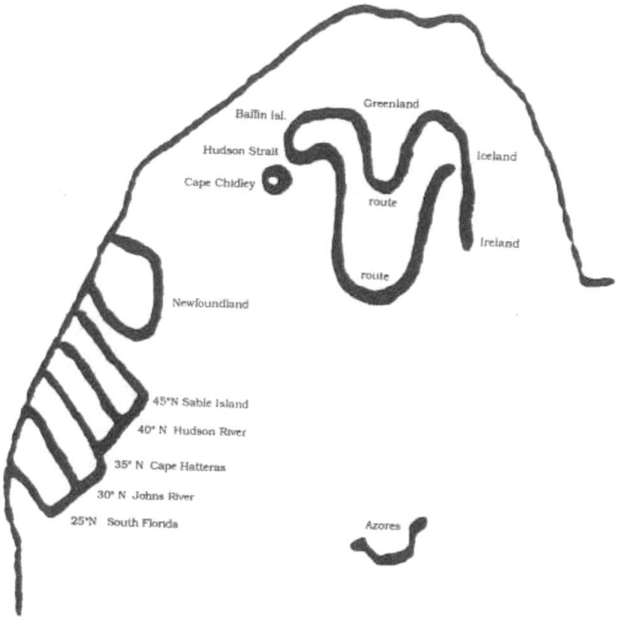


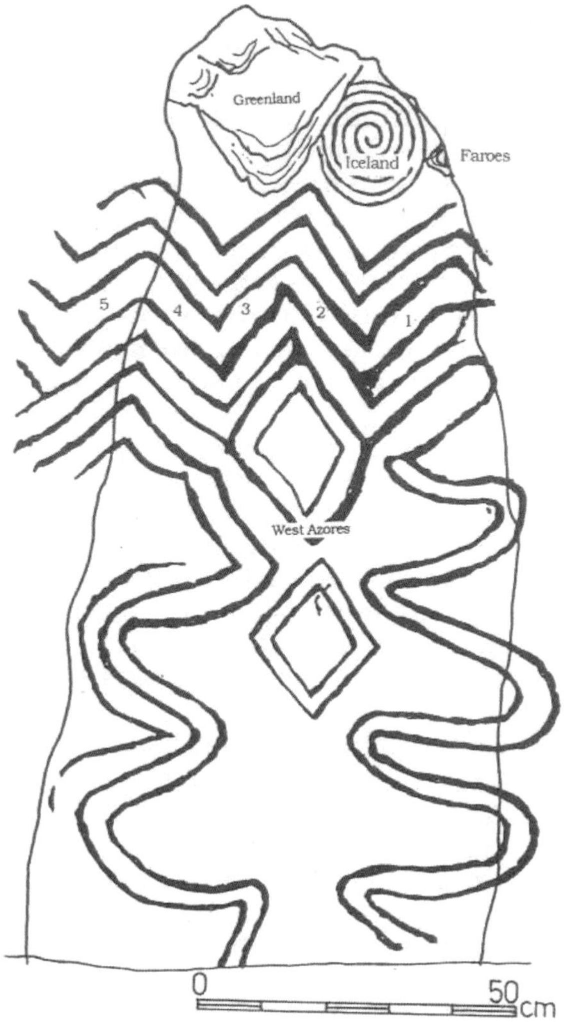


**Fig.9 (Top left):** In the center is South Greenland, with Cape Farvel and the SW Cape, and around it the coastal waters over 0.5DL= 555km. This is also the approximate sailing distance from Iceland to Greenland, and from Greenland to Baffin Island. The dots at the left side are high areas as seen from small boats, which can be reached through Hudson Strait. These areas were important in the search for a sailing route toward the Bering Sea. The position of the stone in the original monument is unknown. (see Fig.6). (Mane Rouillarde, Brittany, Ref.1, c.2200 BC)

**Fig. 10 (Bottom left):** The Upper North Route from Ireland via Iceland to Cape Farvel, and behind the west coast of Greenland (!) to Baffin Island at about 65°N. Cape Chidley is at the other side of Hudson Strait. South of Newfoundland are the important latitudes of the east coast of the U.S. The return route is via the Azores in the middle of the Ocean. (Mane Rouillarde, La Trinite sur Mer, Quiberon, Brittany, c.2300 BC)

**Fig. 11 (Right):** The North Atlantic Ocean, with, at the right, the Old World, and at the left, the New World. On the top, the Faroes, Iceland, and Greenland. In the north, the width of the ocean is indicated over 5DL= 5555km. In the middle, the two important islands of the West Azores. (Barclodiad y Gawres, Anglesey, Wales, c.1900 BC) (Ref.1)

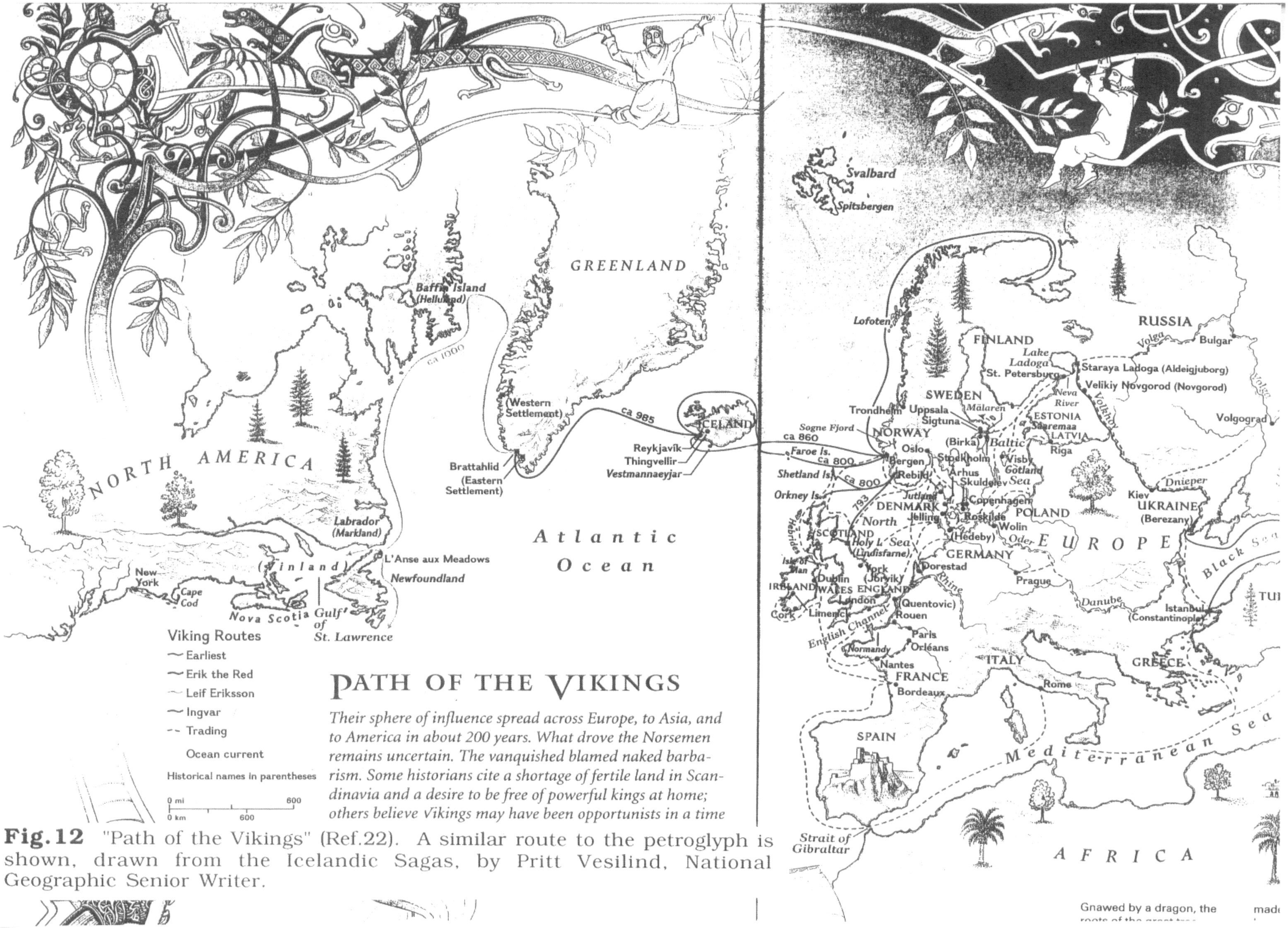


**Fig.12** "Path of the Vikings" (Ref.22). A similar route to the petroglyph is shown, drawn from the Icelandic Sagas, by Pritt Vesilind, National Geographic Senior Writer.

# References

1. Twohig, E. Shee, The Megalithic Art of Western Europe, Clarendon Press, Oxford, 1981
2. Jonge, R. M., de, and IJzereef, G.F., De Stenen Spreken, Kosmos Z&K, Utrecht/Antwerpen, 1996 (ISBN 90-215-2846-0) (Dutch)
3. Jonge, R. M. de, and IJzereef, G.F., Exhibition: The Megalithic Inscriptions of Western Europe (1966)
4. Tour Guide Map, America's Stonehenge, PO Box 84, North Salem, NH 03073
5. Lenik, E.J. and Gibbs, N.L., "The Frost Valley Petroglyph, A Catskill Mountains Enigma", NEARA Journal, Summer1999(ISSN: 0149-25-51)
6. Ferryn, P., NEARA Journal, Vol.XXXI, No.2, p.59 (1997)
7. Fell, B., America BC, Pocket Books, Simon & Schuster, 1994
8. Bailey, J., Sailing to Paradise, Simon & Schuster, 1994
9. Thompson, G., American Discovery, Misty Isles Press, Seattle, 1994
10. Peterson, F.A., Ancient Mexico (1959)
11. Stuart, G.E., "New Light on the Olmec", National Geographic, Nov. 1993.
12. Bernal, I., The Olmec World, University of California Press, London, 1969 (ISBN 0-520-02891-0)
13. Zapp, I and Erikson, G., Atlantis in America, Adventures Unlimited Press, 1998 (ISBN 0-932813-52-6)
14. Mixter, G. W., Primer of Navigation, 4th Edition, D. Van Nostrand Co. Inc., Princeton, N.J., 1960, Lib. Congress #60-12707, pg. 34, Fig.203b
15. Ocean Passages for the World, Third Edition, 1973, published by the Hydrographer of the Navy (first published 1895), Hydrographic Department, Ministry of Deffense, Taunton, Somerset, TA1 2DN
16. MacMillan, M., Green Seas and White Ice, Dodd, Mead &Co., 1948
17. Corliss, W. R., Ancient Man, A Handbook of Puzzling Artifacts, Sourcebook Project, Glen Arm MD 21057,1978(ISBN 0-915554-03-8)
18. Coe, M. D., Principal Adviser & Editorial Consultant, Mysteries of the Ancient Americas, Reader's Digest General Books, USA, 1986 (ISBN 0-89577-183-7)
19. Thirslund, S., and Vebaek, C.L., Vikingernes Kompas, Kopenhagen, 1990.
20. North Atlantic portion of Chart of Magnetic Variation, Epoch 1995.0, Charts of the Earth's Magnetic Field, published by the Defense Mapping Agency, copyright 1996, US Government
21. Portion of "The North", a map Produced by National Geographic Maps, September, 1997.
22. Vesilind, P., "Path of the Vikings", National Geographic, Vol.197, No.5, May 2000, pg.12
23. Lambert, J.D., "America's Stonehenge", an Interpretive Guide", Sunrise Publications, Kingston, N.H., 1996 (ISBN 0-9652630-0-2)
24. Donnelly, I., Atlantis, Harper & Row, San Francisco, 1971 (ISBN 0-06-061960-0) (pg.441)
25. Sanchidrian, Torti, J.L., and others "International Newsletter on Rock Art", No.29, 2001 (pg.17, Fig.1)
26. Malmstrom, V.H., Cycles of the Sun, Mysteries of the Moon, The Calendar in Mesoamerican Civilization, Univ. Texas Press, 1977 (ISBN 0-292-75197-4)

# A Nautical Center for Crossing the Ocean

## The Decipherment of Angular Encoding

(America's Stonehenge, North Salem, New Hampshire, c.2200 BC)

Dr. R.M. de Jonge, drsrmdejonge@hotmail.com
J. S. Wakefield, jayswakefield@yahoo.com

### Introduction

America's Stonehenge (formerly called "Mystery Hill"), is the most important megalithic complex of North America (Refs.1,26). It is situated near a tributary of the Merrimac River in North Salem, New Hampshire (US), about 30 kilometers from the East Coast at 43°N. The Main Site of the complex (Fig.1) has a diameter of c.50 meters, and contains about 15 chambers, some of them connected to each other. Tens of very heavy stones, up to a weight of 50 tons, are incorporated in the structures. It is the center of a larger area (Fig.2), having a diameter of c.250 meters, with numerous stone walls. In these walls, more than twenty big "menhirs" (French term for single large vertical stones) have been placed. These menhirs, as well as the heavy stones in the Main Site, are typical features of the megalithic culture known in Western Europe and NW Africa, dating from roughly 6000 to 1500 BC.

Some of the huge rocks and chambers of America's Stonehenge were used as the foundation of a house once, and the chambers served as a station of the "Underground Railroad" for fugitive slaves. The complex has been studied for its astronomical alignments (Fig.2). Nevertheless, it has been ignored by American archaeology, though they have not always claimed it "fake", as they have with most of the linguistic petroglyphs in the United States. We find this to be a complex megalithic site, with the astronomical attributes of the site design integrated with design features that have encoded geographic meanings. In checking the "astronomical alignment map" at the site, we have found the angles of the groundplan to be accurate, but the symbols indicating the sizes of the menhirs to be sometimes in error. We will note some of the site map mistakes as we discuss these features in the text. **Site angles that encode latitudes are in bold print** to help organize the data and try to keep you awake, because the monument is rather complicated.

We will explain the geographic features we have found encoded in the site design, how the angles of the menhirs reveal the "sailing by latitudes" technology of the times, and how America's Stonehenge was the religo/travel center for crossing the ocean to Europe in the Bronze Age. We think this was a teaching center about oceanic geography and sailing routes, a place to get accurate predictions for the safe timing of oceanic voyages, and a place to make the sacrifices that would ensure safe passages.

You will find that thinking like a man of that time period is a little different

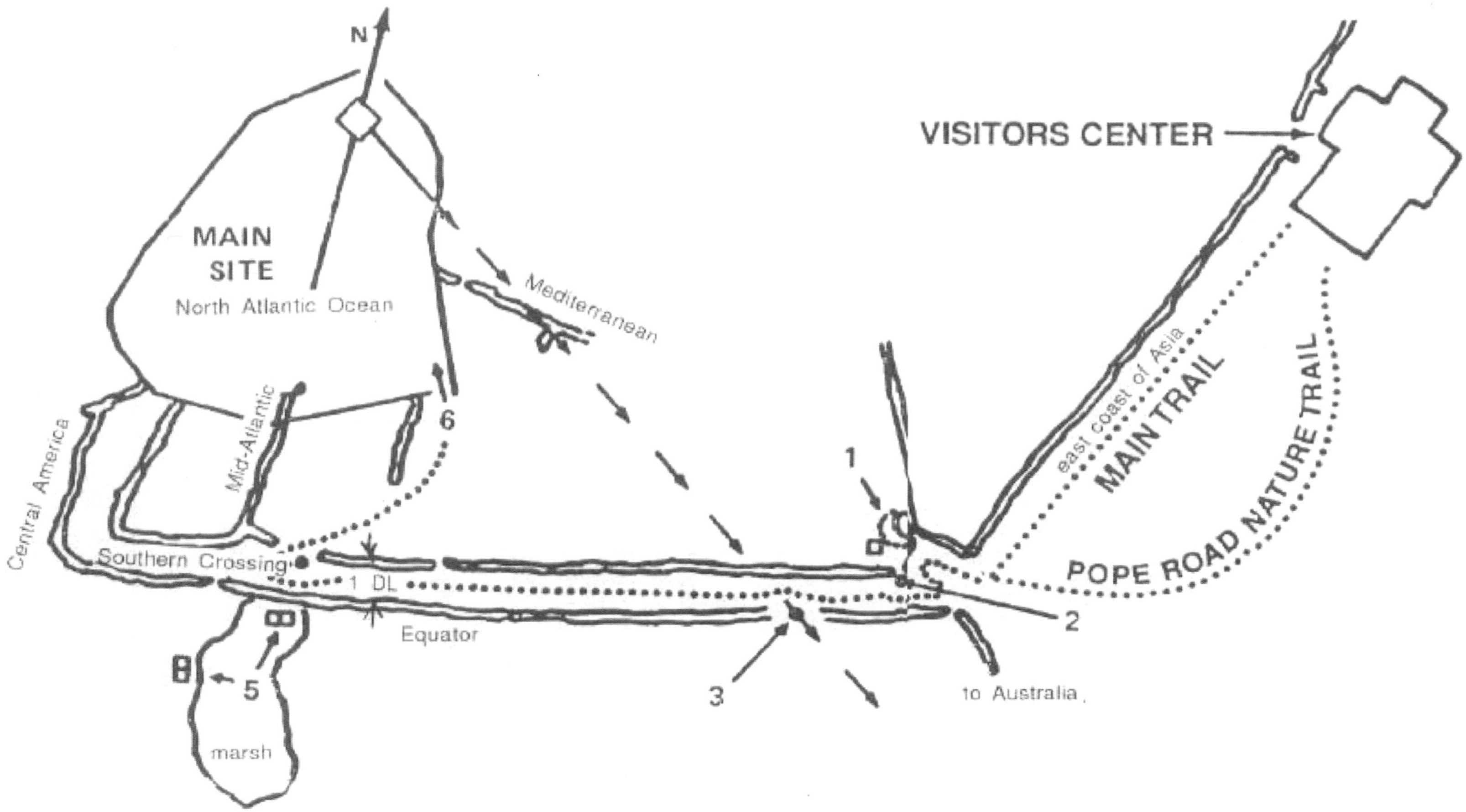


**Fig. 1** The entrance of the Main Site of America's Stonehenge (Ref. 1), with geographic meanings.

Fig. 1

from the way we think of things today. It has taken Dr. de Jonge ten years of study to learn how to see things from their point of view (Refs.2,9). Since there was no written phonetic language (symbols representing spoken sounds) at this time, we have learned about their astronomic knowledge, religion, calendar, and sailing technology by studying the numeric characteristics of their inscriptions and monuments. Dr. Stecchini (Ref.25), feels that "their maps were drawn so that the key positions ... could be memorized, ...and that the lack of a printing press is why we find so many mnemonic devices". Before we can explain the implications of the angles shown in our later exhibits, we need to explain the overall geographic symbolism in the layout of the stone walls found on the site. Then later you will see how this interpretation of the walls is supported by the angular data.

## Some of the Outside Walls

Due south of the Main Site, as shown in Fig.1, is a wall complex with a pattern nearly identical to that of tens of stylized inscriptions in Western Europe (Ref.2), dating from before the discovery of America (c.2500 BC). We know for sure it represents the southern part of the North Atlantic Ocean. The lower south wall is the equator, and the outer west wall is the (stylized) coast of Central America. The outer east wall, which should represent the coast of NW Africa is omitted; it would have been to the right, or east, of the "Madeira-Cape Verde wall" shown at the SE corner of the Main Site (see also Fig.2). The parallel inner walls have been placed at 1 "big distance line" (= 1DL= 10° of latitude) from the equator and the coasts, corresponding to a distance of 1111km. The central North-South Wall (or "true south wall" in Fig.2), divides the ocean into a well-known eastern half, and a less-known western half. It roughly coincides with the Mid-Atlantic Ridge of today (which was already suspected in that time because of islands that had been discovered in mid-ocean, and which is drawn in many megalithic inscriptions (Refs.2,5).

Due north of the Main Site is a single wall (angling off to the upper right in Fig.2, and in detail in Fig.3), which clearly represents the east coast of Greenland (compare it with a map). Stonehenge in England and Loughcrew in Ireland both show that this coast had been discovered by c.3300 BC, a thousand years before the construction of this wall in New Hampshire. The east point of this wall (Fig.3, at the top) is Cape Brewster at 70°N. The lowest point of the wall is nearby the Main Site, with its ceremonial center. This is Cape Farvel, the south point of Greenland, at 60°N. Thus the Main Site of America's Stonehenge is the North Atlantic Ocean, roughly from the Tropic of Cancer in the south to Greenland in the north. This site is a big walk-in scale-model sailing chart of the ocean, with, as we shall see, particular emphasis on sailing routes for getting back, with the wind and current, to the Old World.

## Egypt and Central America

Due east of the Main Site is a single wall, shown in Figs.1&3. This wall represents the Mediterranean Sea. Close to the end we see a short branch

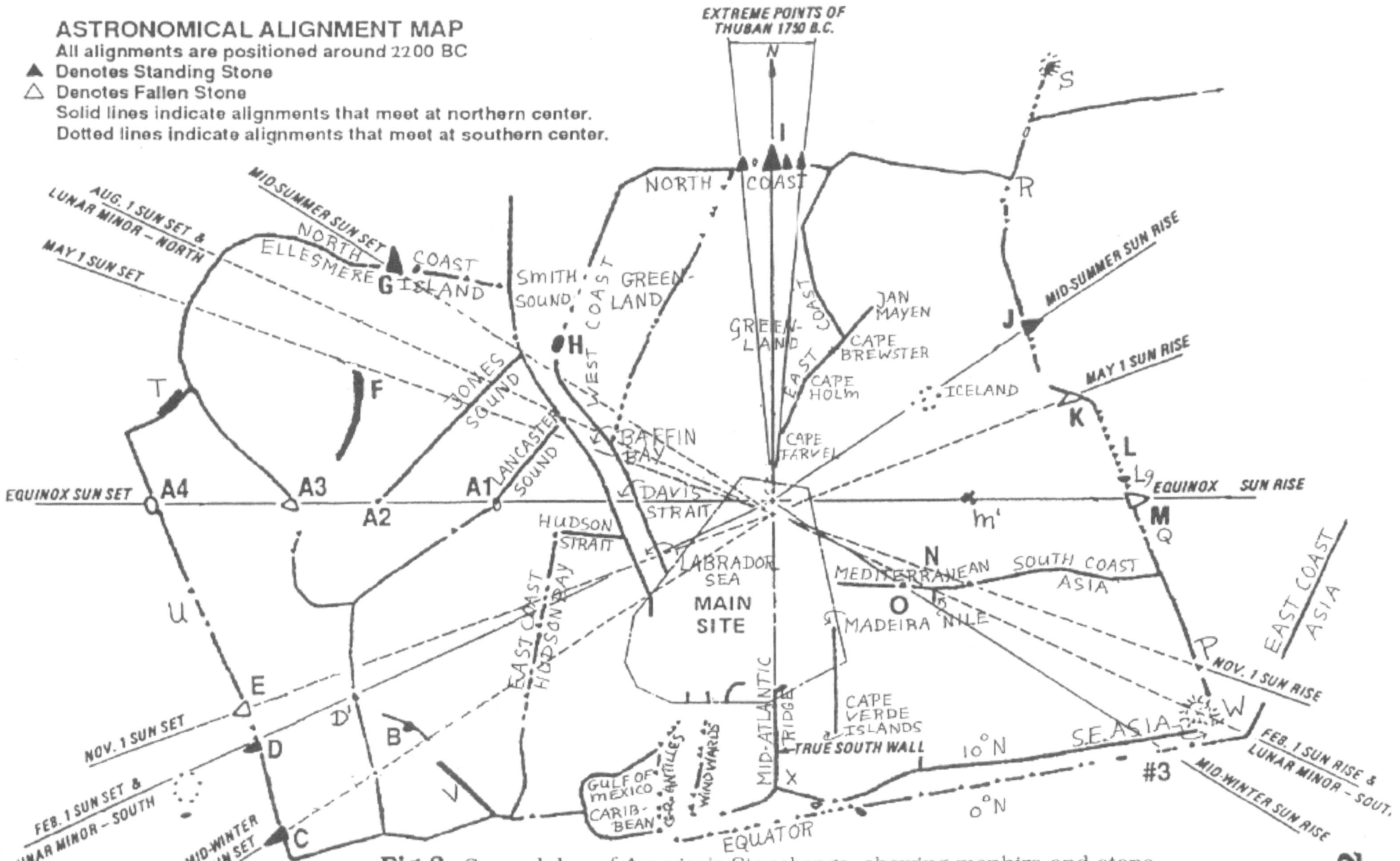


**Fig.2** Groundplan of America's Stonehenge, showing menhirs and stone walls (Ref.1), with geographic meanings. The straight lines are astronomical alignments.

**Fig.2**

to the south. This is the river Nile, the cradle of the greatest civilization then on earth. Not shown on these site maps is the large stone that you will find at the south end of this wall, representing the capital city of Egypt, the center of the Sunreligion.

After the discovery of America (c.2500 BC, Ref.2), many expeditions were undertaken from Egypt to the old civilization of Central America (Refs. 7,11,14-19). From the eastern part of the Mediterranean one sailed to the island of Sicily (the end of the first piece of the wall in Fig.3). Next, one sailed in the western part of this sea to the Strait of Gibraltar (the end of the second piece of the wall). It is known that in those times the Mediterranean was considered to be a two-part ocean, as there were eastern and western portions that were too big to see shores when sailing in the middle of them (Ref.19).

On the Atlantic Ocean, one traveled from Madeira (the starting point of the eastern inner wall on the SE side of the Main Site) via the Canary Islands to the Cape Verde Islands (the south end of this inner wall). Compare this representation with a map of the Atlantic. From there they sailed via the "Southern Crossing" to the NE coast of South America (through the opening in the southern inner wall, where the well is located). They sailed through two big Distance Line segments (indicated by the east and west sides of the mid-ocean line) thus crossing approx. 2222 km before arriving at the South American coast. The tradewinds and currents at these latitudes constantly moved the ships to the west. Below the Southern Crossing we see a marshy area (Fig.1), representing a part of the South Atlantic Ocean. By following the coast on the route below Field VII (Fig.3), one reached the islands of the Caribbean and the Gulf of Mexico (the wall around Field V, see also Fig. 2) (Refs.14-18).

However, the problem is the way back. Because of the tradewind and its current, it is impossible to accomplish the Southern Crossing in the reverse direction. For that reason, like the Spanish Galleons later, they had no choice but to sail to the north, following the Gulfstream along the east coast of North America (the western outer walls of the Main Site). It is not surprising to find megalithic remains at the latitude of America's Stonehenge, which is similar to the latitude and climate of their homelands in Europe.

## The western part of the Main Site: North America

The first piece of wall just outside the lower left corner of the Main Site in Fig.3 is the end of Cuba, where one can cross the Gulfstream to Bimini in the Bahamas. The Florida Strait between Cuba and Florida is shown. After the crossing, the wall continues northeast past the Abacos to the lowest chamber of the Main Site, which represents Bermuda 1000km offshore (see also Figs.4,5, and Photo.3).

The angular wall in the lower left corner of the Main Site (Fig.3), is the coast

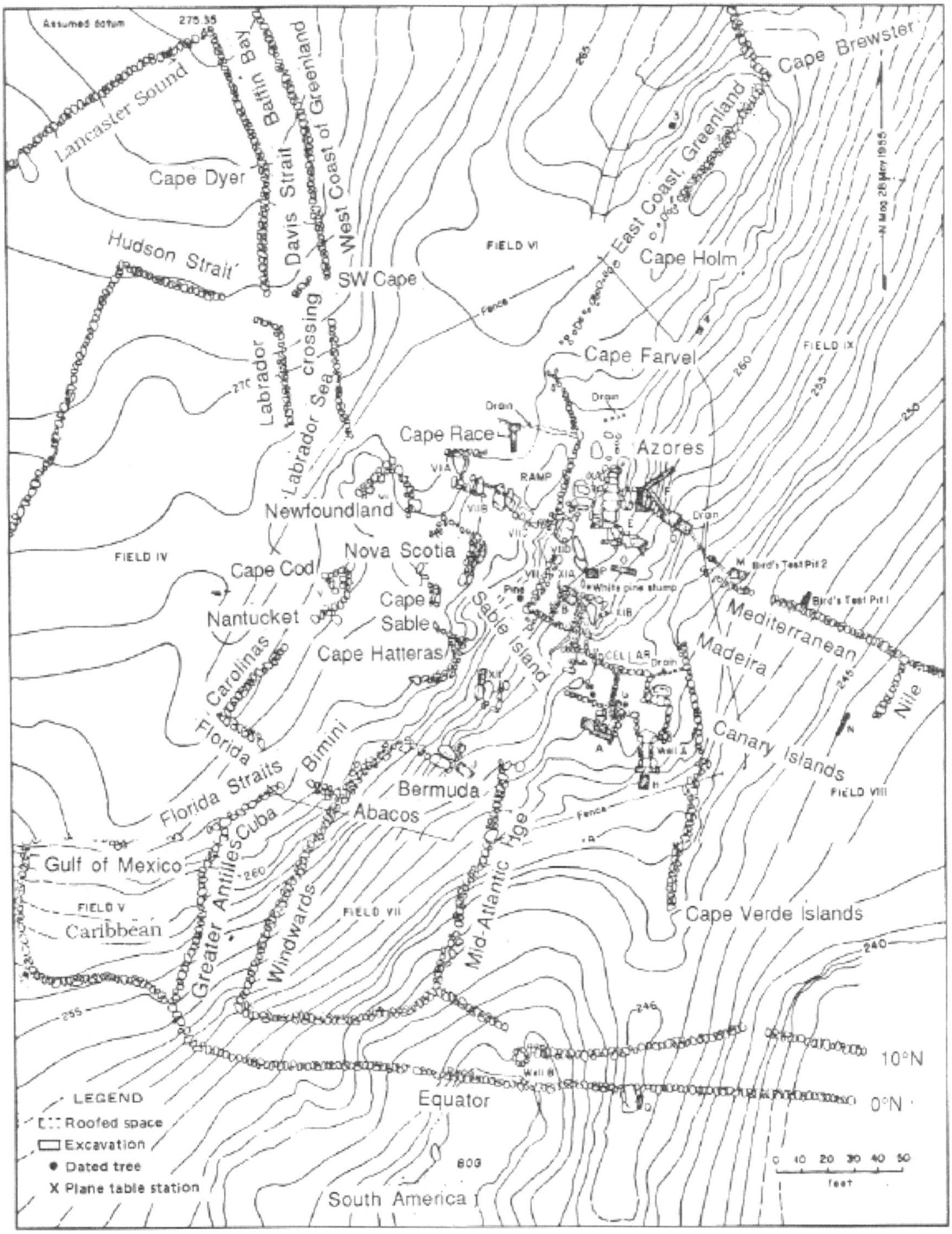


Topographic Map of Mystery Hill, North Salem, New Hampshire. (After *North Salem, N.H., Site Excavations* Report, 1955 by G.S. Vescelius.)

**Fig.3** The Main Site (Ref.26), with geographic meanings.

**Fig.3**

from Florida to Cape Hatteras, North Carolina. The more easterly inner wall (Fig.3, Fig.4 no.14, and Fig.5) shows an enlarged representation of this important cape (Photo.4), that clearly distinguishes between Cape Hatteras South (at 35°N), and Cape Hatteras North (at 36°N), because it is lengthy, and curved back at both ends, like the Cape.

The next northerly wall, or third outer wall (Fig.3, no.V) represents Cape Cod. Follow the stone row with your finger. You will notice at the site a large rock at the tip, probably representing the Peaked Hill Bars visible to sailors rounding the Cape east of Provincetown. The small island of Nantucket, a target to mariners coming from the south, is shown by a stone against the lower wall (there is a small megalithic dolmen out there too). To the right of Cape Cod are rock walls (Fig.3, Fig.4, no.21, and Fig.5) that may have been a chamber (it is thought that big stones were quarried here). The lowest rocky point is Cape Sable (see Fig.5), the south tip of Nova Scotia. Next, the wall shows the south, west, and north coasts of this peninsula. The rocky point at the right side is shallow and foggy Sable Island (Photo.5), later known as the "graveyard of ships", more than 150 km offshore.

The fourth outer wall, which looks like a hat (Fig.3 no.VI, Fig.4 no.17, and Fig.5) shows clearly the NW and NE coasts of Newfoundland, the easternmost land mass of North America. This may have contained a small chamber too. Cape Race is represented by a few big stones at the tip (Fig.5). The corresponding inner wall now finishes at the "V hut" near VII A Fig.3, #18 in Fig. 4, and shown in Photo.6 as the "space under" in Fig.5. This small chamber represents Cape Race also, which on a great-circle sailing chart is the easternmost point of the continent. The large cover stone of this chamber closely resembles the shape of the island of Newfoundland, the easternmost land mass of North America.

From these features we can start to see what was going on at America's Stonehenge. The Main Site has been built to help people who want to return home by sailing across the North Atlantic Ocean to the Old World! We should not forget, that originally the megalith builders were descended from there. It was their homeland! The site terrain slopes down toward the east (Fig.3, Ref.26), so from the high west side of the site, people could survey the layout of the walls to see how this could be accomplished!

## The eastern part of the Main Site: the Islands in the Ocean

The oldest way back (the Upper North Crossing) was via the well-known east coast of Greenland (the wall north of the Main Site). To sail across Davis Strait at the Arctic Circle is a distance of only c.400km. A beautiful petroglyph of this whole return-route from Baffin Island to Iberia can be seen in Chao Redondo in north Portugal (c.2200 BC). However, this is a long and severe course into cold latitudes, which should now be disuaded. For that reason, at the east side of the northern wall (Fig.2) Iceland is only indicated by a small group of 7 stones (if at all), though Iceland is important in the northern route, and fairly common in early megalithic "art" (Ref.9).

**Photo.1** The authors beside the Sacrificial Table in the Processional Walkway (Fig.5), seen from the southwest. Note stone representing Greenland on the left (looked at upside down, it looks like a map of South Greenland).

**Photo.2** Close-up of surface of Sacrificial Table. It represents the North Atlantic Ocean, having an edge with a width of 1DL= 1111km.

The eastern part of the Main Site (Figs.4&5) deals only with the way back via the Azores, which had become much more important. This is the famous, long-known archipelago in the middle of the ocean.

The Ceremonial Center (Photo.1) is by far the most important feature of this site (see Figs.4-6,8,11). There are 3 big chambers, corresponding with the 3 island groups of the Azores: the West Chamber corresponds with the West Azores, the Oracle Chamber Corridor with the Central Azores, and the Cavern with the East Azores (compare it with a map). The Azores Islands, with their discovery c.3600 BC celebrated in the Tumulus of Gavrinis (Brittany), had been revered as the mid-ocean abode of the SunGod for a thousand years, and depicted in many megalithic inscriptions and monuments.

At the right side below the Ceremonial Center is the so-called "northern Pattee area" (Figs.5&7). The two chambers correspond with the two Islands of Madeira. These Islands had been discovered c.4100 BC. The "southern Pattee Area" contains 2 big chambers, corresponding with the western and eastern Canary Islands. However, there is damage reported by the taking of stone by colonists in the last two centuries here, so this ground should be carefully studied. This part of the Pattee Area probably contained 7 small chambers, corresponding with the 7 Canary Islands. These Islands are situated at c.100km from the coast of NW Africa, and they had been discovered c.5500 BC (Ref.2).

As said earlier, the southernmost chamber (#13, Fig.4, Photo.3) of the Main Site represents Bermuda. Probably about a century after the first construction phase of America's Stonehenge, the Islands of Bermuda were discovered from the continent of North America (c.2200 BC, Ref.2). Though they lie at the top of the doldrums of the Sargasso Sea, often called the "Bermuda Triangle", these islands apparently have played an important role in early crossing of the ocean by small vessels. This chamber, like the other western chambers, has its entry in the south, which is the direction people sailed into most of these places represented by chambers, after visiting Central America.

## The Ceremonial Center: The Azores

In the middle of the Ceremonial Center (Figs.5,8,11) is the Processional Walkway. It symbolizes the Atlantic Ocean. The lower Walkway is the southern ocean, and the upper walkway is the northern ocean. As a consequence, these walkways, being waterways, never were covered with stones (Ref.1). Note the orientation of the walls: below and opposite the lower standing rock is a true EW wall, and above the upper standing rock is a real NS wall. Since the Processional Walkway is the Atlantic Ocean, the West Chamber symbolizes the New World, and the Oracle Chamber Corridor and Cavern symbolize the Old World.

The most important travelers came from the center of the United Egyptian

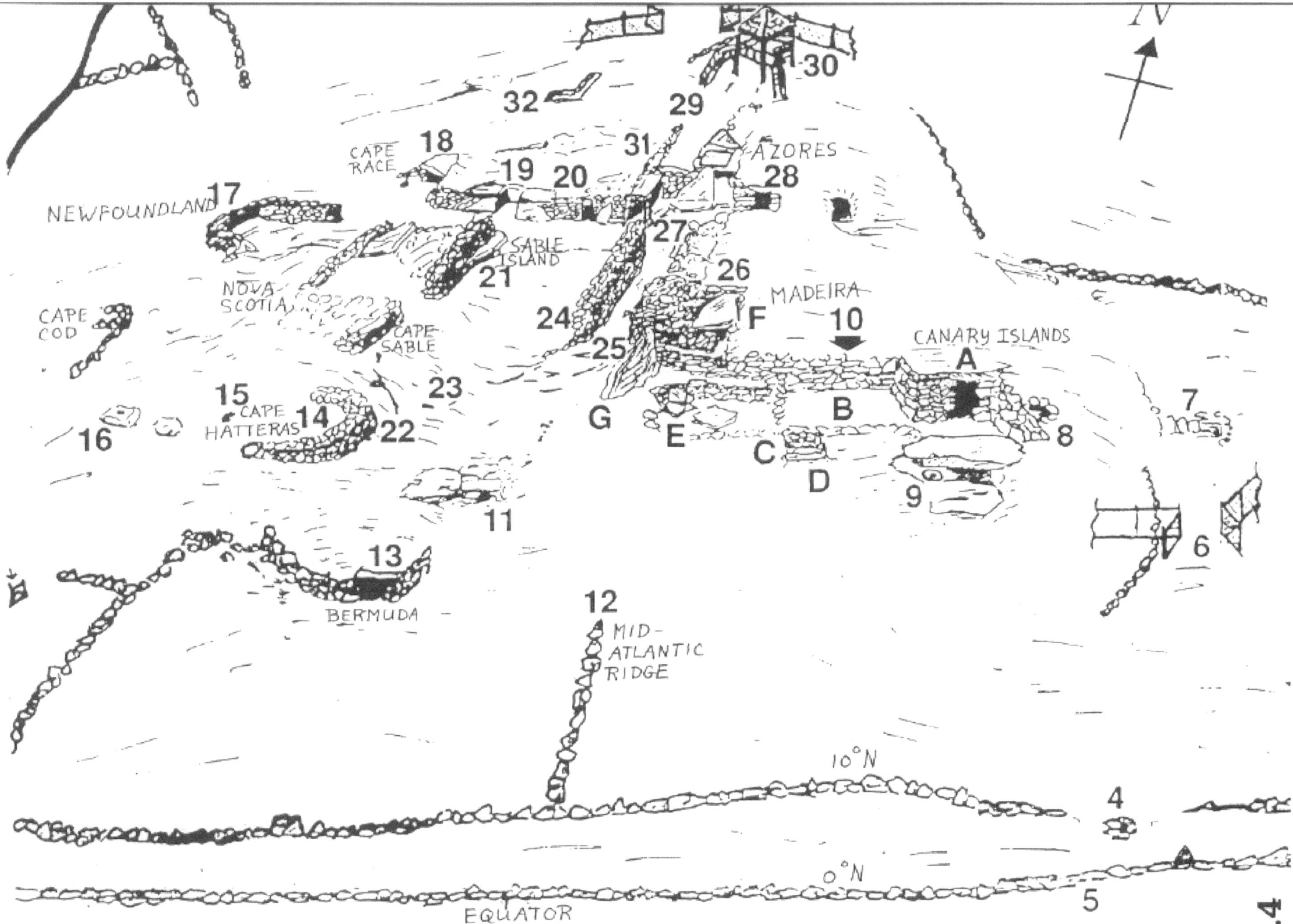


**Fig.4** Survey drawing of the Main Site from the south (Ref.1), with geographic meanings.

Fig.4

Empire (the eastern entry of the Cavern, Figs.8,10,11, "which may have been enlarged by vandals from a window", Ref.26), halfway between the Tropic of Cancer and the Nile Delta, at 26°N. They sailed through the Mediterranean (the Cavern) 10° to the north towards the Strait of Gibraltar at 26°+10°= 36°N. For that reason, the **south wall** of the east side of the Cavern makes an **angle of 10°WNW** (drawn on Figs.8,11). There are five storage niches at the eastern end of this cavern (Photo.7), which could function as repositories for goods from various ports in the eastern Mediterranean. The so-called "closet" (Fig.6, nr.G) at the south side of the Cavern represents the sea area of the Gulf of Sidra and the Gulf of Gabes, which has the same shape as the closet, surrounding the megalithic centers on Malta and Gozo.

From Gibraltar they sailed to Madeira in the Atlantic, 10° of latitude below America's Stonehenge, at 43°-10°= 33°N. For that reason the **south wall** of the west side of the Cavern makes an **angle of 10° WSW**. From Madeira they sailed 18° of latitude to the south towards the Southern Cape Verde Islands, at 33°-18°= 15°N. For that reason the **south wall** of the Oracle Chamber Corridor, which represents these islands, makes an **angle of 18°WSW** (Fig.11). We believe these angles were carefully researched by Payne (Ref.3), as they show in his architectural drawings. Just before you walk out of the south exit of the oracle chamber, you will see a small niche on the east side. This probably represents a then-important port on a river mouth on the African coast.

After the Southern Crossing of the ocean, all sailors landed on the NE point of South America (the Sundeck Chamber, Photo.8). The **wall** below it, which represents the east coast of South America, makes an **angle of 5° SSW, corresponding to the latitude of Cape Sao Roque, the NE point of Brazil, 5°S** (Fig.8). The old carvings at the northern end of this lower left wall show that early visitors of America's Stonehenge understood its orientation. "Cutouts in the north end of the wall indicate that other large stone pillars may once have stood (here)" (Ref.1). Such pillars would have signified entrance to the Americas at about 5°S.

From there they traveled in a WNW direction along the coast to Central America (the West Chamber). The first small door on the drawing (no longer there, as the wall has been rebuilt) led to the small right chamber, symbol of the culture of the north coast of Honduras. For that reason the **walls** of the West Chamber make an **angle of 16°WNW, corresponding to the latitude of the north coast of Honduras, 16°N**. The big door leads to the big left chamber, symbol of the civilization around the Gulf of Campeche. The small central room in between deals with the Yucatan Peninsula. Together, they form the United Central American Empire (the whole west chamber), which was also revealed at Stonehenge in England (Refs.14-18).

It is impossible to return to the Old World from the NE coast of South America (the lower standing rock, Photo.9) against the tradewind and its

**Photo.3** Southernmost chamber of the Main Site (Fig.4, #13), representing Bermuda, seen from the south, with viewing tower (Z) in the background.

**Photo.4** "The Pulpit" (Fig.4, #14), with a low chamber, representing Cape Hatteras.

current. The upper edge of this rock seems to resemble the slanting NE coast of South America. Instead, people crossed north of Cuba to the east coast of North America, just like the Spanish Galleons did later. For that reason, the south wall of the upper Walkway makes an angle of 23°WNW, corresponding to the north latitude of Cuba, at the Tropic of Cancer, 23°N (Fig.8). Over the Sundeck chamber, on this corner (Photo.8), is an upright and a cover, said to have been recently placed there. It is not clear whether there was originally an upper chamber here. If there were, it could have represented the islands of the Greater Antilles (Fig.11).

They sailed northerly along the coast of North America (the west wall of this Walkway, Fig.8). Arriving in New England, all these travelers met in the harbor at America's Stonehenge. In earlier times they would have returned via the south point of Greenland (the upper standing rock) and via Iceland (the north wall of the upper Walkway, Photo.10). But they no longer want to cross the Davis Strait (the exit of the upper Walkway, Refs.10-12). The **upper left wall** (Figs.8,11) makes an **angle of 60°NW, corresponding to the latitude of this crossing of Davis Strait, at 60°N.** On site, you will notice a small chamber under the corner where the 60° wall starts. This represents the important point on Davis Strait, Cape Chidley at the entrance to Hudson Strait at 60°N, where one would depart for Greenland. Note that the edge of the upper standing rock (shown in Photo.1, and in Figs.10,11), forms a coastal map of Greenland (when looked at upside down). Again, in view of the long cold distance across open sea, most people do not want to go directly to Ireland either (perhaps the roof-opening at the north side of the Oracle Chamber Corridor, shown in Fig.6, nr.E).

They want to sail from America's Stonehenge (at 43°N), and catch the Gulfstream off Cape Cod (see Fig.9), sailing to the Azores, Madeira, and the Canary Islands**,** all represented in the Sacrificial Table (Figs.6,8-11, Photos.1,2). These islands are familiar to them, are sited at comfortable latitudes, and provide them sustenance for the continued voyage. That is what America's Stonehenge is all about.

The **west wall** of the upper Walkway makes an **angle of 4°NNE, corresponding to the latitude of Cape Race (47°N), at Newfoundland, 4° north of here (43°+4°= 47°).** It is the easternmost point of North America. The **east wall** of the upper Walkway makes an **angle of 5°SSE, corresponding to the latitude of the Central Azores, 5° south of here.** The **upper east wall** of the Oracle Chamber Corridor also makes an **angle of 5°SSE, corresponding to the latitude of Madeira, another 5°south.** And the **lower east wall** of the Oracle Chamber Corridor makes an **angle of 5°SSE too, corresponding to the latitude of the Canary Islands, another 5° south.** From these islands they can sail to Western Europe, or NW Africa (the upper and lower spaces in the Oracle Chamber Corridor), with most people heading for the Strait of Gibraltar (the Speaking Tube) to the Mediterranean Sea (the Cavern). The **northern wall** of the Cavern makes an **angle of 7°ESE, corresponding to the latitude of Gibraltar, 7° south** of here. As we have said before, the southern

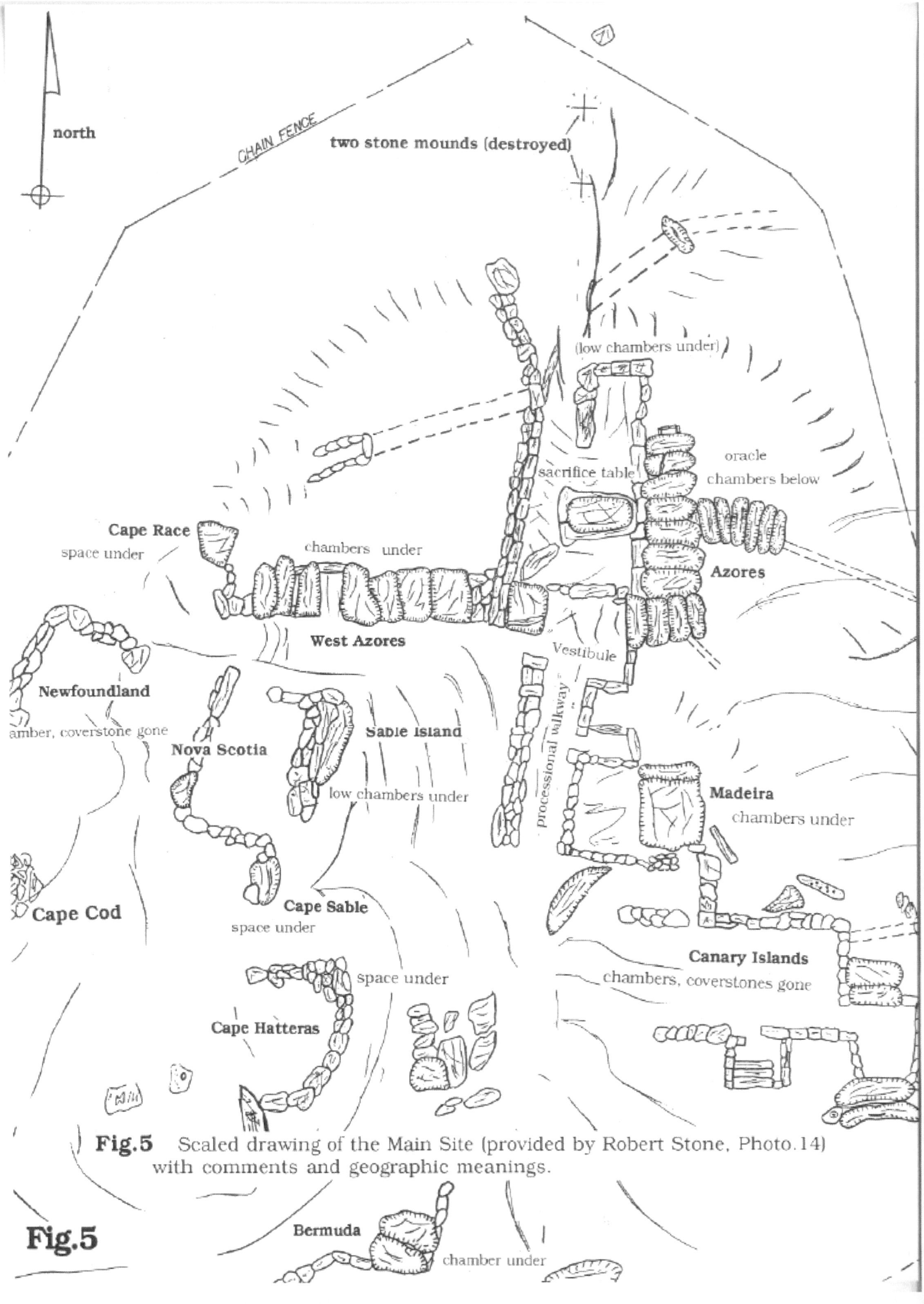


**Fig.5** Scaled drawing of the Main Site (provided by Robert Stone, Photo.14) with comments and geographic meanings.

**Fig.5**

branch of the Cavern is symbolic for Egypt, the great center of civilization.

At the eastern end of the Cavern is the so-called "running deer" carving (Photo.11) (Refs.1,26). Megalithic carvings of animals are extremely rare (Ref.5). We think it is a petroglyph of this crossing to the Azores, and the coastal sailing route from Cape Hatteras via America's Stonehenge to Cape Race.

The surface of the Sacrificial Table (Photos.1,2, Figs.5,6,8,10,11) is a stylized microcosm of the North Atlantic Ocean. Tens of similar tables have been found in the area around Portugal (Ref.7). The edge outside the groove has a width of 1 big distance line (=10°of latitude= 1111km). Within the groove, the surface of the ocean is largely unknown. At the right side below is the Southern Crossing, the exit of the groove. There, the water runs to the South Atlantic Ocean. The upper side of the Sacrificial Table is directed toward the Speaking Tube, because the entire monument of America's Stonehenge deals with the problem of finding the best and safest crossing of the ocean to the Strait of Gibraltar. The Speaking Tube is under the Sacrificial Table (Fig.6, nr.D, also Fig.10). Speaking through the tube distorts the sound of the voice. Probably, powerful personages of the intellectual elite, hidden in the Oracle Chamber, taught the recorded wisdom of the monument, and enriched themselves by selling safe passages.

The west (4°) and east (5°) walls of the important upper Walkway do not run parallel, but are aligned on an important menhir (Figs.3,8,11), making an angle of 9°. This corresponds to the 9° of latitude between Cape Race and the Azores, to the initial sailing direction from America's Stonehenge to the 9 islands of the Azores, 9°ENE, and to the approximate direction of the magnetic north pole, 9.5°NNW. This menhir is in the wall above the Main Site, which represents the east coast of Greenland (Figs.2,3). Coming from the north, the menhir is located at the southern end of the first part of the wall. It represents Cape Holm, situated on the Arctic Circle, at 67° N. When walking the site, we found a large broken menhir (Photo.12), confirming the importance of the wall angle there, which we were previously calling Cape Holm by itself! According to Stonehenge in South England, and the Loughcrew Inscriptions in Ireland, people discovered Greenland here thanks to the SunGod, crossing from Iceland at the Holy Arctic Circle c.3300 BC (Ref.2). The important point here is the Ceremonial Center focus upon this stone, representing a place Holy to them for such a long time.

At the lower end of the second part of the northern wall, two columns of stones were once present (Figs.3,5,10,11). The SE mound was Cape Farvel at 60°N, and the NW mound was the SW Cape of Greenland at 61°N. According to Stonehenge, in England, and Loughcrew, in Ireland, the SW Cape is where the megalith builders gave up their efforts to cross the Atlantic in c.3200 BC (Refs.2,4). For c.700 years the south coast of Greenland was the westernmost land of the then known world. This is why this well-known, historic place is used as the focal point for the crossing of

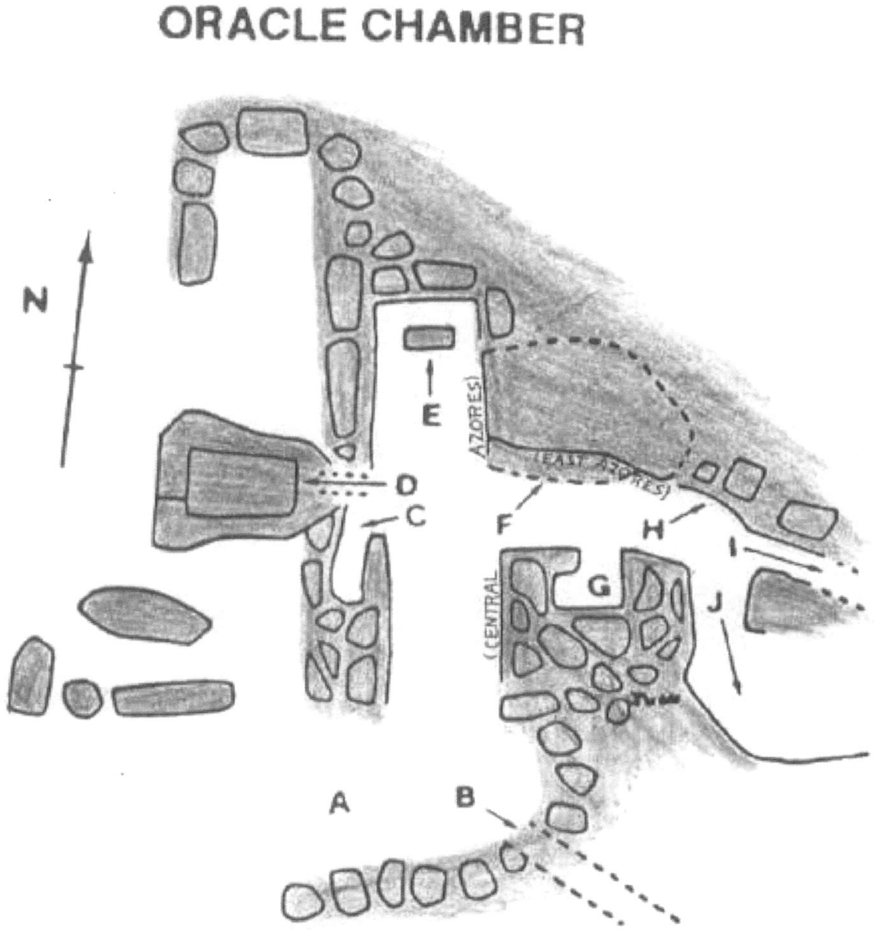


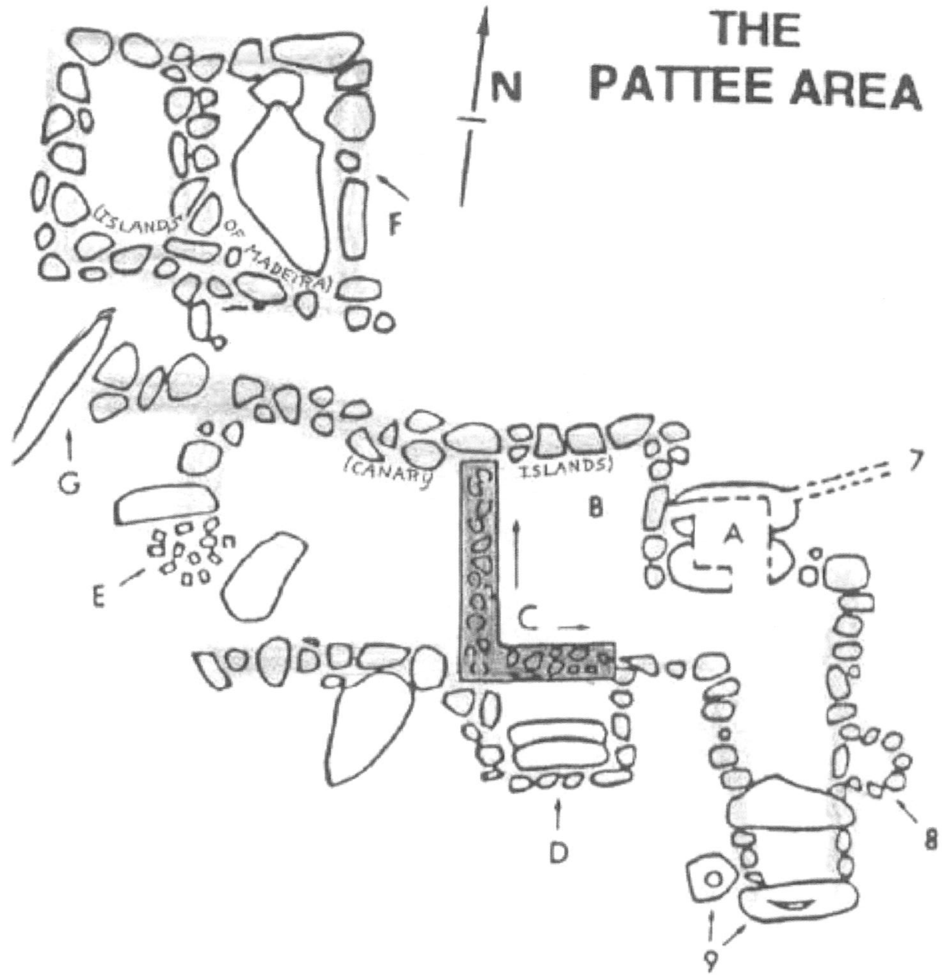


**Fig.6 Left:** Stone drawing of Oracle Chamber Corridor (center) and Cavern (right), with symbolic geographic meanings. A=Vestibule (formerly covered with slabs); B=upper drain; C=Secret Bed; D=Speaking Tube; E=Roof Opening (formerly with sliding slab); F=seat; G=Closet; H=Running Deer carving; I=lower drain and storage niches; J=entry.

**Fig.7 Right:** Stone drawing of the Pattee Area (Ref.1), with symbolic geographic meanings.

**Figs.6,7**

the important NS and EW axes of the complex (Figs.2,11). Any surveyor would start laying out a new chart using a known point as far west as he could be sure about. This is why this is the central viewing point for the alignments with the peripheral menhirs of the site (Photo.13).

## The Menhirs and the Outside Walls

## America's Stonehenge I: The Northern Crossings

America's Stonehenge was constructed to help and to teach people who want to cross the Atlantic Ocean. In principle, these folks want to go directly east. So menhir M, directly east on the EW axis, must be considered the most important (Fig.12). Seen from sighting point X on Fig.12, at the starting point of the true north-south wall, this menhir **M** makes an **angle of 39°, the latitude of the West Azores, 39°N.** (Remember that X is the place where the megalith builders first arrived, near the NE coast of South America (Figs.1-3). Remember that this NS wall, a mid- Atlantic line, points directly up to the chambers of the Ceremonial Center, which symbolize the mid-ocean Azores Islands, and the NS wall points to the south tip of Greenland, which they thought was halfway across the ocean.) Also on Fig.12, note that Menhir **K** is at an **angle of 53°, corresponding to the latitude of Ireland at 53°N**. Menhir **J** makes an **angle of 61°, corresponding to the latitude of the SW Cape of Greenland, 61°N.** These are the three possible destinations on the other side of the ocean (compare with a map). However, as we shall see, the angles of the menhirs on the large scale of the monument primarily provide information about crossing via the West Azores, the same route that was focused upon in the Ceremonial Center of the Main Site. Menhir **K** (Fig.12) makes a complementary **angle of 90°-53°= 37°, corresponding to the latitude of Santa Maria (East Azores), at 37°N**. Menhir **J** makes a complementary **angle of 90°-61°= 29°, corresponding to the latitude of the two easternmost Canary Islands, at 29°N.**

From our analysis so far, we have learned that point X at the start of the North-South wall (Photo.14) is the main viewing point of the site. Note that the complementary angle of the most important menhir of America's Stonehenge, **M, makes an angle of 90°-39°= 51°, which is the latitude of Stonehenge in England, at 51°N.** This is not a coincidence. According to carbon dating at both sites, both the large stone Sarsen Circle and Horseshoe (Phase III of Stonehenge) and the eastern half of America's Stonehenge (Phase I of America's Stonehenge) appear to have been have been under construction at approximately the same time, c.2300 BC. From the meanings of the menhirs and the walls on the eastern side of America's Stonehenge, and including the menhirs in the north, it is clear the two monuments are related to one major event: the discovery of the Americas.

## The Cape Race to Azores Sailing Route

If we place a protractor on menhir M in Fig.13, the menhirs of **I** (at the top of the page) give **the possible places of departure** in the north: **I1=41° (the island of Nantucket at 41°N,** near Cape Cod), **I2=43° (America's Stonehenge**

**Photo.5** The "Mensal Stone", coverstone on small chamber (Fig.4, #21), representing Sable Island.

**Photo.6** The "Vee Hut" (Fig.4, #18), representing Cape Race, Newfoundland, the easternmost point of North America.

**and Cape Sable at 43°N,** a big menhir**), I3=44° (Sable Island at 44°N), and I4=46.5° (Cape Race on Newfoundland at 46.5°N).** Cape Race is the easternmost point on the continent, so we must begin by considering this the most important place for departures. Its latitude is also equal to that of the NE point of Cape Breton Island, 46.5°N. As we know from Fig.12, from X menhir M makes an angle of 39°, which is also the sailing direction across Cabot Strait (c.100km) to the SW Cape of Newfoundland, 39°NE. The wall through M to the south (Fig.13) makes an angle of 21°, corresponding to the theoretical initial sailing direction from Cape Race to the West Azores, 21°ESE. However, because of the long distance, the curvature of the surface of the earth, and the route of the Gulfstream, the problem is more complicated. For that reason we are going to look from the viewing tower in the Main Site for more detailed information.

Viewed from Z, the viewing tower in Fig.13, Menhir K, the biggest menhir of America's Stonehenge (and therefore indicated to be important), makes an angle of 19°, corresponding to the sailing distance to the West Azores: 19 distance lines (19dl= 19°of latitude= 19x111km= 2111km). The short wall through K also makes an angle of 19°, which confirms this distance, which is correct. This route to the Azores involves sailing to the ESE, so let us look at menhirs placed to the ESE of the Z-M axis. Menhir P in the wall through M (see Fig.13) confirms the initial sailing direction with a slightly bigger angle: 22.5°ESE. However, because of wind and current conditions in the Gulfstream one is advised by menhir N (Fig.13) to use an even bigger initial sailing direction (ISD): 24°ESE. The terminal sailing direction (TSD) in the neighborhood of the West Azores given by menhir O is 34°ESE. Both menhirs N and O in the same EW wall have similar V-shaped notches at the upper sides (Refs.1,26). This similarity is indication that both menhirs are related to the same important crossing.

In practice, the chosen sailing direction during the whole voyage was continuously estimated by linear interpolation. (In this case for a 10 day voyage, each day a 1° bigger sailing direction, from 24°ESE to 34°ESE.) We know they used a compass, too, as that has been revealed in petroglyphs. A perpendicular to the imaginary equator of the monument points to magnetic north, 9.5°NNW, which they must have understood. Most of the time they would be aware of true north by the stars or the sun, and they could easily calculate their latitude from data obtained days before. Probably this was combined with a technique called "wayfinding" in Polynesia, where ancient people are known to have followed "starpaths" found by a memorized succession of rising and setting stars on the horizon at night. Once near the West Azores they could sail due east at 39°N, sailing the latitude line of the island group, as the Polynesians are thought to have done in the Pacific.

Other megalithic sites confirm this important route to the Azores. In Dissignac (Brittany, France) the crossing from Cape Race via the Azores to the Strait of Gibraltar was engraved (c.2300 BC, Ref.2). A beautiful inscription of this route to the Iberian Peninsula is shown in Chao Redondo,

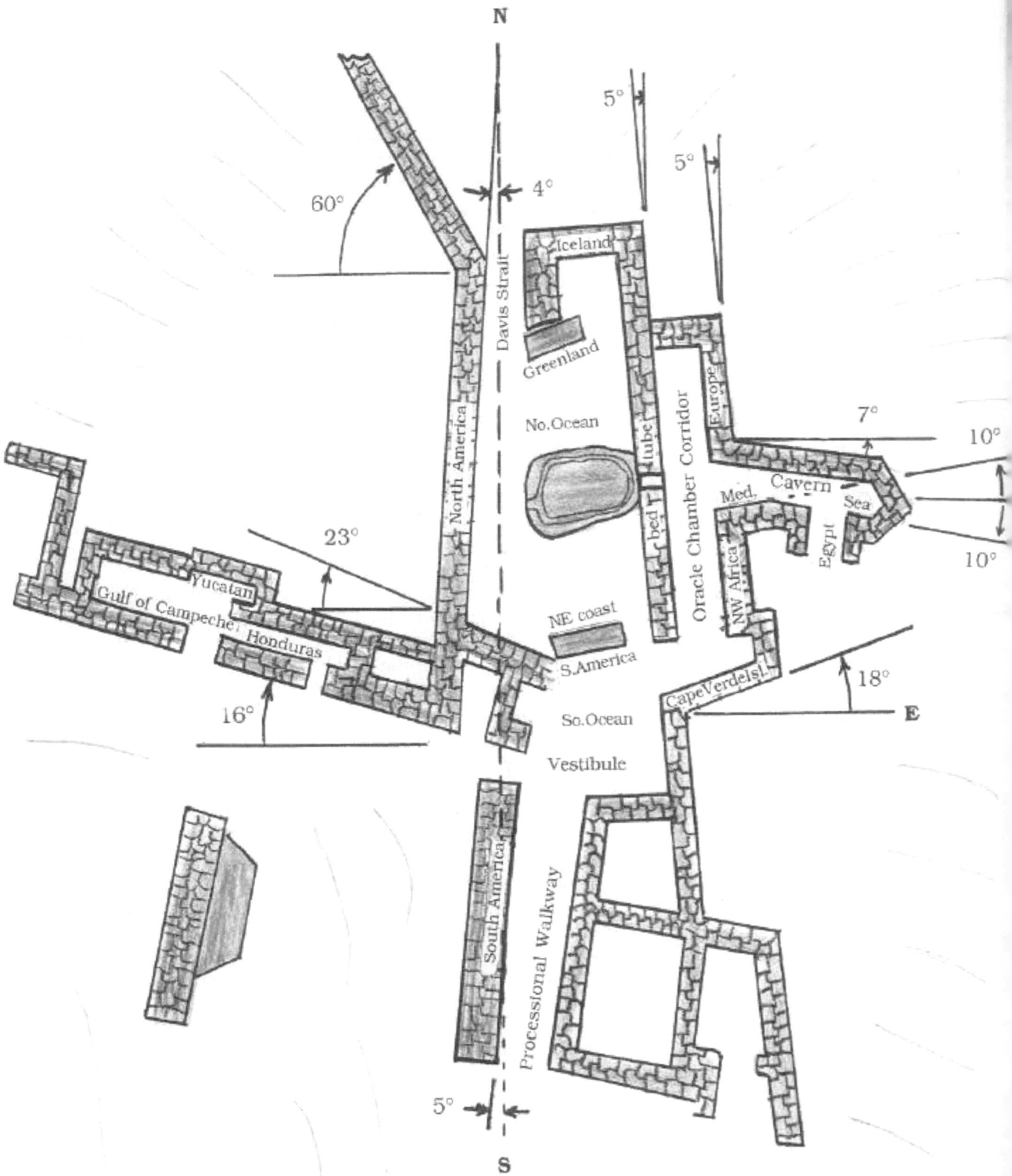


**Fig.8** Drawing of the Ceremonial Center of America's Stonehenge (after Payne, Ref.3), with wall angles and geographic meanings.

North Portugal (c.2200 BC). Stonehenge III in south England shows that this route was one of the "gates" to and from America (c.2000 BC).

Originally, America's Stonehenge was built for people coming from the south, who wanted to cross to the West Azores. However, the obvious point of departure, Cape Race, Newfoundland, at 47°N, is a long way about for ships which initially departed from the Gulf of Mexico. In addition, south of Newfoundland the current inconveniently flows southwest along the coast (Fig.9), and the cold Labrador Current forms a dangerous fog "wall" full of icebergs where it hits the Gulfstream. The more southerly location of America's Stonehenge at 43°N is a latitude where one can catch the Gulfstream running east off Cape Cod, and cut the corner to the Azores with a current boost, while missing the "wall". All the detailed alignments built into this site show that the megalith builders used the best routes known today, and did not simply sail the 39° latitude line to the West Azores.

## Sailing from the Azores to Madeira

From the West Azores they sailed along the other islands of this archipelago to the easternmost island, Santa Maria. We have already shown, from point X in Fig.12, how menhir **K** makes a complementary **angle of 37°, corresponding to the latitude of Santa Maria at 37°N.** If the protractor is moved slightly down from X to Y (which is on the imaginary equator in Fig.16), the most important menhir **M** makes an **angle of 33°, corresponding to the latitude of Madeira at 33°N.** So M also represents travel destination Madeira. In the wall through M, 9 menhirs have been placed, corresponding to the 9 islands of the Azores. The small menhir above M, plus M correspond to the two islands of Madeira, so Menhir M must also represent the main island of Madeira (Ref.2). The two menhirs near Q correspond to the two eastern Canary Islands.

When we go to the viewing tower (point Z, Fig.16), **menhir J** makes an **angle of 33°, which strongly confirms the latitude of Madeira, 33°N.** Between the viewing tower and menhir J are 7 stones, corresponding to the sailing distance to Madeira: 7 distance lines (7dl= 7°of latitude= 7x111km = 777km). At the end of the wall through M is the "Watch House" W. It makes an angle of 28° to the ESE, equal to the initial sailing direction (ISD) from Santa Maria to Madeira, 28°ESE. Finally, menhir O (Fig.13) gives the terminal sailing direction (TSD) in the neighborhood of Madeira, 34°SE.

## Sailing from Madeira to the Canaries

From viewing point X of Fig.12, menhir **J** makes a complementary **angle of 29°, corresponding to the latitude of the two eastern Canary Islands, 29°N.** The wall through J makes an angle of 17° (Fig.13), which encodes the total sailing distance from the West Azores via Madeira to these two eastern Canary Islands, of 17 distance lines (17dl= 17°of latitude= 17x111km= 1888km). When the protractor is placed at Y in Fig.16, the two menhirs of **Q** (representing the Canaries, see above paragraph) make an **angle of 29°, confirming the latitude of the two eastern Canaries at 29°N.** Oddly, the

**Fig.9** **Above:** Sketch from Ulu Burun shipwreck off Turkey, showing trading vessel of 1300 BC. In its holds were found six tons of oxhide-shaped copper ingots, a half-ton of tin, and amber from the Baltic (Ref.33).
**Below:** Great Circle Sailing Chart of the North Atlantic, showing the problem of finding the Azores from American Stonehenge (Ref.32).

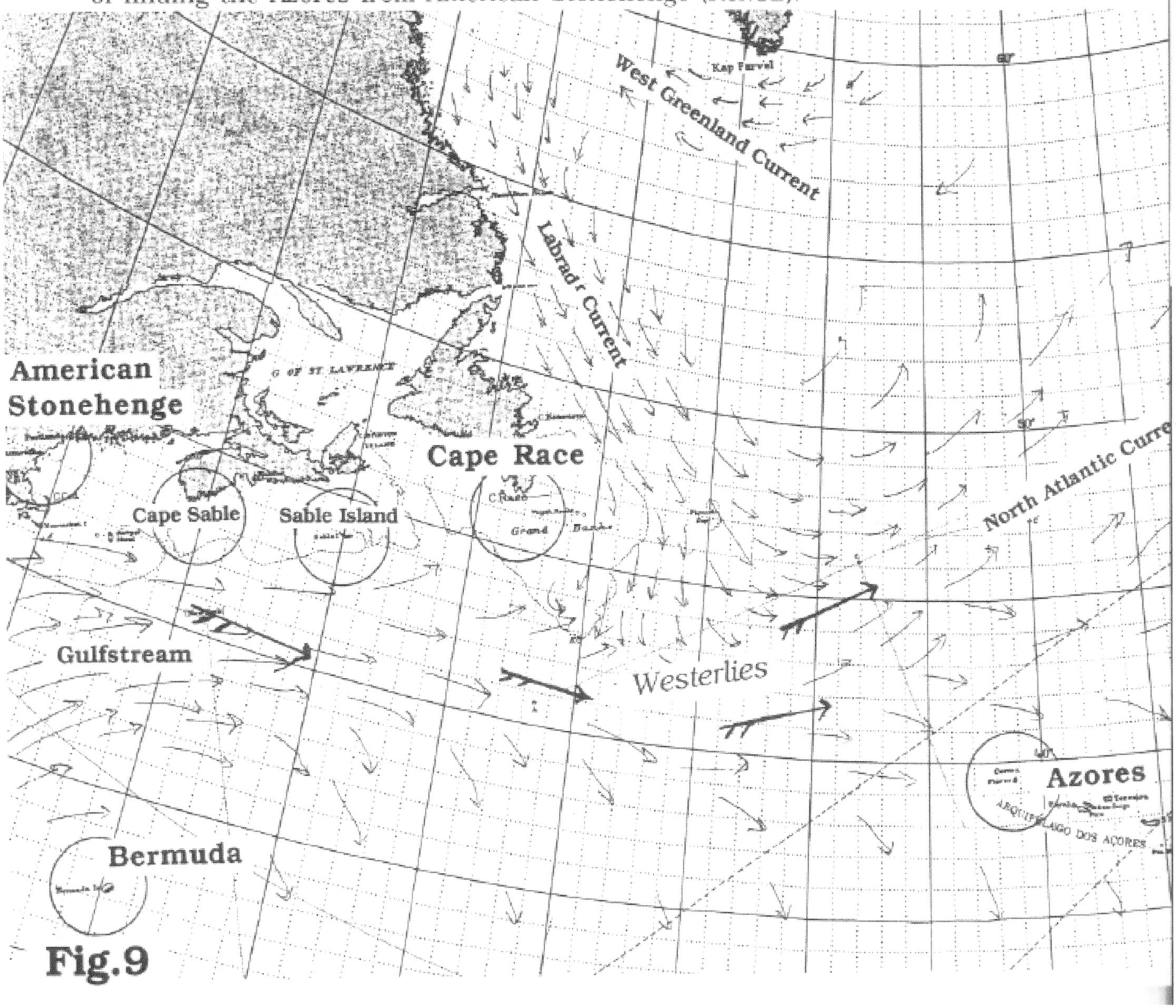


**Fig.9**

circle of rocks in the brush at the far west may also be a representation of the Canaries, as from Z in Fig.13, the large stone makes an angle of 29°. But, as we will show, the menhirs erected later in the west have meanings in another context.

From the viewing tower Z in Fig.16, the Watch House **W** makes an **angle of 28°, encoding the latitude of the Canary Islands at 28°N.** Here at W a chamber is placed instead of a menhir, probably because the Canaries are the beginning and the end of a visit to the New World. In addition, it was for a long time the western home of the SunGod, because before c.4500 BC, these were the westernmost islands of the then known world. The history of discovery of the islands in the Atlantic started with the Canaries. The 2 menhirs of Q in Fig.12 make an angle of 5°, encoding the sailing distance from Madeira to the Canary Islands, of 5 distance lines (5dl= 5°of latitude= 5x111= 555km), which is correct.

Finally, we move the protractor to menhir I2, representing America's Stonehenge itself (Fig.16). Menhir **J makes an angle of 33°** (between J and I2), again **confirming the latitude of Madeira at 33°N.** Also viewed from I2, the Watch House W makes an angle of 53° to the SE, the sailing direction (ISD) from Madeira to the eastern Canary Islands, 53°SE. Note that the important viewing points for the crossing to the West Azores are X (the latitudes) and the viewing tower Z (the sailing directions and distances). For Madeira Y is added, and for the Canaries, I2 is further added.

## Sailing Directly from Cape Race to Ireland

In the monument the crossing to Ireland is only given as an option. For this crossing, menhir M (Fig.13) provides the points of departure, at the northern menhirs of I. The easternmost point of North America, at 46.5°N is Cape Race, menhir I4. From point X in Fig.12, we have seen how menhir **K** made an **angle of 53°, encoding the latitude of Ireland at 53°N**. Menhir J makes a complementary angle of 29° to the ENE, encoding the initial sailing direction from Cape Race to Ireland, 29°ENE. Also, Q' in Fig.12 shows an angle of 29° to the ENE, showing this ISD of 29°ENE.

When the protractor is moved to Y along the imaginary equator in Fig.16, menhir **J** also makes an **angle of 53°, confirming the latitude of Ireland at 53°N**. Note how the menhirs of Q make an angle of 29° to the ENE (Fig.16), confirming the ISD from Cape Race to Ireland, 29°ENE.

From the viewing tower (Fig.12), the uppermost wall crossing **R** makes an **angle of 53°, which again confirms the latitude of Ireland at 53°N.** Just below this wall crossing (also near J) are three menhirs, which indicate that the west coast of Ireland extends over three degrees of latitude. The 5° angle of the menhirs of Q on Fig.12, previously used as a sailing distance to the Canaries, might here indicate the TSD in the neighborhood of Ireland,5°ESE. But since they can sail the latitude line of 53°, exact TSD is not very important. The Watch House W angle of 28° (Fig.13), which previously

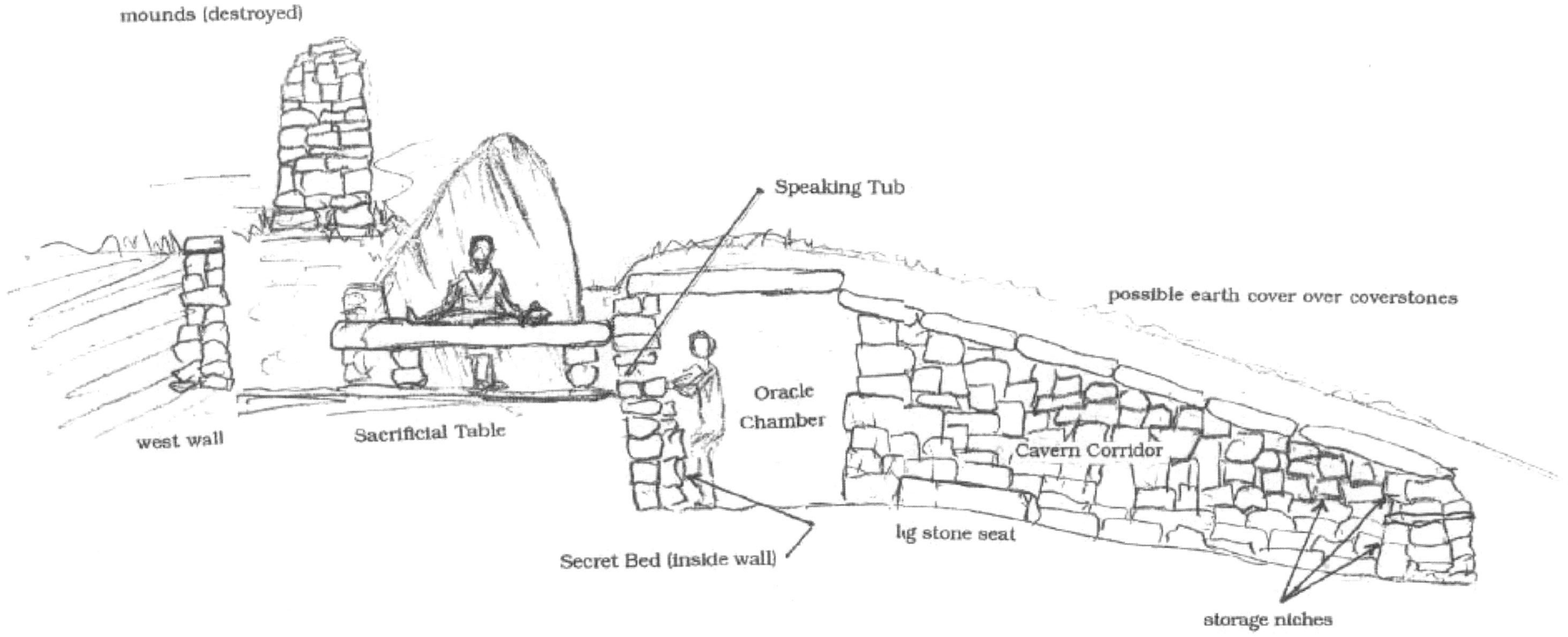


**Fig.10** Cross-section of Ceremonial Centε, Oracle Chamber and Cavern, showing location of Speaking tube, Secre Bed, Stone Seat and Storage Niches.

**Fig.10**

revealed an ISD from the Azores to Madeira, now encodes the sailing distance from Cape Race to Ireland, 28dl (28dl= 28°of latitude= 28x111km= 3111km), the correct distance.

Probably, the direct crossing on the Gulfstream from Newfoundland to Ireland was not a common route. We believe Ireland is not shown in the Main Site, though the roof opening in the Oracle Chamber Corridor might be a hint, as discussed previously. Two major drains (the Gulfstream) from the "Vee Hut" (Cape Race, Figs.4,5) are important indications of this route. The builders could have placed a bigger menhir near R had they wanted to make Ireland important. Nevertheless, the direct cross-Atlantic route to Ireland is engraved on a menhir in Navalcan, Toledo, Central Spain, dated c.1900 BC (Fig.13, Chapter 13; Ref.5).

## Sailing from Cape Chidley to Greenland and Scotland

The crossing via the upper north is shown in the monument as an option, too. Seen from X in Fig.12, menhir **M** makes a complementary **angle of 51°, encoding the latitude of Belle Isle Strait, at 51°N.** This advises one to sail north more safely, by going west (inside) of Newfoundland. Big menhir **K** makes an **angle of 53°, corresponding to the latitude of Cape St. Charles,** the east cape of the mainland of North America, at 53°N. Menhir **J** makes an **angle of 61°, encoding Cape Chidley , and the latitude of the crossing from Cape Chidley to the SW Cape of Greenland, at 61°N** (Photo.15). The wall corner at W makes an angle from X of 8°, encoding the sailing distance of this crossing, 8 dl (= 8°of latitude= 8x111= 888km).

When the protractor is placed at **Y** in Fig 16, menhir J makes an **angle of 53°, again showing the latitude of Cape St. Charles at 53°N.** The upper wall crossing and menhir **S** in Fig.16 make an **angle of 60°, the latitude of Cape Farvel, the south cape of Greenland at 60°N.** The wall crossing at **Q'** makes a complementary **angle of 67°, the latitude of Cape Holm on the Arctic Circle, 67°N.** Continuing on Fig.16, W now shows an angle from Y of 4°, showing the sailing distances from Greenland to Scotland, three journeys of 4dl each (4dl= 4°of latitude= 4x111= 444km). Finally, menhir M makes a complementary angle of 57°, the ISD from the Faroes to Scotland, 57°SE, at the latitude of 57°N, the end of the voyage.

From the viewing tower, point Z, Fig.13, menhir **N** makes a complementary **angle of 66°, the latitude of Cape Dyer at 66°N**, important for the shortest crossing of Davis Strait, as discussed before (c.400km), and the **NW point of Iceland, both at 66°N.** Menhir **P** makes a complementary **angle of 67°, which encodes the latitude of Cape Holm on the Arctic Circle, 67°N.** The menhir opposite the small southern **wall crossing,** and the crossing itself (Fig.18) make **angles from Z of 64° and 62° respectively, corresponding to the latitudes of the SE coast of Iceland and the Faroe Islands at 64°N and 62°N.** Again, the 28° angle to the ESE of W (Fig.13), shows the ISD from Iceland to the Faroes, 28°ESE. The complementary **angle of W, 62°, confirms the latitude of the Faroes at 62°N.** Menhir L' (Fig.16) makes an

**Photo.7** "Storage niches" (two others are further left) at the east end of the Cavern (Fig.6), representing the eastern part of the Mediterranean.

**Photo.8** "Sundeck Chamber" (Fig.4, #27), probably representing Cape Sao Roque and the NE coast of South America.

angle of 4°, confirming the three sailing distances from Greenland to Scotland, all three 4 distance lines, as discussed. Finally, from Z (Fig.16), menhir **J** makes a complementary **angle of 57°, showing the ISD to Scotland, 57°SE, and also the latitude of Scotland at 57°N.** Menhir **S** (Fig.12) also shows this **angle of 57°, which confirms 57°N again.** Note that beyond Cape Brewster, the northern wall in Fig.2 has been extended to the islet of Jan Mayen, discovered c.2900 BC, as illustrated by petroglyphs. Menhir **K** makes a complementary **angle of 71° (Fig.13), corresponding to the latitude of Jan Mayen, at 71°N.**

When we place the protractor in the stone mound **M'** (see Fig.18), the menhirs of I confirm the most important latitudes at the end of the crossing of the Upper North: **I1=57° (Scotland, 57°N); I2=60° (the Shetland Islands and Cape Farvel, Greenland, both 60°N); I3=62° (the Faeroes, 62°N); and I4=64° (the SE coast of Iceland, 64°N).** Menhir I1 also provides the ISD from the Faroes to Scotland, 57°SE. The front side of stone I2 shows two large super-imposed images of South Greenland, in haut-relief (Ref.26). (Note that from the stone mound **M'**, the big menhir K makes an **angle of 43°, again showing the latitude of this departure point of America's Stonehenge at 43° N,** which may explain the size of this menhir.)

As a route to reach America from Europe, this crossing via the Upper North was known after c.2500 BC (Refs.2,3,7,9,10). A separate glyph was made for this route in the Dissignac, Brittany petroglyphs, and a beautiful inscription of this route is shown in Chao Redondo, North Portugal (c.2200 BC). America's Stonehenge was built as a nautical base for people in the Americas who wanted to cross the Ocean to the West Azores. The original point of departure, Cape Race Newfoundland at 47°N, turned out to be a long way about and dangerous. Menhir S, which gave us important Northern Crossing latitudes, is far outside the site; apparently in view of the severe weather conditions in the upper north, one is being disuaded from sailing that far north. Iceland, which is usually so important in the Northern Route is also weakly shown, perhaps the stone pile above M' (Fig.2), and we think it is represented by the central stone at the north end of the ceremonial center area (see Photo.10). The location of America's Stonehenge on the coast of New England at 43°N (the big menhir I2 indicates big importance) was shown to be a better departure point, and so became a popular nautical support center for cross-ocean sailings (Refs.7-12). However, as a return route, it is seen in this monument that Northern Crossings became less popular, because there is so much explanation of other routes focused on more southerly approaches to the Azores.

## America's Stonehenge II: The Southern Crossings
## Sailing from Cape Race to the Azores

Seen from the main viewing point X, menhir **A1** (Fig.12) makes an **angle of 47°, corresponding to the latitude of Cape Race, Newfoundland, 47°N.**(Though not indicated this way on the map, A1 is the largest menhir on the site.) Cape Race is the most easterly cape of North America, and therefore

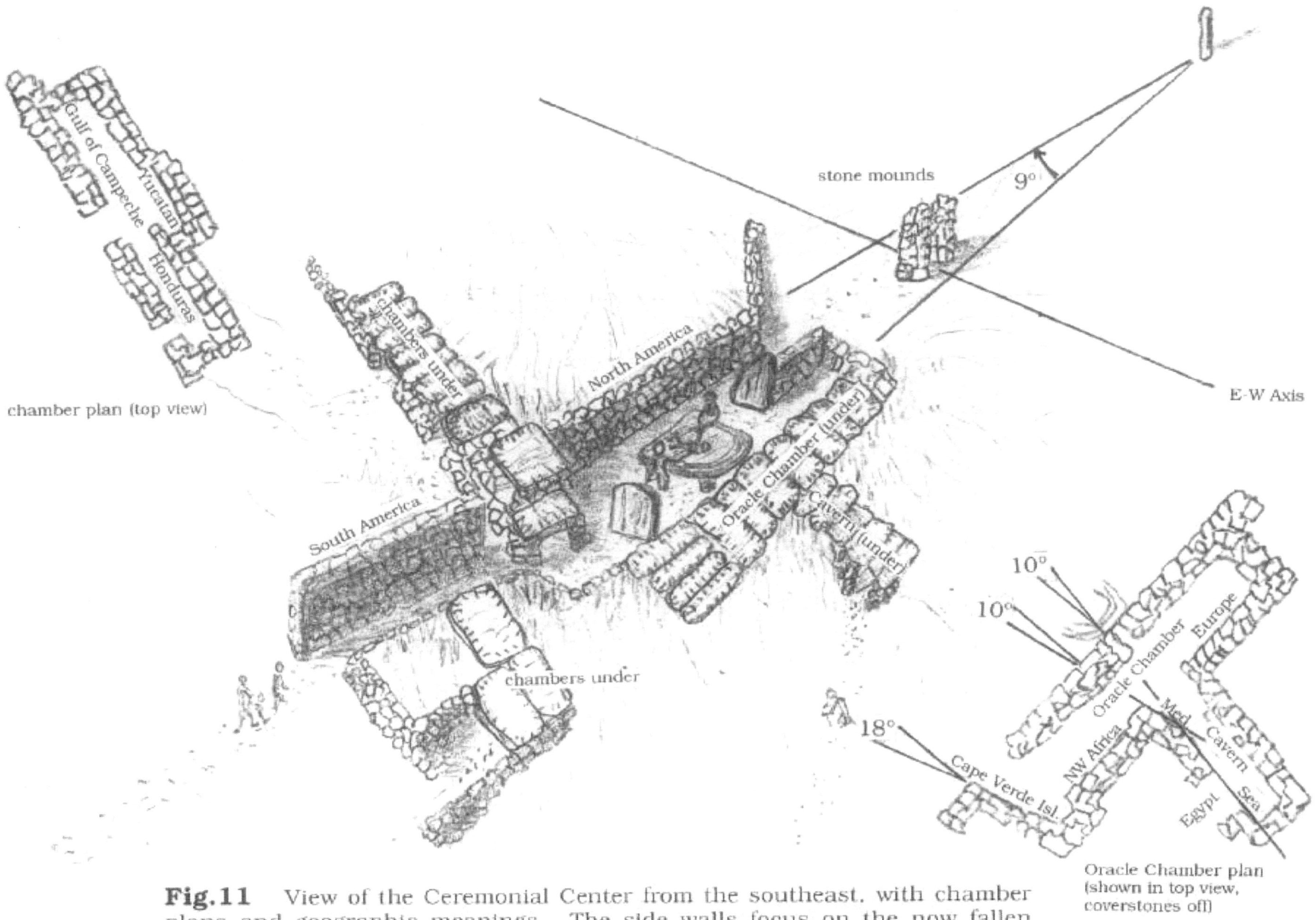


**Fig.11** View of the Ceremonial Center from the southeast, with chamber plans and geographic meanings. The side walls focus on the now fallen menhir at the upper right (also in Photo.12), representing Cape Holm, Greenland (at 67°N). Some of the coverstones, lower on the sketch, have been removed by quarrying.

**Fig.11**

has been the most important point of departure. The complementary **angle of 43° corresponds to the latitude of America's Stonehenge, at 43°N.** The surface on the top of this boulder A1 shows a very deep groove, resembling the Merrimac River near this monument, running to the coast (the eastern edge of the stone) (Ref.26). Moving now to the western side of the complex on Fig.13, note that the wall above A1 points to the big menhir in the north, I2, that we previously found represents America's Stonehenge. This **wall** makes **angles of 39°, the latitude of the West Azores, at 39°N, and 51°, the latitude of the megalithic monument Stonehenge in England, at 51°N**, again reinforcing the primary importance of these two places to the builders of America's Stonehenge. A menhir that does not show on the drawings and is not lettered at the end of this wall, makes an angle of 21° from point Z of Fig.13, which is the ISD to the West Azores, 21°ESE. This crossing from America's Stonehenge via Cape Race to the West Azores is the primary route recommended by the monument.

However, there are explanations of alternative southern sailing routes to the West Azores, which have been later discovered. Menhirs G and H (from point X, Fig.12) provide the most important places of departure. Menhir **G** has an **angle of 35°, which corresponds to the latitude of Cape Hatteras South, 35°N.** Menhir **H** makes an **angle of 25°, the latitude of the south point of Florida, 25°N.** (Seen in person, menhir H is far smaller than it is drawn on the site map, and therefore is less important than one would think, looking at only the map.) Maybe it is significant for its angle of 65° from X (Fig.20), relating to the crossing from Cape Mercy to West Greenland, at 65°N (see Photo.16).

## Southern Crossings

For the Southern Crossings, only the viewing points X (Photo.14) and Z (Photo.13) turn out to be important. Viewing point Y and menhir I2 do not play a role in these routes. From the meanings of the menhirs and walls on the western half of the site, it is clear that this half of the site represents a second phase of construction at the monument. This America's Stonehenge II was built about a century later than the first half, just after the discovery of Bermuda c.2200 BC.

## Sailing from Cape Hatteras to the Azores

From the large menhir I2 (America's Stonehenge) a wall leads west (Fig.12) then down to menhir H, the "Eye Stone" because it has a petroglyph of an eye on it (Photo.16, Ref.1). This eye appears identical with one at Vermont Calendar Site 2, near South Woodstock, Vermont (Ref.7). This stone **H** has an **angle of 36° (seen from Z), corresponding to the latitude of the Strait of Gibraltar at 36°N**, the likely goal of many expeditions. This latitude of **36°N is also the latitude of north Cape Hatteras.** Note that the wall above H makes an angle of 21° to the ENE, indicating that the correct ISD from Cape Hatteras to the West Azores is 21°NNE. Seen from the viewing tower Z on Fig.18, the top of thick wall T (not so thick in reality) is 11°, the TSD in the neighborhood of the West Azores, 11°ESE.

**Photo.9** Lower standing rock in the Ceremonial Center (Figs.8, 11), representing Cape Sao Roque, the NE point of South America.

**Photo.10** Stone at north end of Ceremonial Center (Figs.8, 11), representing Iceland.

The wall coming down to menhir H is more complete in reality than shown on the map (the gaps are smaller). This wall points through A1 to menhir B on Fig.12. Note that B has an angle of 10° from point X, indicating the TSD in the neighborhood of the West Azores is 10° ESE. From menhir A1 a **wall** runs downward at an **angle of 35°, which corresponds to the latitude of Cape Hatteras South, 35°N.** It also encodes the distance from Cape Hatteras to the West Azores, 35dl (35dl= 35° of latitude= 35x111= 3888km), which is correct. From the viewing tower Z in Fig.13 menhir C makes an angle of 35°, which confirms this distance. The ISD to the Azores of 19° ESE (Fig.13) given by the large menhir K at 19° was the sailing direction well known before the discovery of Bermuda. The TSD was given by the 11° angle to Q', 11°ESE (Fig.18). Either of these sets of ISD and TSD would work, helping a crew to stay with the Gulfstream.

## Sailing from Cape Hatteras to Bermuda and the Azores

The wall from A1 bifurcates, indicating two different routes to the West Azores! By the upper leg, we go to **A3,** which has an **angle of 32° from X on Fig.12, representing Bermuda at its latitude of 32°N.** Taking the south leg goes to **D'** which has an **angle of 25° from Z on Fig.13, encoding the tip of Florida or the Bahamas at 25°N.** On Fig.12 you can see that the end points of these walls make angles of 18° and 9° respectively (also from A1, reversed), encoding the ISD from Cape Hatteras to Bermuda of 18° ESE, and the distance to Bermuda of 9dl (9dl= 9°of latitude, = 9x111= 999km).

The small menhir A2 (Fig.12) makes an angle of 37°, corresponding to the total sailing distance from Cape Hatteras South via Bermuda to the West Azores: 37dl (= 37°of latitude= 37x111= 4111km). This is a long distance! Above this stone a **wall** runs at an **angle of 47° (Fig.13), which refers to Cape Race at 47°N**, only 19dl from the Azores. This wall has a complementary **angle of 43°, referring to the favorable location of America's Stonehenge at 43°N.** This shows that if you sail up the American coast to America's Stonehenge, you get a shorter sailing distance to the Azores!

This wall rising from A2 is connected by walls to menhir G. As we have seen before, menhir **G** (from X, Fig.12), makes a complementary **angle of 35°, corresponding to the latitude of Cape Hatteras South, 35°N.** However, from the viewing tower Z of Fig.12, menhir **G** makes an **angle of 32°, which strongly confirms the latitude of Bermuda at 32°N!** (Again we see that menhirs are used for multiple encodings, frustrating our desire to specify a location for each one, as they sit in walls that are intended to make geographic sense.) From G a wall runs in a western loop to menhir A3, then a branch wall runs south through A4 to E. Menhir E makes an angle of 9° from X on Fig.12, providing the distance from Cape Hatteras South to Bermuda, 9dl (9dl= 9°of latitude= 9x111= 999km) which is correct. Rising above A3 is a thick wall F. The end of F wall (on a straight line between G and A3) makes an angle of 18° on Fig.12, equal to the ISD from Cape Hatteras South to Bermuda, 18°ESE. The F wall is actually four feet wide, and about four feet tall, telling us this route is highly important.

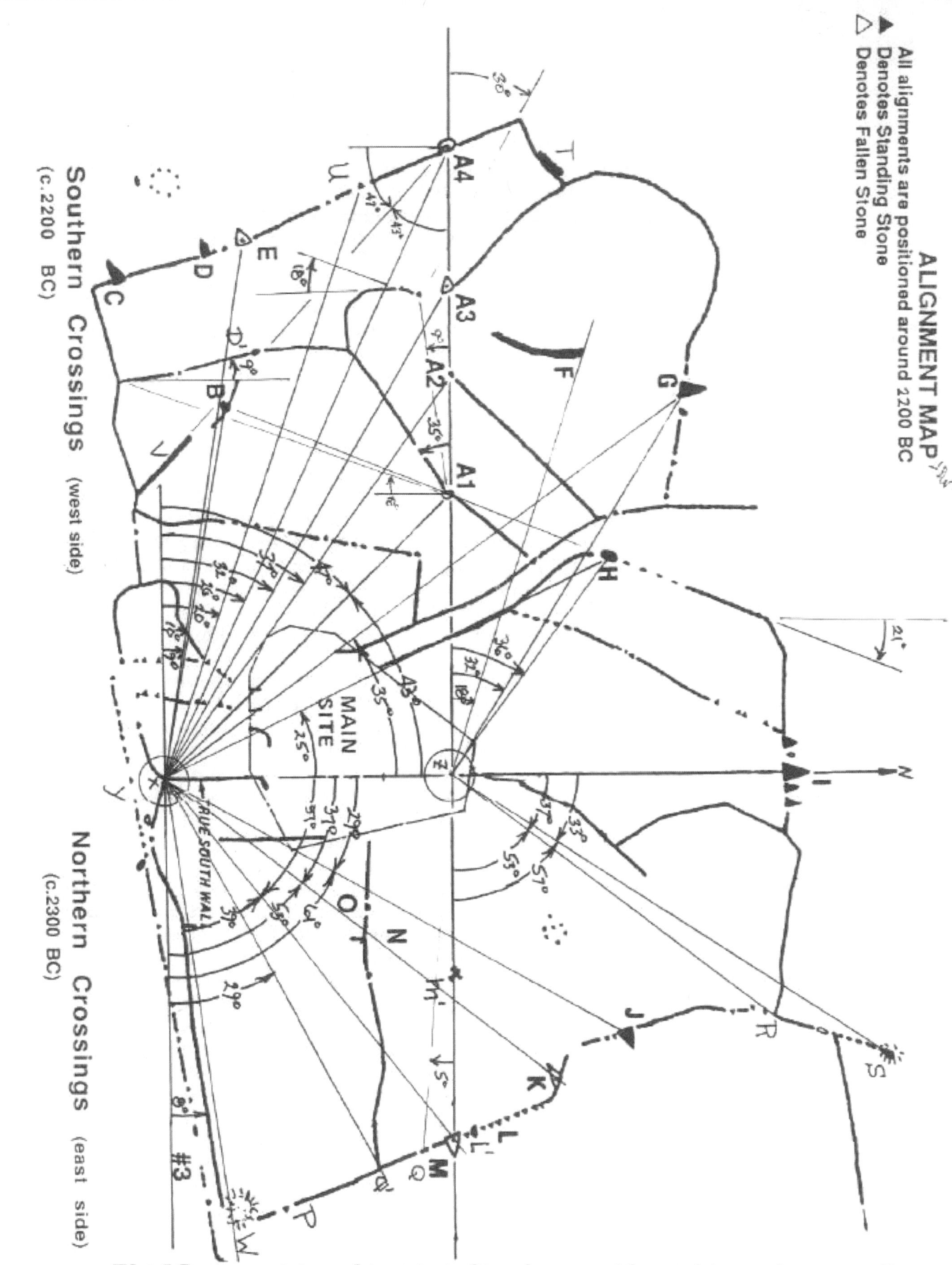


**Fig.12** Groundplan of America's Stonehenge, with menhirs and stone walls. Site angles of menhirs from viewing points X and Z.

**Fig.12**

Menhir **C** (Fig.13) makes an **angle of 35°, again confirming the latitude of Cape Hatteras at 35°N.** Menhir **B** nearby makes an **angle of 32°, confirming the latitude of Bermuda again at 32°N.** A short wall runs through B. The central end of it points to menhir A4, which makes an angle of 26° on Fig. 12, showing the less important TSD in the vicinity of Bermuda, 26°ESE. A beautiful inscription of the crossing from Cape Hatteras via Bermuda and the Azores to the Iberian Peninsula was found in Chao Redondo, North Portugal c.2200 BC. It confirms that in prehistoric times they used these islands as mid-ocean stopover points.

## Sailing from the Bahamas to Bermuda and the Azores

From X on Fig.12, menhir **A4** makes an **angle of 26°, encoding the latitude of Bimini and the Abacos, the NE islands of the Bahamas at 26°N**. Via walls, menhir A4 is connected to G and A3, both of which have provided the latitude of Bermuda, confirming this route and its importance. The thick wall V points (via B and D') to menhir A4, at 43° and 47° angles (from A4, Fig.12), which are a beautiful confirmation of the latitudes of America's Stonehenge and Cape Race again. The angle of menhir B from Z in Fig. 13, 32°, is also the ISD from the Bahamas to Bermuda, 32°NE, as well as confirming the latitude again. The 25° angle of D' also provides a less important TSD in the vicinity of Bermuda of 25°ENE.

Menhir B, which has given us the ISD to Bermuda, is at the same level (or latitude) as Watch House W. From A4 Fig.17, W makes an angle of 12°, which probably provides the sailing distance from Great Abaco to Bermuda:12dl (12dl= 12°of latitude= 12x111= 1333km). From Z in Fig.13, W makes an angle of 28°, encoding the distance from Bermuda to the West Azores: 28dl (28dl= 28°of latitude= 28x111km= 3111km).

**Wall V** has its end point at an **angle of 38°, encoding the latitude of the Central Azores at 38°N.** Menhirs D and D' make angles of 25° and 26°, the ISD from Bermuda to the West Azores, 25° to 26°ENE. (The Z-D line could be extended westward to a circle of 8 stones and a menhir, together symbolizing the 9 islands of the Azores.) The northeast wall of Field V in Fig.3 (above Z', see Fig.13), makes an angle of 32° too, so persons traveling from the Caribbean are advised they can sail directly to Bermuda (this wall is nearly pointed at the Bahamas and chamber #13 representing Bermuda in the Main Site, see Fig.3).

In Europe, petroglyphs are known (Kercado, Chao Redondo) in which the crossing from the Bahamas to Bermuda is recommended. As we have just seen, in America's Stonehenge this route is indicated by a stone row in the Main Site. However, this wall is nothing more than an extension of the western inner wall. Other parts of the double walls point to a crossing from American Stonehenge via Cape Cod to Bermuda. Yet, wind and current charts show that the Bahamas to Bermuda route is not unfavorable. We have the impression that, as time went on, this Southern Crossing from the Bahamas became popular for craft returning directly from Central America.

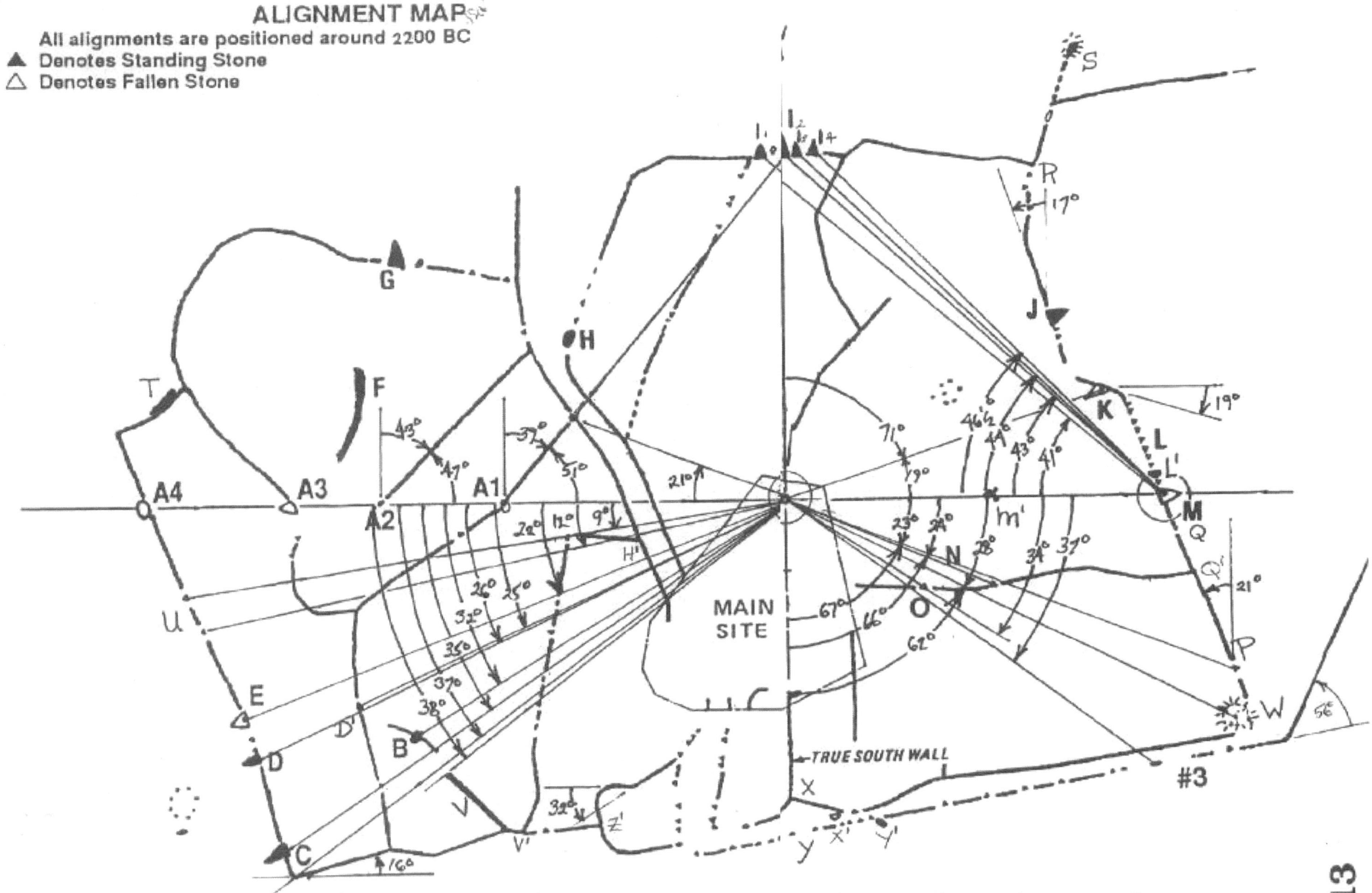


**Fig.13** Groundplan of America's Stonehenge, with menhirs and stone walls. Site angles of menhirs from viewingpoints M and Z.

**Fig.13**

The mounds and underwater stone blocks and "piers" at Bimini indicate these islands were then, as they are today, a refuge on the east side of the rough Gulfstream, which have behind them one of the few deep passages across the shallows of the Bahama Banks (Refs.11,-15, and others).

## The Gulf of Mexico and Central America

From point X on Fig.12, menhir **A4** makes an **angle of 26°, showing the latitude of the Gulf Coast of Florida where it starts above the Everglades, at 26°N.** Above A4 the wall runs to a sharp bend at 30°, showing interest in the northern latitudes of the Gulf of Mexico up to the important level of 30°N, the mouth of the Mississippi. The small menhir **A2** (37° from X) represents the center of **waterways of the US, where the Ohio joins this river near Cairo, at 37°N,** and the huge menhir **A1** (47° from X) symbolizes the **south coast of Lake Superior, at 47°N**, with its enormous copper deposits, near the start of the Mississippi (Refs.7,11,13).

On Fig.17, note a small menhir **X'** which is to the right and below X; it makes an **angle of 23°, corresponding to the latitude of the north coast of Cuba on the south side of the Florida Straits, and the Tropic of Cancer, at 23°N.** This menhir is also a tribute to the SunGod, associated with the holy 23° latitude. From menhir A4 (Fig.17) a **wall** runs downward to menhir E at an **angle of 23°, confirming the 23°N latitude**. From A4 menhir **D** makes an **angle of 22°, corresponding to the latitude of the crossing from the SW Cape of Cuba to Cape Catoche, Yucatan, at 22°N.** (Actually, when seen on the site, E is a standing stone, and D is a fallen stone.)

Menhir **D** (see Fig.13) makes an **angle of 26°, which confirms the latitude of south Florida, at 26°N.** The NS wall points due south, the sailing direction to Cuba. Menhir **E** makes an **angle of 22°, which confirms the latitude of the crossing from Cuba to Yucatan, 22°N.** The wall below D (Fig.17) makes an angle of 13°, the sailing direction from Cuba to Yucatan, 13°WSW. The lower one of two U menhirs makes an angle of 12° (from Z, Fig.13), confirming this sailing direction, 12°WSW.

From the south cape of Florida (menhir A4 or D), one can also sail directly to Yucatan, though this is against the wind and current, and was probably done infrequently and by waiting for favorable conditions. Menhir C is on the wall below A4 and D. From Z in Fig.13, C gives the right sailing direction, 35°SW. Below menhir C is a bend at 37° SW (Fig.13) and a wall under an angle of 16°. The site explains, that a sailing direction of 37°SW can be used when the destination is the north coast of Honduras at 16°N. From the important menhir I2 (America's Stonehenge) the western **wall crossing Z'** in Field V, the Caribbean (see Figs.17&2), makes an **angle of 16°, which confirms the importance of the north coast of Honduras at 16°N.** From viewing tower Z (Fig.17) this **wall crossing** makes an **angle of 30°, encoding the Nile Delta, at 30°N.** This shows again that economic and political contact between these civilizations was considered highly important (Refs.7,11,13-18).

**Fig.14** America's Stonehenge (Mystery Hill) is 20 miles from the Ocean today, and six miles from the Merrimack River. The finding of beaches far inland in Maine has shown that this coast has risen (100m since 10,500 BC) because the weight of the mile high ice sheet has melted off (Ref.27). Careful estimates by Paine show that at 2000 BC the sea level was 40m higher than today (Ref.3), so there were salt water harbors around Mystery Hill. A partial reason for the eventual decline of America's Stonehenge as time went by may have been its increasing distance from the sea.

**Fig.14**

Near Y, below the imaginary equator, there is a menhir **Y'** (Fig.17). From X this menhir makes an **angle of 16°, encoding the latitude of the Cape Verde Islands, at 16°N.** The furthest west point in the Cape Verde Islands is even today labeled on charts as the "Cape of the Sun", probably referring to the ancient spreading of the Sunreligion. These islands are important as the starting point for the Southern Crossing. From the frequent megalithic referral to the this latitude of 16°, we believe one of their most important destinations was a culture at 16°N, on the north coast of Honduras, or the eastern coast of today's Guatemala. From the viewing tower Z (Fig.17), and via the east end of the eastern inner wall (which we previously labeled as the Cape Verde Islands), menhir Y' makes an angle of 15°, corresponding to the latitude of Brava, the SW Cape Verde Island, 15°N, and also of Cape Gracias a Dios, the NE cape of Honduras, 15°N.

## Sailing to Cape Sao Roque, the Antilles, and Yucatan

This ocean crossing is an easy one, with trade winds on the stern all the way. It has been reported that in 1989 even "some one-inch grasshoppers from Africa made it to the Windward Islands of the Caribbean in a dust storm" (Ref.20). Approaching the coast, some researchers suspect that sailing craft turned south and across a shallow Amazon basin, to mining-based cities that developed on islands (Ref.34) and in the Andes (Ref.23), but we are dealing here with the many which turned north through the opening of the inner wall between X and Y (Fig.2). They then had a choice of sailing up the east coast via the island chain of the Antilles, or sailing further west along the coast to Central America and the Gulf of Mexico. The winds on all these Caribbean routes are at your back or on the stern quarter.

From the viewing tower, point Z (Fig.18), menhir **I1** makes a complementary **angle of 5°, revealing the latitude of Cape Sao Roque, the NE point of South America, at 5°S.** It is the official landing place for voyagers from the Old World. Menhir **I2** makes a complementary **angle of zero degrees, corresponding to the latitude of the mouth of the Amazon River, on the equator.** Menhir **I4** makes a complementary **angle of 5°, corresponding to the latitude of the north coast of French Guyana, at 5°N.** This landing place was often used to cut the corner on the way to Central America (Ref.16). Menhir **X'** makes an **angle of 11°** from the equator when viewed from Y, which corresponds to the latitude of the start of the Antilles Route (sailing from island to island to the north), the **latitude of Trinidad and Tobago at 11°N.** The sailing direction from the SW cape of Cuba to Cape Catoche, Yucatan, is also 11°WSW.

The small chamber that can be seen at **U** has an **angle of 20° from X (Fig.12) perhaps encoding the tip of Yucatan,** where the resort of Cancun is now located on the island of Cozumel, near the **Mayan seacoast trading city of Tulum at 20°N.** Above Z' (Fig.17) is the wall of the Gulf of Mexico. When walking the site, you will notice there are exceptionally large rocks in the wall here, as shown in Fig.3. This indicates interest in correctly recording the location of the Mississippi River, their entry point to the interior of the

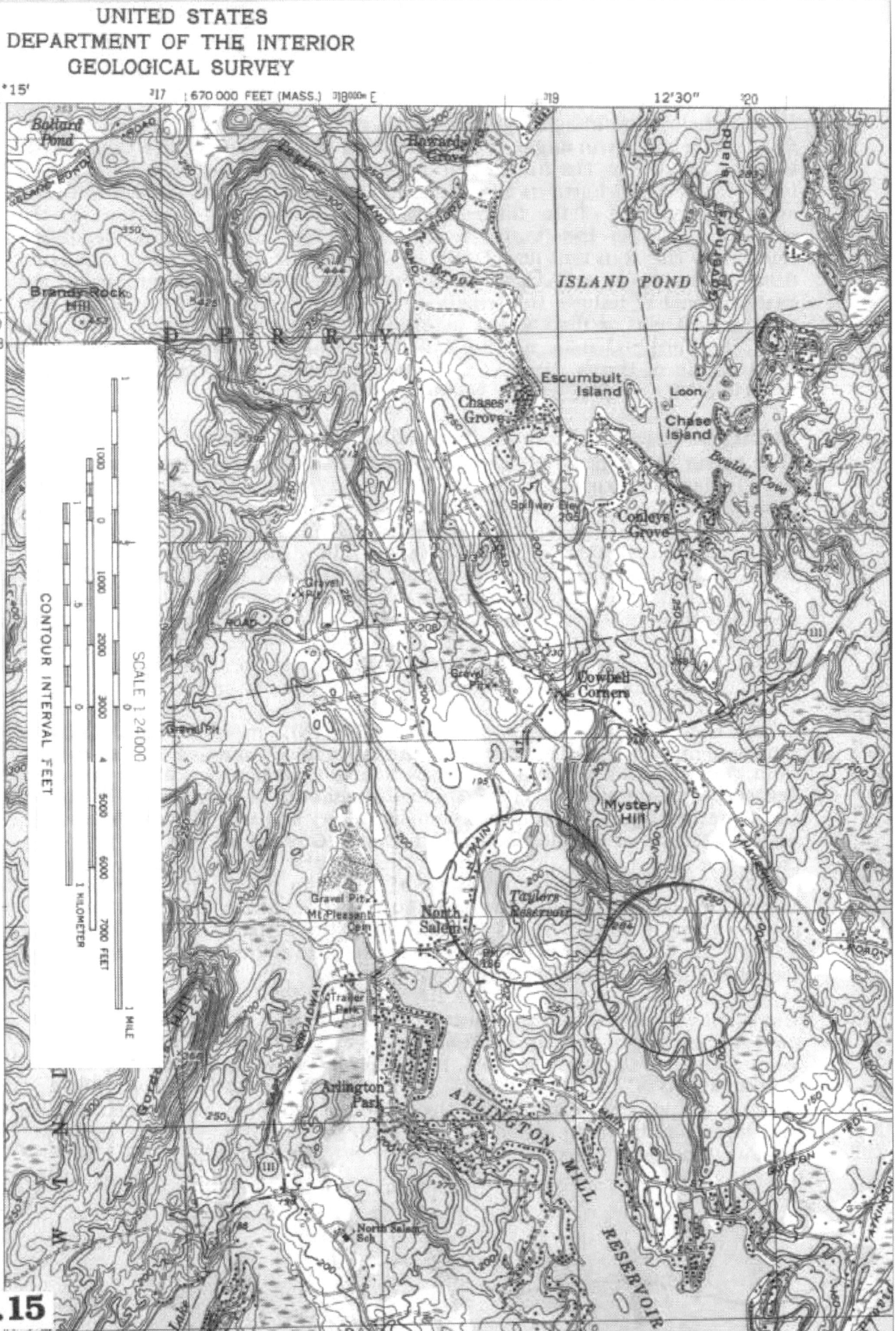


**Fig.15** Topographic map of t surroundings of America's Stonehenge (Mystery Hill) (Ref.30). The circl areas should be carefully examined for early port structures.

**Fig.15**

continent (and the copper trading at Poverty Point), at 29°N, as measured from Z (Fig.17).

## The Landcrossing of Panama to the Pacific

From X (Fig.18), menhir **E** makes an **angle of 9°, corresponding to the latitude of the isthmus of Panama, apparently important for the land-crossing to the Pacific, at 9°N** (Ref.2), though archaeological evidence points to sites in Lake Nicaragua (statuary) and on the pacific coast of Costa Rica (huge stone balls) (Ref.15). From Y menhir **C** also makes an **angle of 9° above the equator, confirming this latitude of 9°N.** From viewing tower Z, menhir **X'** makes an **angle of 8°, which identifies the Pearl Island Archipelago at 8°N** in the Gulf of Panama. From viewing point I2 in the north, menhir **Y'** makes an **angle of 8° too, confirming these islands at 8°N.**

## Other Important Features: The Religion

The megalith builders were followers of the Egyptian SunGod religion. Long before the building of America's Stonehenge (c.2300 BC), the SunGod religion was already the official state-cult in Egypt (from c.3200 BC onwards). From 3600 to 3200 BC the Azores (the Ceremonial Center) were the westernmost islands of the then known world. From 3200 until 2500 BC the south point of Greenland (the double stone mound previously standing at the viewing tower Z) was the westernmost land. Since the SunGod traveled to the west every day, these areas were often called "the islands of the supreme god Ra".

In Europe, from the time of the discovery of the Azores (c.3600 BC), the Azores had a religious meaning. The finding of this mid-ocean home of the SunGod was a milestone in prehistory. The Azores were considered sacred for a thousand years. At America's Stonehenge, the Ceremonial Center now itself represents the Azores, replicating them further west.

As we have seen, the west and east walls of the Processional Walkway are directed to a point in the north, representing Cape Holm (Greenland) on the Arctic Circle (Fig.11). The discovery of Cape Holm was also a milestone in the mission history of the megalith builders of c.3300 BC. Since the Arctic Circle is the northernmost line where the sun still shines on midwinter day, this is one of the two holiest days of the SunGod Ra, the highest god of Egypt. Since the Ceremonial Center represents a holy place (the Azores), focused on a sacred place (Cape Holm), the America's Stonehenge Ceremonial Center has an important religious meaning.

In this early mythology, below Ra are two other main gods, Horus and Osiris. The Old World (The Oracle Chamber Corridor and the Cavern) is the land of the sungod Horus. The New World (the West Chamber) is the land of the moongod Osiris. These two lands are separated by a huge sea, which we now call the Atlantic Ocean (the Processional Walkway). This is the work of the earthgod Maat, symbolized by the Sacrificial Table. To keep the two kingdoms of heaven in position, from time to time probably some children

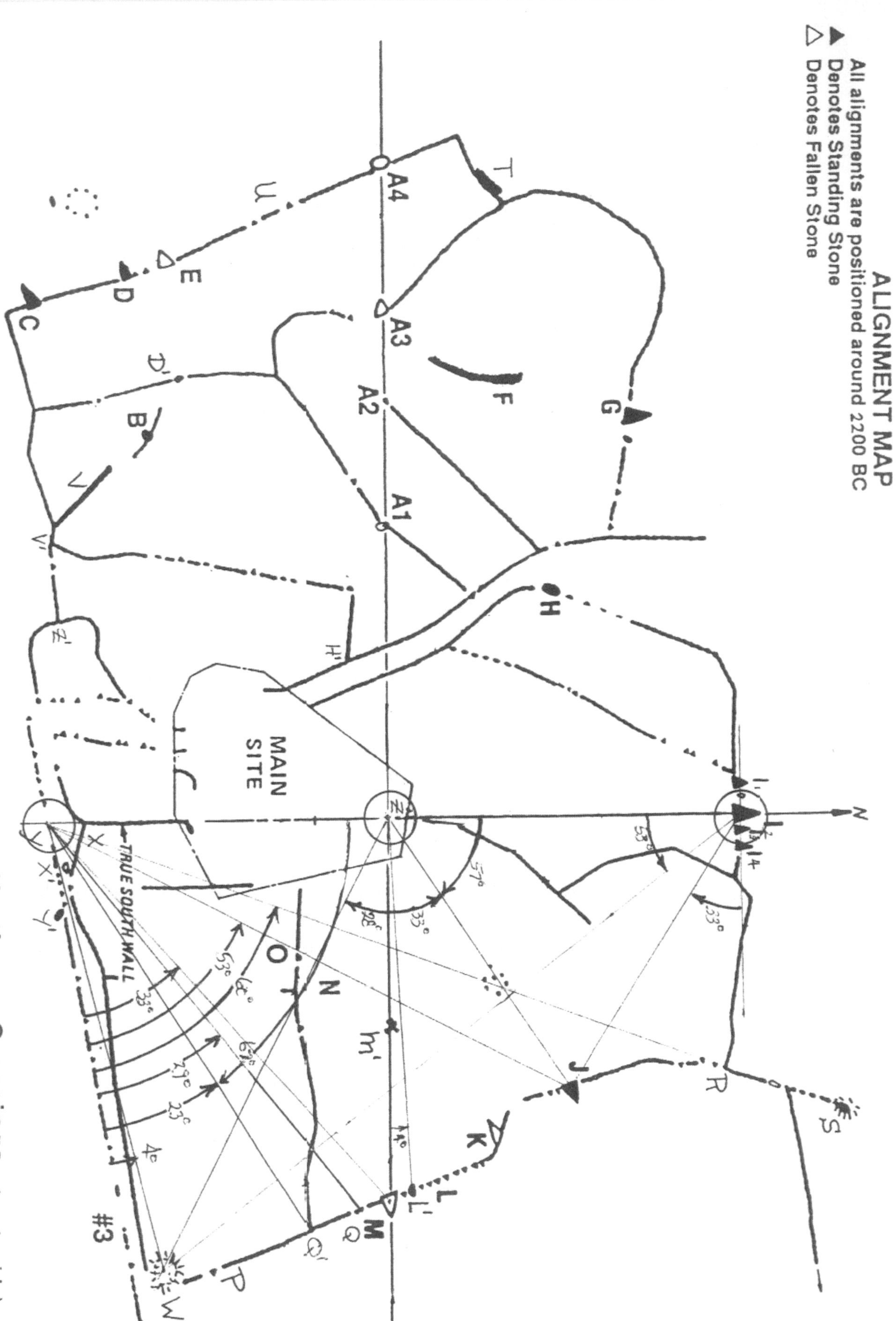


**Fig.16** Groundplan of America's Stonehenge. Site angles of menhirs from viewing points Y, Z, and I2.

**Fig.16**

had to be sacrificed, which happened in the name of god on the Sacrificial Table. We have no evidence for this, but it is likely according to historical records (Ref.35). Child sacrifice was still going on 2,000 years later in Phoenecian Tophets, and the Christian god was telling Abraham to sacrifice his son. People believed that from the blood of children, the most valuable sacrifice, the gods were propitiated, and so the members of the expeditions would survive. It is recorded in Egypt that the SunGod Ra has said that "From the realm of the dead in the west you can return to the east, to the land of the living, and you will receive eternal life". These hieroglyphics should be given a new look for broader meaning in this context.

At the west side of the Oracle Chamber Corridor is the "secret bed" (Fig.6, nr.C), just below the Speaking Tube. Geographically, this bed represents the coast of NW Africa, situated on the holy Tropic of Cancer, which also runs through the center of the Southern Egyptian Empire. Probably here, lying in this sacred place, a priest of the site spoke through the Tube under the Sacrificial Table, helping the leader of the prospective expedition make his final decisions about the course to take, and the timing of the perilous crossing to the Old World.

Simultaneously with their geographical meaning, the major menhirs have astronomical, calendrical, and religious meanings. Previous investigators have carefully studied the astronomical and calendrical aspects of the site, such as calendrical control of the Holy Days, which we accept as accurate, and can build upon, in this new context. The menhirs of I in the north and the equinox menhirs on the E-W axis, M, and A1 to A4 are clearly standing in honor of the supreme god Ra. The whole site of America's Stonehenge has been built around these menhirs. The midsummer and midwinter menhirs G, J, O, and C are the focus of honor for the sungod Horus. The Watch House W and menhir D were raised in honor of the moongod Osiris. Note that the alignment from the viewing tower to the Watch House W coincides with a small wall at the right side of menhir O which ends in another menhir. As said before, this represents the center of the Egyptian Empire at 23°N, as does menhir N, at the holy 23° angle (Fig.13).

## The Southern Walls to the East

As explained earlier, the southern walls under the Main Site mark the south border of the North Atlantic Ocean at 0° and 10°N, respectively (Figs.1&2). Geographically, they are extended towards the far east, in our opinion to SE Asia. (Compare the length of the Mediterranean Sea east of the Main Site with a map. Note that the Red Sea, the Persian Gulf, the Arabic Sea, and the north side of the Gulf of Bengal are all above 20°N). Archaeological evidence is showing that early voyages to the Far East must have not been uncommon (Ref.21). The single wall at the end toward the SE (Fig.1) points to the continent of Australia (not visible). The Watch House with its entrance to the south was built because of the then-recent discovery of this continent (c.2700 BC), as shown in Dissignac (Brittany). It also marks the corner of the site. This means that in America's Stonehenge, a crossing of the Atlantic

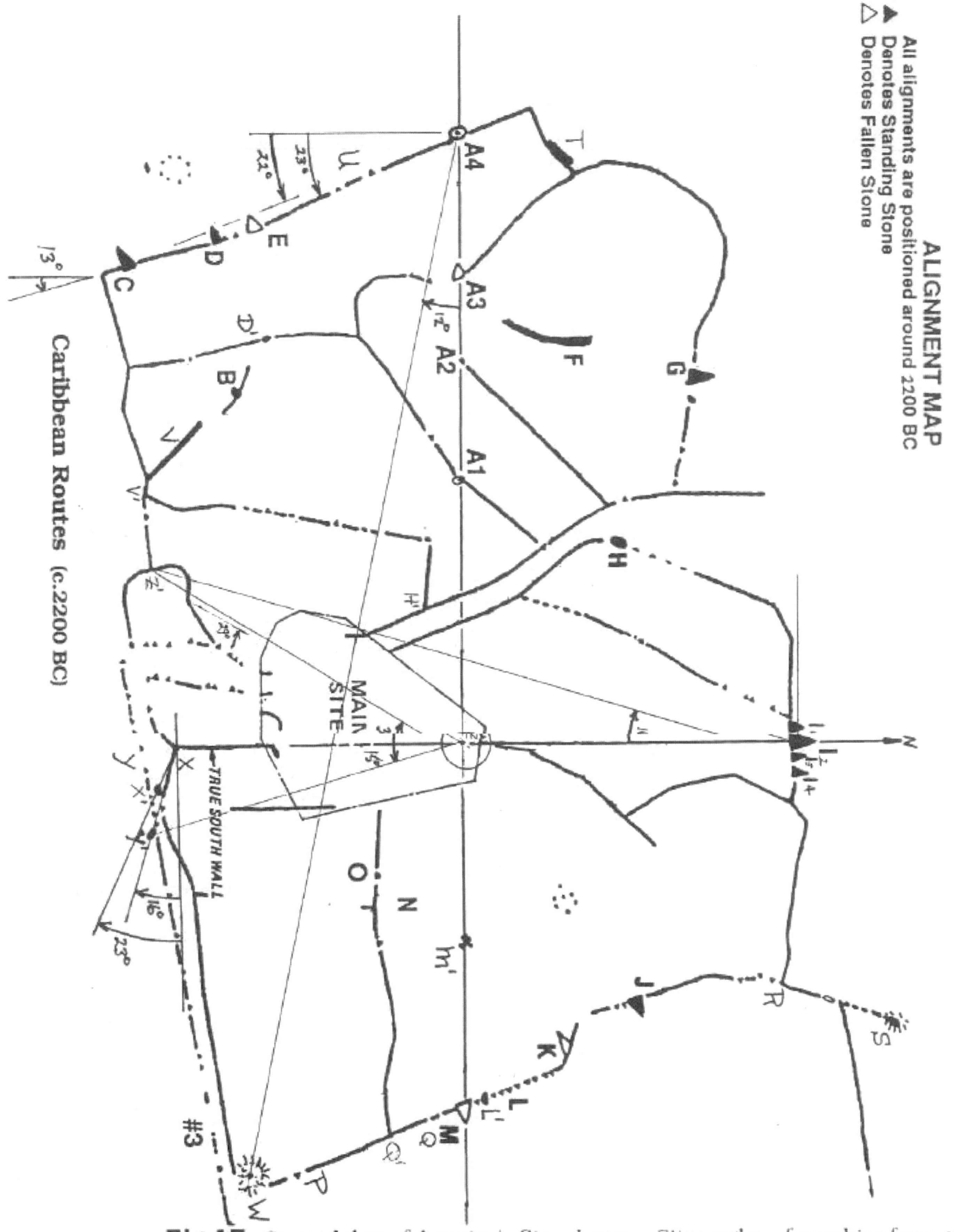


**Fig.17** Groundplan of America's Stonehenge. Site angles of menhirs from a variety of viewing points.

**Fig.17**

is recommended for destinations as far as SE Asia. The long wall toward the northeast points along the stylized Pacific Coast of Asia to the Bering Sea (not visible). This **wall** makes an **angle of 56°** with the imaginary equator (Fig.13), **encoding the latitude of the east cape of the Kamchatka Peninsula at 56°N,** where the Aleutian Route starts towards the Alaska Peninsula (North America). Originally, this was the location of the first crossing to America by megalithic people. According to the inscriptions of Dissignac (Brittany, France), and Stonehenge in South England, America was discovered via the Bering Sea (c.2600 BC).

## The Northern Walls: Hudson Bay and Baffin Bay

Northwest of the Main Site a double row of walls can be identified with the second building phase (Figs.2&3, c.2200 BC). The lower part resembles the Labrador Sea, with at the right side a western inner wall at 1 big distance line (1DL= 10°of latitude= 1111km) from the coast. At the end of the right side is the SW Cape of Greenland, and at the left side is the entry to Hudson Strait (look at a map). The small piece of wall in between is the literal crossing of Davis Strait.

From the start of the central NS-wall X (Fig.18) to H', the **left wall** makes an **angle of 60°, corresponding to the latitude of Cape Chidley, at the entry of Hudson Strait at 60°N.** The wall from H' makes an angle of 6°, showing the length of Hudson Strait, 6dl (= 6°of latitude= 666km). At the left side, the end of this **side-wall** makes an **angle of 50° from X, corresponding to the latitude of the then existing south point of James Bay, at 50°N.** Following the wall south, we recognize the east coast of enormous Hudson Bay, and beyond, James Bay. This near-vertical **wall** makes an **angle of 10°, confirming that a latitude of 60-10= 50°N** could be reached via Hudson Bay. It also indicates that at that time, the width of Hudson Bay was c.10 distance lines (10dl= 10°of latitude= 1111km). Sighting from the viewing tower Z in Fig.18, the end of this **wall at V'** makes an **angle of 50.5°, correcting the latitude of the south point of James Bay to a slightly higher value of 50.5°N.** The present south point is at 51°N (further north), but in that time the land was much lower, due to deformation of the earth's crust by the weight of the ice during the Ice Age. This all confirms again the very ancient date of the second building phase. At c.2200 BC the sea level was approximately 35 meters higher at America's Stonehenge, which was a protected seaport, with a freshwater lake and protected island above it (Fig.14, Ref.3).

The second portion of the double row of walls (Figs.2&3) is Davis Strait, with at the left side, the coastline of Baffin Island, with its prominent **Cape Dyer at the holy Arctic Circle at 67°N.** Note that a **menhir A'** (Fig.18) at the point where the wall from A1 joins the Davis Strait wall (actually noted to be a high menhir when walking the site, as we would have predicted) has an **angle of 67°** from viewing tower Z (this is also a solar and lunar alignment line). The third part is Baffin Bay, and finally, the fourth part is Smith Sound. At the right side is the west coast of Greenland. The first side-wall on the left (running toward A1) is Lancaster Sound, and the second wall on

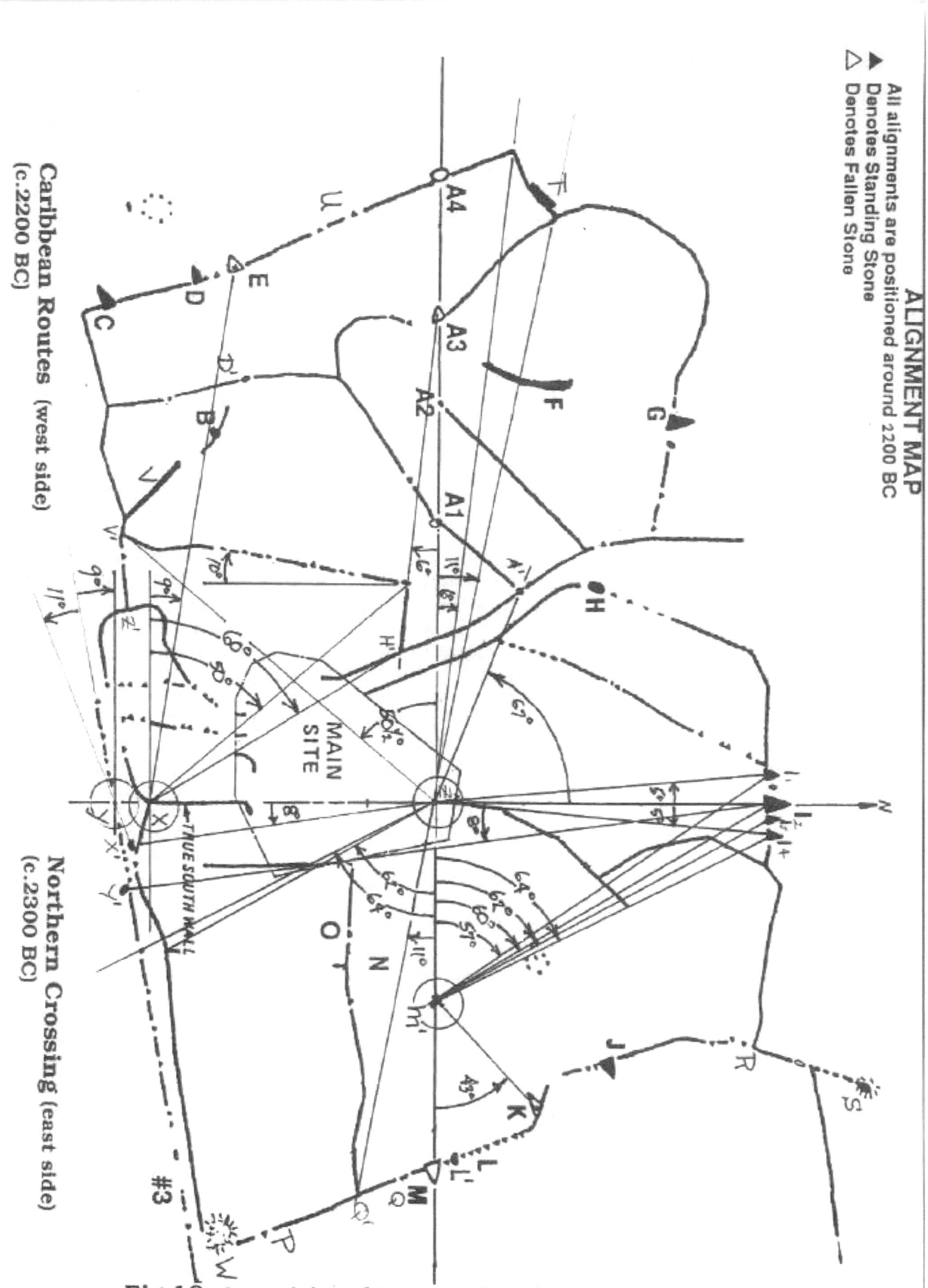


**Fig.18** Groundplan of America's Stonehenge. Site angles of menhirs from viewing points X. Y. Z. and M'.

**Fig.18**

the left side (toward A2) is Jones Sound, coming to a dead end. The third wall (running to G) is the route along the northern coast of Ellesmere Island along the Northern Ice Sea.

These details, and the extent of these amazing far north walls at America's Stonehenge suggest that people at this time were very interested in the possibility of a northern passage to the Bering Sea, which was at the limits of their explorations. The inscriptions of Dissignac (Brittany, France, Ref.2) also show that this was the last major geographical problem people were dealing with around this time period.

## Geographic Groundplan

A frustrating feature of some megalithic sites is the multiple symbolisms used. At this site, the menhirs have encoded multiple geographic, as well as astronomic meanings. It appears, though, that some meanings are so strongly indicated, that they have been used as "primary meanings". Basically, America's Stonehenge has a geographic groundplan, and the primary meanings of the menhirs are given by the latitudes encoded in the angles, as seen from the main viewing point X (see Fig.20). Probably, there are more phases in the development of the site over time which we have not discerned yet. For example, note that menhir **A2 (37°** from X) and the unnamed menhir in the wall below **A1 (39°** from X) appear to represent **Chesapeake Bay (37°N) and Delaware Bay (39°N),** which with their river systems, were probably colonial centers to the south, during the later periods of use of America's Stonehenge.

## Dating

America's Stonehenge dates back from the time directly after the discovery of America via the Atlantic c.2497 BC (Refs.2,7,9,10,11,26). However, the monument appears to have two building phases, which succeeded one another closely. The first, c.2300 BC, is before, the second is just after the discovery of Bermuda in c.2200 BC. As said before, the wall structure due south of the Main Site points to the most early date: c.2300 BC. The walls between the Main Site and the Visitor Centre, in which the builders refer to the original discovery of America via the Bering Sea, are without doubt of the same age. However, all these walls are interconnected with the Main Site and the eastern half of the location (walls and menhirs). So this part is the first building phase, America's Stonehenge I.

The southernmost chamber of the Main Site (Figs.3-5, Photo.3) represents Bermuda. The double walls in the northwest (Figs.2&3) point via Cape Cod (and America's Stonehenge) exactly to this chamber, as discussed previously. The right wall in the northwest is the west coast of Greenland (the first part is a western inner wall), and the left wall is the newly discovered coast of North America. Note that the double wall does not fit geographically to the old single wall north of the Main Site, so this is the joining area between the two building phases. Bermuda is involved in all of the routes left of the Main Site, constructed c.2200 BC in the second construction phase. If Bermuda

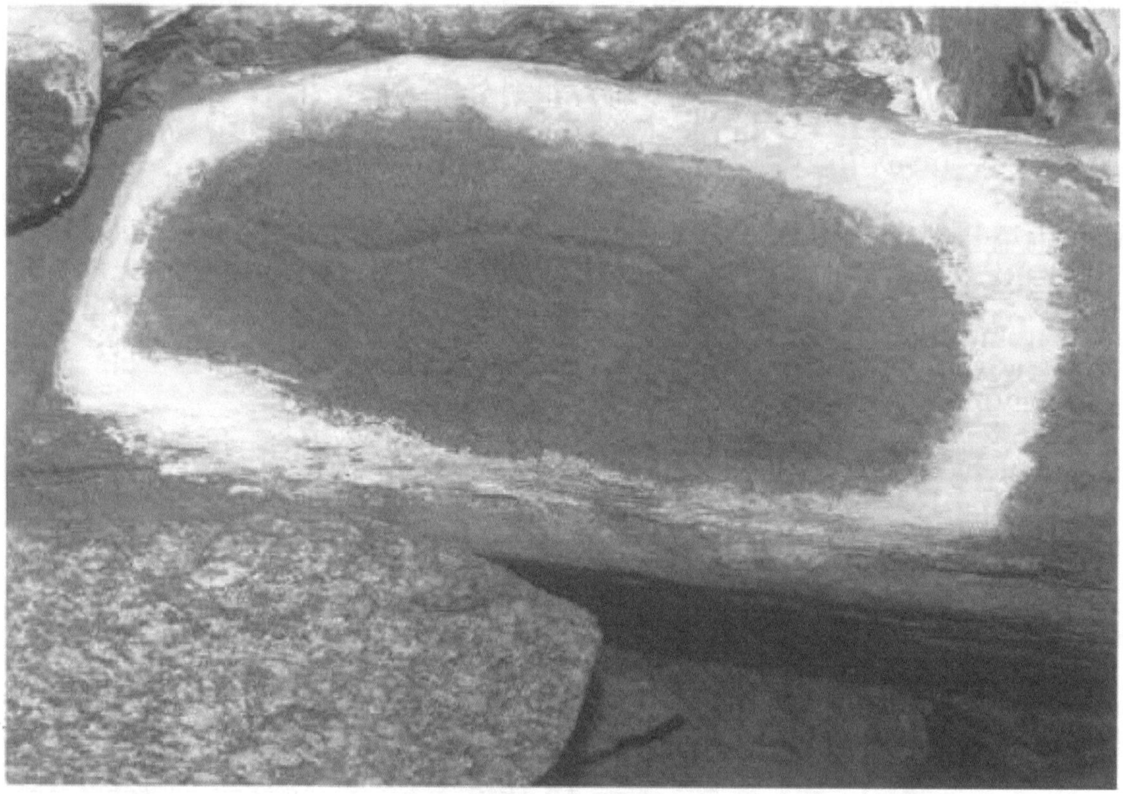

**Photo.11** "Running Deer" stylized petroglyph of the crossing from America's Stonehenge to the Azores, and the coastal sailing route from Cape Hatteras via America's Stonehenge to Cape Race. (East end Cavern, Fig.6)

**Photo.12** Dr. de Jonge pointing to the large fallen menhir north of the Main Site. representing Cape Holm, Greenland, on the Arctic Circle, at 67°N. (The site map did not show this menhir, but we thought there should be one, and there it was!)

(32°N) had been discovered earlier, America's Stonehenge would not have been built at such a high latitude (43°N). It appears that in later times, the sailing routes via Bermuda were the most popular ones. The dates mentioned here are compatible with the corrected Carbon-14 dates from the monument, the oldest of six samples being at 2000 BC (Ref.26, pg.92). Our dates also match the situation of the land level rise (see Ref.3). At c.2200 BC the sea level was about 40 meters higher, with a protected anchorage wrapping around the site (Fig.14, Photo.19).

A very intriguing question is at what time the site was definitely abandoned. In 1975 a broken 14 foot menhir was found in the center of a stone circle, in the woods north of the Main Site (Photos.17,18). The stone was found to be inscribed with roman numerals "followed by a triangle, the Celtic symbol for days, reading in translation by Barry Fell "DAY XXXVIIII". Barry Fell called the stone the "Beltane Stone" with a calendrical explanation, and concluded that the Celts still occupied the site as late as the century following the introduction of the reformed calendar of Julius Ceasar in 45 BC (Ref.7). If Fell is right, that means that America's Stonehenge was in use for 2500 years. Pieces of the menhir are lying in the Visitor's Center. We have recommended that a replica of the inscribed menhir be erected in its original position in the stone circle where it was found. We suspect that if the shadow of the menhir is carefully watched, one might discover a peaked stone in the circle that would reveal the best "day" of the year for setting sail to the Azores, which are located at "39°N". So the carved triangle on the stone also means "degrees", and the three X's correspond to the 3 island groups of the Azores, and the nine symbols correspond to the 9 islands of the Azores!

Ms. P. Underwood describes, in her book The Walking People (Refs.24,31), how after centuries of wanderings, this indian people arrived in a wooded area near the Ohio (The Winding River). They had to build enormous mountains for the Ohio Mound Indians (the Sun Disk People), who exploited them. Finally, they fled across the Appalachians to the east, where they settled down between the "Southern People" and the "Stone Hill People". This last group lived around America's Stonehenge, called "Stone Hill". According to oral history it was a strange people. They did inconceivable,and complicated things with stones. They had a lighter hue, were not hostile, and had beautiful metal artifacts, like the Ohio Mound Indians. Clearly they felt superior to the Walking People. They considered America's Stonehenge as their sacred place. When they walked on the site in procession, they always approached the monument in a northerly way (this confirms our approach from the south along the NS axis!). The Walking People heard that they were not as numerous as they had once been. On one of the few occasions they met, they warned the Walking People never to disturb their sacred hill. Rather soon after their arrival, the whole Stone Hill People disappeared with unknown destination. From this story it can be concluded, that, probably, this last event happened about 500 AD, so America's Stonehenge was used for nearly 3000 years!

**Photo.13** The viewing tower (Z), north of the Ceremonial Center, important for both geographical and astronomical alignments.

**Photo.14** The authors with Robert Stone (right) at the start of the NS wall (X), July 18, 2000. Mr. Stone is owner of America's Stonehenge, and founder and former president of the New England Archaeological Research Association (NEARA).

## Possible Use of the Site

A place used by mariners from around the world for almost three thousand years must have had a name. Maybe it was "Stone Hill". Probably, today's owner, Bob Stone, would be comfortable with that. It is thought that during its period of use, most of "Mystery Hill" was a bare glacially-stripped rock hill, with none of the trees it has today (Ref.26). The southern walls formed the original entry to the site, because the megalithic people arrived in the Americas by the Southern Crossing, and it would have made sense to them to enter this geographic site in the same manner (Fig.19). Next, they walked to the western walls of the Main Site to discuss their position along the East Coast of North America (the orientation). The West Chamber was the reception room because it symbolizes both the West Azores and the New World. Here travelers would receive information about Central America. Visitors who wanted to cross the ocean via Cape Race could receive information about that in the V-hut (nr.18). Those who were planning to travel via Bermuda were sent to the south chamber (nr.13). At scheduled dates and times the Oracle Chamber Corridor and the Cavern (nr.28) were opened for information on how to reach the Azores and the appropriate sacrifice ceremonies. Visitors with questions about Madeira and beyond were sent to the northern Pattee Area (nr.26). Staff and management of the site may have held office in the southern Pattee Area as well as in the Cavern. Information about the Northern Crossings could be obtained there. The Watch House chamber (earlier shown to represent the Canaries) at the start of the entry was the gate-keeper, as the name implies, and the Davis Strait western walls were the corridor out of the site.

The ground of the site is uneven, though generally sloping east. It is thought that the site was not wooded when it was occupied or in use. Today it is difficult just to keep the alignment corridors free of trees. The wooden viewing tower now at Z is necessary to sight the alignment stones for astronomical alignments. The megalith people must have had a similar wooden structure here. There had to have been a similar viewing tower at X for the geographic alignments, and we have recommended building one there. It probably had an extension to Y. The two viewing towers on the north-south line, and removal of all trees between them and the menhirs, would allow clear viewing and would probably replicate the in-use site situation.

In the Main Site rock holes indicate that two stones have been used as holders for wooden posts, probably used for measuring the angles of the menhirs from the most important viewing points. How this might have been done should be studied by someone. A great many interesting stones, such as quarries, dolmen, walls, and menhirs are being found, even at a considerable distance beyond the site. The island in the adjacent lake should be examined carefully, especially for remains of habitation. The steep hillsides, particularly SW and SE of the Main Site should be examined, especially at elevations that would have been shoreline, mooring, or pier areas at the estimated harbor locations recently described, that are now dry

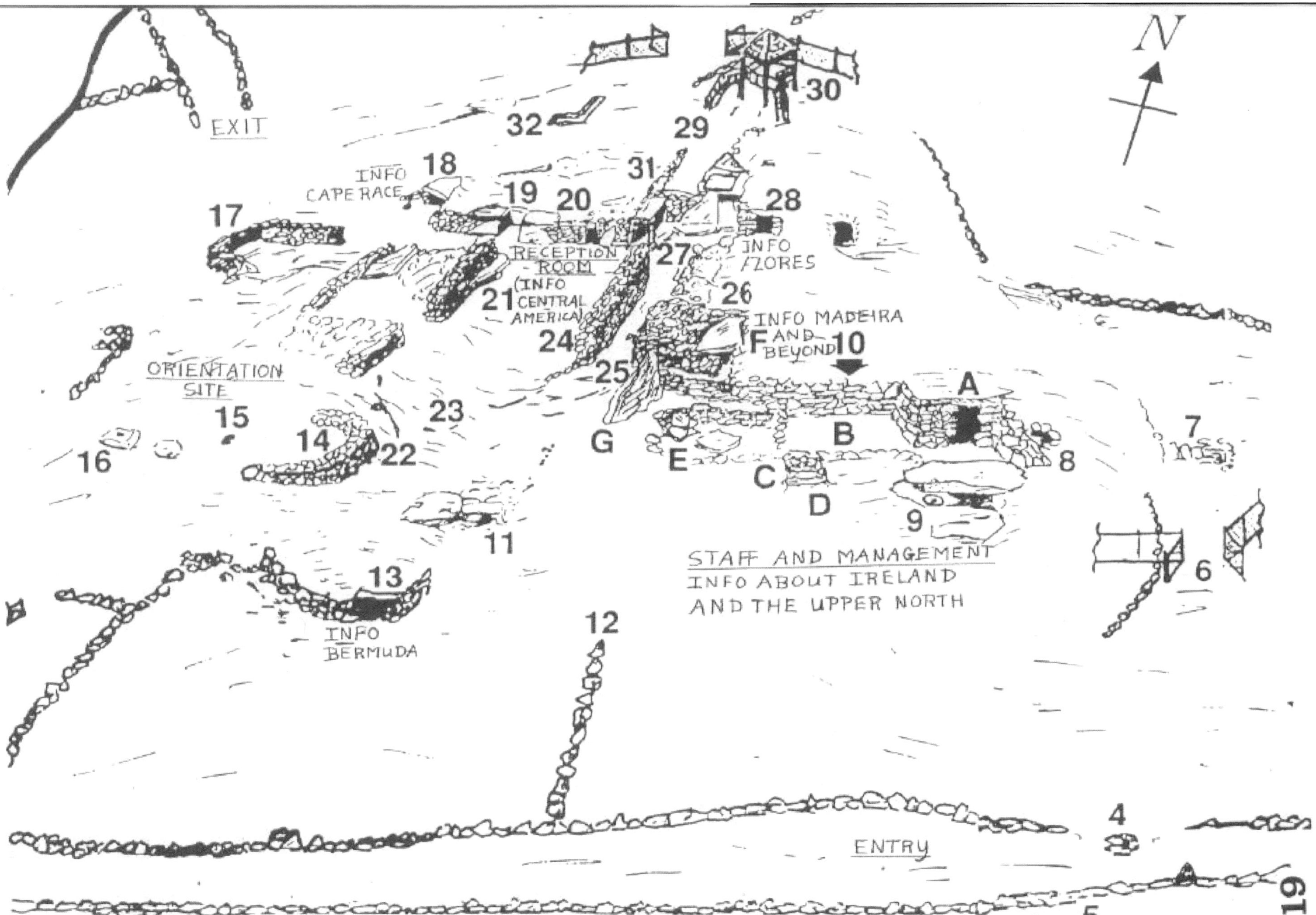


**Fig.19** Possible use of the Main Site of America's Stonehenge as a teaching center.

Fig.19

(Figs.14.15; Photo.19; Ref.3). The entire area should be carefully explored, and precise photographic and location surveys of all lithic features should be done.

In the walls of America's Stonehenge a surprisingly high number of upright triangular stones are included. In fact, they have the same shape as the surface of the Main Site. Obviously, these stones are also stylized images of the North Atlantic Ocean because this is the focus of America's Stonehenge, and another example of the use of complex multiple symbolisms, or multiple levels of meaning used by megalithic people.

Walking the site, there is tour emphasis given to the channels cut here and there in the bedrock, to facilitate the draining rainwater off the site (Fig.5, Refs.1,8). Since this is an ancient nautical center, with expert sailors working on the site, one might ought not be surprised by a dislike of water in the bilges or underfoot!

## The de Jonge Rules of Decipherment

We admire the cleverness of these monuments, built by people with no way of writing their spoken languages. We have been trying to learn to think like they did, so their complicated monuments would become clear to us. We need your help to decipher more of these sites, and confirm our findings. Here are a few "principles" we have learned to use, that will help you in deciphering megalithic sites:

**1)** Look for encoded latitudes in the number of stones, and the alignments of the site. Particularly look for the latitude of the site itself. Number of stones can be related to number of islands (or island groups) in the ocean. Degrees of latitude are usually given by lines (angles) pointing NE or NW, though there are exceptions. Angles of 23° and 67° have religious meanings.

**2)** It is usually impossible to write down that a certain menhir represents a certain place. It would certainly simplify explanations of these sites if this were true, but most menhirs have a number of different meanings. This use of multiple symbolisms is true at America's Stonehenge, and most other megalithic sites as well. Menhirs here can represent astronomic dates, latitudes of places, initial and terminal sailing directions, geographic distances, compass readings, or all of these.

**3)** Big menhirs are more important than small menhirs, and menhirs are more important than walls. Big wall rocks are more important than walls made of small rocks. Thick walls are more important than thin walls. In every case, think about the labor involved quarrying and moving the weight of the stones. Long walls are more important than short walls. Chambers are much more important than menhirs or walls. Always think of the labor involved. Big monuments are always easier to understand than small ones.

**4)** For the megalith builders, it is a natural thing to have more than one viewing point on a site. For example, Stonehenge has at least three. In American Stonehenge there are at least four, with all the most important ones on the true north-south axis, and others on the east-west axis.

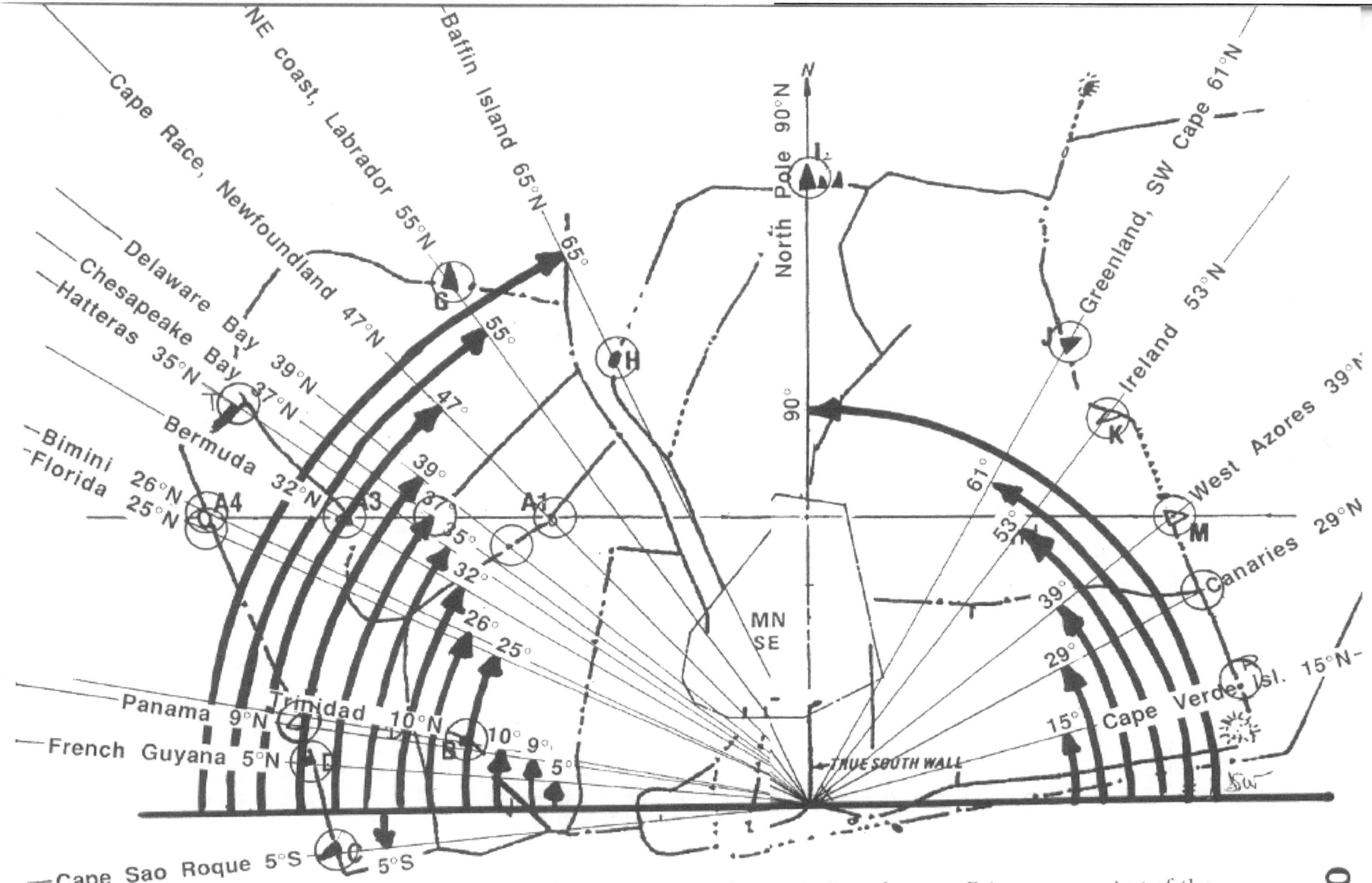


**Fig.20** Groundplan of Arrica's Stonehenge. Primary meaning of the menhirs, seen from wing point x.

Fig.20

**5)** The walls are important for the whole story, because they reveal how to move across the site. To record geographic information, you have to move from one menhir to another menhir. Setting this up so it works both astronomically and geographically is not an easy process, so do not expect the maps to always look perfect to you.

**6)** Look for the numbers 23, 30, and 26.5, because they relate to the latitudes of the centers of the Southern, Northern, and the United Egyptian Empire, respectively. After c.2500 BC, the numbers 16, 18 and 17 can be important, because they can relate to the latitudes of the centers of the Southern, the Northern, and the United Central American Empire, respectively.

**7)** Monuments can be oriented on sunrise or moonrise, on midsummer day, midwinter day, mid-spring day, or mid-autumn day, all of which have religious and cultural meanings. Monuments can be oriented on geographical north (like America's Stonehenge), but also west, south, or east. The history of the megalith builders is often involved in a monument.

**8)** Sailing Directions should be in the right direction. A line pointing to the NE cannot represent a sailing direction to the SE. However, a line pointing to the NW can represent a sailing direction to the SE, the opposite side.

**9)** The use of complementary angles is frequent. It is clear they often took care to use both sides of an angle, and both angles or either can be important.

**10)** Be careful in mixing the geographical alignments with the astronomical alignments in reading geographic meanings. The builders mixed them, but we do not understand their intentions yet. Much work remains to be done.

## Discussion

We do not think America's Stonehenge has been demolished in recent centuries in such a terrible fashion as suggested in the literature (Refs.1,7,26). The damage was primarily done to large chambers, not the menhirs. Perhaps it has helped that the site has been privately owned by people who have been interested in the antiquity of the site, and have, for the most part, protected it as a historic curiosity. Sure, the western border of the Main Site seems to be rather disturbed, while the southern Pattee Area has had damage too. Quite a few capstones appear to have been removed here (Fig.5). However, we are happy to conclude that since what remains makes sense, the site is less disturbed than it has appeared to be. The complexity of the existing site, with its geographic layout, and angular accuracy superimposed and combined with astronomic phenomena, is astonishing. All these coincidences do not occur in nature by accident, but have been built by the human mind. Thousands of years of compounded knowledge and effort by the brightest minds of society are reflected here.

This amazing integration of astronomical site design with geographic encoding opens an interesting opportunity for interdisciplinary research. What meanings might have been intended by the builders when certain star, sun, or moon risings or settings use the same menhir that describes certain

**Photo.15** The authors at menhir J (July 18, 2000), representing the SW Cape of Greenland, at 61°N. The edge resembles a coast map of Greenland (upside down).

**Photo.16** Menhir H, with "eye of Baal" (Ref.7), showing the shortest crossing of Davis Strait at 65°N (Fig.20).

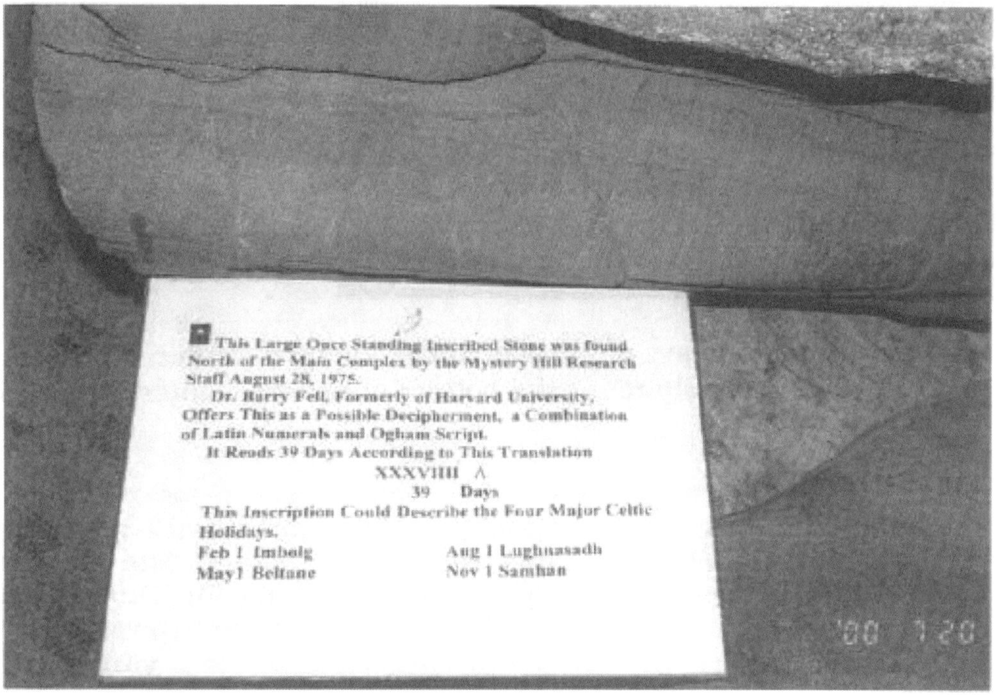


**Photos.17&18** "Day 39 Inscription" on 14 foot broken "Beltane Stone", on display in the Visitor's Center, referring to the latitude of the West Azores, at 39°N. The 9 symbols correspond to the 9 islands of the Azores (c.50 BC)

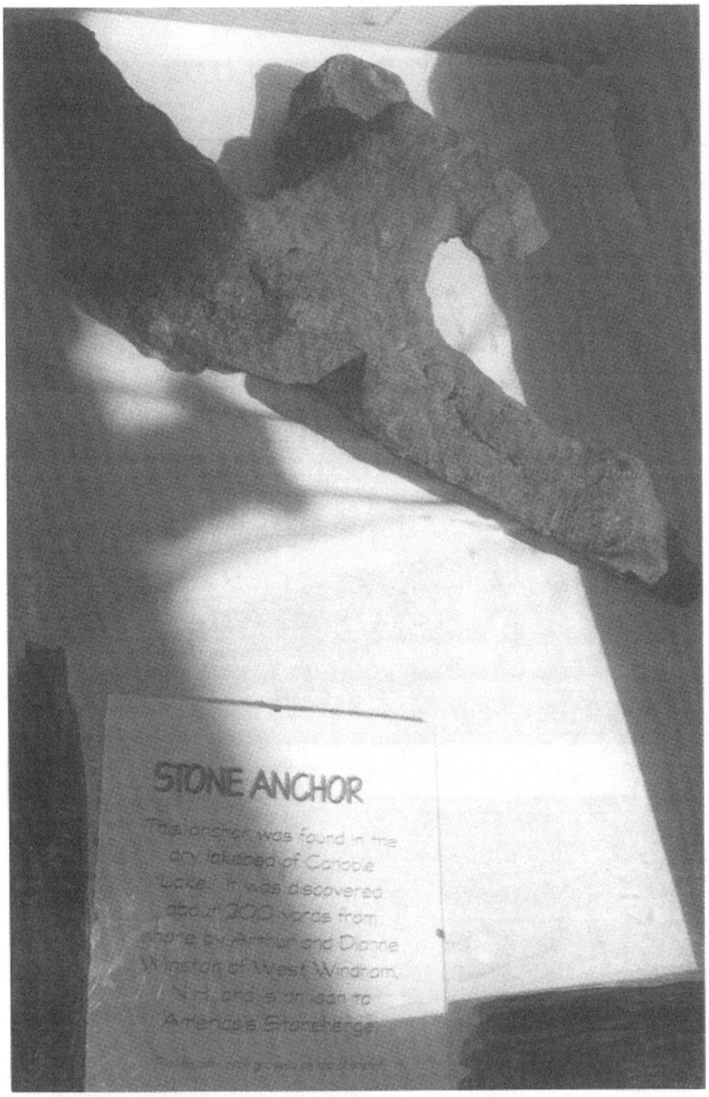


**Photo.19** Stone anchor, on display in the Visitor's Center. It was found "200 yards from shore" at the bottom of nearby Canobie Lake to the SE (Fig.14).

sailing routes or sailing destinations? Much more work is needed here, including more carefully done large area site mapping. Site measurements should be part of the mapping, particularly looking for Thom's Megalithic Yards of 2.77 feet. Mining engineer Whitney "measured every wall ... and found no place where the measurements were in feet, yards, or inches. Whoever built that place either didn't give a damn about standard linear measure, or he didn't know it existed" (Ref.26). This reference also reports that "Frank Glynn ... discovered that all measurements ... conformed to an ancient Egyptian unit known as the cubit". Which cubit, the Royal Egyptian Cubit of .5775m, or the Egyptian geographic cubit of .462m? This wonderful site needs and deserves a lot more work.

A test of any scientific theory, like the de Jonge Rules of Decipherment, is whether it can be used reliably as a predictor. Without intending to do so, we did this experiment upon ourselves. Dr. de Jonge figured out most of this decipherment at home in Holland, using the Site Brochure provided to site visitors. From the site drawing, we knew that a place of major importance (Cape Holm, where they first discovered Greenland at the Holy Arctic Circle) was represented by only a minor change in the angle of a wall. This was curious, because their methods, as we understand them, would usually represent a place like this with a large stone. Lo and Behold, when Dr. de Jonge walked the site for the first time (see Photo.14), there was a large fallen menhir, not included in the Site Plan, right where we thought there should be one! Reinoud did not actually get excited, because he is used to having this sort of thing happen.

Generally speaking, humans live in the present moment, and have more difficulty understanding the past, the further back it gets. Combined with the comfort of existing beliefs which color perceptions, such as "Columbus discovered America", it is hard to find a receptive audience for new ideas that cannot be "witnessed". Our findings are dependent upon the liklihood of a great many coincidences in the number of symbols, stones, or angles of menhirs. We have found so many coincidences, that they are playing to us from one prehistoric site to another like a symphony. These results, springing from Dr. de Jonge's insights obtained on his premise of latitude use, need further study and confirmation by other researchers, so this material can be added to the increasing knowledge of the prehistory of man on earth.

# References


1. "Tour Guide Map", America's Stonehenge, PO Box 84, North Salem, NH, 03073, and website "America's Stonehenge"
2. Jonge, R. M., de, and IJzereef, G.F., De Stenen Spreken, Kosmos Z&K, Utrecht/Antwerpen, 1996 (ISBN 90-215-2846-0) (Dutch)
3. Payne, M. H., America's Stonehenge as Architecture, NEARA Journal, Vol. XXXII, No. 2, pg.93 (1998)
4. Richards, J., Stonehenge, English Heritage, 1992
5. Twohig, E. Shee, The Megalithic Art of Western Europe, Clarendon Press, Oxford, 1981
6. Hawkins, G. S., Stonehenge Decoded, Barnes & Noble, 1993 (ISBN 0-88029-147-8)
7. Fell, B., America BC, Pocket Books, Simon & Schuster, 1994 (pgs.145,200)
8. Whittail Jr., J. P., "Mystery Hill Revisited", NEARA Journal, Vol.XXXII, No. 2, p.91 (1998)
9. Jonge, R.M., de, and IJzereef, G.F., "Exhibition: The Megalithic Inscriptions of Western Europe" (1996)
10. Ferryn, P., "5000 Years Before Our Era: The Red Men of the North Atlantic", NEARA Journal, Vol. XXXI, No. 2, pg.59 (1997)
11. Bailey, J., Sailing to Paradise, Simon & Schuster, 1994 (ISBN 0-684-81297-5) (pg.397)
12. Lenik, E.J. and Gibbs, N.L., "The Frost Valley Petroglyph, A Catskill Mountains Enigma", NEARA Journal, Summer 1999 (ISSN: 0149-2551)
13. Mallery, A.H., and Harrison, M.R., The Rediscovery of Lost America, The Story of the Pre-Columbian Iron Age in America, Dutton, NY, 1979 (ISBN 0-525-47545-1)
14. Thompson, G., American Discovery, Misty Isles Press, Seattle, 1994
15. Zapp, I. and Erikson, G., Atlantis in America, Navigators of the Ancient World, Adventures Unlim. P., 1998 (ISBN 0-932-813-52-6) (pg.357)
16. Peterson, F.A., Ancient Mexico, 1959
17. Stuart, G.E. "New Light on the Olmec", National Geographic, Nov. 1993
18. Jairazbhoy, R.A., Ancient Egyptians and Chinese in America, Rowman & Littlefield, Totowa, N.J., 1974 (ISBN 0-87471-571-1)

19. Aubet, M.E., The Phoenicians and the West, Politics, Colonies and Trade, Cambridge Univ. Press, 1996 (ISBN 0-521-56598-7) (pg.143)
20. The Seattle Post-Intelligencer, 9/23/00 (pg. A4)
21. Oppenheimer, S., Eden in the East, The Drowned Continent of Southeast Asia, Phoenix, London, 1998 (ISBN 0-75380-679-7)
22. White, P., "The Oz-Egyptian Enigma", Exposure Magazine, Vol. 2, No.6, 1996
23. Allen, J.M., Atlantis, The Andes Solution, Windrush Press, Gloucestershire Press, 1998 (ISBN 1-900624-19-2)
24. Kayworth, A.E., Legends of the Pond, Branden Books, Boston, 2000 (ISBN 0-8283-2053-5)
25. Tompkins, P., Secrets of the Great Pyramid, Harper & Row, London, 1971 (ISBN 0-06-090631-6) (with Dr. Stecchini)
26. Lambert, J.D., America's Stonehenge, an Interpretive Guide, Sunrise Publ., Kingston, N.H., 1996 (ISBN 0-9652630-0-2) (pgs.31,33,48,55)
27. Bauer, J.L.,"Glacial Retrostatic Land Movement", NEARA Journal, Vol.XXXI, No.2, pg.73, 1997
28. Boutre, M., Ancient Celtic New Zealand, De Danann Publishers, Auckland, 1999 (ISBN 0-473-05367-5), and website under title
29. Winkler, I., and Stone, R., "Construction and Use of America's Stonehenge", NEARA Journal, Vol. XXXIII, No. 2, 1999
30. USGS, Salem Depot Quadrangle (upper left corner)
31. Underwood, P., The Walking People, A Tribe of Two Press, PO Box 913, San Anselmo, Ca. 94979
32. Great Circle Sailing Chart of the North Atlantic Ocean, rev.1970, Defense Mapping Agency, Hydrographic Center, Wash. D.C. 20315
33. Curtsinger, B., "Oldest Known Shipwreck", National Geographic Magazine, Vol.172, No.6, Dec. 1987
34. Honore, P., In Quest of the White God, G.P. Putnam's Sons, New York, 1964 (transl. from German, publ. 1961)
35. Irwin, C., Fair Gods and Stone Faces, St.Martin's Press, New York, 1963

# The Discovery of Bermuda

## The Decipherment of American Megalithic Petroglyphs

(The Devil's Head Petroglyphs, Harmony, Maine, c.2200 BC)

Dr. R.M. de Jonge, drsrmdejonge@hotmail.com
J.S. Wakefield, jayswakefield@yahoo.com

## Introduction

The Devil's Head Petroglyph is situated on a hill of that name (Fig.1) just SE of the village of Harmony, central Maine (US), only 25km east of the well-known Embden inscriptions on the Kennebec River (Ref.16). To get to the site, you walk an hour up the marked snowmobile trail that runs east and steadily up from the road junction just east of the bridge. The petroglyph (Fig.2), is applied on nearly flat granite bedrock underneath a huge overhanging rock on the south side of the hilltop. The lower photo in Fig.1 shows the authors sitting in this rockshelter, with the petroglyph between them. The top photo shows Dr. de Jonge on top of the shelter, with Great Moose Lake in the distance. The place has a very nice ambience; a beautiful perspective combined with a very protected, hidden feeling.

The petroglyph (Fig.3) is c.52cm long and c.21cm wide. It is a carved ovoid, about a half cm deep in the granite, with three carvings inside. The ovoid has been described as a "cartouche". This word "cartouche" refers to a similar-looking ovoid figure developed at about this date in Egypt, which always contained characters for pharaonic names, a fact of great help in the decipherment of Egyptian hieroglyphics. Here, this somewhat look-alike figure has been called a cartouche, though it does not contain Egyptian symbols, or possess the bottom knot of the Egyptian cartouche. In this article, though a misnomer, we will continue to call this shape a cartouche, because it has recently been labeled as this in the NEARA literature (Ref.1).

The shape of this cartouche (see Figs.3&5) shows a great similarity to the left half of a well-known type of megalithic inscription from Western Europe, of which a few tens are known, as shown in Fig.7 of the Introduction (Refs.2,12). Most all of them date from before the discovery of America, c.2500 BC. They represent the stylized North Atlantic Ocean. In this case, the cartouche enfolds the important, but unknown, western part of the ocean, while the space to the right of it, which is east of the meridian through Greenland, is the well-known eastern part of the ocean.

## Newfoundland

Look at the upper inscription within the cartouche (Fig.3, also Figs.4&5). Most of this carving is 0.9cm deep. The first part of the dark slanting line is the east coast of Labrador, from Cape Chidley in the north to Cape St. Charles in the south. Cape Chidley is the NE point of North America, and Cape St. Charles is the east point of the mainland of North America. Left, near the bottom of the black line, another line attaches, representing Belle

**Fig. 1** The Devil's Head Petroglyph is inscribed on nearly flat granite bedrock in a rockshelter on the top of Devil's Head, near the village of Harmony, in central Maine (US). In the **lower photo,** the petroglyph is between the authors, with white sand in it, so it is easier to see. In the **upper photo**, Dr. de Jonge is looking at other petroglyphs carved into the top of the roof block. To the southeast, behind him, Great Moose Lake is visible.

Isle Strait, and, following it, the north side of St. Lawrence Bay. The first part of the line to the south is the west side of St. Lawrence Bay, along the east coast of New Brunswick. This line extends to Maine, where the petroglyph is located, and runs down via America's Stonehenge on the coast at 43°N (Refs.3,6), and via Cape Cod at 42°N due south of this petroglyph, to Delaware Bay at 39°N. Below this line we just followed to the left (to the west) are two small, deeper lying patches, shown as circles in Fig.4. The right patch is the island of Newfoundland (new-found-land), a habitation center of the Red Paint People (Refs.5,23). The left patch is St. Lawrence Bay. Below this bay a small triangle may be seen which would represent the peninsula of Nova Scotia (New Scotland). Newfoundland, however, is the easternmost area of North America, and Cape Race is the easternmost point.

## Bermuda

The central inscription of the cartouche is a vertical line that is carved 3.9 cm deep. You can see that it runs south from the direction of Nova Scotia. It is the meridian, today at 65° eastern longitude, running through the Bermuda Islands, due south of Nova Scotia. Bermuda is at 32°N, indicated by a small side branching on the NE side (most clear in Fig.5, where white sand has been put into the carving). This small archipelago is situated 1000km from the east coast. It was discovered from America at c.2200 BC, and this petroglyph encodes its location. The center vertical line continues to run down to about the Tropic of Cancer at 23°N, now pointing SSE in the direction of the Lesser Antilles.

## Puerto Rico

The lowest inscription of the cartouche is a deeply inscribed (5.1 cm) cross, pointing to the north. The horizontal line is the 18th latitude line through Puerto Rico, the easternmost island of the Greater Antilles (Ref.2), where recently very ancient stone steps have been found underwater by Scuba divers. This line extends from Hispaniola in the west to the Virgin Islands in the east. The vertical line is the meridian through Puerto Rico, which is located slightly west of Bermuda, as indicated on the petroglyph. We are familiar with two petroglyphs in Europe, one in Portugal (Chao Redondo), and one in Brittany (Kercado), that indicate both Bermuda and Puerto Rico. The inscription of Puerto Rico here at Devil's Head has been carefully made, to indicate the accurate east-west position of the recently discovered Bermuda Islands.

## The Western Part of the Ocean

The cartouche is the stylized western part of the Atlantic (Figs.3-5). The vertical line which is the left side of the cartouche is a meridian along the east coast of Hudson Bay and James Bay, down to Cape Hatteras, the east cape of the US, at 35°/36°N. As clearly visible in the petroglyph, this point is located below Delaware Bay at 39°N (the bottom end of the top glyph), and above Bermuda at 32°N (the branch on the middle glyph). The bow-shaped top of the cartouche runs from Hudson Bay/Hudson Strait via the shortest crossing of Davis Strait at the level of the Arctic Circle (67°N), to the SW

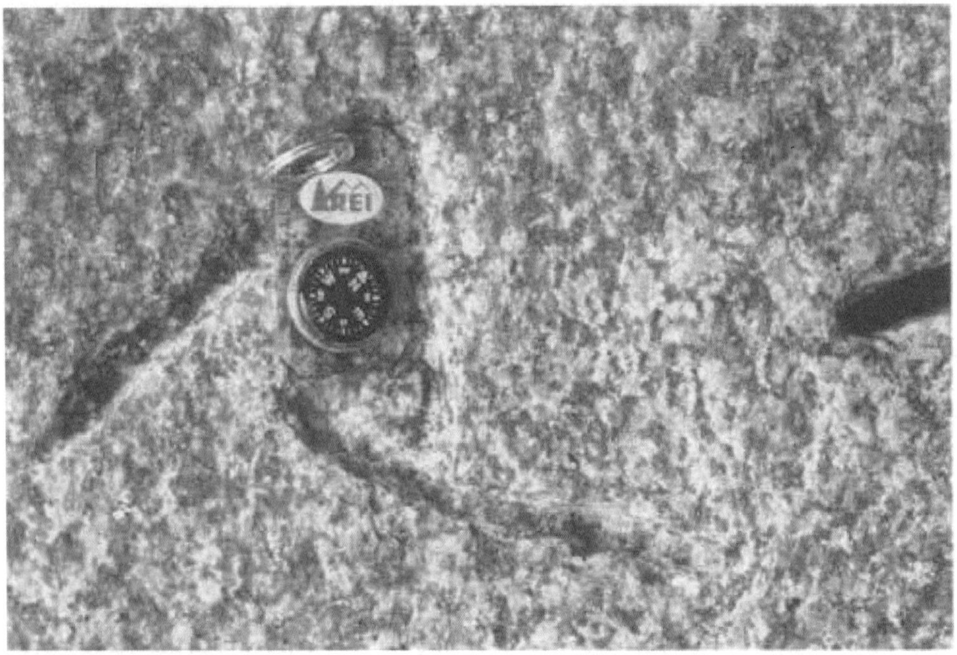


**Fig.2** Photo of the Devil's Head Petroglyph, Maine. **The top photo** (with white sand in the inscription) shows how it could have been carved by people sitting in the shelter. A bed of boughs would have been nice to level the area, and make it more comfortable. **The lower photo,** with no sand, shows the depth of the inscription, and the axis direction of 30°SSW compared with the magnetic north pole.

Cape of Greenland at 61°N. Note that the carving has a wide spot, or "knob" at the top of the cartouche, which is an acknowledgement of the "Holy Arctic Circle" at the crossing. The bow continues to a shallow indention at its upper right corner. This reveals the two important southwest corners of Greenland: its SW Cape at 61°N, and Cape Farvel (cape farewell), the south cape of Greenland at 60°N.

The long vertical line running downwards, which is the right side of the cartouche, is the meridian running south from Cape Farvel. This imaginary line divides the ocean into a western part, which is of great importance to those sailing east from America now, and an old known eastern part, explored a thousand years before.

## The Crossing to the Old World

All these details show that we are dealing with a megalithic petroglyph dating from shortly after the discovery of America c.2500 BC. The carving illustrates Cape Farvel, which continued to be of interest for only a few hundred years after that (Ref.2). The megalith builders arrived in America via the Southern Crossing between Africa and South America, riding the tradewind and the current, a distance of more than 2000km. To return from Central America, they were forced to ride the Gulfstream north, up the American east coast. Note that the end wall of the rock shelter at the bottom of the petroglyph (Figs.1&2), is slanting up from the right (Figs.3-5), thus representing the slanting NE coast of South America. Together with the bedrock sloping downwards, they illustrate that it was impossible to return via the Southern Crossing, against the prevailing winds and currents. In the beginning, they returned via Greenland and Iceland, but they hated this long and cold route across the Upper North. Soon they found they could cross between Cape Cod and Cape Race to the West Azores, a distance over open sea of c.2000km (Refs.2,3). Note that the West Azores (which are at the same latitude as the lower end of the upper glyph, Delaware Bay, 39°N) are almost due east of the Devil's Head petroglyph in Harmony, Maine (at 45°N). For this reason, and because the river valleys of New England had become colonized (as demonstrated by Barry Fell, Ref.6), America's Stonehenge Part One was built on the coast at 43°N (c.2300 BC, Refs.3,22), which is the oldest part of the most important megalithic monument of North America.

About a century later they discovered the Bermuda Islands at 32°N. This Devil's Head petroglyph in Maine testifies to this discovery, c.2200 BC. Now the sailing routes on the Atlantic suddenly changed! From Delaware Bay (the lowest point of the top glyph and the left cartouche line), one could sail via Bermuda to the West Azores. The distance between Cape Hatteras (left side of cartouche) and Bermuda (middle glyph) is 1 big distance line (=10° of latitude) equal to 1111km. The distance from Bermuda to the meridian through Cape Farvel (line, right side of cartouche) is estimated to 1DL= 1111km too as shown by the petroglyph. This is not exactly correct, and it turns out to be too optimistic (it was a little further). However, it indicates that the new route was preferred!

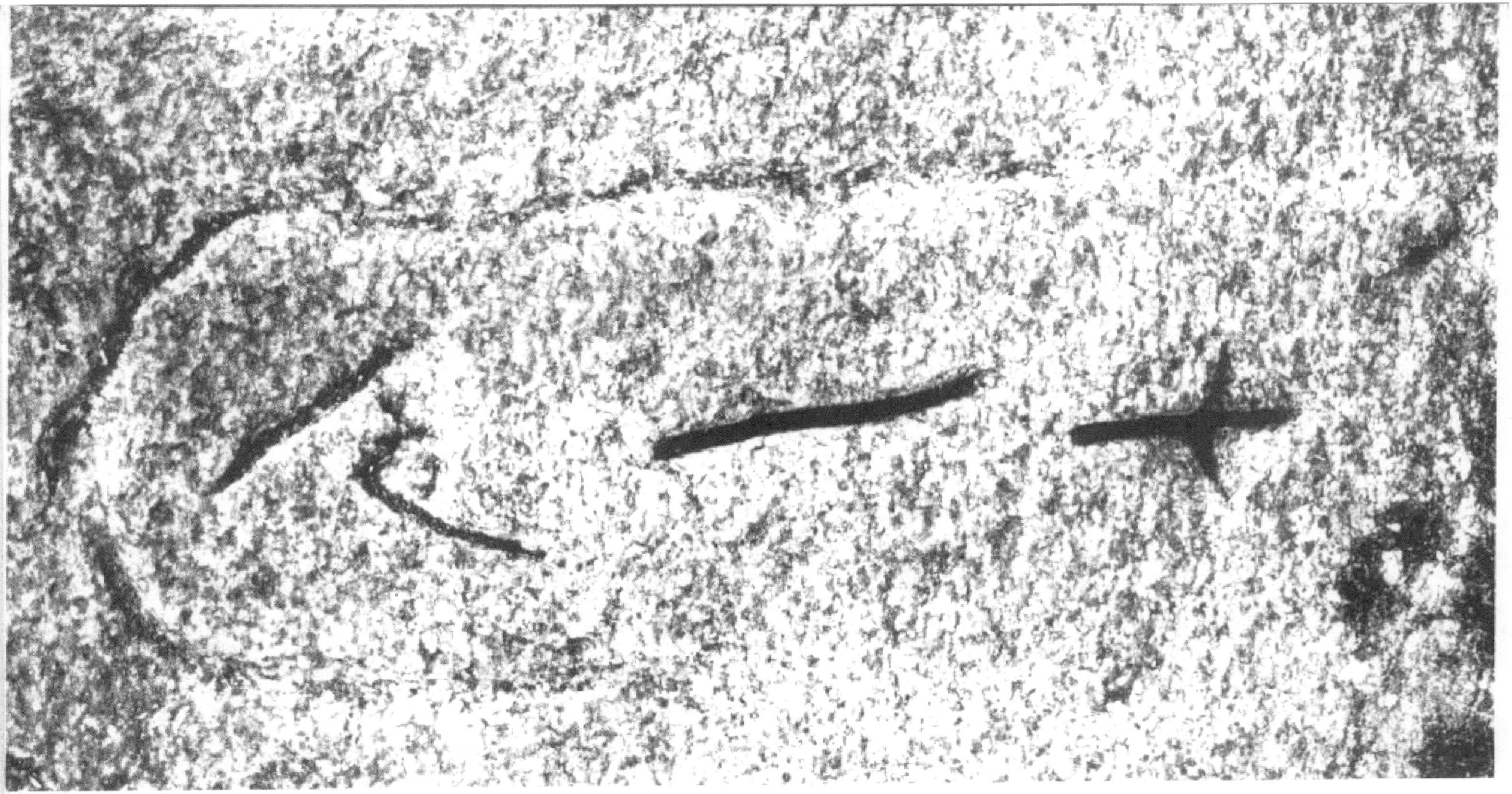

**Fig.3** The vertical cartouche is the western half of the North Atlantic Ocean. The upper glyph at the top is the east coast of Labrador, from Cape Chidley to Cape Race, with a tail running down the New England coast. The middle glyph is a meridian line, with Bermuda shown as a small branch (see Fig.5). The lower glyph is Puerto Rico, where the meridian line crosses the 18°N latitude line. According to the makers of this petroglyph, the western part of the ocean has a width of 2DL. =2222km (c.2200 BC).

As a result of the discovery of Bermuda, c.2200 BC, America's Stonehenge II was built on the coast of New Hampshire. This second phase was an important extension of the original, hundred-year old or so, monument. It now became possible for ships from Central America to sail directly from the Bahamas to Bermuda, up the left edge of the Bermuda Triangle. Note that this Devil's Head petroglyph in Harmony, Maine, is exactly at 45°N, and 45° SW of here is the center of the Olmec civilization at the southern border of the Gulf of Campeche (Refs.7-11,13-15). Details of the glyph suggest that the advised route to Bermuda is at the latitude of the Holy 23°N line through the Florida Straits along the north shore of Cuba, then via Bimini, not Florida, which is on the wrong side of the dangerous Gulfstream. The 4 mph northward current of the Gulfstream is nearly always running into strong north winds, causing rough seas and crossings from Florida to Bimini that are often legendary even today. The east side of huge shallow Andros is mostly mangrove swamp, while the shallow Bahama Banks and the treacherous reefs to the south later sank many Spanish Galleons. The Islands of Bimini, on the edge of the Banks, mark a sailing route across the banks to the deep Tongue of the Ocean in the east. According to the details of the petroglyph, the distance from Bimini to Bermuda is slightly longer than from Cape Hatteras to Bermuda (1DL= 1111km), which is correct. In the second half of the 2nd millenium BC, this crossing would become the most important one for travelers from Central America!

## The Sailing Route from Cape Hatteras to Bermuda

We were surprised to find other inscriptions under the pine needle duff on the top of the rockshelter. As far as we know, these petroglyphs have never been reported. By letting the mouth of the plastic sand bottle jiggle along in the carvings, trailing white aquarium sand, we made the glyphs much more visible. The largest inscription, Fig.6, top photo, is shown in detail below. The wave line runs from the left (west) side, indicating Cape Hatteras South at 35°N, to the right (east) side, indicating Bermuda (the dot), at 32°N. It is the shortest crossing to Bermuda (see also Figs.7&8). The line goes up and down 8 times, corresponding to a distance of 8dl= 8°of latitude,= 888km. Indeed, the distance from Cape Hatteras South to Bermuda is slightly more than 888km, as the petroglyph correctly shows. The wave line makes an angle of 18°ESE to the lower latitude line, equal to the correct initial sailing direction from Cape Hatteras South to Bermuda, 18°ESE. Note that the edges of the stone in the photo (Fig.6, top), indicate the approximate geographic position of Bermuda. The right edge of the stone in front of Dr. de Jonge is the East Coast of the U.S., while the (high) top edge is Florida, and the left edge is the Tropic of Cancer at 23°N.

Above the wave line (Fig.6, bottom) two horizontal lines are visible, connected by a vertical piece of line. The lower one is the latitude line through Devil's Head at exactly 45°N, and the vertical line is the meridian through Devil's Head, today at 70° eastern longitude, also through Cape Cod further south. According to the petroglyph (Fig.7), the distance from Cape Hatteras South to this meridian equals 4 dl,= 4° of latitude,= 444km, which

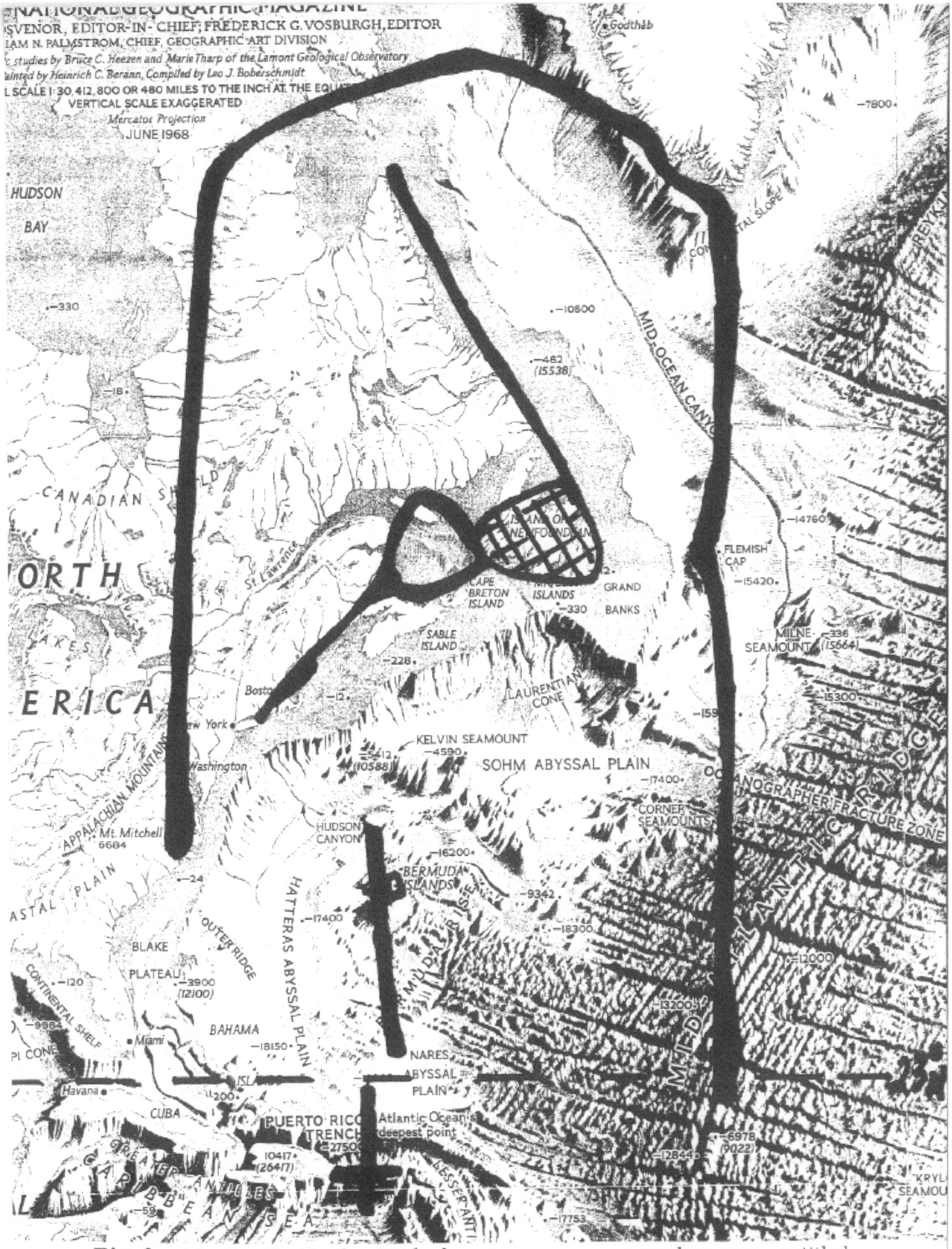


**Fig.4** The Devil's Head petroglyph as it appears on a modern map, entitled "The Atlantic Ocean Floor", National Geographic Society, 1968.

is correct. In the upper right corner we see a slanting piece of line to the northeast. This is the sailing route above 45°N, along the coast of Cape Breton Island in the direction of Newfoundland. The long vertical line to the south is the meridian between Cape Breton Island and Newfoundland, today at 60° longitude. This line confirms that, when sailing via Bermuda to the Azores, orientation on the island of Newfoundland was not important anymore. The meridian runs east of Bermuda (the dot), as the glyph correctly indicates. As a consequence, the highest horizontal line must be the latitude line at 50°N, coinciding with the north coast of St. Lawrence Bay, along southern Labrador.

When Cape Hatteras South, the place of departure, is indicated at 35°N, below the latitude line through Devil's Head at (35+10=) 45°N, the lowest latitude line will be at (35-10=) 25°N (because the spacing equals one DL, or 10° each side). At the far left (west) side, this line finishes at the south tip of Florida, at 25°N, as illustrated. As a consequence, the vertical line is the meridian along the west coast of Florida. The sharp tip above is Cape Hatteras North at 36°N, which was important for the direct crossing to the West Azores at 39°N. However, the petroglyph bears the message that the new route via Bermuda (at 32°N, the dot) is now strongly preferred. Note, that in the photo (Fig.9), at the right side of the glyph, (and above the IIW/LCC box) a small line-shaped carving (with white sand in it) is visible, representing the archipelago of the Azores in the Ocean. This can also be seen at the bottom of the upper photo, in Fig.6. The meridian from Cape Hatteras to the south (Fig.8) finishes in a slanting piece of line pointing southeast. This indicates the important NE islands of the Bahamas, which, because of their special geographic position, are important departure places for Bermuda when sailing from Central America (Refs.7-11,13,14,20).

This petroglyph about the shortest sailing route to Bermuda was made directly after the inscription inside the shelter about the discovery of Bermuda (Ref.5). The use of distance lines, the composition, and the type of weathering of the carving confirm that we are dealing with an ancient megalithic petroglyph of c.2200 BC. Both petroglyphs strongly support each other's meanings, and both confirm the importance of crossing the ocean to the Azores in prehistoric times.

## Crossing the Gulf of St. Lawrence

In Fig.13 (bottom photo, curve in lower portion of photo) a simple geographic inscription is shown of the sailing route from the east coast of Cape Breton Island (right), via Iles de la Madeleine across St. Lawrence Bay, to the entrance of the St. Lawrence River (left), at the other side of Gaspe Peninsula (see Fig.11). This sailing route was important right from the beginning of the time period of the copper trade (Refs.5-7). It was much shorter than following the coast. Clearly visible are the NW part of Cape Breton Island, and Prince Edward Island. This ancient petroglyph dates back from the megalithic time period, probably from the second half of the 3rd millennium BC.

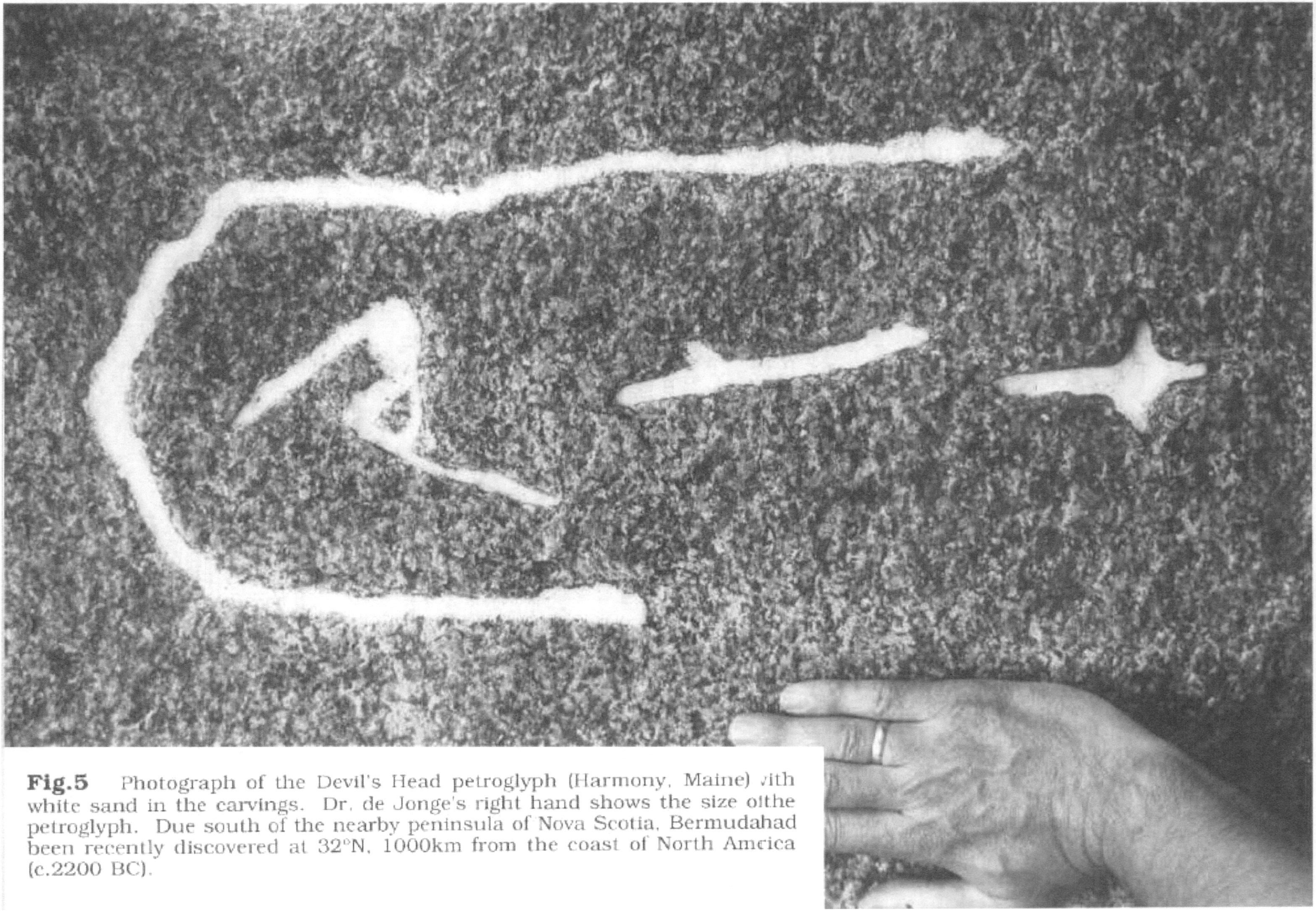

**Fig.5** Photograph of the Devil's Head petroglyph (Harmony, Maine) ;ith white sand in the carvings. Dr. de Jonge's right hand shows the size oithe petroglyph. Due south of the nearby peninsula of Nova Scotia, Bermudahad been recently discovered at 32°N, 1000km from the coast of North Amcica (c.2200 BC).

## The Megalithic Petroglyphs

Devil's Head Mountain was probably named by Puritan colonists after this incomprehensible petroglyph in the rockshelter on its top. The Indian people of the region, the Wabanaki (Ref.6), did not make the petroglyph, never having felt the urge to cross the ocean to Europe. The close siting of the petroglyph near the Embden inscriptions on the Kennebec River (Ref.16) suggests that sailors had quite a lot of time to explore the Maine hinterlands, if they had not in fact settled the area. It also suggests that the Embden copper trading center was important right from the start of the first megalithic settlements in New England, c.2400 BC. It may be that in the course of history and for unknown reasons, most of the megalith builders left the region of New England. Fell documents through pictographic evidence, movement of Celts to British Columbia (Ref.17), while Mallory reports evidence of sudden mass death, and postulates that people crossing from Europe to escape the Black Death brought it with them, wiping out the (mixed?) Norse and Celtic populations along the Ohio River (Ref.18). From linguistic evidence (Ref.13 and others) it appears the survivors melded into the aboriginal population throughout North America over the 400 years or more prior to the arrival of Columbus.

## Iberia

Figure 10 is a typical megalithic petroglyph on a menhir of NW Spain, dated from after the discovery of America (c.2500 BC). Viewing the stone's face, we are looking along the tall surface of the stone from Iberia to the west. The semi-circles are the islands in the Ocean on the return route from America, at the latitudes of the Iberian Peninsula. The Azores consist of the 2 islands of the East Azores (right), the 5 islands of the Central Azores (left below), and the 2 islands of the West Azores (above, several lines). The open ends mean "the land continues", not the land of the islands, but the land of the continents on each side, of course. This side of the rock may have been selected for the glyph by the artist because the natural features at the top illustrated the American coast (top, single line). All lines of the petroglyph provide the distances involved: Bermuda is 1 DL from the Bahamas and the American coast, the West Azores are 3DL from Bermuda, Madeira is 1DL from the Azores, and the Iberian Peninsula is about 2DL from the Azores. At the right side below, the coast of NW Spain is shown with 6 sea inlets, the most southern one being the mouth of the Minho River (the dot), important for the location of this menhir. (Pola de Alande, NW Spain, after the discovery of Bermuda, c.2100 BC).

## The Gulf of St. Lawrence and Newfoundland

Above the big megalithic petroglyph, and related to it, a small geographic inscription is visible, resembling the letters "OK" (Figs.6&9,top). The O represents the shape of the Gulf of St. Lawrence, and the K symbolizes the meridian through the east coast of Newfoundland, followed by the start of the two sailing routes to Ireland (ENE) and the Azores (ESE), respectively.The short horizontal line above the K indicates the 50th latitude line. The letters O and K also symbolize the two destinations, the Azores and Ireland

**Fig.6 Top:** Dr. de Jonge looks at the big petroglyph on the rooftop of the sailing route from Cape Hatteras to Bermuda (c.2200 BC). Above it is the OK inscription showing St. Lawrence Bay and the start of the two sailing routes from Newfoundland (500 BC - 0 AD).
**Bottom:** The sailing route from Cape Hatteras South (left) to Bermuda (the dot, right), directly after the discovery of Bermuda (c.2200 BC).

and/or the Azores and Madeira. The inscription also has the meaning of the word "okay", which is common knowledge, stressing that not only Cape Hatteras in the big petroglyph below it, but also Newfoundland was considered to be a good starting point for crossing the ocean. If you give credibility to this second meaning in letters OK, the inscription cannot be older than the 1st millenium BC, and it probably dates back from the second half of this millennium.

At the left side of the "big empty box" (Figs.9,13) is a similar inscription, resembling the letters CII. The C stands for the shape of the west side of the Gulf of St. Lawrence, the first I for the meridian through the west coast of Newfoundland, and the second I for the meridian along the east coast of Newfoundland (see Fig.9). The letters are placed in a rectangular box with a geographic meaning, related to the sizes of St. Lawrence Bay and Newfoundland. The long side corresponds to a distance of 1DL =10° of latitude= 1111km in the EW direction, and the short side corresponds to adistance of (0.5 DL=) 5° of latitude,= 555km in the NS direction. The three letters also symbolize the three island groups of the Azores, the principle destination of a sailing voyage from Newfoundland. It may have been dated, because the letters CII are also Roman numerals, and could mean the year 100+1+1= 102 AD. Could these people have been Christians, who counted the years after the birth of Christ? The inscription shows that Newfoundland was still an important starting point of the crossing of the ocean in this time period (Refs.6-8). It was copied from the older OK inscription of roughly the same size, because they liked the double meaning of this carving. However, it was engraved on another part of the overhanging rock, and with a different orientation, though the left edge of the stone still represents the East Coast of North America. Probably, the people that made the petroglyph considered themselves different from the megalith builders who made the other petroglyphs.

Below the "big empty box" (Figs.12,13) is again a similar inscription, resembling the symbols CI3 II. Again, the C stands for the shape of the west side of the Gulf of St. Lawrence, and the I for the meridian through the west coast of Newfoundland. The 3 now indicates the start of the two sailing routes from the east coast of Newfoundland, having a distance of 4° of latitude= 444km (the 3-shaped zigzag), to Ireland (ENE) and the Azores (ESE), respectively. This part is a combination of the OK and the CII inscriptions! The two symbols II at some distance are the two meridians east of Newfoundland through the SW Cape and Cape Farvel (South Greenland). The petroglyph stresses the geographic meaning of the CII inscription of 102 AD, but it has its own additional geographic meaning. It is important that the first group of three symbols CI3 is to the southeast of CII. These newer glyphs symbolize the three island groups of the Azores, with the open ends of the C and the 3 (meaning "the land continues") directed to the I (the Central Azores). The second group of two symbols II, almost a U of "Unity", symbolize the two islands of Madeira, the fourth island group east of the Azores. These are the most important destinations if one

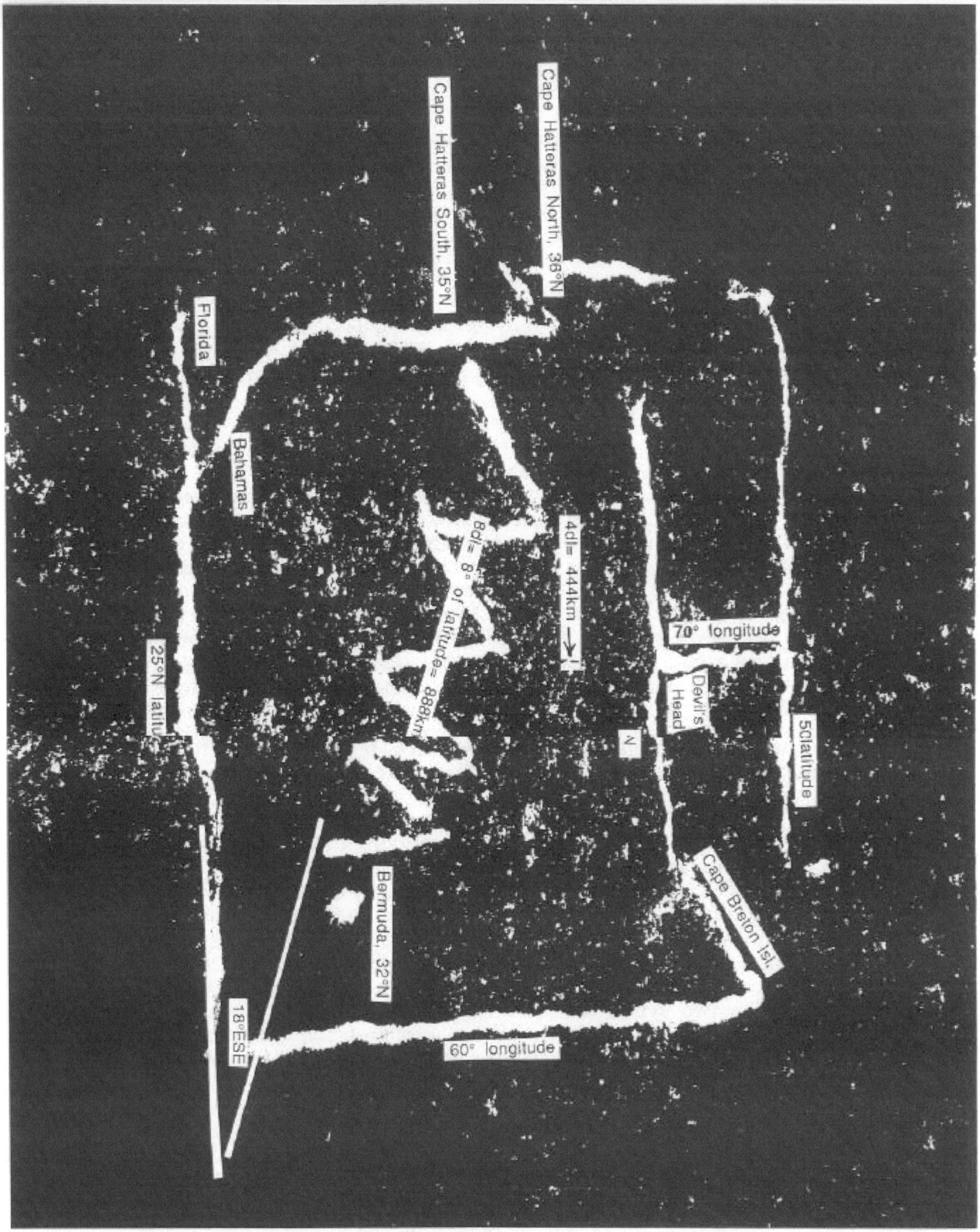


**Fig.7** Petroglyph of Fig.6 (bottom) with geographic meanings. The sailing route from Cape Hatteras South (left) to Bermuda (the dot, right). The distance is 8dl, =888km. The sailing direction is 18°ESE. The lower end of the vertical piece line above it is Devil's Head Mountain, at 45°N (c.2200 BC).

departs from Newfoundland (CII). The petroglyph CI3 II has been made after the first one of 102 AD, actually in the year 100+1+4+2 =107 AD. Both badly weathered petroglyphs were made here in central Maine at 45°N, because voyagers here were thinking about voyages to the Azores, which they oriented, for their safety, on Newfoundland, the easternmost land of North America.

## From Bermuda to the Azores

At the right side of the big megalithic petroglyph (Fig.9), a box was carved with the symbols IIW, followed by the probable letters LCC below them. When the top side of the box is extended to the left, it hits the dot in the neighboring petroglyph, which represents Bermuda. If the left side of the box is extended upwards, it hits the start of the small line-shaped carving, representing the West Azores. So, the text in the box relates to the sailing route from Bermuda to the West Azores, and we have to start reading the text at the left side above. Well, this is the easy part of the meaning of the petroglyph. The problem is: what is the meaning of the letters? If the information relates to the sailing distance, the II may stand for "two big distance lines" and the W may stand for "4 small distance lines", in total "20+4"= 24dl. However, this number does not correspond to the sailing distance, but to the Initial Sailing Direction from Bermuda to the West Azores, ISD= 24°ENE. If the same information is read backwards, IIW may have the meaning of "4+2"= 6dl. This number does not correspond to a sailing distance either, but would be a correct Terminal Sailing Direction in the neighborhood of the West Azores, TSD= 6°ENE.

We suspect that this use of multiple symbolisms in a complex glyph is a mneumonic device, transcribed from a deerskin, a very carefully thought- out shorthand means of helping to accurately remember, record, and teach navigational information. When we read the complete message, we may combine the letters in the following way: IIW= "20+4" and LCC= "50+200", which is a total of 274 small dl= 274x11.11= 3044km (a small dl= 0.1dl =11.11km). This is the correct sailing distance from Bermuda to the West Azores. When we read the whole message backwards, we might get the right date of the petroglyph, largely in the usual notation CCL= "250" and WII="6", which is 256 AD. Note that this boxed petroglyph is influenced by the earlier boxed petroglyph CII of similar size, of 102 AD (Fig.9). Its sides are equal to the length of the CII petroglyph, so it is also a big unit of surface area typically related to the large ocean: 1 DL squared= 1111x1111km squared. The six letters of the petroglyph also symbolize the crossing via the six island groups of the ocean: Bermuda, the West, Central, and East Azores, Madeira, and the Canaries. Because W= 4, the complete glyph symbolizes the 5+4= 9 islands of the Azores, too. Like the CII inscription, this is a dated, geographical petroglyph. Contextual information from further research and other decipherments will help sort out the intended nuances.

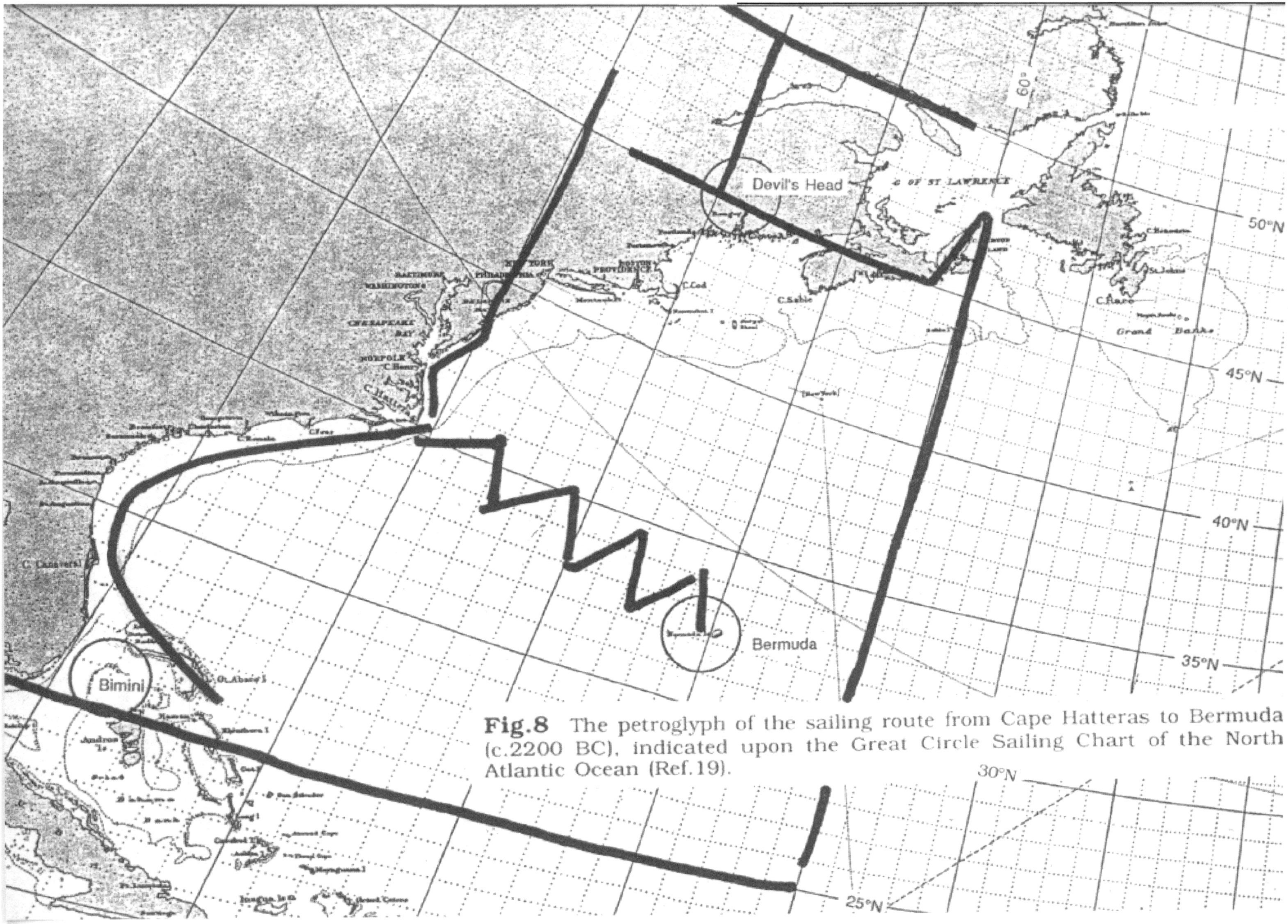


**Fig.8** The petroglyph of the sailing route from Cape Hatteras to Bermuda (c.2200 BC), indicated upon the Great Circle Sailing Chart of the North Atlantic Ocean (Ref.19).

## The Azores

In Fig.13 (bottom photo) we see a petroglyph resembling E.E.P.II I'. The petroglyph is oriented on the old inscription of the sailing route across the Gulf of St. Lawrence. If we look at a map (Fig.12), the petroglyph seems to run along the NE coast of Newfoundland in the direction of the Azores. We cannot decipher it, but in view of the previous carvings, we can make a reasonable guess. The letters may represent an abbreviation of the words "Extremely Eastern Province" (of North America), or something similar to this in a related language, pointing to the archipelago of the Azores. The symbols II I are big distance lines, showing that the distance from Cape Race to the West Azores is barely 2DL= 2222km (in reality 2100km), and further to Madeira is well over another 1DL= 1111km (in reality 1400km). The dot at the right side above the last distance line is the Strait of Gibraltar. The carving may be related to the copper trade, like the older curved carving below it. Probably, the glyphs II I' also carry the meaning that it was carved in the third century AD.

## The Crossing of the Ocean

In Fig.13, (top photo), we see a "big empty box", followed by the inscribed message 8AM Cinture, or something close to that. The big empty box that separates the letter carvings was placed at the right (east) side of the CII petroglyph (Newfoundland), so it must have the meaning of "the ocean" (see Fig.11). It is a copy of one of the oldest megalithic petroglyphs from Europe of the "North Atlantic Ocean", dated from the 5th millenium BC (Ref.12). It has sides equal to the length of the CII petroglyph, so it is again a big unit of surface area typically related to the large ocean: 1 DL squared = 1111x1111km squared. The whole petroglyph, including the text, was placed at the right side and as a continuation of the CII petroglyph, which they apparently revered quite a lot, as people continued to carve here, extending the ideas previously carved.

Some readers have suggested that "J.A. McIntyre" is a modern graffiti, which might be correct, if the carving had sharp edges, which it does not. All these petroglyphs are typically a centimeter wide, and a millimeter deep. They were just visible. Consider the possibility that since in the Roman languages of Europe, CINTURE means "belt" or "girdle", the translation might be something like "ocean 8 belt". The North Atlantic has an 8-shaped sailing route pattern, used to cross the ocean: the lower part being the westerly Southern Crossing, the middle being the easterly routes to the Azores, and the top of the 8 being the Upper North route. As a consequence, a possible translation for "Box" 8AM Cinture is "8 shaped conveyor belt", referring to the wind and current belts which were used to cross the ocean for several millennia. Possibly "8AM CINTURE also has a second meaning: dated "8th century".

## The Canary Islands

In Figure 13 we see the CANY petroglyph, surely an abbreviation of "Canary Islands". From left to right the size of the letters increase, because all the

**Fig. 9** **Above:** A survey of seven prehistoric petroglyphs on top of the roof block of the shelter (2200 BC - 1200 AD). It appears that people visited the site over a considerable period of time. At the right side the boxed petroglyph about the sailing route from Bermuda to the Azores (256 AD).

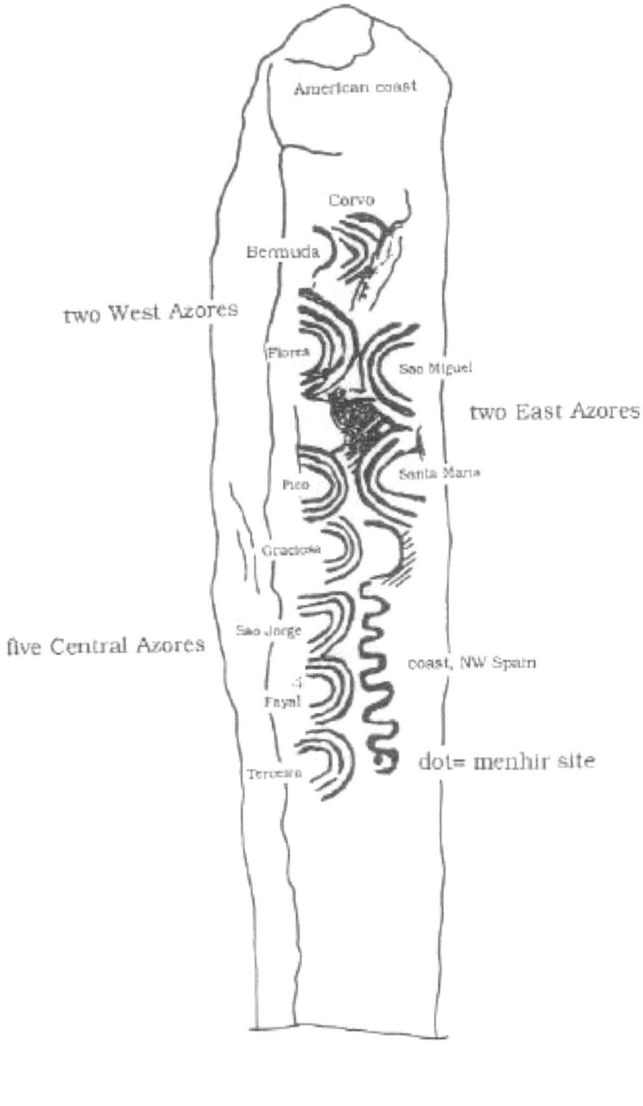


**Fig. 10** **Right:** The Azores in the middle of the ocean consist of the 2 islands of the East Azores (right), the 5 islands of the Central Azores (left, below), and the 2 islands of the West Azores (above). The open ends mean "the land continues" (not the islands, but the continents to each side, of course). From the menhir site (the dot), in NW Spain, there are 6.5 waves, (13 up/ down legs)= 13dl= 13x111km= 1444km, which is the correct distance to the Azores. (Pola de Alanda, NW Spain, after the discoveries of America and Bermuda, c.2100 BC) (Ref.12).

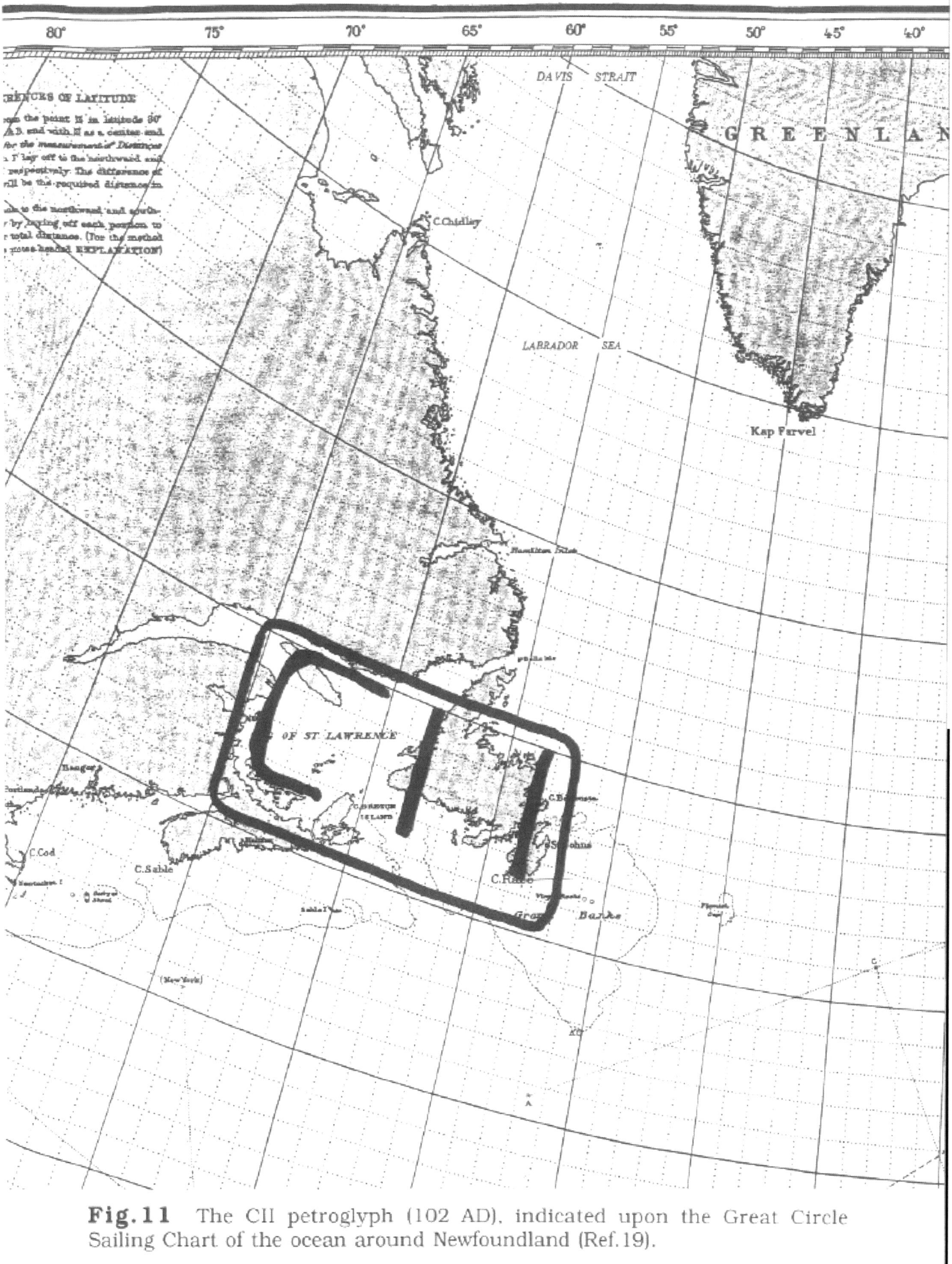


**Fig.11** The CII petroglyph (102 AD), indicated upon the Great Circle Sailing Chart of the ocean around Newfoundland (Ref.19).

# GREAT CIRCLE SAILING CHART

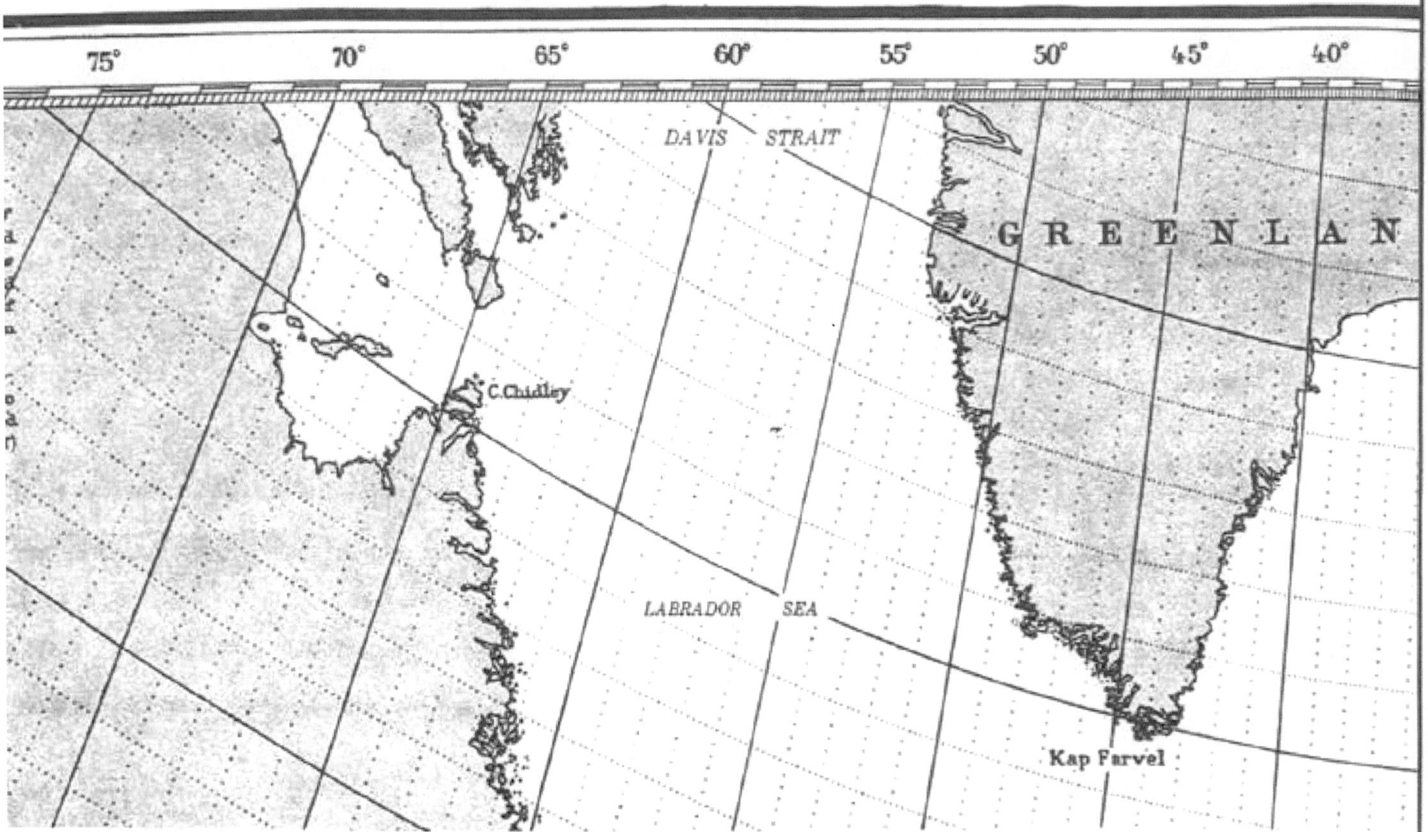


**Fig.12** The CI3 II petroglyph (107 AD), indicated upon the Great Circle Sailing Chart of the ocean around Newfoundland (Ref.19).

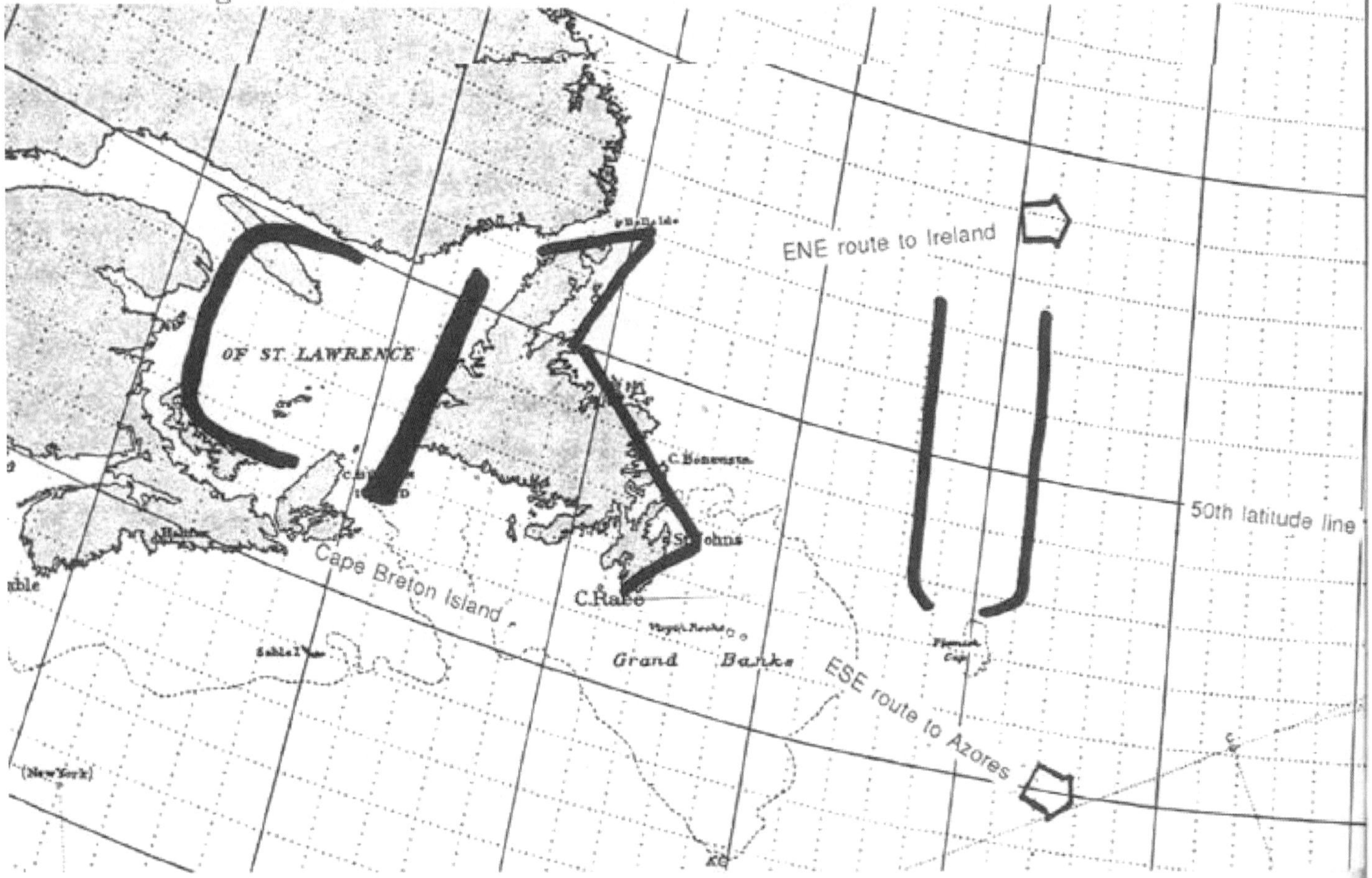

petroglyphs below it are related to the sailing route to this eastern archipellago. The four letters indicate that this island group can be reached via 4 other island groups. The C stands for the West Azores (the open end to the right; the land continues to the east). The A stands for the Central Azores (the A has the shape of the highest mountain , on the main island of Pico). The N stands for the East Azores (the open end downwards; the land continues at a lower latitude). The Y stands for the islands of Madeira (the two arms for the islands of Madeira and Porto Santo, the single leg for the route to the Canaries). It appears that the artist studied all the inscriptions, and understood the meaning of them. The tangents to the CANY glyph cut the upper corners of the "empty box". This confirms that the glyph deals with this "canny" crossing of the Ocean. Again, it is a very weathered petroglyph. For these reasons, we date the CANY petroglyph shortly after the one below it, 800-1200 AD.

## Discussion

We must stress that these vague, badly weathered petroglyphs in the hard granite on the rooftop are definitely prehistoric. However, we have specialized in megalithic petroglyphs, and are not very familiar with carvings from the historical period. All these carvings show that people have visited the site over a considerable period of time. Note that prehistoric people had no difficulty reading each other's petroglyphs. The idea that these carvings cannot be understood is a modern hype. We hope some of you may contribute in helping to decipher them.

Other petroglyphs, though perhaps difficult to see, may be present in the immediate surroundings. A survey of lithic features and horizon alignments should be done on this hilltop to see if it has a stone circle, menhir, or other signs that it might have been a nautical, astronomical, or otherwise important megalithic site that would have attracted people up here. We are sure a full exploration and excavation of the immediate surroundings of the site will be rewarding.

Because petroglyphs have not been taken seriously by American Archaeology, which denies pre-columbian oceanic travel, there is not a high-quality record of US petroglyphs, such as Twohig provides for Europe (Refs.12, 21). The Devil's Head petroglyph and the related inscriptions on top of the rockshelter are of great importance for their insights into the prehistory of America. For that reason we recommend its complete protection for posterity as an official Prehistoric Monument of the State of Maine.

**Fig.13** **Top:** To the left and below "the big empty box" (the Ocean, 8th century AD), two petroglyphs of St.Lawrence Bay and Newfoundland (102 and 107 AD, respectively). **Middle:** The petroglyph "BOX" 8 AM CINTURE may mean "8-shaped ocean conveyor belt", but also "8th century". **Bottom:** The sailing route from Cape Breton Island (right) across St. Lawrence Bay to the St. Lawrence River (left) (2500 - 2000 BC). Above it a possible abbreviation of "Extreme Eastern Province" (the Azores), and three distance lines (3rd century AD).

# References


1. Carlson, S., www.neara.org/CARLSON/
2. Jonge, R.M. de, and IJzereef, G.F., De Stenen Spreken, Kosmos Z&K, Utrecht/Antwerpen, 1996 (ISBN 90-215-2846-0)(Dutch)
3. Tour Guide Map, America's Stonehenge, PO Box 84, No. Salem, N.H. 03073
4. Lenik, E.J. and Gibbs, N.L., "The Frost Valley Petroglyph, A Catskill Mountains Enigma", NEARA Journal, Summer 1999 (ISSN 0149-2551)
5. Ferryn, P., NEARA Journal, Vol.XXXI, No.2, p.59 (1997)
6. Fell, B., America BC, Pocket Books, Simon & Schuster, 1994
7. Bailey, J., Sailing to Paradise, Simon & Schuster, 1994
8. Thompson, G., American Discovery, Misty Isles Press, Seattle, 1994
9. Peterson, F.A., Ancient Mexico (1959)
10. Stuart, G.E., "New Light on the Olmec", National Geographic, Nov. 1993
11. Bernal, I., The Olmec World, University of California Press, London, 1969 (ISBN 0-520-02891-0)
12. Twohig, E. Shee, The Megalithic Art of Western Europe, Clarendon Press, Oxford, 1981
13. Gruener, J., The Olmec Riddle, An Inquiry into the Origin of Pre-Columbian Civilization, Vengreen Publications, Rancho Santa Fe, 1987 (ISBN 0-942185-56-0)
14. Xu, Mike, Origin of the Olmec Civilization, University of Central Oklahoma Press, 1996 (ISBN 09648694-2-x)
15. Smith, G.E., The Migrations of Early Culture, A Study of the Significance of the Geographical Distribution of the Practice of Mummification as Evidence of the Migrations of Peoples and Spread of Certain Customs and Beliefs, The University Press, Longmans, Green & Co., Manchester, 1915
16. Strong, R., "Did Glooskap Kill the Dragon on the Kennebec?", NEARA Journal, Vol XXXII, No.1, pg 38
17. Fell, B., Saga America, Times Bks, New York, 1980 (ISBN 0-8129-6324-5)
18. Arlington M., and Harrison, M.R., The Rediscovery of Lost America, The Story of the Pre-Columbian Iron Age in America, E.P. Dutton, New York, 1951 (ISBN 0-525-47545-1) (pg. 28)
19. Great Circle Sailing Chart of the North Atlantic Ocean, Defense Mapping Agency, Hydrographic/Topographic Center, Washington DC
20. Hancock, G., Fingerprints of the Gods, Random House, NY 1995 (ISBN 0-517-88729-0)
21. Whittall, McGlone, et al. Ancient American Inscriptions: Plowmarks or History?, Early Sites Research Soc., 1993, (ISBN 01095365 XXX)
22. Lambert, J.D., America's Stonehenge, an Interpretive Guide, Sunrise Publications, Kingston, N.H., 1996 (ISBN 0-9652630-0-2)
23. Schuster, A.M.H., Archaeology, May/June, 2000, Pg.60

**Fig. 1** The Three Rivers Petroglyph (c.1500 BC) on the Neversink River, Catskill Mountains of New York looking upstream to the East, July 16, 2000.

# The Three Rivers Petroglyph

## A Guidepost for River Travel in America

(Frost Valley, Neversink River, New York, c.1500 BC)

Dr. R.M. de Jonge, drsrmdejonge@hotmail.com
J.S. Wakefield, jayswakefield@yahoo.com

### Introduction

In 1995 a high quality petroglyph was discovered on a boulder along the Neversink River in the Catskill Mountains in southeast New York (Figs.1&2, Ref.1). It shows a spiral with three "legs" attached to it, indicating to us, that it might be an ancient megalithic inscription. In the summer of 2000 we visited the site to investigate the petroglyph.

The boulder with the petroglyph on it is located on the north bank of the West Branch of the Neversink River near the town of Frost Valley, New York (Fig.3, Ref.1). The petroglyph has a size of 80x70cm. It has been carved in a fine-grained metamorphic sandstone, with a vee-shaped groove about a centimeter deep and a centimeter wide. The boulder itself has a length of c.160cm, a width of c.110cm, and a height of c.100cm. It appears that the stone once was standing on the bank, but was undercut by the river, fell off the bank, and now lies on its backside in the river. High waters and floating ice and debris in Spring months must flow over the stone, and damage the inscription. It looks as though it is also being walked on.

### Geographic Inscription

We will show that this is a stylized megalithic inscription with a pure geographic meaning. The spiral has five spaces between its coils (Fig.2), so it indicates that the whole inscription relates to a circular area having a radius of about 5 megalithic distance lines (=5dl= 5°of latitude= 5x111 km= 555km) (Ref.3). The center of the spiral is the place where the petroglyph is located, which is on the West Branch of the Neversink River.

The petroglyph is located halfway between the Hudson River in the east, and the Delaware River in the west. The distances to these rivers are both c.50km. Notice how a small tributary of the Hudson River, Esopus Creek (Fig.3), now a popular whitewater "tubing" river at Phoenicia, NY, runs in a near-perfect circle around Panther Mountain, which is the now-exposed hardened sedementary plug of a buried meteor impact crater (Ref.5). These are thought to be some of the oldest mountains on earth. Above the upper end of Esopus Creek is small Winnisook Lake with a summer camp on it. This little lake drains southwest, starting the west fork of the Neversink River. Coming down the river about five miles, the petroglyph is at the foot of a straight stretch, prominently on the right bank as the river curves left (Fig.1 shows the view upriver). The Neversink flows south to Port Jervis into the Delaware River (Fig.9). It appears that the Neversink River was a "transportation corridor" through the Catskill Mountains.

**Fig.2 Above:** The authors study the petroglyph. **Below:** When filled with white aquarium sand, the inscription is easier to see (c.1500 BC).

From our other work, such as the study of the petroglyphs at Loughcrew, Ireland, we know that megalithic inscriptions often use the shape and features of the stone itself as parts of the story being told. We start with the assumption that the pointed end of the stone is the top (Fig.8), which later makes sense in a number of ways as we shall see. The rock narrows at the top, because the land north of the Catskills, which includes the Adirondacs, narrows between Lake Champlain in the east, and Lake Ontario and the St. Lawrence in the west, which is within the 555km of the spiral. Note that the top of the stone has an upper edge in the center, representing the Adirondack Mountains literally.

## The Delaware River

The right leg of the inscription, pointing downwards, represents the Delaware River (see Fig.4) from Port Jervis to the sea, because the shape of the leg closely resembles the shape of this river. Its end is the entrance of Delaware Bay, south of Philadelphia. When we follow the spiral from the right leg upwards and left toward the center leg, then the Delaware River is followed from Port Jervis, where the Neversink joins it, to the Susquehanna River near Binghamton (Windsor, Deposit), NY. The shape of the spiral is more or less similar to the shape of this part of the river.

## The Susquehanna River

Doing this, one arrives at the middle leg of the figure, which represents the Susquehanna River, because the shape of this leg is similar to the shape of this river. The long "foot" of the middle leg is Chesapeake Bay, because the shape of the foot resembles the shape of the bay. The three sharp peaks to the left point to the major river entrances in the Chesapeake: the Potomac, the Rappahannock, and the James. These places are about 555km from the stone site in the Catskills, as indicated by the spiral.

## The Ohio River

When the spiral from the middle leg is followed upward, again the Delaware River is followed from Port Jervis, where the Neversink River joins it, to beyond Binghamton (Windsor, Deposit), NY. The shape of this part of the spiral is exactly the shape of this part of the river. The "nose" of the spiral corresponds to the major river bend, and it points to the beginning of the third leg.

The big upward bow in the petroglyph before the third leg is their route across the Allegheny Mountains to the start of the Ohio (Allegheny branch) River in the west (Fig.4). The shape of the bow does not have a geographic meaning, but does represent the crossing at an altitude that is required between these river basins. In view of the site location of the petroglyph near small creeks in the Catskills, they probably crossed the land between the headwaters of the Susquehanna and the Allegheny Rivers near Coudersport, Pa., a distance of less than 20 kilometers.

As a consequence, the third leg is the Ohio River (Fig.5). This enormous

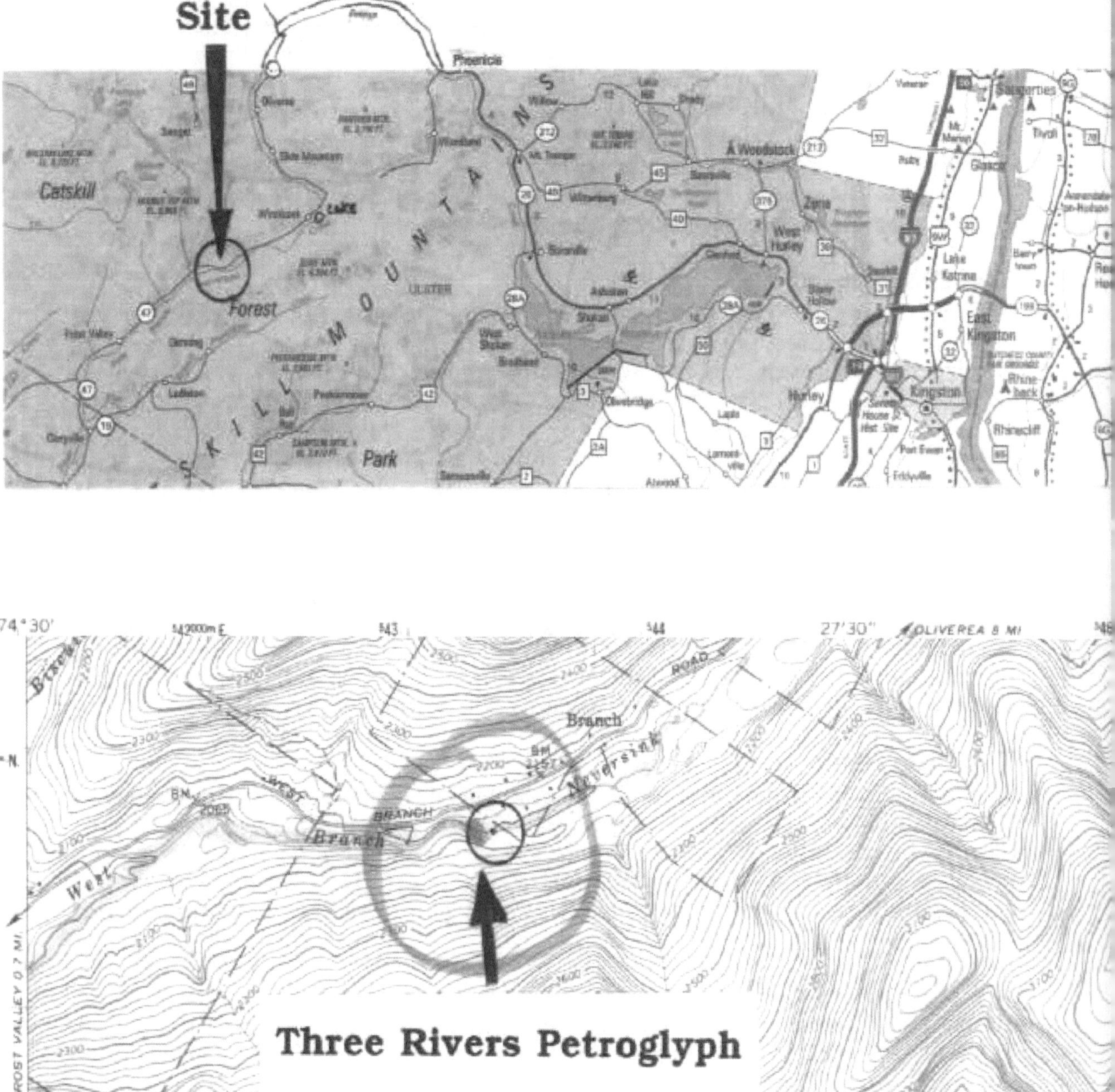


**Fig.3** Maps of the site. **Above:** The Three Rivers Petroglyph on the West Branch of the Neversink River, 45km west of the Hudson River (1cm=3.3km). **Below:** Portion of USGS Quadrangle topographic map. The Three Rivers Petroglyph at 42°N, altitude 2100ft.= 640m (1cm=0.24km).

river is much more important to them than the Delaware or the Susquehanna. For that reason the third leg is made of round forms, just like the spiral! Before Pittsburgh the river jogs, and then a side river joins in from the southeast, the Monongahela River. The next part of the leg is a continuation of the Ohio. The first sharp peak points to the entrances of the first rivers of reasonable size on the other side of the Ohio, the Muskingum and the Hocking. The next point is the continuation of the Ohio itself. The final sharp peak points to the Kanawha and Big Sandy, flowing west from the Appalachian Mountains. Again, these places are about 555km from the stone site in the Catskills, as indicated by the spiral. But, as so often in megalithic petroglyphs, there are multiple symbolisms. The right side of the stone is now the Atlantic coast between Chesapeake Bay and the Hudson, and the left side is Lake Erie (Fig.5).

But also concerning the Ohio, there are multiple symbolisms. In this case there is besides the first meaning, a second, and even a third meaning. This is the reason why the stylized left leg been shaped from round forms, like the spiral, and unlike the two eastern rivers. Now the first big curved leg (Fig.6) represents the courses of the Kanawha and Big Sandy, with the points being the Wabash, Missouri, and the Mississippi. The right side of the stone is now the Atlantic coast, from Florida to Cape Hatteras, and the left side the Mississippi.

But there is even a third meaning (Fig.7). The curved leg also symbolizes the drainage of the Cumberland and the Tennessee. The Mississippi flows all around the Cumberland and Tennessee, as shown in the petroglyph. The three sharp peaks point to the river entrances on the other side of the main stream, the Arkansas and the Red Rivers, and the third is the Mississippi to Poverty Point and the Gulf of Mexico. Here the left edge of the stone is the Rocky Mountains. These distances are far greater than the 555km of the spiral, so they have made this leg curved, to show "that the open spiral continues here".

## The Three Rivers Petroglyph

The stone can also be seen another way, giving a superb artistic second meaning (Fig.2): "If you are on foot (the spiral with two legs below), and by boat (foot of the second leg in the boat), and you are heading west (the nose points to the left leg), then this map (the petroglyph) is meant for you!!" Note that all this sophistication, all this information, is worked out in one single, uninterrupted line! Close investigation shows only minor errors (if any), such as the shape of the line just before the start of the first leg, which is the Neversink River, literally.

This beautiful megalithic petroglyph is a beautiful map of the waterways of America that form a transportation network from the megalithic culture in the northeast. America's Stonehenge (Mystery Hill, Ref.14), the megalithic center in New England, is only 350km away (ENE), which is within the radius of the spiral. The very short piece of straight line in the center of the

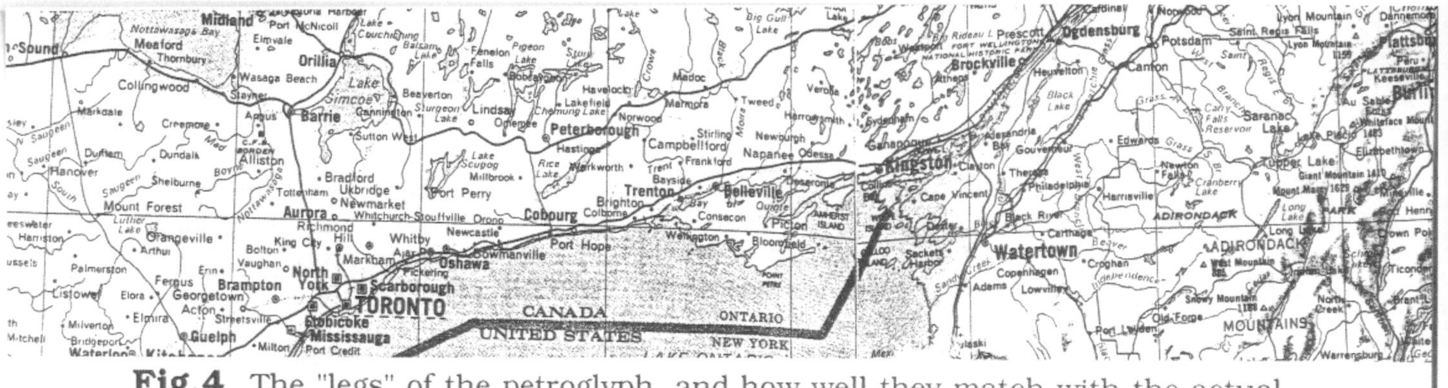

**Fig.4** The "legs" of the petroglyph, and how well they match with the actual courses of the Delaware, Susquehanna, and Allegheny (Ohio) Rivers on an Eastern US map.

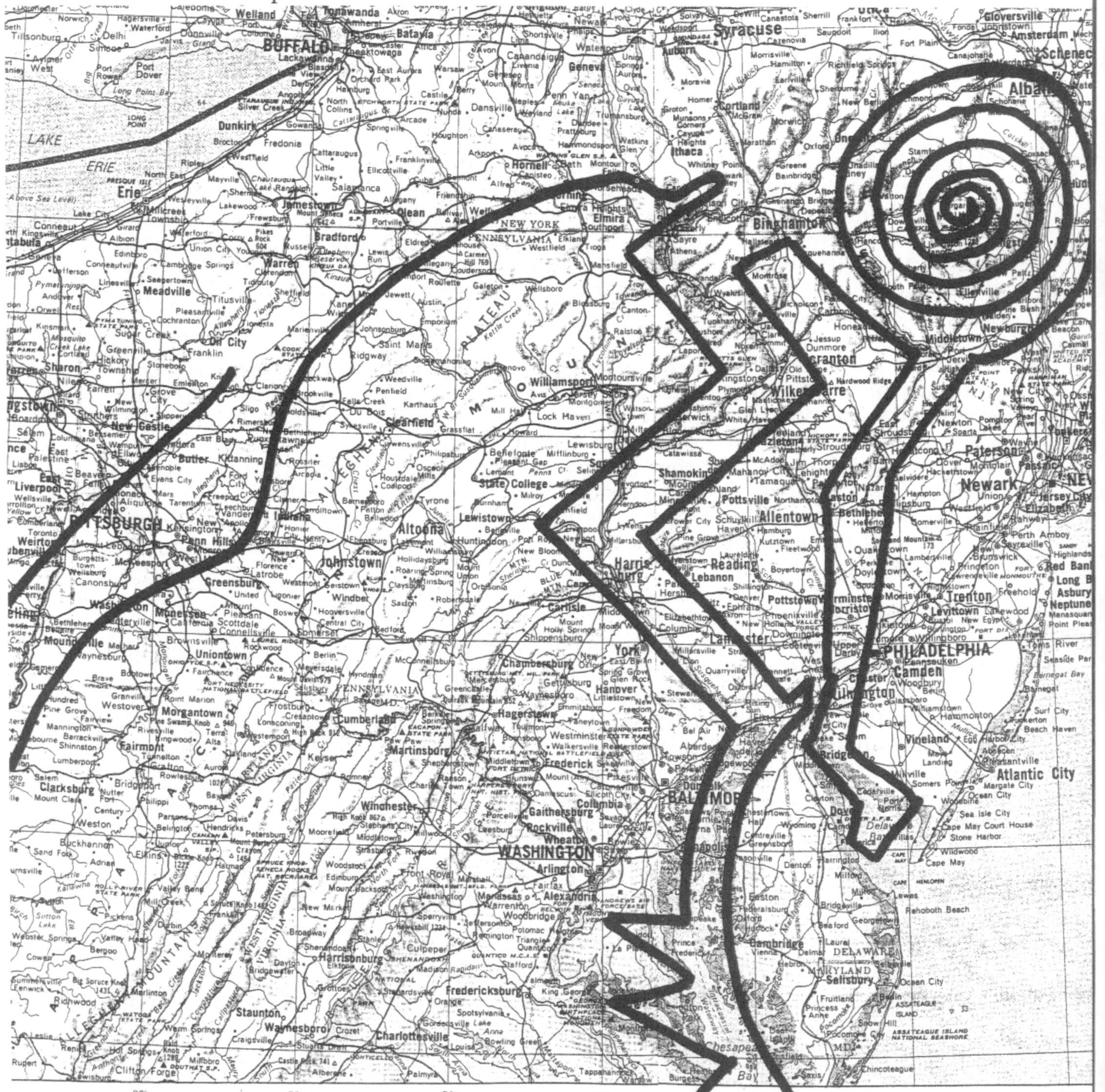

coils (Fig.2) might be an indication of the approximate direction where the makers of the petroglyph came from: c.65°NE. Because of the three legs, we feel this inscription should be named the "Three Rivers Petroglyph". It is a proof that in this time period this transportation corridor through the Catskill Mountains was used to reach these three rivers from the Hudson. If the stone were standing upright on the bank (Fig.8), the figure would confirm this with "traveling feet". The map in Fig.9 shows an overview of the finding spots of ancient stone structures and artifacts in the region. This map suggests that usually, in order to reach the Delaware from the Hudson, the creeks between Kingston and Port Jervis were used. In that case the West Branch of the Neversink River was primarily used to reach the upper course of the Delaware and the Susquehanna. However, because of the Allegheny Mountains, this was also the only way to reach the important Ohio River. So this petroglyph is a prehistoric version of the "Gate to the West" tought in American colonial history classes.

When you find a map along the road, there is usually a junction nearby. Here too! To sail the Delaware, you just follow the Neversink to Port Jervis, and go downstream beyond to the sea. However, if you are heading for the Susquehanna or the Ohio, you might have taken the first, second, or third small creek to the right, below the stone site, and carry your skin boat across the land for a portage of about 1.5km to Beaver Kill Creek or Willowemoc Creek. Either of these would bring you to the Delaware near the village of Hancock, NY, 80km northwest of Port Jervis. This portage looks perhaps difficult on the topographic, but should be examined, because it would give a reasonable explanation for the site location of this petroglyph. Old marker stones which might be present near the junctions with these small side creeks should be looked for.

Notice on a map (Fig.9) how close (15km) the Delaware approaches the Susquehanna River near Binghamton (Windsor, Deposit), NY, 110km west of this stone site. Tributaries of the Susquehanna are close to those of the Ohio River, 280km to the west, along the New York/Pennsylvania border. The distance between the headwaters of the Susquehanna and the Allegheny (Ohio) River near Coudersport, Pa., is less than 20km. These short passage areas might provide important corroborating archaeological sites, and should be examined, particularly the railroad pass at Gulf Summit (order the "Gulf Summit NY USGS Quadrangle). A portage only 2 miles in length (3km), is located further south, near Altoona, Pa., called "Horseshoe Bend" (Ref.10). The towns of "Summit" and "Portage", and the "Allegheny Portage Railroad National Historic Site" are located there today. An ancient name for the Ohio is "the Winding River" (c.500 AD, Refs.12,13). This might be a reason, too, for the round forms of the left leg of the petroglyph.

## Latitudes

By correctly sizing the distances of the petroglyph, the spiral again demonstrates the use of latitudes, so the community of the man who made the inscription was aware of the astronomic, mathematical and geographic

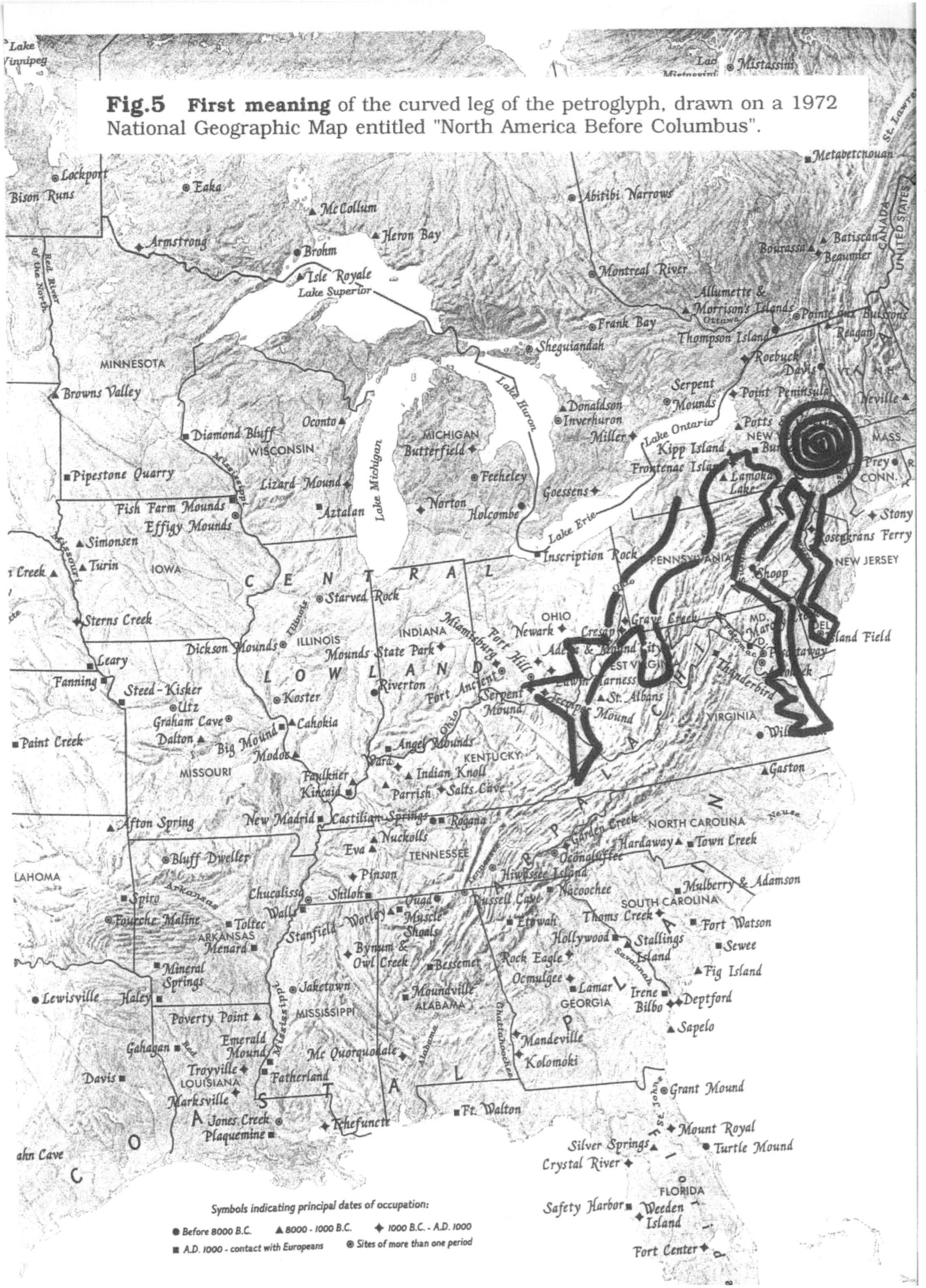


**Fig.5 First meaning** of the curved leg of the petroglyph, drawn on a 1972 National Geographic Map entitled "North America Before Columbus".

knowledge of the megalithic people, and undoubtedly had a background in sailing at sea. The Atlantic coast is only 150km away, and the Atlantic entrance to Chesapeake Bay, shown in the petroglyph, is exactly 5° of latitude south of this site, as shown by the 5 coils of the spiral. The center of the spiral (meaning the spot where the petroglyph is located), makes an angle of 42° with the lower edge (bottom) of the stone (see Fig.8), encoding the latitude of this site, which is 42°N. We agree with a previous study (Ref.1), that the making of the carving was started at this center. The inscription thus reveals that it was designed in the megalithic fashion (first identifying its latitude).

The petroglyph is located at the end of a straight halfmile long section of the Neversink River (Figs.1,3), which points 20°WSW in the direction of the shortest distance to the Delaware River, 50km, and also is the correct direction to the place where the Ohio joins the Mississippi. This is near Cairo, Illinois, 1300km away from this site. This place is 5° of latitude south of the petroglyph, at 37°N, as indicated by the spiral (42°-5°=37°). Clearly, this junction, illustrated in the 3rd leg of the petroglyph, was considered to be the center of the waterways of North America.

Copper from Michigan was brought down the St. Lawrence (47°N) and traded for valuables brought from Europe by traders, who returned to the Old World via Cape Race at 47°N (42°+5dl of the five-wide spiral=47°). It is not uncommon for multiple symbolisms like this to have been built into megalithic inscriptions, indicating that they were very carefully thought out before the carving was done. No doubt latitudes were sometimes difficult to figure in the overcast skies and inclement weather that occur in the Catskill mountains. This carver did a valuable service to his community by creating this inscription, probably because several expeditions had badly lost their way.

## Neighboring Archaeological Sites

Southeast of this petroglyph site (90km away), on the east side of the Hudson, in Putnam County, N.Y., there has been found a great concentration of stone "chambers". In fact, so many in such a small rocky area, that today's researchers are thinking there was a colony here. Such a colony nearby would certainly be supportive of the need for a transportation corridor to the west through the Catskills. Martin Brech, who lives in the area, gave us a tour of some of the chambers (Fig.10). Notice the standing menhir with celestial alignments in the bottom photo. This is one of the remaining "hundred" chambers that Mr. Brech and others are studying. He said he thinks there were 200 chambers once, but housing development is proceeding quickly in the county, and reducing their number. Unfortunately he reports that local archaeologists are being paid fees from developers for signing "Declarations of Non-Significance" on Environmental Impact Statements, so the corbelled chambers (at least 1st millenium BC, Ref.11), some with inscribed Ogam dedications to Baal, can removed from housing sites. The "professionals" still cling to the "colonial root cellar" explanation.

**Fig.6 Second meaning** of the curved leg of the petroglyph, on the map of Fig.5.

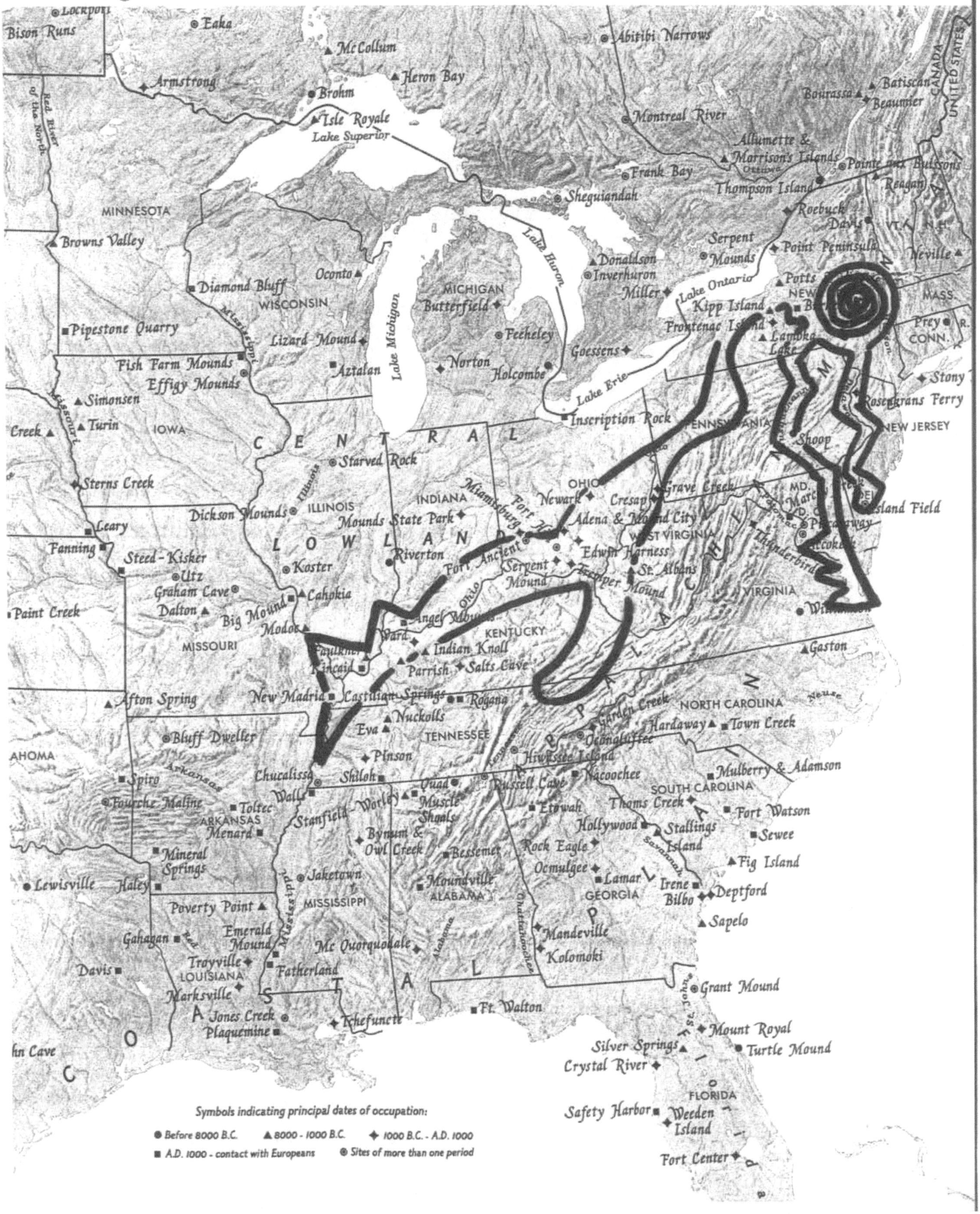

We also went to see the Cobble Hill Mound in South Granville, and finally found it, covered with trees, behind a nursery business. The nice nursery owner told us "there were two more, smaller mounds, over there, but we just bulldozed them, to make more room for the nursery"!

For an intriguing review of megalithic stones in the river valleys of the East Coast, see Trento's books (Refs.7&8). The transportation corridor from Kingston on the Hudson to Port Jervis on the Delaware (about 60km south of the petroglyph, Fig.9), was replaced at some later date by a 104 mile (165km) road, called the "Old Mine Road", which has stone cairns along the way. "Along the (adjacent) ridge ... there are numerous cuts and tunnels carved entirely by hand tools ... some of these tunnels measure 6 feet wide, 4 feet high, and 500 feet deep" (1x2x150m)."There is no trace of gunpowder or the subsequent blasts, (but there is) evidence of pick and wedge mining. (There are) radial lines of stone piles, hundreds of conical stone piles, all constructed on base boulders... (apparently) the debris of an ancient Old World People, (leaving behind) inscribed amulets, engraved pebbles, aligned rocks.. (Ref.7). Trento shows two photos of these mine tunnels and a rough map in his new book (Ref.8). This mining for copper, iron, and lead was probably going on in the 1st millenium BC.

Trento discusses the Hawley Stone (Fig.11, right), found near the Lackawaxen-Delaware River confluence (30km NW of Port Jervis, 70km SW of the petroglyph). This stone is not yet deciphered. It probably dates from the second half of the first millenium BC. His comments on the Delaware are helpful: "the River travels the same route today that it did some 200 million years ago, (and) before sucked dry by New York City's municipal water needs, (it was) much deeper and more navigable ... there have been inscribed stones found in abundance in the Delaware River Valley. As usual, regional archaeologists, unable to explain the meaning of the etched symbols, have simply put the artifacts aside" (Ref.7).

Figure 11 (left) is a photo of a small metal urn (found by a ten-year old) washed out in a backyard close to Binghamton, NY, in 1973. This is very near the portage we suggest near Windsor/Deposit NY (see Fig.9). On page 79 of his book Saga America (Ref.4), Dr. Fell discusses this urn "decorated with Phoenician themes in the upper part which depicts goddess Astarte, with Egyptian-inspired ornament below, apparently African rain-dancers. The style, of mixed derivation, recalls the work of Cypriot Phoenicians around 600 BC, when such objects were manufactured for distribution and sale, mainly to semi-civilized barbarians, by traders from Carthage. The urn was discovered in an excavation carried out by the Middlebury Archaeological Research Center near the junction of the Susquehanna and the Chenango Rivers, New York. Phoenician inscriptions had earlier been recognized in the Susquehanna region by Phillip Beistline and Dr. William W. Strong" (Ref.4).

**Fig.7** **Third meaning** of the curved leg of the petroglyph, on the map of Fig.5.

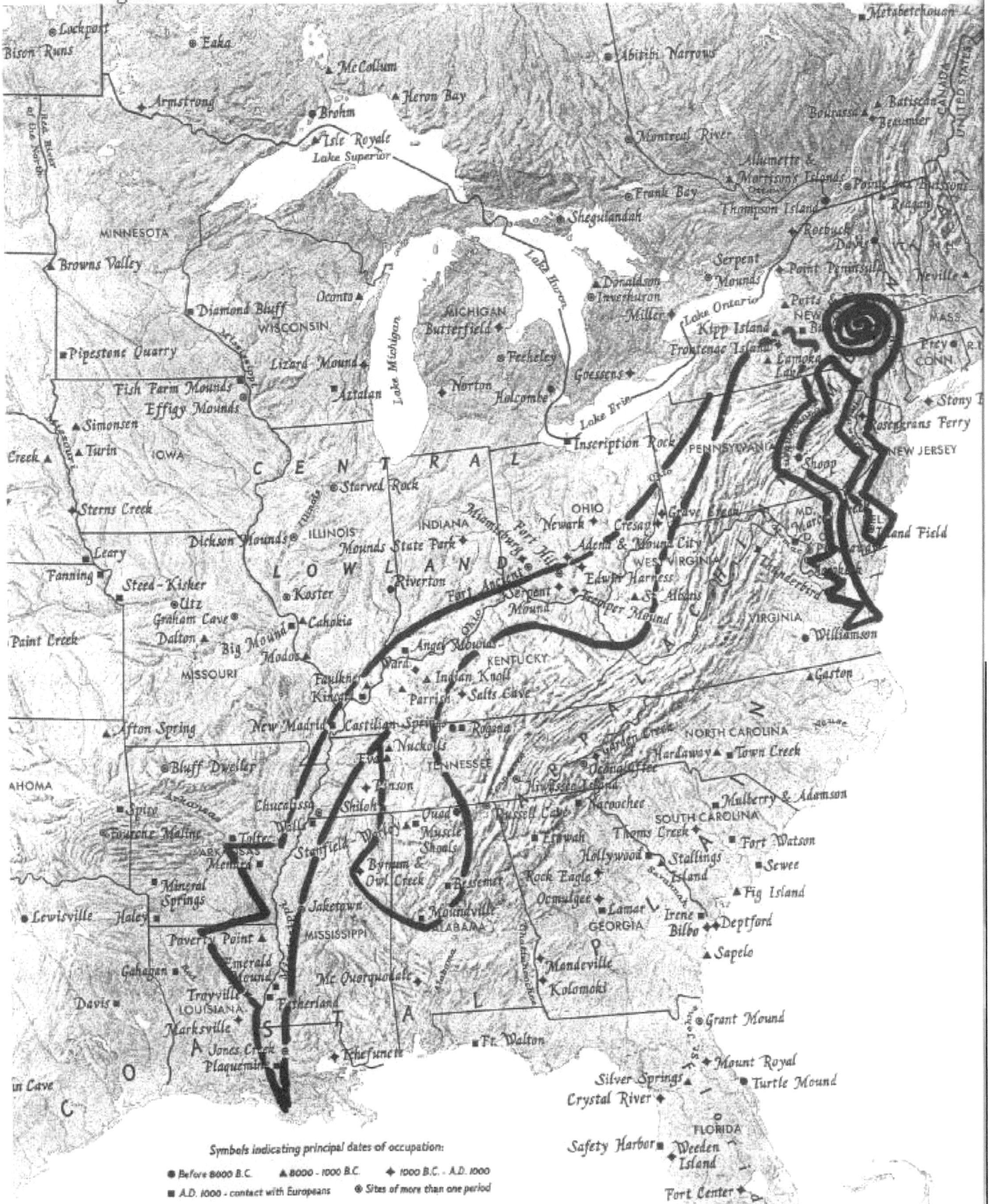

Near Elmira (75km west of Binghamton, and 185km west of the petroglyph), on a tributary of the Susquehanna, is an ancient set of earthworks called Fort Hill. Trento explains, "that it is an embankment, 270 feet in length, 6 to 9 feet wide at the top, and higher than his head" (80x3x2m), maybe older than the 1st millenium BC (Refs.7,8).

Other relics have been found near the probable portage near Coudersport.One of these is the Genesee Stone (Ref.6). This stone was found on the Genesee River before 1975. "Barry Fell deciphered the stone ... identified it as the Iberic alphabet and Punic language." The authors Warren Dexter and Donna Martin illustrate their book with maps of rivers and relic site locations. Prominent is the Genesee/Allegheny/Susquehanna junction near Coudersport. "The Genesee flows northward across New York State into Lake Ontario. At its beginning is a short portage between it and the western branch of the Susquehanna River as well as a short portage to the Allegheny River" (pg.53, Ref.6). "Punic inscriptions cover the "Susquehanna Stones" in Pennsylvania" (probably 300 BC - 200 AD). In the same valley, the late Dr. W. Strong collected 500 inscribed stones comprising Celtiberian, Phoenician, and Basque grave markers, from a Bronze Age settlement, about 800-600 BC (Ref.11).

Mertz reviews a cardinally-oriented boat burial found in a mound on Vincent Island, in the Pigeon River, Tennessee, as reported by the Thomas Report to the Smithsonian, which probably dated from the 1st millenium BC (Ref.25). She goes on to say that "Huntingdon stands today on a site which in Revolutionary War times was known as Standing Stone. Here at this spot an ancient stone marker once towered on a promontory. Amazed explorers described the stone as an obelisk - fourteen feet high, broad of base, tapering to six or seven inches at the top, covered on all four sides with undecipherable letters, believed by some who first saw it to have been Egyptian Hieroglyphics. This stone is only one of a great many recorded standing stones that earliest travelers and explorers found marking the trail from the Atlantic to the Ohio by water - a trail of standing stone markers found as far west as Lancaster, Ohio. Diaries of various early wanderers through this then wild region - David Zeisberger, Conrad Weiser, George Groghan, Hechwelder, Gist and Peter Kalm - all faithfully testify to these standing stone markers having been there before their own time" (Ref.10).

## The Mysterious Ohio Connection of the 3rd Leg

Though most archaeologists claim that there was never an Iron Age in America, steel executive and engineer Arlington Mallery has shown otherwise (Ref.2). In his excellent book "The Rediscovery of Lost America" are photos of both Viking and Celtic style iron furnaces along the Ohio River and in Virginia. His photos show remains of iron implements, some cladded, such as shovels, and there is a photo of a group of 60 pound iron bars lined up against his garage. Such colonies of a Pre-Columbian iron industry on the Ohio would date from the 1st millenium BC. He found skeletons stacked 3-5 deep, and believed that Black Death had reached these settlements. Also,

**Fig.8 Above:** The top of the boulder now points downstream, to a rapid in the middle of the small Neversink River. **Below:** The spiral makes an angle of 42°, showing the local latitude, 42°N. The upper edge on top of the stone represents the Adirondack Mountains (c.1500 BC).

Caesar's defeat of the Celtic fleet off Brittany in 51 BC could have sent waves of refugees west or isolated overseas settlements through loss of the fleet.

A lot more study of the mysterious so-called Adena and Hopewell peoples is needed, who are credited with the Serpent Mound, the enormous astronomic circles, squares and octagon earthworks that were up to a quarter of a mile in diameter every two miles along a 14 mile stretch near Chillicothe, Ohio, and covered four square miles with walled avenues at Newark, Ohio. Carl Munck has found mathematical relationships between these huge monuments and other megalithic monuments, as well as evidence for a spherical mathematics, and a global longitude/latitude grid system based on a Giza meridian, which appears to be a very fertile area for more study (Ref.19). The global scale of these findings is reinforced by Stecchini, who states that "all the measures of the ancient world constituted a rational and organic system ... the figures I have succeeded in establishing so far, suggest that Pre-Columbian American units agree with those of the Old World" (Ref.24). A former Director of the US National Park Service has written a historical perspective on these monuments, with his opening chapter entitled "The Great Dying" (Ref.18). There is a Cherokee legend of their defeat of the "Moon Eyes" (blue eyed) people at Fort Ancient, Tenn, and there are many related colonial era stories, such as a colonist in Virginia saving himself from death by speaking "Welsh" with his captors. Nine ancient hilltop dry-stone wall structures (6"x6'x600') in the southern tip of Illinois are laid out E-W at 38°N, in the same pattern as the Azores, which are at this same latitude. These structures indicate the ongoing importance of the Azores and the SunGod to people who had their roots in the Old World (Refs.9,21). These structures may date from the first millennium AD.

European readers may find hard to believe the quantity and quality of artifacts that have come out of the ground in the US that are unreported,largely unknown, undated, unstudied, and ignored by professional archaeology in the US. Hopefully these objects have not been lost, but have only disappeared into private collections. Fascinating examples of this state of affairs are the Richardson and Wilmington tablets (Fig.15) found with a mica plate (10"x13") under a mound in Ohio in 1879, reported in 1881. The mound had a diameter of c.43 feet, and a height of c.7 feet. Similar mounds were nearby (Ref.9). Corliss calls these the "most interesting relics from the Mound Builder's Period that have ever been found" (c.500 BC - 200 AD), with well-dressed male and female persons, animals, and blocks of strange glyphs (Ref.9). Evidence in the Old World is a problem too. Copies and Greek translations of a column of the Temple of Neith, telling the story of Atlantis, which were given by Egyptian priests to Greek Historian Crantor (310 BC) are thought to have been burned in the Library of Alexandria (Ref.25). The column may now lie under the City of Cairo.

Two 6"x8" golden metal plates were reported found by Joseph Smith "in a hollow cavern" atop a mound or hill on his farm in the early 1800s. You may visit the mound, though the tablets are gone, now called the Angel of Moroni

**Fig.9** The location of the 3 Rivers petroglyph, the metal urn, the Hawley Stone, the Old Mine Road, and the Windsor-Deposit area. The little shield figures are hand-chipped tunnels identified by Trento, and the other figures are stone piles, walls, menhirs, chambers or dolmen (Ref.7).

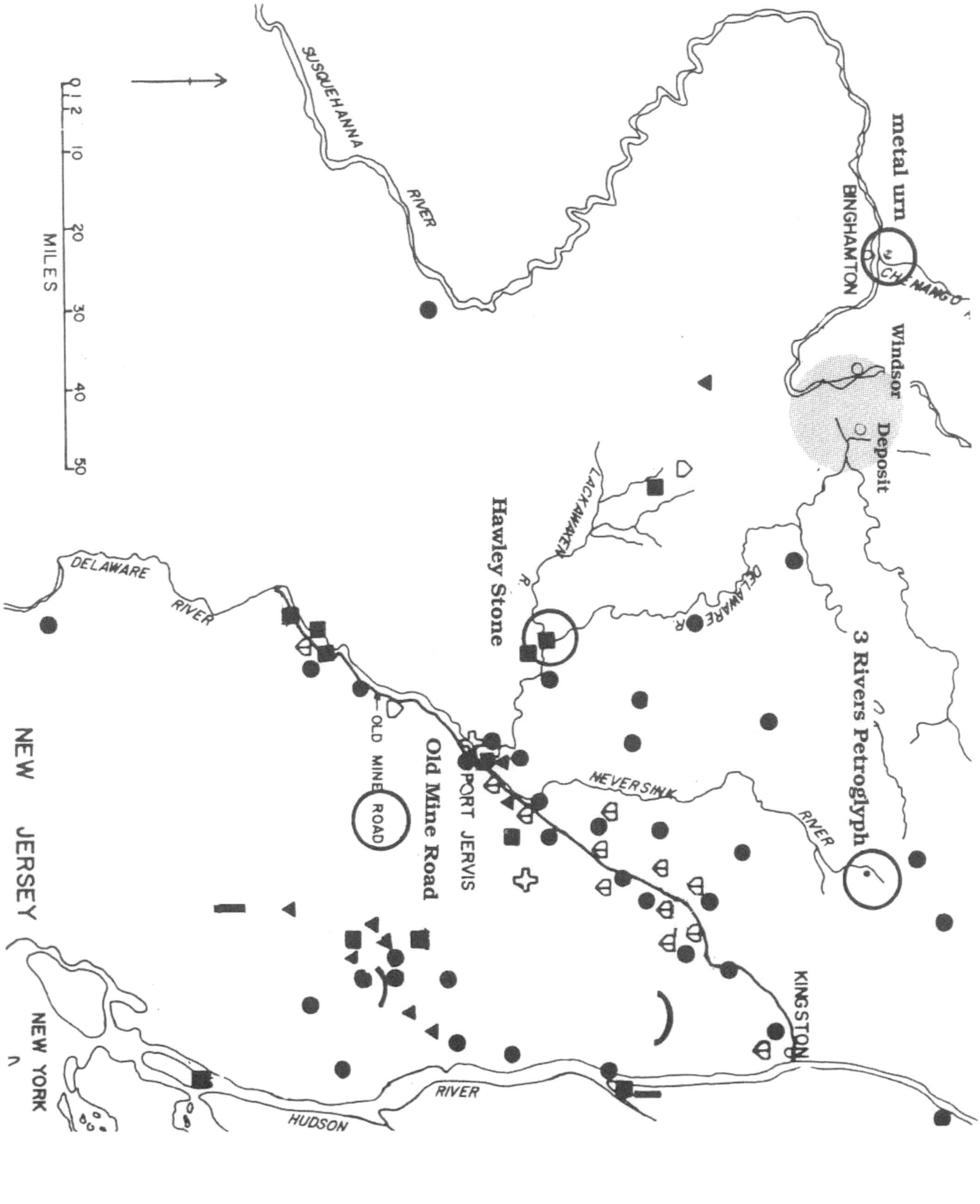

Monument at the Joseph Smith Farm in Palmyra, NY, near I-90 south of Rochester, NY. Three lines of the "hieroglyphic" inscriptions that Smith copied from the plates are reproduced and discussed as having been identified as Egyptian, Chaledeac, Assyriac and Arabic in Trento's book (Ref.8). Smith wrote about these tablets, which formed the basis of his Book of Mormon, which he said told a tale of ancient Israelites who came to America in 600 BC. Archaeologists have not declared these plates to be fakes.

## Dating

To study the geology of the creekside boulder, the authors removed for study two small pieces of river rock from the streamside nearby, which looked similar to the stone of the inscription. Both the petroglyph and the samples appear to be stream-eroded sedementary rocks, not glaciated rocks. They are not igneous because there is even a mudstone inclusion in one of them, and there are faint stratification lines in the other (Fig.13). They are fine-grained metamorphic sandstone, meaning they have been fused under heat and pressure, to almost a quartzite. Remember that Panther Mountain, at the head of the creek, has been shown to be a plug of hardened sedementary rock itself. This might be some of it (Ref.5). The rocks show iron oxide weathering, fairly deep into the stone (1-2mm). As can be seen on the sample, and by the abrasive damage on the petroglyph, the oxidized surface is much less hard than the deeper stone. Notice on the stone how the groove is lighter-colored where it is cut through the oxidised layer on one of the samples. Yet the glyph groove walls are dark, showing the patina of re-oxidization over considerable time.

The person who made the Three Rivers Petroglyph on the Neversink was undoubtedly skilled in metal tool use, because the details of the carving have been beautifully done. The "pecking" method is clearly seen in Figure 12. The fellow chose a fine grained hard rock with a smooth, water polished surface. We know that people from the end of the "Stone Age" had knowledge of the qualities of stone. The inscription was probably carved with a bronze or iron tool. A substantial amount of copper and bronze artifacts have been found in the surroundings in New England (Ref.11). Some early iron had metal impurities in it that serendipitously gave some early tools and weapons the qualities of steel. The author's groove (Fig.13), with a steel tool, took quite a lot of impact force. Since this petroglyph shows the river routes to Virginia and Ohio, it might have been carved by an iron tool manufactured in the USA.

The petroglyph dates back from the time period when route-finding and transportation was by boat, via the rivers. In America there are more maps like this one, but they are not deciphered yet. At Walnut Island in the Susquehanna River (SE Pennsylvania, 280km south of the petroglyph) two "coilshaped figures" were recorded (below Safe Harbor Dam, but above Holtwood Dam) prior to the construction of a power dam and the subsequent inundation of the site (Refs.1,8).

**Fig. 10** **Above:** The authors with Martin Brech (left), in the "California Hill Chamber" in Putnam County, N.Y., July 15, 2000. **Below:** Note the menhir at the left side (behind a tree) outside the chamber, which is lined up with other megaliths on astronomic and geographic lines (1st millenium BC).

The spiral is a very old megalithic symbol, in Europe used often in the fourth and third millennia BC. Probably, the meaning was still known in the second millennium BC, because their consistent use in inscriptions continues for a long period. Some old petroglyphs of rivers are known in Ireland, and some newer, more stylized ones are found in the rest of Europe (Refs.3,15), but these are not very common, especially the use of parallel groves. Indication of river entrances by sharp peaks has not been previously seen. The last mentioned features, and the way the petroglyph is performed, point to a late megalithic date. It is for sure that the petroglyph dates after the megalithic discovery of America, c.2500 BC (Ref.3). Referring to the closely related cup and ring petroglyphs, Barry Fell (Ref.11) states, that in general: "large numbers of (these) examples are known from regions where metalic ores occurr", and he adds: "Petroglyphs of this kind may be among America's oldest legacies of visits by people from the Old World". It is for sure that the petroglyph dates from before 1000 BC, because spirals in stone having a geographic meaning, were simply not carved by the cultures that came after the megalith builders. Lenik and Gibbs (Ref.1) confirm that "the surface of the rock including the glyph and the flake scars are patinated" (Figs.12-14). Also because of the likely use of the petroglyph to aid travelers between inhabited areas, we think it has a late megalithic date of c.1500 BC.

## Discussion

An article in the NEARA Journal ("A Catskill Mountains Enigma", Ref.1) by Edward Lenik and Nancy Gibbs carefully and usefully researches the site. Due to unfortunate Pre-Columbian professional prejudice, the article concludes a priori that since it must have been created by a pointed metal tool, it must have been created in the last 400 years, probably by an indian. It has been more recently said that the petroglyph was carved by "hippies in the 60's". But the petroglyph is large, and carefully executed. It shows forethought, and a large amount of work and serious intent. Its location in the river, where it appears to be under water during the snow meltoff in the Spring, and the dark weathering and patina of the old chip and groves argue that it is not a recent carving. Because of the spiral and the legs, and the geographic meanings of both, we are convinced it is a megalithic petroglyph from the 2nd millenium BC, carved by the megalithic culture of New England, who had bronze, and later, iron tools. We agree that "the aesthetic achievement and technical excellence present in the spiral design suggests that making it was a time consuming, exacting, and important activity".

The edges of the stone should be investigated accurately. The right edge may show details of the Hudson River, or the Atlantic coast, and the left edge may display elments of either Lake Erie, or the Mississippi and the Rocky Mountains which run largely parallel. In the immediate area of the stone are stone walls and other lithic features which should be surveyed. We are dismayed that the figure appears to be damaged by being walked upon, or hit by ice or flotsam in the river (Fig.14). In view of the antiquity, high quality, and historic value of the petroglyph, we recommend immediate protection of the site.

**Fig.11 Left:** Metal urn found near the confluence of the Chenango and Susquehanna Rivers, Binghamton, NY (c.600 BC, photo by Trento, Ref.7).
**Right:** Hawley Stone. The inscribed block found near the Lackawaxen-Delaware River confluence. The inscription (chalked for photo) has yet to be deciphered (probably c.200 BC, photo by Dr. V. Leslie, Ref.7).

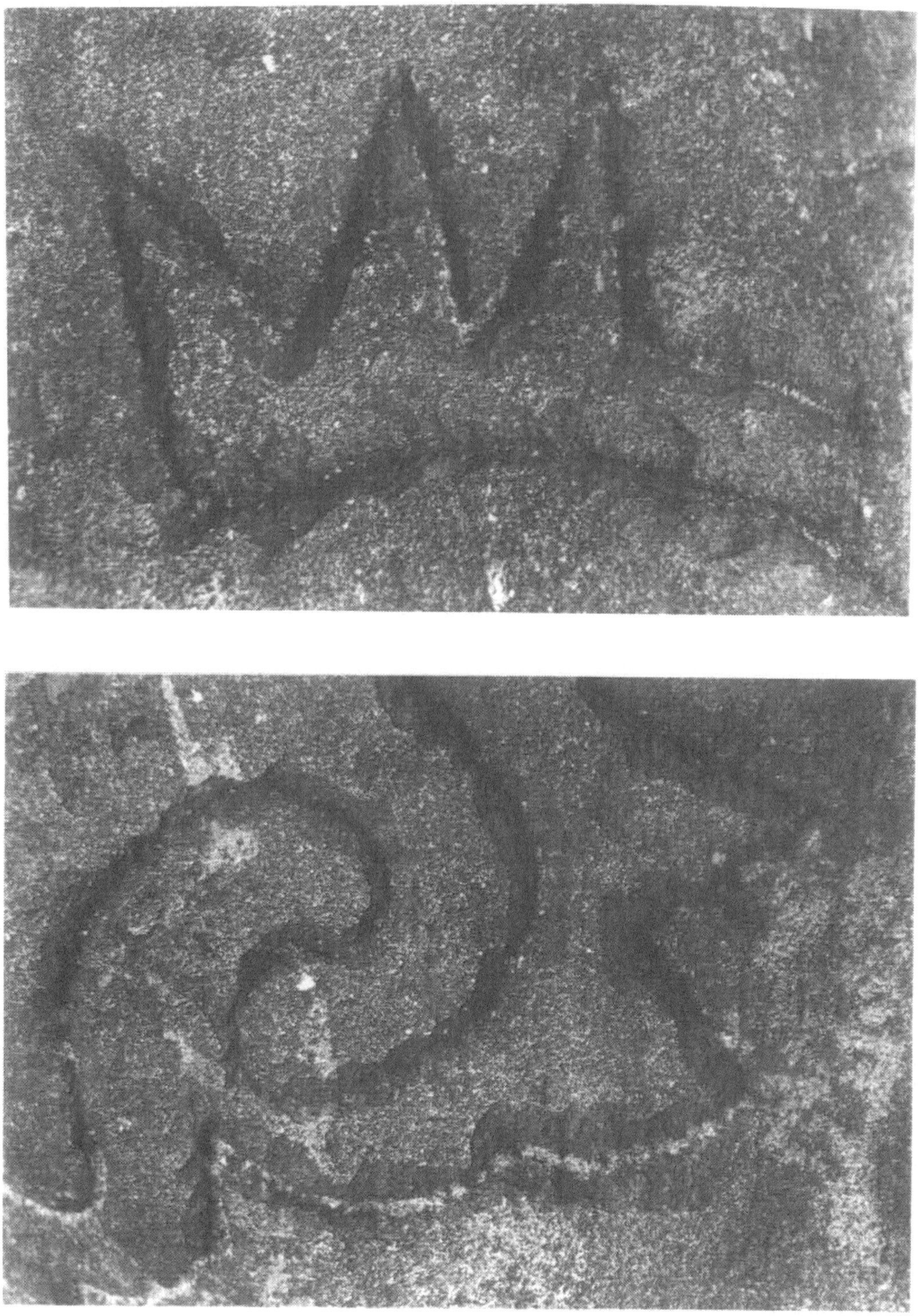

**Fig.12** Close-ups showing the "pecking" style of the petroglyph, and the dark patination of most of the deep grooves (c.1500 BC).

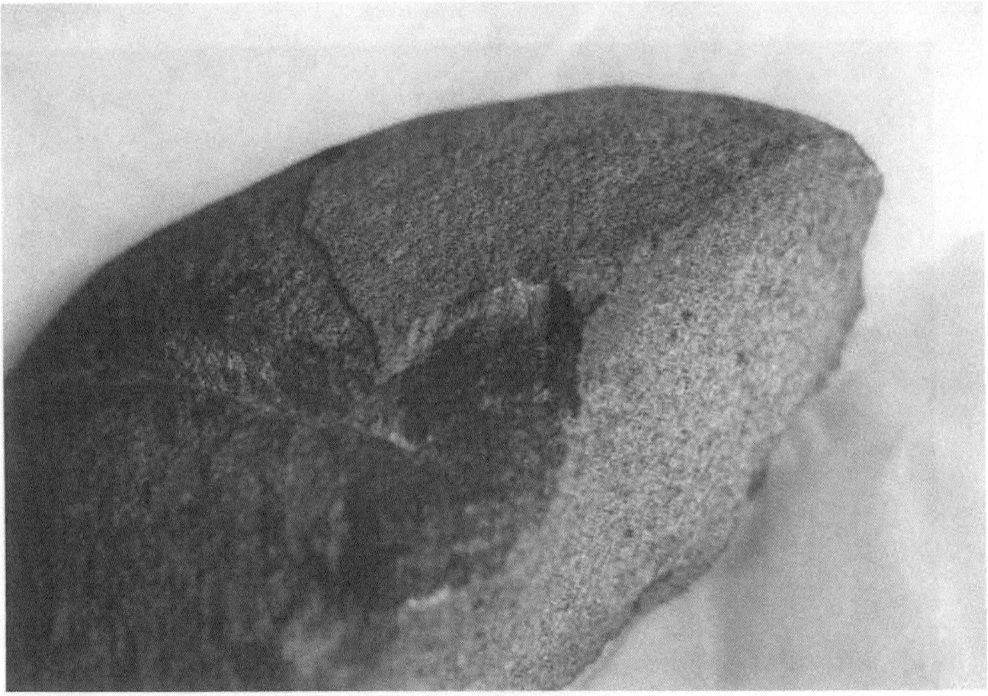

**Fig. 13 Below:** Fresh groove cut by authors in a similar fine-grained metamorphic sandstone sample from the site. Note brightness of the fresh cut compared with the discoloration of the old chip, which is darkly weathered, like the petroglyph. The depth of the surface oxidation of this old rock can be seen at its right edge, where the rock was split.

**Fig.14** **Above:** Dr. de Jonge shows the difference between a dark old chip made by the carver (left forefinger, c.1500 BC), and a recent light orange abrasion damage (right forefinger). **Below:** Note the bright-colored damage to the top of the spirals, probably from being walked on.

**Fig.15** **Right, upper**, the Richardson Tablet. This decorated 3 7/8"x 4 7/8" piece of Waverly Sandstone was found clasped in the left hand of a skeleton in a charcoal and ash filled vault under Sparks Mound in Wilmington, Ohio, Jan. 31, 1879. **Left,** the Wilmington Tablet, found in the center of a pit surrounded by a circular vault of stones filled with charcoal and ashes, also nearly under the center of the same mound on Feb. 12, 1879. **Right, below,** a portion of the upper left Wilmington Tablet "reproduced 2 1/2 diameters" (Ref.9, pg.588, from Welch & Richardson, "American Antiquarian" 4:40-48, 1881). "Mazes and labyrinths are thought to have a history of about 5,000 years" (Ref.21).

# References


1. Lenik. E.J. and Gibbs.N.L.. "The Frost Valley Petroglyph. A Catskill Mountains Enigma". NEARA J.. Summer 1999 (ISSN: 01492551)
2. Mallery.A. and Harrison.M.R. The Rediscovery of Lost America. the Story of the Pre-Columbian Iron Age in America. E.P. Dutton. New York. 1951 (ISBN: 0-525-47545-1)
3. Jonge. R.M.. de. and IJzereef. G.F.. De Stenen Spreken. Kosmos Z&K. Utrecht/Antwerpen. 1996 (ISBN: 90-215-2846-0) (Dutch)
4. Fell. B.. Saga America. Times Books. New York. 1980 (ISBN 0-8129-6324-5).
5. Guterl. F.. "Panther Mountain Crater". Discover. August 2000. pg. 52.
6. Dexter. W. . and Martin. D.. America's Ancient Stone Relics. Vermont's Link to Bronze Age Mariners. Academy Books. Rutland. Vt.. 1995 (ISBN 1-56715-050-0)
7. Trento. S.. M.. The Search For A Lost America. The Mysteries of the Stone Ruins. Contemporary Books. Chicago. 1978 (ISBN 0-8092-7852-9)
8. Trento. S. M.. Field Guide to Mysterious Places of Eastern North America. Henry Holt & Co.. New York. 1997 (ISBN 0-8050-4449-3)
9. Corliss. W. R.. Ancient Man. A Handbook of Puzzling Artifacts. Sourcebook Project. PO Box 107. Glen Arm. MD 21057. 1978 (ISBN 0-915554-03-8)
10. Mertz. H.. Atlantis. Dwelling Place of the Gods. 1976 (ISBN 0-96000952-3-3)
11. Fell. B.. America BC. Pocket Books. Simon & Schuster. 1994
12. Kayworth. A.E.. Legends of the Pond. Branden Books. Boston. 2000 (ISBN 0-8283-2053-5)
13. Underwood. P.. The Walking People. A Tribe of Two Press. PO Box 913. San Anselmo. Ca 94979
14. Lambert. J.D.. America's Stonehenge. an Interpretive Guide. Sunrise Publications. Kingston. N.H.. 1996 (ISBN 0-9652630-0-2)
15. Twohig. E. Shee. The Megalithic Art of Western Europe. Clarendon Press. Oxford. 1981
16. Ferryn. P.. "5000 Years Before Our Era: The Red Men of the North Atlantic". NEARA Journal. Vol.XXXI. No.2. pg 59 (1997)
17. Bailey. J.. Sailing to Paradise. Simon & Schuster. 1994
18. Kennedy. R.G.. Hidden Cities. The Discovery and Loss of Ancient North American Civilization. Penguin Books. N.Y..1994 (ISBN 0 14 0255273)
19. Munck. C. P.. Whispers from Time. and The Code. 1997. from Radio Bookstore Press. PO Box 3010. Bellevue. Wa 98009
20. Welch. L.B.. and Richardson. J.M.. American Antiquarian. 4:40-48. 1881
21. Coleman. L.F.. NEARA Newsletter 5. 1970 (pg.68)
22. Shackleton. R.. American Antiquarian. 15:295-304. 1893
23. Lepper. B.T.. Archaeology. Nov/Dec 1995. (pg.52)
24. Tompkins. P.. Secrets of the Great Pyramid. Harper Colophon Books. Harper & Row. New York. 1971 (ISBN 0-06-090631-6)
25. Zapp. I.. and Erikson. G.. Atlantis in America. Navigators of the Ancient World. Adventures Unlimited Press. Illinois. 1998 (ISBN 0-932813-52-6)

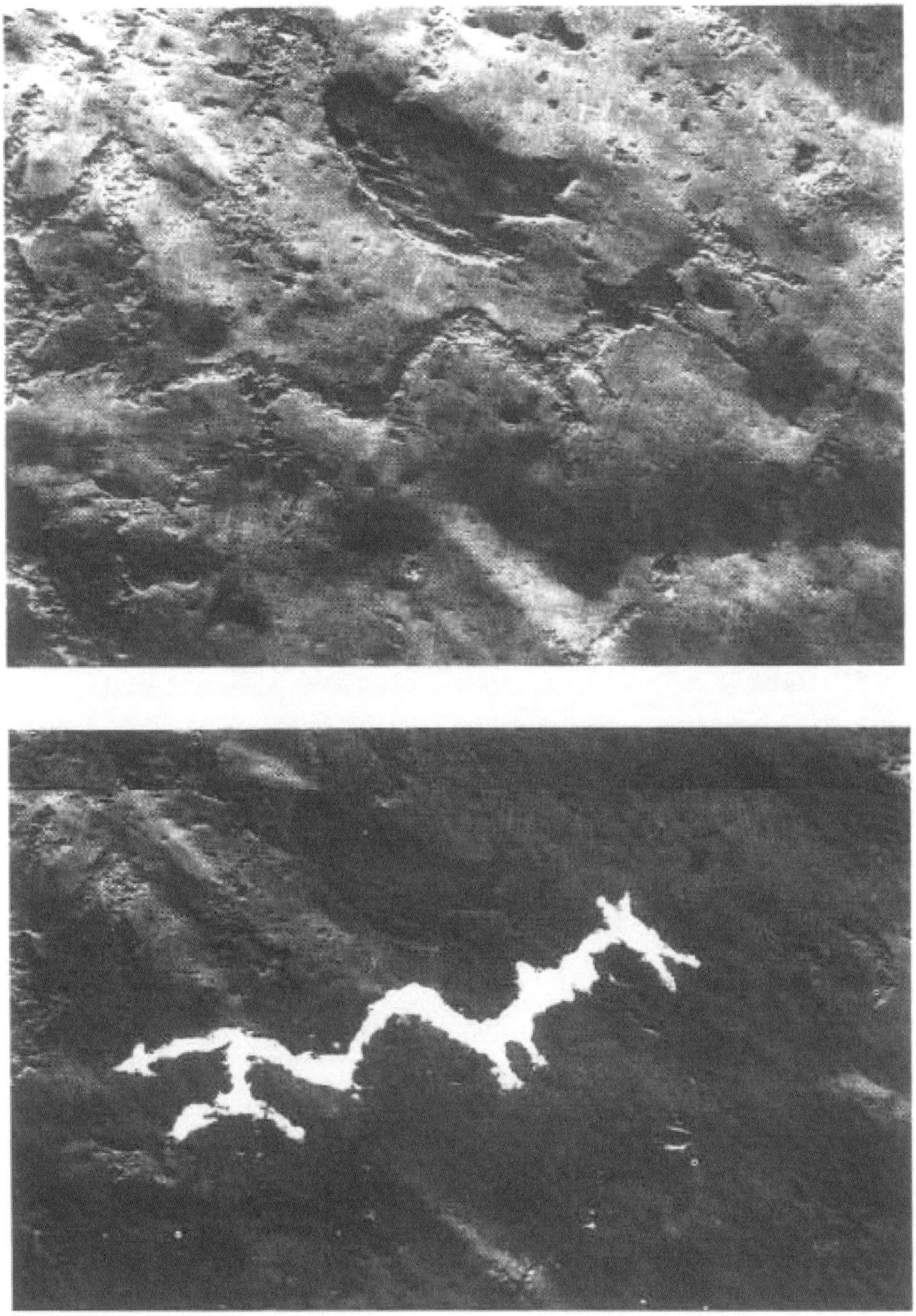

**Fig.1** The Embden Dragon Petroglyph, with and without white sand. It is a stylized carving of the Kennebec River, running from Moosehead Lake (right) to the Ocean (left). (Embden, Central Maine, c.1500 BC).

# The Embden Dragon Petroglyph
## A Copper-Trading Route of the Bronze Age
(Kennebec River, Embden, Maine, c.1500 BC)

Dr. R.M. de Jonge, drsrmdejonge@hotmail.com
J.S. Wakefield, jayswakefield@yahoo.com

### Introduction
The dragon (Figs.1&2) is one of over a hundred prehistoric petroglyphs (Ref.20), carved on a pier-like rock on the west bank of the Kennebec River, south of the bridge to the town of Embden. There are quite a few canoes, humans, and animals of various kinds. The whole group is engraved on the south side of the smooth, black surface of the rock, which appears to be hard basalt. The dragon has a length of c.2 feet, and the actual carving is c.1" wide and c.1/10" deep. It has been applied horizontally, parallel to the present July waterline, a meter or so below. The dragon looks to the river, here c.200 yards (meters) wide. Like so many prehistoric petroglyphs, it has, among others, a geographic meaning.

### The Kennebec River
The dragon petroglyph represents the Kennebec River System (Fig.3), and the head of the dragon represents Moosehead Lake, because it has more or less the shape of the river and the lake. Moosehead Lake is probably a name as old in usage as the petroglyph itself. For that reason, it is likely that the head of the dragon actually represents the head of a moose. Spencer Bay and Lily Bay can easily represent the upper and lower jaws of the moose, and the bays of Indian Pond, the two ears of the beast. These big animals no doubt have been a common sight there for the last several thousand years, and probably since the Ice Age. Guidebooks today describe Moosehead Lake as having the greatest density of moose in Maine. A petroglyph of a moose about a yard away is shown in Fig.8.

The unusually long neck down the breast of the figure is the long first part of the Kennebec River, from the present dam in Indian Pond where the river narrows, to the front legs of the figure at the villages of Moscow and Bingham in the south. The thick right leg is Austin Stream, leading to Austin Pond, and the smaller, left leg is the stylized Moxie Creek, leading to Moxie Pond. These are the two tributaries on the east side of this section of the river. The actual course of the creek differs here from the shape of the petroglyph (Fig.3), but it is also the part which is least important for the carvers, who turn out to be traders, as will be explained. Above The Forks, at the confluence of the Dead River, 30 miles north of Embden, is the Kennebec River Gorge, where there are 4 miles of class IV rapids (Ref.15).

The upper section of the body of the dragon is the Kennebec River from Austin Stream to the villages of Anson and Madison, where the river narrows again. These narrows define the sections of the river, probably due to their

**Fig.2** **Above:** The petroglyph of the Kennebec (c.2 feet) goes up and down five times, showing the total length of the river: 5x 1/2 dl= 275km (c.1500 BC).

**Right:** The petroglyph was carved on the west side of the river, near Embden, Maine. The width of the Kennebec is c.200 yards.

difficulty of passage, due to currents or rapids at some times of the year. About halfway, near what would be the heart of the dragon, is the location of these petroglyphs at Embden.

The central section of the body is the Kennebec from Anson/Madison to the present town of Waterville. This section is stylized, because the loop in the river from Norridgewock to Waterville (the belly) has not been represented accurately. The lower section of the body of the dragon is the Kennebec from Waterville to beyond the town of Augusta, near the villages of Gardiner and Randolf, where the river narrows again.

The tail of the dragon is the Kennebec River again from Gardiner and Randolf to the Atlantic Ocean at Popham Beach, the tail taking the shape of the landforms around and south of Bath, as well as representing the river mouth and the shape of the surrounding waterways. The lower part of the arrow is the inlet to Wiscasset and Boothbay Harbor, and the upper part is the inlet to Crooks Corner, including the bay of Orrs Island.

The haunch of the dragon, the short carving downwards to the rear legs, (Fig.2), does not correspond to a creek, but points to "Whitefield Lake", east of Randolf. The left hind leg (forwards) is the waterway from there via "Somerville Pond" and Sheepscod Pond to Lake St. George in the northeast. The right hind leg (backwards) is the salt water inlet to Damariscotta Lake, all the way to the Ocean in the south. This emphasis on the hindlegs of the dragon may show us that these complex waterways of the Maine coast were greatly frequented by Bronze Age sailors, and were familiar to the artist. This is reinforced by the eight Phoenician triangular stone anchors (with shank holes and carved recesses for shank pins) that a diver recently found in the harbor of Pemaquid, at the south end of Damariscotta Lake (Ref.25). "With the aid of a tow truck, three of them are now in the diver's flower garden"!

The dragon petroglyph consists of five zigzags (five sections) to its head (Fig.2), which means that the length of the Kennebec River up to Moosehead Lake (Indian Pond) is 5 "half distance lines" equal to 5hdl= 5x55= 275km, which is correct (1dl= 1°of latitude= 111km, Ref.13). The unit of "half a distance line" is nowhere indicated, but all sailors who got this far knew the Kennebec was not twice as long, and could understand the proportions of the standard units. The "distance line" is a megalithic unit of length, used in Europe since c.4700 BC, as shown by the eastern cairn of Barnenez, Brittany, France (Ref.13). It is equal to the ancient Egyptian "moira", proven to be in use in the Archaic Period (c.3000-2800 BC), but probably much earlier (Ref.28).

## Copper From Lake Superior

The hind legs of the dragon have a strange shape. Note also that the arrow on the tail of the beast points due west, to the region of the Great Lakes. When looking at a map of the U.S., you see that the haunch and the hind legs

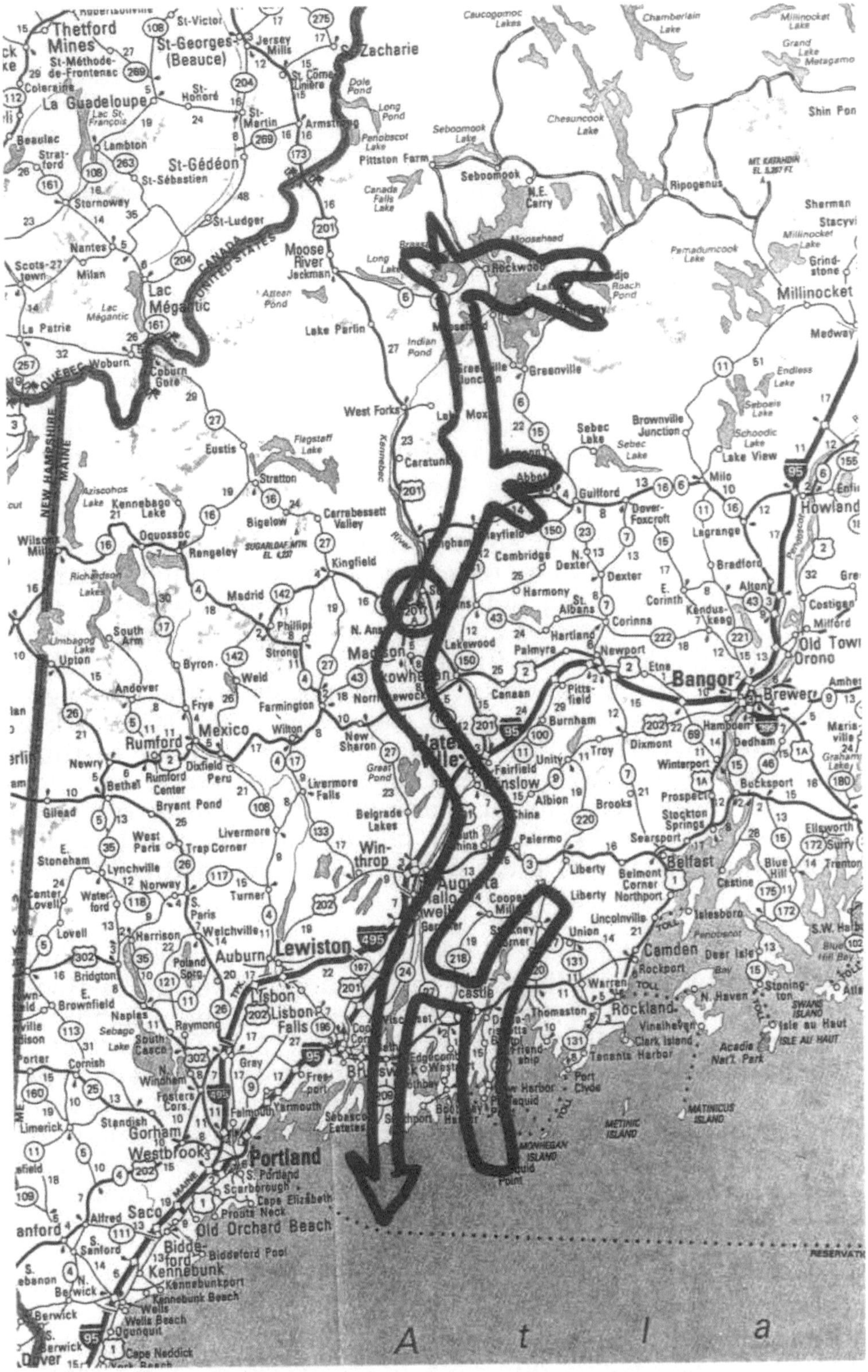


**Fig.3** Map of central Maine, with overlay of a drawing of the Embden Dragon on the Kennebec River, running from Moosehead Lake via Embden, where the dragon is carved (at the center of the small circle), to the Ocean in the south. (1cm=10.5mi)

closely resemble Nipigon Lake and Lake Superior. This area is famous for the huge amounts of copper which were shipped from there in prehistoric times (Refs.3,4,7,18,19). Even the latitude of this lake may be indicated, as the 5 sections of the dragon at 45°N correspond to 5 x 1/2dl= 2.5° of latitude, and Lake Superior is located 2.5° north of Embden, at 45°+2.5°= 47.5°N. This is the latitude of the south part of Isle Royale, and of Copper Harbor on the Keweenaw Peninsula, at the eastern end of Mineral Range, the northern tip of Michigan. The Amerindian name for Isle Royale was "Minong", close to the English word "mining". The five sections of the dragon also correspond to 5 x 1/4DL= 5 x277= 1388km, the distance from Embden to these sites. The head of the dragon resembles an oxhide ingot (Refs.3,4). It has this shape, because during the extraction process (the melting and smelting) impurities had to be removed in the directions of the four corners of the plate. Samples of minature oxhide ingots, called "reels" (6"x4") have been found in prehistoric mounds in Ohio, West Virginia, Indiana, and Kentucky, and also outside North America (Refs.3,4). Some of these are now located in Harvard's Peabody Museum in Boston.

The prehistoric copper mines around Lake Superior provided an average of 1,000 to 1,200 tons of ore per pit, yielding about 100,000 pounds of copper each (Refs.7,18,19). There were 5,000 mines, mostly along the Keweenaw Peninsula and along Lake Superior above the St. Mary's River. On the northern shore, the diggings extended 150 miles, varying in width from four to seven miles, through the Tap Range, to include three Michigan counties, Keweenaw, Houghton, and Ontonagon. At Isle Royale ("King's Island"), the mining area was 40 miles long and averaged five miles across. This is the only area in the world where pure, native copper exists in substantial quantities. It has been estimated that in the Bronze Age somewhere between 250,000 and 750,000 tons of copper were removed from the Upper Peninsula and Isle Royale. Frank Joseph's extensively-documented second chapter (Ref.18) is entitled "Missing - half a billion pounds of copper". He estimates 10,000 men working the mines for a thousand years seems credible, as does the conclusion that they were not slaves. All they left behind were their tools, literally hundreds of thousands of them. Ten wagonloads of stone hammers were taken from a single location near Rockland. The mauls were mass-produced in various sizes and types to serve different tasks. Most hammers were five to ten pounds, grooved to fit a wooden handle tied around the middle. According to W.H. Holmes (Ref.19), we may assume that a considerable trade existed in native copper, in raw material (ore), and in extracted, purified copper. To produce the easier-to-carry oxhide shape, the copper was melted into ingots of standard size and weight. This was probably done near the mining area. The grandiose mining enterprise that began suddenly around 3000 BC, terminated just as abruptly c.1200 BC.

Fell shows in his book America BC (Ref.4) pieces of metallic copper found near Lake Superior, and how they were used in the manufacture of tools and weapons that closely resemble those of the Old World. The presence of

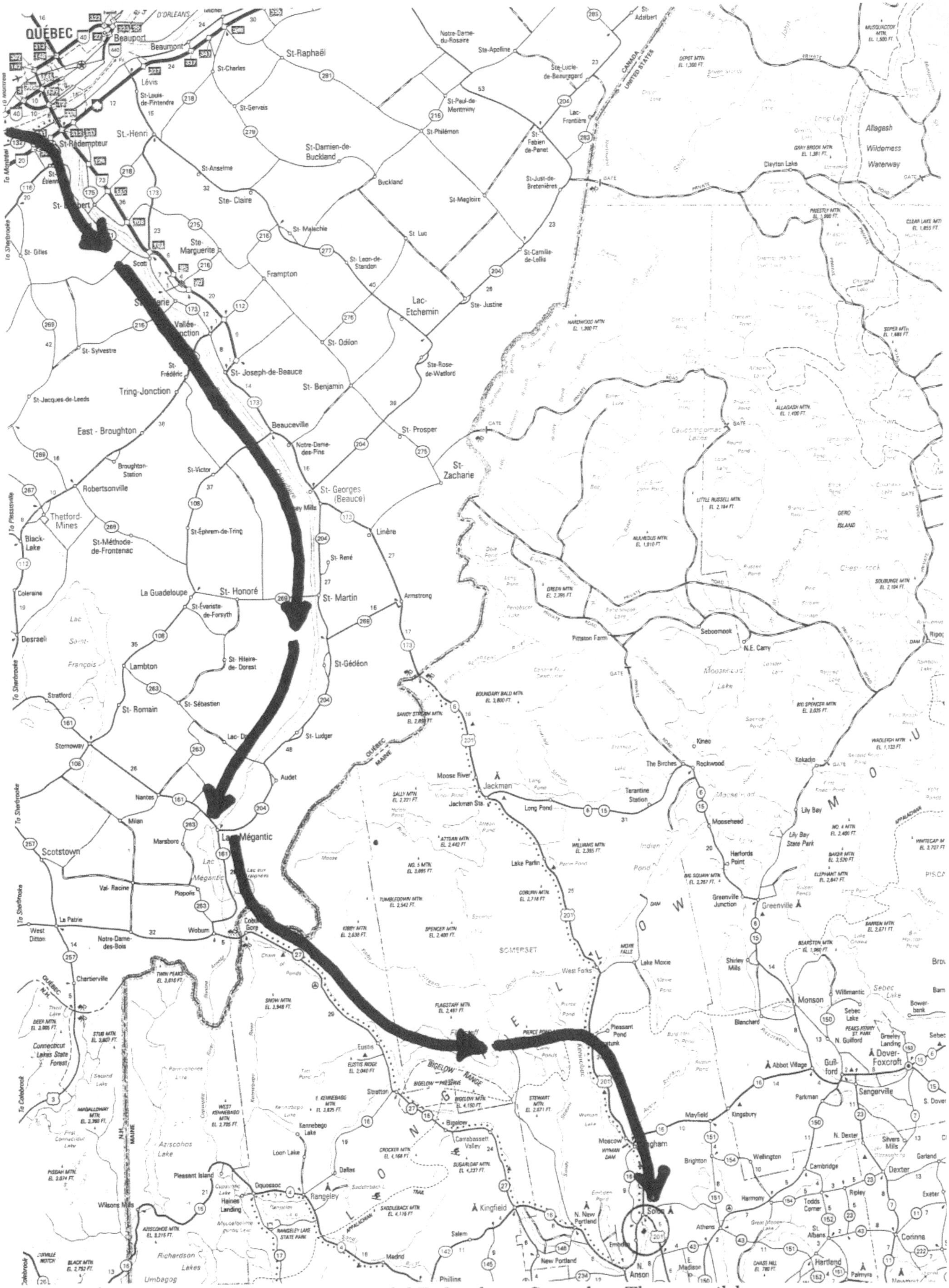


**Fig.4** Map of NW Maine, and SE Quebec, Canada. The possible copper-trade route of the Bronze Age from the St. Lawrence River in the north, via the Riviere Chaudiere and Flagstaff Lake, to the Kennebec River at Embden in the south, where the petroglyphs are located. (1cm= 5.9mi)

silver nodules shows that these artifacts were made by cold-working, not from molten metal. The engineer Mallery (Ref.31) points out that more than 100,000 Pre-Columbian copper artifacts from the Great Lakes Copper Culture have been found in the surrounding area. The accepted theory was that they all were hammered out of pieces of crude copper. However, this turns out to be wrong. Many of the heavier pieces are clearly cast, and it could be shown that the copper had been melted in open crucibles. He discovered numereous castings in nearly every museum. His conclusion was that the vast majority of the ancient copper tools were cast in molds.

## Transport to the East

The dragon goes up and down, like the waves at the surface of the water (Figs.1&2), so it represents a sailing route via a waterway. On the rock, the animal looks to the east showing its direction. The ships contained copper ingots, because the head of the dragon has their shape, and the hind legs show they were fully packed. The 5 sections of the dragon indicate the distance from Lake Superior to Cape Race, Newfoundland (5x 1/2DL= 5x555= 2777km). Basically, the dragon can be seen to represent the copper transport route from Isle Royale to the east, to the Atlantic Ocean.

In 1954 Canadian field geologists discovered a rich petroglyph site near Petersborough, Ontario (c.1700 BC). It features ancient ships, interspersed with human and animal figures, resembling those of Embden, although there are also differences. According to Fell (Refs.4,5) the petroglyphs literally illustrate that Scandinavian traders were exchanging woven textiles for copper ingots from Lake Superior, supplied by the Algonquians. Fifty miles north of Peterborough is the huge Algonquin Provincial Park, having a surface area of c.3000 square miles. The Peterborough site is northeast of Toronto, between Georgian Bay and Lake Ontario, along a route through large shallow lakes with portages. Today this route is known as the "Trent/Severn Waterway" (Fig.11), which eliminated the longer route via Lake Huron, Lake St. Claire, Lake Erie, and Niagara Falls. These petroglyphs (Fig.10) prove that this route was used for transport of copper to the east. In an Indian Reservation near Brantford, Ontario (125 miles SW of Peterborough), a bronze case axe of a Scandinavian type was found, dated from the Late Bronze Age (c.700 BC), now in the Museum of the American Indian in New York City (Refs.4,5). This bronze axe confirms that the Scandinavian people visited or lived in this part of North America, and used this material in prehistoric times.

The haunch and the hind legs of the dragon also closely resemble Lake Nipissing and the North Channel/Georgian Bay, respectively, while the rest of the body may represent the Ottawa River and the St. Lawrence. This may have been an alternative route to transport the copper to the east. Octave Du Temple ( a foremost authority on early Michigan, Refs.7,18) showed, that the Atlantic Coast was easily accessible from the Upper Peninsula mining regions during the centuries of the European Bronze Age. Approximately 3,500 years ago, the post-glacial Great Lakes were in the Lake Nipissing

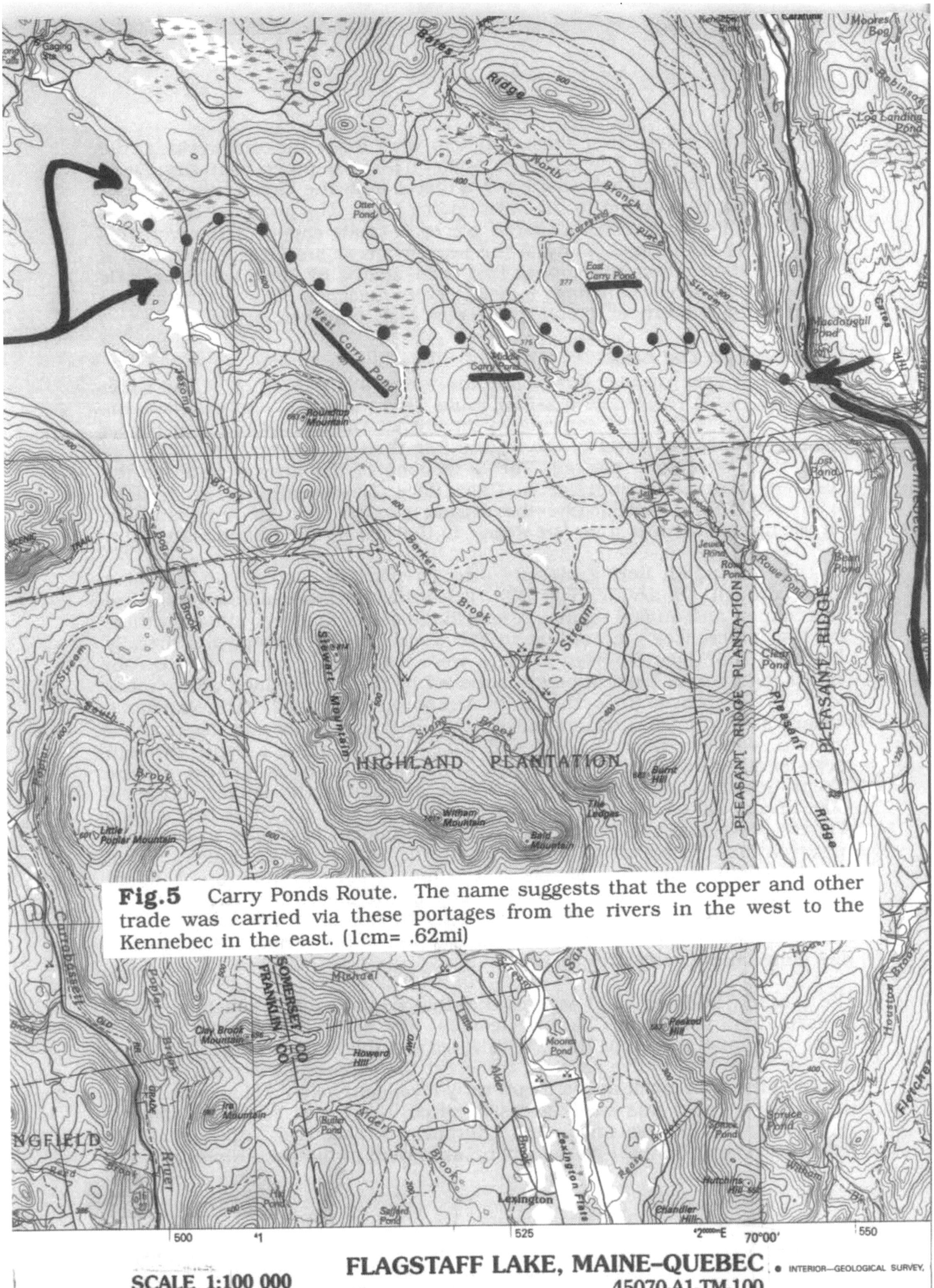


**Fig.5** Carry Ponds Route. The name suggests that the copper and other trade was carried via these portages from the rivers in the west to the Kennebec in the east. (1cm= .62mi)

Stage. Lakes Superior, Michigan, and Huron were all at 605-foot elevations above sea level. At this time, it was possible to travel east directly to the ocean, via North Bay (Lake Nipissing) and the Ottawa River, and thence out the St. Lawrence to the sea.

"Anta" (or "onta") and "anti" are old names for copper, now hidden in the present geographical names of Ontonagon in Michigan (copper vein), Antigo in Wisconsin (copper place), Deseronto in Ontario (copper quay), Toronto (copper rock or copper god, Thor=Norse god), Ontario (copper river), Ottawa (copper stream), Anticosta (copper coast), and Atlantic Ocean (copper sea) (Ref.3).

## Transport from Quebec, and down the Kennebec

The copper traders from the west sailed along the St. Lawrence River to the City of Quebec. We think some of the ingots were sold there to traders from Maine. These local traders sailed the rather big Riviere Chaudiere to Lake Megantic (see map Fig.4), and then to the current American-Canadian border at the small portage of Coburn Gore (a portage of 4km). From there, they traveled today's Scenic Route down the Dead River, through the Chain of Ponds to big Flagstaff Lake. From Flagstaff Lake "the Dead(!) River offers the longest stretch of continuous whitewater in the East. The 16 mile trip begins at Grand Falls and runs through Class IV and V whitewater to The Forks" (Ref.15). Therefore from Flagstaff Lake the traders chose (see Fig.5) to go up short Pond Stream out of Flagstaff Lake to big West Carry(!) Pond, then portage the trails (1.5 mi & 1.0 mi) to Middle and East Carry Ponds, and then paddle down Carrying Place(!) Stream to the Kennebec thirteen miles upstream of Embden. The Embden dragon petroglyph (Fig.2) goes up and down 5 times, corresponding to the approximate distance from Quebec to Embden, 5hdl= 5x55= 275km. Just a few miles south of Embden on the Kennebec are two towns named North Anson and Anson. These names might be derived from anti=copper from Central America, and soom=copper in the Micmac language (Refs.3,4). In nearby Madison, a hoard of Bronze daggers was found in 1924, only 8 miles from the Embden glyphs (Ref.4). The metal composition of these daggers should be studied. These routes need to be paddled, and examined for any signs of ancient petroglyphs, copper artifacts, or stone works by local history buffs. What would it be like to carry a sack of copper ingots along the route of these portages?

The dragon petroglyph is on a small peninsula of smooth rock that formerly extended about 80 feet into the river from the west bank (Figs.2&9). Unfortunately loggers had trouble with log rafts hanging up on this rock, so they blew off the "outer 50 feet" with dynamite (Fig.9). Below the petroglyph rock is a large pool, which is about four feet deep (c.1m), right to the vertical rock river's edge. This flat rock ledge is several feet above water, remarkably like a natural dock. This location is just downstream from the river's first rapids (Ref.1). The glyphs are directly across the river from a campground that "has been dug by state archaeologists and produced artifacts in great quantities from many different periods, some very old.

**Fig.6** Petroglyph of a sailing ship (length c.6"), with and without white sand. (Embden, Maine, 2000-1500 BC)

There is another prehistoric site just a few miles upriver on the same (west) side as the petroglyphs" (Ref.1). Might there be any artifacts related to Europe, Africa, Egypt or Central America? An inventory of the artifacts, and thorough excavation of these known sites along the river would be interesting.

Probably for a long time period, Embden was a center of traders between the St. Lawrence River and harbors near the mouth of the Kennebec. The Wabanaki Indians gave the Kennebec River its name, meaning "long level water" (Ref.21). Among the most valuable commodities of the Bronze Age were copper and tin. Many known copper artifacts, such as chisels, wedges, knives, spear points, rings and bracelets have been hammered into shape from masses of native copper, and not smelted (Ref.3). However, we agree with Mallery (Ref.31) that most copper was cast, and that ingots for shipping were made from molten copper, a technique that was well-known in this time period (c.2500-1200 BC). The copper from this area is the only copper in the world with a relatively high silver content. It would be interesting if any copper fragments turned up of similar composition to the material from Lake Superior (Refs.3-5,7,17,18). The soil itself in the Kennebec campgrounds should be analysed at various depths for copper.

The Hopi Indians considered themselves the first inhabitants of America. One of their oldest legends is that white men came across the Atlantic and returned to the east "using stepping stones routes" (via islands, Ref.35). In many Greek and Roman texts, like those of Plato and Diodorus, it is written that the Phoenicians knew of America, and traded with this continent, as early as 1000 BC. In reality, this happened much earlier. In the Old World there was an enormous demand for copper. Most of the seafaring ships from the east returned from America's Stonehenge in New Hampshire (c.2300 BC), via the south point of Nova Scotia and Sable Island to the Azores, the archipelago in the middle of the Ocean (Ref.13). Both the shape of the copper ingots (the head of the dragon) and the mouth of the Kennebec (the arrow-shaped point of the tail) are shown in the petroglyph.

Evidence for other copper trading routes should be explored. What about the land crossing from Lake Ontario to the Mohawk River in the State of New York, in order to reach the Hudson? The Kennebec River petroglyph (Figs.1&2) goes up and down 5 times, corresponding to the distance from Embden to Lake Ontario, 5dl= 5x111= 555km. Further east, another likely crossing from Lake Champlain to the Hudson is 5hdl (=5x55), or 275km from Embden.

The dragon indicates the copper trade route from Isle Royale to Embden, and in this meaning, the front legs of the dragon indicate the Hudson and the Connecticut Rivers. The distance from Isle Royale to these rivers is c.4x 1/4DL=1111km, as the dragon indicates. The Connecticut River runs nearly straight north-south through many of the "stone chamber" sites in New England, and its mouth is across from Orient, NY, at the tip of Long Island.

**Fig.7** The "Dancing Man" (renamed "the lewd guy" by author's wife) with and without white sand (height c.16"). Note that the body of the man is similar to the man of Peterborough, Ontario, in Fig.10. (Embden, Maine, 2000-1500 BC)

The Connecticut can be reached after a portage from the large Lake Memphremagog, on the Quebec/Vermont border. The Yamaska runs north from there via the St. Francis River to the St. Lawrence east of Montreal. On a hill on the western shore of the lake is a large inscribed stone, referred to as the "Memphremagog Stone", which is profusely covered with vowlless Ogam with dipthongs. Dr. Barry Fell and a local Wabanaki linguist agreed that the inscription gives the length and depth of certain areas of Lake Memphremagog (Ref.17). Who might be more concerned about shallows in the lake than skippers of heavily laden vessels carrying copper ingots?

The big arrow at the end of the Kennebec River petroglyph (Fig.3) points due south, to Cape Cod and Nantucket Island. The 4 or 5 segments of the Kennebec River petroglyph provide the distances from the mouth of the river to these places, 4hdl=4x55= 220km and 5hdl=5x55= 275km. The big ships of the Old World arrived off Nantucket and the Cape, sailing north on the Gulfstream. The hind legs of the petroglyph (Fig.3), represent the coast of the Gulf of Maine, and the shape of these ships, fully packed. These sailing vessels are also shown by other petroglyphs on the site (see Fig.6). They went to America's Stonehenge in New Hampshire, the nautical center of New England, for the preparations of the long crossing of the Atlantic Ocean (Refs. 8,13). Whittall (Ref.4) has shown that 12 different ancient copper and bronze artifacts from New England have virtually identical shapes compared to corresponding artifacts from the Iberian Peninsula.

More sailing ships are engraved in downeast Maine, on a rock seen now only on a low tide, near Machiasport on the ocean shore (Refs.1,20). Sailing to the east, sailors headed for Cape Race, Newfoundland, which is the most easterly point of North America. The distance to Cape Race is c.1388km, corresponding with the 5 segments (5x 1/4DL= 5x277= 1388) of the dragon. From Cape Race the ships sailed via the Azores, Madeira, and the eastern Canaries to Gibraltar. The serpent goes up and down 5 times (Fig.2), corresponding to this distance, 5DL= 5x1111= 5555km. In this meaning of the petroglyph, the arrow points to Embden, the start of the haunch is Cape Race, the end of section 2 is the Azores after a crossing of 2DL= 2222km, the left front leg points to Madeira, the right front leg points to the Canaries, and the head is the Strait of Gibraltar!

"Keftiu or Phoinikes" (Red Men) appear on Egyptian wall-paintings carrying ingots of copper on their shoulders, brought by Phoenician seamen after three-year round trips from the "Isles of the Sea" (Ref.26). On the Kennebec, due north of Cape Cod, they could trade for copper with the valuable articles from the Mediterranean. Thus the sea-ships wishing to supply the warring states in the Middle East with copper for Bronze weapons, were saved from sailing all the way to Quebec to get copper, because ancient traders brought the high-value metal right to the coast of New England. The traders in Embden could also obtain articles from the civilization in Central America, previously visited by the ships. Embden is located in New England at exactly 45°N, and the center of the Olmec Culture is in the southern Gulf of

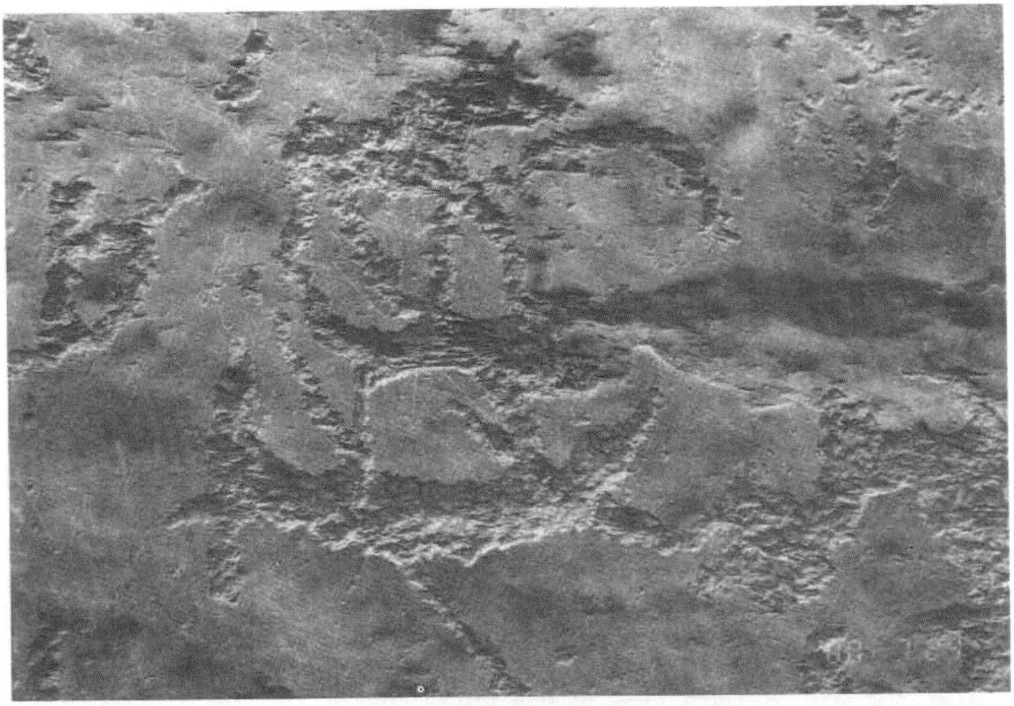

**Fig.8** Moose, canoe, and woman giving birth (total height c.28"), with and without white sand. (Embden, Maine, 2000-1500 BC)

Campeche, 45°SW of Embden (Refs.10-12). The Kennebec petroglyph goes up and down 5 times (Figs.2,3), corresponding to the latitude of the Gulf of Campeche, 5hDL= 5x555km= 2775km/111km/°= 25°of latitude south of Embden, at 45-25= 20°N. When the dragon petroglyph is viewed from the north, the rear legs of the beast may even represent the coast of the Gulf of Campeche. The many stories told in the Embden petroglyphs indicate that people spent significant time here waiting for traders (Figs.7&8). The famous trading triangle of New England rum/African slaves/Caribbean sugar in Colonial times that is taught in American schools lasted a hundred years. It was preceeded by a Caribbean/New England copper/European triangular trade route in Bronze Age prehistory that lasted at least 1000 years.

In the neighborhood of Embden are a number of interesting archeological sites, some related to the dragon petroglyph. Only 15 miles east of Embden are ancient geographic petroglyphs on Devil's Head Mountain (Ref.29). The two oldest carvings relate to the discovery of Bermuda, 600 miles from the American coast (c.2200 BC). On Monhegan Island, 25 miles east of the mouth of the Kennebec, is an old Celtic Ogam inscription (c.700 BC), with the message: "Cargo platforms for ships from Phoenicia". Petroglyphs of Iberic Punic script are found at York Harbor, 60 miles SW of the mouth of the Kennebec, dated to 300-400 AD. Scuba diver Norwood Bakeman found ancient anforeta jugs (Spanish olive jars, c.300 AD) on the sea bottom at Castine Harbor, at the mouth of the Penobscot River, 70 miles NE of the mouth of the Kennebec (Ref.20).

## Serpents and the Copper Trade at Peterborough

We studied the petroglyphs of Peterborough (Refs.4,5) as well as those of Bohuslan, Sweden (Ref.6), where they are beautifully painted in red ink, and have come to the conclusion that they are similar (Ref.1). The choice of the subjects, with many canoes, humans and animals, carved on flat rocks polished by the Icecap during the Ice Age, is typical for these petroglyphs. According to Fell (Ref.4,5), in Peterborough, Scandinavian traders were exchanging woven textiles (illustrated in the petroglyphs) for copper oxhide ingots (also illustrated) from Lake Superior, supplied by the Algonquians. The Peterborough site is between Georgian Bay and Lake Ontario, along a route through large shallow lakes with portages, today known as the "Trent/Severn Waterway" (Fig.11). Fell goes on to say: "Some 5,000 abandoned mines have been found on the shores of Lake Superior, dated between 2000 and 1000 BC, while it appears that millions of pounds of metallic copper were extracted and cannot now be accounted for. The copper must have been shipped overseas, and now the discovery of the Peterborough site seems to confirm that." Note that "Serpent Mounds" Canadian National Park (see Fig.11) (with interred copper and Conch shells) is located on the waterways near Peterborough. An ancient myth holds that an extinct super-race, the 'Mound Builders' raised the earthworks (Ref.19).

There are a number of dragons among the Peterborough petroglyphs. In Fig.10 we see several serpent-dragons, accompanied by old Norse Script,

**Fig.9** **Above:** The Embden Dragon Petroglyph, symbolic for Orm, the god of trade, and two manifestations of the Norse god Thor. (Embden, Maine, c.1500 BC)

**Below:** Local schoolteacher & friend in kayaks tell us how the outer fifty feet of the petroglyph rock was blown off by loggers. This quiet water near the pier-like rock shore (behind the photographer) would have made a nice docking area.

called Tifinag. Translation of the Tifinag text by Fell (Refs.4,5) produces a story from old Norse mythology. Figure 10, right, where the Norse God Thor uses his hammer, reads as follows: "the Hammer has struck, the serpent-dragon weakens". Figure 10, left, where he wears his glove, reads: "Hammer and Glove, woe is their power to the serpent-dragon writhing". The names of some villages near Peterborough confirm this translation. Thirty miles NW is Tory Hill (hill of Thor), and 25 miles NE is Ormsby (Serpent village). Thirty five miles SW is the village of Moira, the ancient Egyptian name for the one degree distance line.

Like so many ancient stories, this old Norse mythology is Egyptian in origin. The dragon or serpent is derived from the Egyptian goddess Maat, the goddess of law and order in the universe. The story goes that she was defeated by the god Horus, here represented by "Thor with the Hammer", who was helped by the god Osiris, represented by "Thor with the Glove". It is part of a very ancient tradition of Sun/Moon, Horus/Osiris, Hammer /Glove, bull/bird, father/son mythologies. In Egypt, the kings were the substitutes of Horus and Osiris. For that reason it is not a mere coincidence that the old town at the entrance to the Trent/Severn Waterway into the St. Lawrence has been called Kingston (or "Kingstown"), only 90 miles from Serpent Mounds and these petroglyphs.

The petroglyphs of Peterborough, Embden, and Machiasport are all at 45°N. There might be a religious reason for this. On the rock in Embden are again two glyphs of Thor in the immediate vicinity of the dragon petroglyph (Fig.9). The names of villages nearby may show the influence of this Godking: 15 miles west of Embden is Kingfield, and 20 miles NE is Kingsbury. The serpent itself now represents Orm, the Norse version of the goddess Maat. Orm (Maat) was also the god of trade. The head of the dragon looks like a copper ingot, and the whole petroglyph deals with the international copper trade. The dragon was also the standard of the Phoenicians (3000-200 BC), and at the prow of the Phoenician galleys stood a figurehead of the goddess Astarte ("a sister of Orm and Maat"), holding a cross (possibly a compass card) in one hand, and pointing the way with the other (Ref.30). Fig.14 shows a copper helmet from South France (Ref.32, c.800 BC), with 23 burls on top (and 24 on the side), corresponding to the latitude of the Tropic of Cancer, at 23°N. So this helmet, possibly made of American copper, is for a soldier of the army of the King, substitute of the SunGod!

The title of Strong's article about the dragon (Ref.1) reads: "Did Glooskap kill the Dragon on the Kennebec?" She tells us that she met Mr. Mike Sockalexis, a Penobscot Indian: "He had painted an oil painting of Glooskap killing the water monster. I knew that Glooskap was a culture hero of the Indian tribes of the entire northeast, including the Canadian Maritimes: Penobscot, Passamaquoddy, Maliseet, and Micmac. They are often referred to as the Abenaki or Wabanaki. Mike showed me his copy of a book written by an ancestor of his in 1893, Joseph Nicolar's 'Life and Traditions of the Red Man' (in which this mythological story was told extensively, Ref.9)."

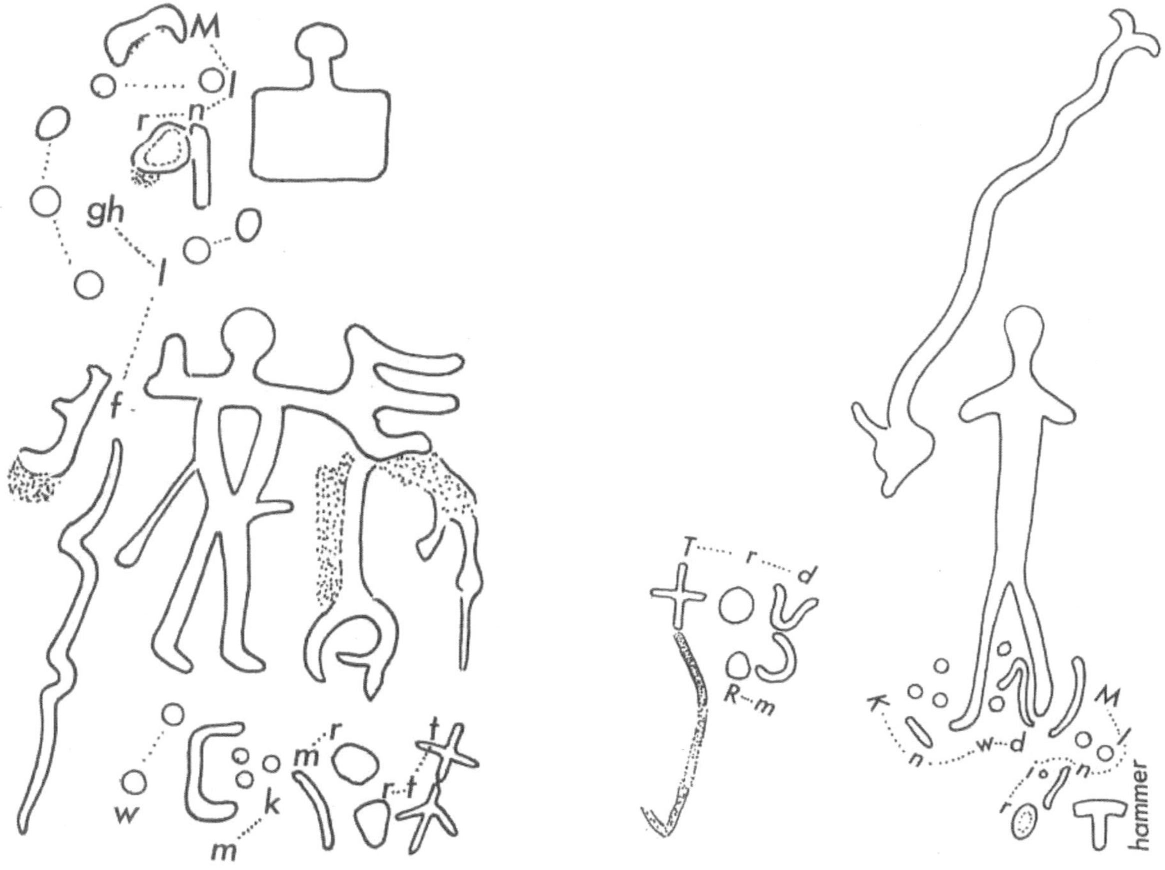


**Fig.10 Left:** The Norse god Thor using his Hammer (above) and his Glove (center right) to defeat the Serpent-dragon. The Tifinag text reads "Molnir, Glofi, ve maki Orm rittit", which means "Hammer and Glove, woe is their power to the Serpent-dragon writhing". The mythological story of the Indian culture hero Glooskap killing the dragon is c.3700 years old.
**Right:** The Norse god Thor using his Hammer (below right) to defeat the Serpent-dragon. The Tifinag text reads "Molnir knudh, traudh Orm", which means "The Hammer has struck, the Serpent-dragon weakens". Peterborough, Ontario, Canada, c.1700 BC, Refs.4,5).

We quote Mrs. Strong, because we are convinced that the Norse god Thunor the Thunderer, the later god Thor, at least three times depicted in the petroglyphs of Peterborough is exactly the same person as the Indian culture hero Glooskap. His weapons are a "glove", literally "Glofi" in old Norse, very close to "Gloos-", and a hammer or copper/bronze(?) axe, which he uses to "cut" the dragon in half, is in modern Swedish literally "Kapa" or "Kap"! Thus the Indian culture hero Glooskap or "Glofikapa" or "Glovecut" is the old Norse god Thor, and this frequently illustrated story comes all the way from Scandinavia, and is at least c.3700 years old.

## European Petroglyphs of the Copper Route

Brittany is situated at the latitude of Newfoundland, at 48°N. This must have been interesting to the people of Brittany (as seen in Fig.13) when copper and other goods were shipped from Newfoundland. At the top of Fig.12, we see a petroglyph in Brittany which shows the important coasts of the Gulf of St. Lawrence (the two curves). Below, at the right side of a nearby stone, is the distance from Brittany to Newfoundland, 4DL= 4x1111km= 4444km. On the left side we see the width of the continent of North America, a large and impressive 4DL. These stones from c.2000 BC confirm that people in Europe were very interested in the hinterland of North America at the level of Brittany.

Figure 13 is a stylized petroglyph (c.1900 BC) of the North Atlantic Ocean. The Southern Crossing and Upper North route are shown here by the edges of the stone. In the north the direct crossing from Newfoundland to Ireland is engraved. In the center we see the crossing via the Azores (in the middle) to Iberia, at about 40°N, the latitude of this stone itself, located near Toledo at the river Tagus in central Spain. These return routes had become very important in the Bronze Age.

Northern Bohuslan, 70 miles north of Goteborg, is the most important petroglyph area of Scandinavia (Refs.6,39,40). On the smooth, granite hills near the coast, which were polished during the Ice Age, about 1500 rock carving sites contain more than 40,000 carvings of the Bronze Age, the majority dated between 1800 and 500 BC. The petroglyphs include approximately 8,000 images of boats, 2,500 humans, 1500 animals of various kinds, 27,000 cupmarks, and a large number of wagons, circles, feet, and other images. In Fig.14, upper left, we see a beautiful example of a seagoing vessel, with three warriors that is commonly reproduced. They are accompanied by two flying birds, and a warm sun. The sun image shows they believe in the SunGod, which will make them invincible! This petroglyph is characteristic for Bohuslan.

Figure 14 also shows a special petroglyph from Gerum near Tanum, in the center of Northern Bohuslan (Refs.6,40). Three men hang on ropes, and turn fast around a pole. A fourth, horned man is standing on top of the pole, with his arms in the air. While this event is unknown in Europe, similar illustrations can be seen in and around Mexico, where it is called the "Tree

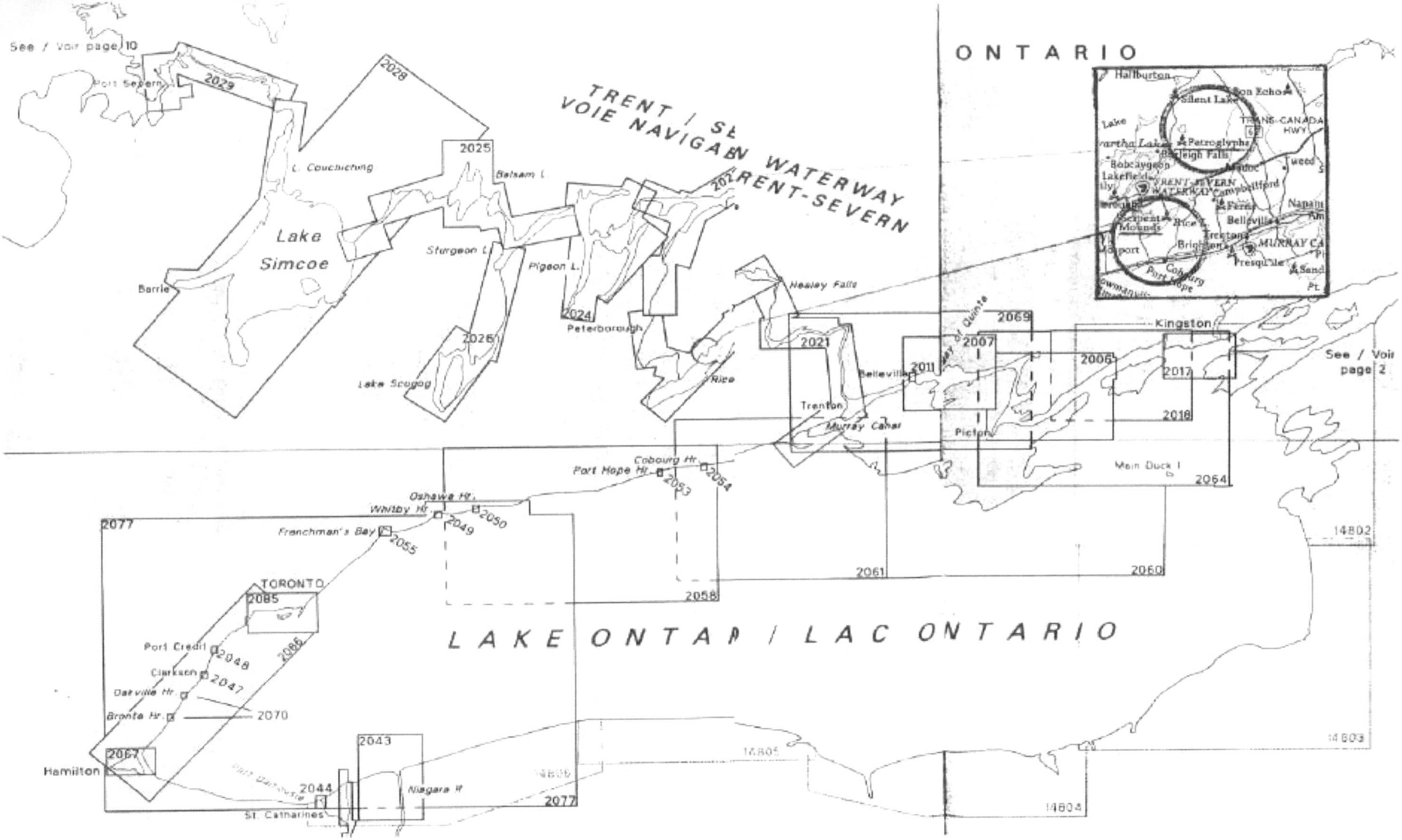


**Fig.11** The Trent/Severn Waterway, ing from Georgian Bay in the west, via Lake Simcoe to the eastern p: Lake Ontario (Kingston). The Petroglyph Park and the Serpent Mounds are located on this waterway (see upper right corner, Ref.27).

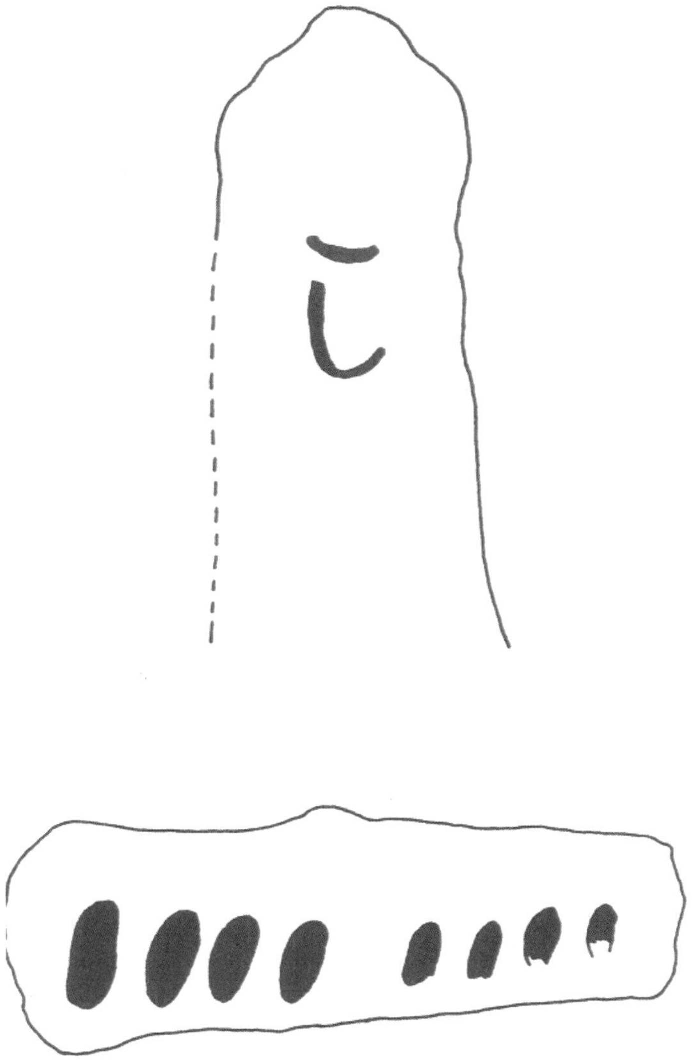

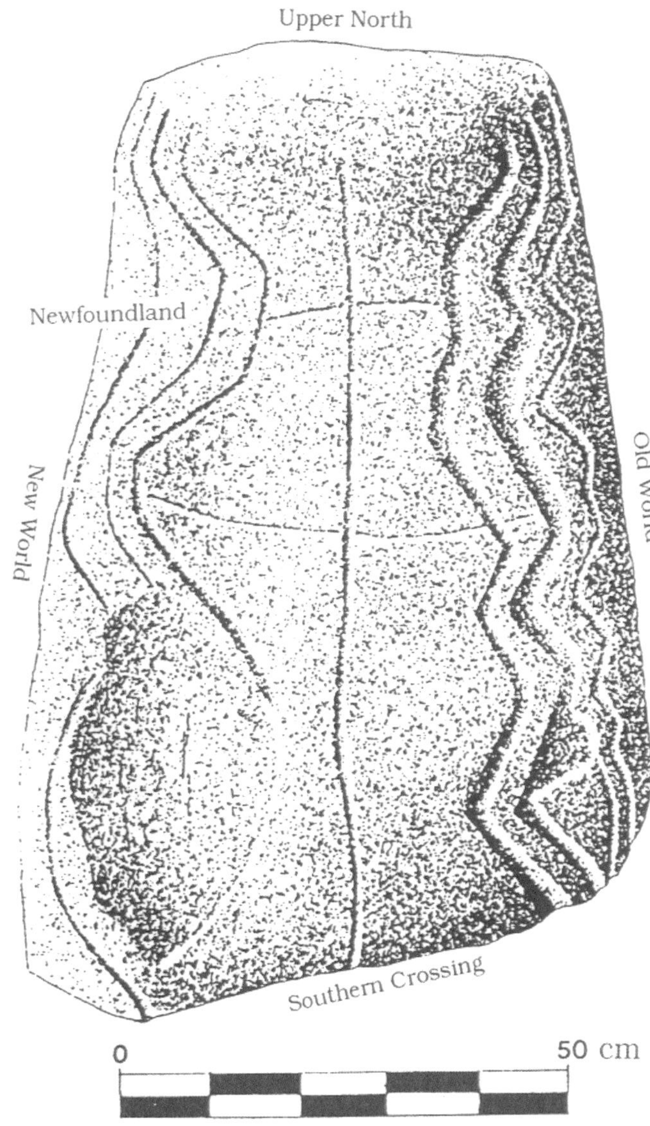


**Fig12 (Left), Above:** The two curves are the coasts of the Gulf of St. Lawrence behind Newfoundland, the approach to Peterborough and the copper mines of Lake Superior. **Below:** The four dots at the right side illustrate that the width of the Ocean between Brittany and Newfoundland is 4DL= 4444km. The four larger dots at the left side show that the width of the continent of North America is even more, and difficult to cross. (Butten-er-Hah, 48°N, Ile de Croix, Brittany, Ref.41, c.2000 BC)

**Fig.13 (Right):** Stylized petroglyph of the North Atlantic Ocean with the Old World in the east and the New World in the west, and the imagined Mid-Atlantic Ridge (vertical) in the middle. The crossing from Cape Hatteras (via Bermuda and the Azores) to Iberia is the center horizontal line. Below Iberia the protruded coastlines of the Canaries and the Cape Verde Islands are clearly indicated. Higher on the stone the horizontal crossing from Newfoundland, the east point of North America, to Ireland has been engraved. Left and right, the coastal waters are indicated over 2x 1/2DL= 2x555km= 1111km. So the total width of the Ocean as shown is over 4DL= 4444km, which is correct.
(Navalcan, 40°N, Toledo, Central Spain, Ref.42, c.1900 BC)

of Voladores" or "Tree of the Flying Men" (Ref.33). In Huaxteca (Mexico) is an ancient Totonakes-petroglyph of the same tradition, and in front of the step-pyramid El Tajin in Vera Cruz the ceremony was and is still regularly performed. The original tradition has been accurately described by Clavijero (Ref.34), and a nice drawing of "the flywheel" appears in Beltran's book (Ref. 38). This petroglyph proves prehistoric contact between Bohuslan and Central America between 1000 and 750 BC.

## Dating and Origin

The Kennebec River petroglyph is, like most petroglyphs, difficult to date. However, the petroglyph is one of a large group at the site, which together with the inscriptions of Machiasport (SE Maine), resemble those of the Peterborough site. Compare the great similarity of the body drawings of Figs.7&10. According to Fell (Refs.4,5), the Peterborough site is of the Bronze Age: "astronomical details included in the inscriptions suggested a date of about 1700 BC". The rock carvings from Bohuslan, Sweden, are accurately dated between 1800 and 500 BC (Refs.6,39,40). The Embden petroglyphs resemble only the older ones of Bohuslan, from before 1000 BC. We think the dragon with the moosehead is basically a Scandinavian petroglyph. Also the other petroglyphs point to a Scandinavian origin. However, there are also differences. The Swedish inscriptions appear to be carved deeper in the rock, and the humans generally have thicker calves. Compared with the Peterborough site, the Embden petroglyphs do not possess Tifinag script.

The discovery of America via the Atlantic is recorded in Egyptian history at 2497 BC. A petroglyph of a dragon on the east coast of North America is definitely after this date. According to Mr. Carter (Ref.16) and Mrs. Strong (Ref.1): "...only Celtic dragons (have) arrows on their tails." If this is true, then there were Celtic influences at the time the petroglyphs were made. According to Fell (Ref.4), Celtic elements were already present in New England c.1700 BC, and regular Celtic contacts with Indian groups and immigrants to the Northeast began c.1500 BC. On the basis of all this information, we think that most of the petroglyphs at Embden were made between 2000 and 1500 BC (some may be older), and that a date of c.1500 BC is most probable for the Kennebec Dragon petroglyph. It is possible and even probable, that after c.1500 BC, most ships from Central America returned from Cape Hatteras, or the Bahamas via Bermuda to the Azores, and that the trade at Embden more or less collapsed (Ref.13). The copper mining itself collapsed precipitously at c.1200 BC (Refs.18,23).

## Serpents/Dragons and other Copper Trade Routes

A lot of copper was shipped via the Mississippi to the south. Fifty miles west of Milwaukee is ancient Aztalan (c.1500 BC). In the Aztalan Archaeological Park (21 acres), there are at least two temple-mounds: the "Pyramid of the Moon", a square earth mound, 20 feet across at the base, 11 feet high, and the "Pyramid of the Sun", with a 15 foot summit. The scene was dominated by 16-foot high walls that run for more than a mile around and through the entire complex. There were 150 guard towers 20 feet high, which gave Aztalan an unusual appearance for an American site. The immediate area swarmed with more than 70 effigies of many shapes and sizes, of which only a handfull survive. Originally, there were a total of c.10,000 effigy mounds in Wisconsin. From Aztalan the copper was transferred via the Rock River to the present town of Davenport at the Mississippi, c.140 miles SW of Aztalan, where in 1874 a famous stele was excavated. It contains an important religious message in Egyptian hieratic, Libian, and Iberic Punic script (c.700 BC), illustrating its influence. Via the Mississippi River the copper was shipped to Cahokia, near present day St. Louis, c.300 miles south of Aztalan.

Dragons and Serpents are associated with the midwestern prehistoric copper trade, as documented by Frank Joseph in his books (Refs.18,19). For example, near Rock Lake (three miles west of Aztalan) is Lake Sinissippi. "The word, in Algonkian dialect, means "Serpent Ore" ... there was a colossal earthwork, called the Sinissippi Cross at the south end of the lake. Four hundred and twenty feet across ... the arms were constructed almost precisely at right angles" (Ref.19). "We know it was oriented to the four cardinal directions, its designers having used the North Star... Hydrological history charts ... confirmed that Lake Sinissippi lay at the precise midpoint between Isle Royale and Cahokia for anyone traveling between these two locations via the rivers that connected them (Ref.18). (Monk's Mound, at Cahokia Mounds State Park, Illinois, one of Cahokia's hundred or so mounds, was 100 feet high, and covered 14 acres (5.7 hectares.) "The caduceus ... was the symbol of the Pelasgians, those pre-Greek Sea Peoples identified as Bronze Age voyagers from Atlantis. Hesiod, the early mythographer of Greek prehistory, wrote that "the Pelasgians claimed their birth from Ophion, the great primeval serpent" (Ref.18). According to Algonkin legends, the great serpent always came up the Mississippi searching for copper, for which it had an enormous appetite. ...A gargantuan effigy mound skirted the south shore of Rock Lake ... described by the Wisconsin Archaeologist Magazine as a "water spirit", the 84 foot-long earthwork portrayed some monstrous being... Before it, too, was obliterated, its erect head was 10 feet wide rising from a 24 foot-long body with a 60 foot-long tail... On the mythic level, the Menomonie (Indians) told of Michibissy, a deadly lake monster installed by the ancient miners to guard the waters around Isle Royal in the Upper Great Lakes for the protection of its rich copper deposits. The creature is so strongly identified with these prehistoric civilizers that Michibissy represents the surviving spiritual energies of the dead "marine men", who sank back into the sea whence they came" (Ref.19).

Frank Joseph is one of the first people in the modern world to describe some petroglyphs as "rock maps": "One was found on Beaver Island, in upper Lake Michigan, an archaeologically rich site. Here monolithic observatories and sophisticated garden beds have been excavated. The Beaver Island petroglyph strongly resembles the configuration of lower Michigan itself. Another rock map of Michigan's lower peninsula was found on the Lake Huron shore opposite Drummond Island. Still another representation of

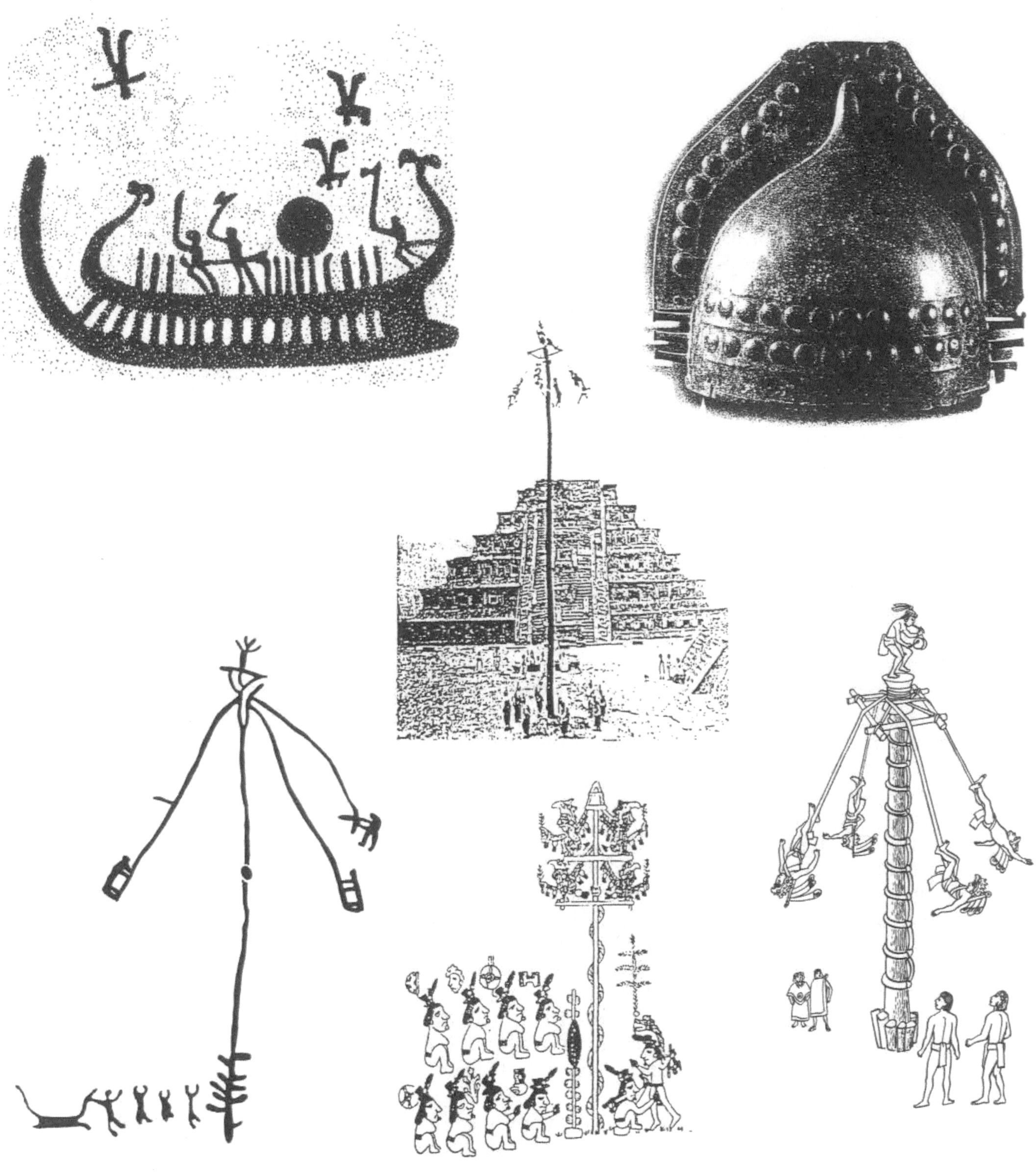

**Fig.14** **Above left:** "Sunboat" with three warriors (Dingle,Bohuslan. Sweden, c.800-700 BC). **Above right:** Copper helmet from South France, with 23 burls on top corresponding with the Tropic of Cancer at 23°N, so this is a helmet for a soldier of the Sun-King! (Musees de Nice/Lauros-Giraudon, South France, c.800 BC, Ref.32). **Below, left:** "Tree of Voladores" flywheel game-ceremony with men dressed as birds spinning down on ropes. This petroglyph reveals contact with Central America (Gerum, Tanum, Bohuslan, Sweden, c.1000-750 BC, Refs.6,40). **Below, center & right:** three Mexican images, including the drawing by Beltran (Refs.33,34,38).

Michigan was retrieved in 1989 from the waters off Keweenaw Peninsula.

The map is shaped from a rectangular four-ton copper nugget ten feet long, four feet wide, and twenty inches thick. Underscoring its supposed function as a map is the specimen's smooth surface and its obvious resemblance to the very place in which it was found. This copper billboard appears to have fallen overboard during transport between 3000 BC and 1200 BC, when mining flourished along the Keweenaw Peninsula" (Ref.18).

Warren Dexter and Donna Martin, in their excellent out-of-print book (Ref.22) discuss the copper shipping site at Poverty Point, La. (c.1600 BC): "Mitchell M. Hillman, an Assistant Curator for the Louisiana Office of State Parks, has found quite a number of spots of copper on the surface both north and south of Poverty Point, for a distance of five to fifteen miles, on both sides of the river. He found no copper below the surface, only on the ground surface. There are no deposits of copper in this region naturally (Ref.22). Old World sailors who reached South America via the Southern Crossing called the big archipelago on their way to the Mississippi the "Antilles", or copper islands, though today we have forgotten why this was so.

Classical writings need be re-read in light of this new perspective. Book III from the 40 Volume Library of History of Diodorus Siculus (Agyrium, Sicily, 80 BC-20 BC), for example says that the "Atlanteans, dwelling as they do on the edge of the Ocean ... subdued the ... regions to the west and the north ... introduced the year as the basis of the movement of the sun and the months. ... The letters, as a group, are called "Phoenician" because they were brought to the Greeks from the Phoenicians, but as single letters the Pelasgians were the first to make use of the transferred characters, and so they were called 'Pelasgic'. ...The gods, they say, were born among them ... Oceanus source of the gods ... Atlas was the first to publish to mankind the doctrine of the sphere (spherical geometry) ... Chronos gave the name Crete after his wife ... the Titans had assembled their united forces together and had crossed over to Crete to attack Ammon ... and a great war had arisen ... In a great battle which followed Dionysus was victorious and slew all the Titans." (Ref.43)

# References


1. Strong, R., "Did Glooskap Kill the Dragon on the Kennebec?", NEARA Journal, Vol. XXXII, no.1 (pg. 38)
2. Internet site www.neara.org
3. Bailey, J., Sailing to Paradise, Simon & Schuster, 1994 (pgs.31,51,147)
4. Fell, B., America BC, Pocket Books, Simon & Schuster, 1994 (pg.128)
5. Fell, B., Bronze Age America, Little, Brown; Boston, 1982 (pg.188)
6. Coles, J., Images of the Past, A Guide to the Rock Carvings of Northern Bohuslan, Bohuslans Museum, 1990 (ISBN 91-7686-110-4)
7. Drier, R.W., and Temple, O.J., du, Prehistoric Copper Mining in the Lake Superior Region, Calumet, Michigan, 1965
8. Zapp, I. and Erikson, G., Atlantis in America, Navigators of the Ancient World, Adventures Unlimited Press, 1998 (ISBN 0-932813-52-6)
9. Nicolar, J., The Life and Traditions of the Red Man, Saint Annes Point Press, Fredericton, N.B., 1979
10. Peterson, F.A., Ancient Mexico, 1959
11. Stuart, G.E., "New Light on the Olmec", National Geographic, Nov, 1993
12. Bernal, I., The Olmec World, University of California Press, London, 1969 (ISBN 0-520-02891-0)

13. Jonge, R. M. de, and IJzereef, G.F., De Stenen Spreken, Kosmos Z&K, Utrecht/Antwerpen, 1996 (ISBN 90-215-2846-0) (Dutch)
14. Ferryn, P., "5000 Years Before Our Era: The Red Men of the North Atlantic", NEARA Journal, Vol. XXXI, No. 2, pg.59 (1997)
15. Brochure, Raft Maine, an Association of Professional Whitewater Outfitters, PO Box 3, Bethel, Me. 04217; www.raftmaine.com
16. Carter, G., Ellsworth American, "Before Columbus" (Ref.1)
17. Fell, B., Saga America, Times Bks, 1980 (ISBN 0-8129-6324-5) (p.142)
18. Joseph, F., Atlantis in Wisconsin, New Revelations About Lost Sunken City, Glade Press, Lakeville, Mn.,1998 (ISBN 1-880090-12-0)
19. Joseph, F., The Lost Pyramids of Rock Lake, Glade Press, St.Paul, Minn., 1992 (ISBN 1-880090-04-X)
20. Cahill, R., New England's Ancient Mysteries, Old Saltbox Publishing, Salem, Mass., 1993 (ISBN 0-9626162-4-9)
21. "The History of the Kennebec", menu publ. by Kennebec Tavern, 2000
22. Dexter, W., and Martin, D., America's Ancient Stone Relics, Vermont's Link to Bronze Age Mariners, Academy Books, Rutland, Vt. 1995 (ISBN 1-56715-050-0)
23. Drews, R., The End of the Bronze Age, Changes in Warefare and the Catastrophe ca.1200 BC, Princeton University Press, Princeton,, NJ, 1993 (ISBN 0-691-04811-8)
24. Jewell, R., Ancient Mines of Kitchi-Gummi, Jewell Histories, Fairfield, Pa., 2000, (ISBN 0-9678413-0-5)
25. Stapler, W., "Ancient Pemaquid and the Skeleton in Armor", NEARA Journal, Summer 1998, Vol XXXII, No.1, pg.35
26. Mertz, H., Atlantis, Dwelling Place of the Gods, (ISBN 0-9600952-3-3)
27. Canadian Maritime Chart Index of the Trent/Severn Waterway, 2000
28. Tompkins, P., Secrets of the Great Pyramid, Harper Colophon Books, Harper & Row, N. Y., 1971 (ISBN 0-06-090-631-6)/L.C. Stecchini
29. Carlson, S., www.neara.org/CARLSON/
30. Bayley, H., The Lost Language of Symbolism, Citadel Press (ISBN 0-8065-1100-1) (pg.311)
31. Mallery, A.H., and Harrison, M.R., The Rediscovery of Lost America, The Story of the Pre-Columbian Iron Age in America, Dutton, NY, 1979 (ISBN 0-525-47545-1)
32. People of the Stone Age: Hunter-Gatherers and Early Farmers, Weldon Owen Pty Lim., Australia, 1994
33. Evers, D., Magie der Bilder, Praehist. Skand. Felsgravuren, Pulsar Verlag, Warmsroth, 1995 (ISBN 3-9802716-6-8) (German)
34. Clavijero, F.J., Los Voladores, Mexikanische Jesuiten Geschichte, 18 Jh (German)
35. Hitching, F., Earth Magic, Casell, London (1976)
36. Hawkes, J., The Atlas of Early Man, St. Martin's Press, 1993 (ISBN 0-312-09746-8)
37. Donnelly, I., Atlantis, Harper & Row, San Francisco, 1971 (ISBN 0-06- 061960-0) (pgs.246,441)
38. Beltran, A., Ancient Sun Kingdoms of the Americas, World Publishing Company, New York, 1957
39. Hygen, A.S., and Bengtsson, L., Rock Carvings in the Borderlands (Bohuslan and Ostfold), Warne Forlag, 2000 (ISBN 91-86425-02-1)
40. Evers, D., Die Ware Entdekkung Americas, Weissbach, 2000 (ISBN 3-930036-45-2) (German)
41. Twohig, E. Shee, The Megalithic Art of Western Europe, Clarendon Press, Oxford, 1981
42. Art et Symboles du Megalithisme Europeen, Revue Archeologique de L'Ouest no 8, Nantes, 1995 (ISSN 0767-709-X) (French)
43. Siculus, D., Books II, 35-IV, 58, Loeb Classical Library, Harvard University Press, London 1935, reprint 1994 (ISBN 0-674-99334-9)
44. Trento, S., Field Guide to Eastern North America, Henry Holt & Co., New York, 1997 (ISBN 0-8050-4449-3)
45. Vastokas, J.M., and Vastokas, R.K., The Sacred Art of the Algonkians, A Study of the Peterborough Petroglyphs, Mansard Press, Peterborough, 1973

# The Orient Tablet

## An Egyptian Expedition to America

(Eagle Neck, Orient, Long Island, N.Y., c.850 BC)

Dr. R.M. de Jonge, drsrmdejonge@hotmail.com
J.S. Wakefield, jayswakefield@yahoo.com

### Introduction

In about 1888 an inscribed tablet was found in a shell midden at Eagle Neck, Orient, at the eastern tip of Long Island, NY (Figs.1&2, Ref.1). Besides unknown symbols, it showed images of a man, a boat, a mammal, a tree and a bird. For that reason it was supposed to be a collection of Indian petroglyphs recording a hunting trip. That is why it presently belongs to the Museum of the American Indian in New York (cat.#09/8603, Ref.34).

In 1975 Barry Fell started studying the inscriptions (Ref.2). It turned out to be a bilingual tablet. One side, shown in Fig.1, is inscribed with old Libyan script, and the other side, shown in Fig.2, is carved with old Egyptian script, in the "hieratic", or "informal" style. Because of the unhealthy archaeological climate in the United States, Fell discusses the possibility of a forgery, which is very unusual outside the American continent. Scientifically speaking this discussion is correct, of course, and his argument is very convincing: "It has never been considered spurious and, indeed, the Libyan inscription (Fig.1), which matches the Egyptian (Fig.2), could not have been forged for old Libyan was not deciphered until 1973". In his famous book America BC (Ref.1), Barry Fell gave a decipherment of the majority of the symbols of both sides of the tablet. We will complete the translation here, and we will explain the geographic and religious meanings of this tablet.

### The Libyan Side

We numbered the Libyan characters in Fig.1 as Fell deciphered them (Refs.1,2), in the right order 1 to 14. The phonetic rendering, largely in the Libyan language, is: "Ta(1) sailing ship(2) d(3)-p(4)-t(5) T(6)-m(7)-i(8)-r(9) W(10)-a(11)-s(12) Empire(13) Bull-god Horus(14)". This rendering is the same as that of Fell, except for "sailing ship(2)", "Empire(13)", and "Bull-god Horus(14)".

Inscription 2 is not a Libyan character for "ship" ("d-p-t"), but just a drawing of a boat. Inscription 1 is carved above the boat as if a sail, so the right translation of inscription 2 will be "sailing ship". Inscription 13 is not a Libyan character either, but the Egyptian (!) hieroglyph for "Dominions" (Ref.1). In other words, these appear to be a special kind of "dominions". In view of the meaning of inscription 14, we will see that the translation of "Empire" is much better. Inscription 14 is not a drawing of a bear, as suggested by Fell, but an image of a bull. Bears do not have such long tails, and the figure shows similarity with horned cattle, while the head suggests a male.

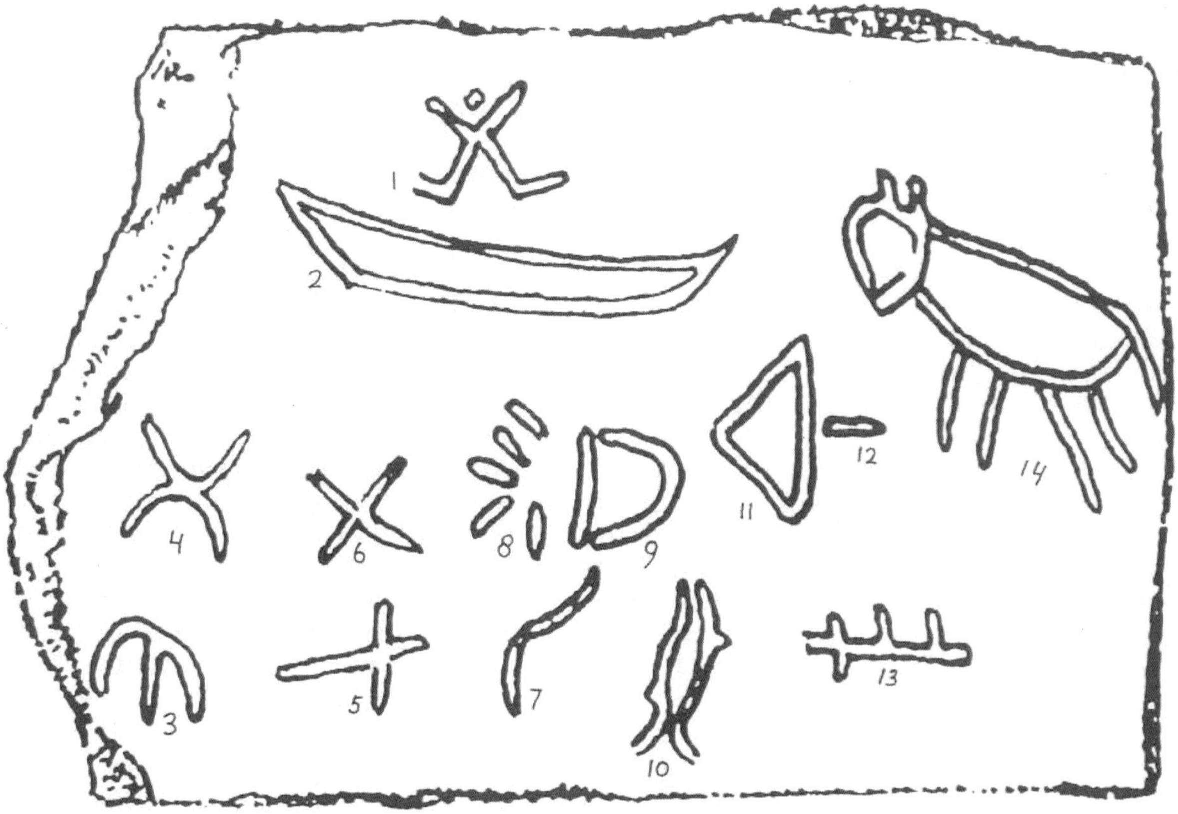


**Fig. 1** The Libyan text on the front side of the Orient Tablet from Long Island, NY, c.850 BC (Refs.1-3): "This sailing ship is a vessel from the Egyptian Empire of the Bull-god Horus". The 14 characters mean: Cape Verde at 14°N. (The religious meaning is: "We belong to the empire of our king, substitute of the god Horus.") (Museum of the American Indian, New York)

Because of the two languages of the tablet, it dates from the 22nd, 23rd, or 24th (Libyan) dynasty of Egypt between 950 and 715 BC. According to Fell, it was probably the 22nd dynasty (c.850 BC), "at which epoch it is likely that Libyan Pharaohs would encourage long distance voyages". The most famous Pharaoh in the preceeding era was Ramses II ("the Great"), 1301-1235 BC, from the 19th dynasty. The first of his five official titles was: "Horus, strong bull, loved by the goddess of the law" (Ref.4). Similar titles were often used during several centuries, often much longer, without any change.

Fell's translation of the Libyan text was: "This ship is a vessel from the Egyptian Dominions". In light of the above, a better translation of the Libyan side of the tablet is: "This (1) sailing ship (2) is a vessel (3,4,5) from (6,7,8,9) the Egyptian (10,11,12) Empire (13) of the Bull-god Horus (14)", or, without numbers: **"This sailing ship is a vessel from the Egyptian Empire of the Bull-god Horus".**

The bull, being the symbol of the strong, powerful king, can also be seen on the tablet of King Narmer (Menes) from Hierakonpolis (Fig.3), pharaoh of the 1st dynasty of Egypt (c.3000 BC) (Ref.4,12). This famous tablet, now in the Egyptian Museum in Cairo, has beautiful inscriptions in so-called haut-relief. A sculpture of a sacred bull of the Iberian Celts was found in the Lawrence area of the Merrimac River valley, Massachusetts, only c.200km from Long Island (Ref.1). It was actually thought to have been a Bison. It probably dates from about the same time period as the Orient Tablet.

## The Egyptian Side

We numbered the Egyptian hieratic characters or group of characters in Fig.2, as Fell deciphered them (Refs.1,2), in the right order 1 to 9. The phonetic rendering of the hieroglyphs, largely in the Egyptian language, is: "Isw-w-t(1) Rsw-t(2) i-iri(3) oho(4) r-m(5) mso(6) R-n-p(7) Sea-Fowl-god Osiris(8) W(9)". This rendering is the same as Fell, except for the inscriptions R-n-p(7), Sea Fowl-god Osiris(8), and W(9).

Inscription 7 was omitted by Fell, but the meaning of the symbol (which looks like the new growth of a plant) is given in "America BC" six pages before, in his discussion of the Davenport Stele from Iowa (see Ref.1, pg 264). It simply means R-n-p in Egyptian, or "New" in English. According to Fell, the presence of inscription 8, the sea bird, is "incongruous", as is the "bear" on the other side. However, both figures are engraved in the same style, without doubt by the same people at the same time. Fell should have considered the "incongruous animals" as religious ideograms! Now we have seen that carving 14 on the other side is a bull, representing the god Horus, so we should expect the animal of inscription 8 to be the other important god Osiris (Ref.5). It may be true that Osiris is not commonly depicted as a sea fowl in Egypt, but it is known that this popular god is pictured in many different ways. He is also known as the god of the sea, and according to a myth, he drowned and washed ashore on the coast of a distant land (Ref.4). Outside Egypt, Osiris and related gods are often pictured as birds or

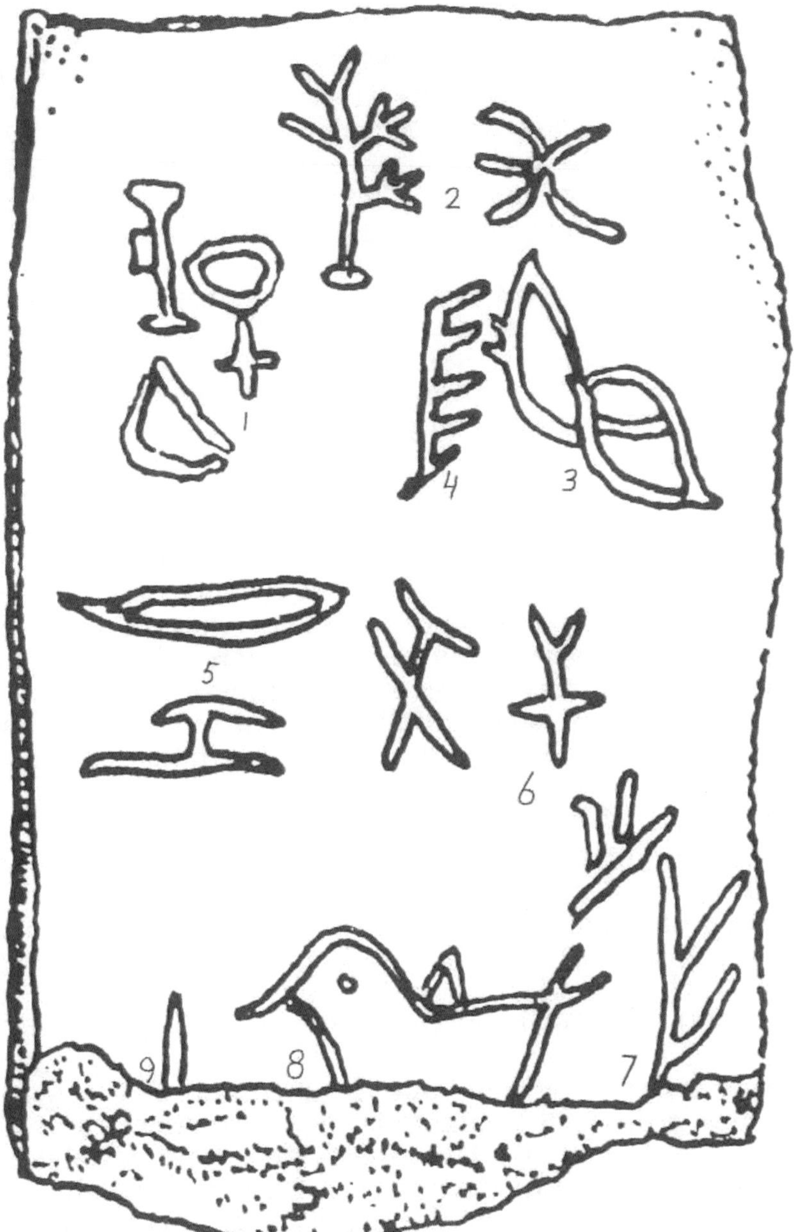

**Fig.2** The Egyptian text on the back side of the Orient Tablet from Long Island, NY, c.850 BC (Refs.1-3): "A ship's crew from Upper Egypt made this tablet because of their expedition to the New Land of the Sea Fowl-god Osiris (America)". The 16 characters mean: Honduras at 16°N. (The religious meaning is: "We made this tablet, because we are heading for the empire of our new king, substitute of the god Osiris." (Museum of the American Indian, New York).

birdmen, notably in Africa, Central and South America, and on the islands of the Pacific (Refs.9,20,21).

One edge of the tablet is damaged, and as a result some inscriptions on the Egyptian side (Fig.2) are barely visible. Inscription 9 is also omitted by Fell, but we think that the shown part is enough for a rendering. Again, the meaning of the symbol is mentioned in Fell's own book America BC, five pages before, also in the discussion of the Davenport Stele from Iowa (see Ref.1, pg.265). Is simply means "W" in Egyptian, or "Region" in English, which, in the context of the sentence, should be replaced by "Land" (of one of the "Dominions" of Egypt; see the Egyptian(!) inscription 13 at the other side).

Fell's translation of the Egyptian hieratic text was: "A ship's crew from Upper Egypt made this stele in respect of their expedition". We think the best translation of this is: "A ship's crew (1) from Upper Egypt (2) made (3) this tablet (4), because of (5) their expedition (6) to the New (7) Sea Fowl-god Osiris (8) Land (9)", or: **"A ship's crew from Upper Egypt made this tablet, because of their expedition to the New Land of the Sea Fowl-god Osiris".**

In protodynastic Egypt another bird, the falcon, often symbolized the main sky god (Fig.3, Ref.9). Many birds are seen as characters in the hieroglyphic writing in Egypt, for instance in the text accompanying the seagoing vessels arriving for the first time in Punt (Central America), in the year 2497 BC, which is in the 13th year of government of King Sahura, depicted on his pyramid at Abusir (Ref.4,13). The five titles of Ramses II (c.1250 BC), discussed earlier, contain six hieroglyphs of birds out of a total of 38 (Ref.4 ). On the cover and inside his book Sailing to Paradise, Jim Bailey shows a picture of a Greek boat of the early Iron Age, c.750 BC, only a century after the Orient Tablet, showing a symbol of the sacred bird on the prow (Ref.9, see also Figs.4,5,7).

## The Geographic Meaning

Let us look again at the complete text of the Orient Tablet: **"This sailing ship is a vessel from the Egyptian Empire of the Bull-god Horus". "A ship's crew from Upper Egypt made this tablet, because of their expedition to the New Land of the Sea Fowl-god Osiris".** The Egyptian text is clearly a continuation of the Libyan text, and not the other way around, so the Libyan text with the bull is the front side of the tablet, and the Egyptian text, with the bird, is the back side. "The New Land of the Sea Fowl-god Osiris" is the New World, now called America. In this time period, centers of civilization in Central America were by far the most important (Refs.6-8,10,15,16). It is highly probable that these sailors arrived in America via the Southern Crossing, on the trade winds between Africa and South America (Ref.5). (Again, it is clear that the translation "sailing ship" on the front side is better than Fell's translation "ship".) Next, they sailed along the north coast of South America to Central America. They could have returned from the Bahamas or Cape

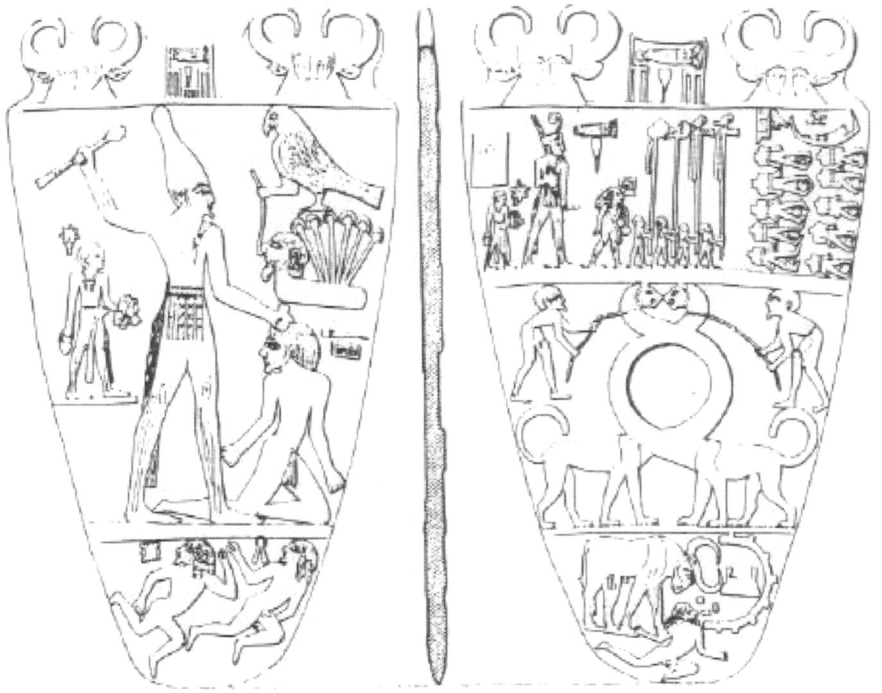

**Fig.3** **Above:** The Narmer Palette, commemorating the reign of the first king of united Egypt. On top of both sides, the heads of sacred cattle. In the lowest register at the back side (right), the conquering power of the king, symbolized by a bull (Egyptian Museum, Cairo, c.3000 BC, Ref.30).
**Below:** The sacred Apis bull with the sundisc between the horns, symbolizing the Sunreligion. (British Museum, London, c.1000 BC, Ref.31).

Hatteras via Bermuda to the Azores. But this tablet is found in Long Island, just south of Cape Cod and America's Stonehenge (Refs.17-19,40). Apparently they decided to take the opportunity to pay a visit to New England, the Celtic center of North America, before returning home, probably from Cape Sable via Sable Island to the Azores.

We have shown (Ref.5) that Bronze Age people encoded the latitudes of places important to them in their petroglyphs and monuments. Although it is already the Iron Age (c.850 BC, Ref.14), note that there are 14 characters on the Libyan side with the bull, which encode the latitude of Cape Verde, the westernmost point of continental land of the Old World at 14°N. On the back side with the bird, there are 3+2+2+1+2+3+1+1+1 =16 characters, corresponding to the latitude of the culture along the north coast of Honduras at 16°N (Refs.5,9,15,16). There are 30 characters (14+16 =30) in total, corresponding to the latitude of the Nile Delta, the center of Lower Egypt, the site of the beginning and end of this sea voyage at 30°N! The latitudes confirm that the Libyan text is the front side, and the Egyptian text is the back side, because the ship first left the Old World at 14°N, and landed in the New World at 16°N. Note also, that there are 9 groups of characters at the back side (Fig.2), equal to 9 words in the Egyptian text, corresponding to the 9 islands of the Azores which are so important for the return route to the Old World.

The Orient Tablet is rectangular, which is a logical shape to use. However, the makers of the tablet had to choose a certain length/width ratio. Probably, the (original) length of the tablet corresponded to the distance from Cape Verde to the north coast of Honduras (Cape Gracias a Dios), which is 65dl =65°of latitude =65x111 =7222km. The width of the tablet probably corresponded to the distance from the equator to Long Island, where the tablet was found, which is 41°of latitude= 41x111=4555km. So, the original length/width ratio of this tablet probably was 7222/4555 =1.59, which is equal, or at least very close, to the present ratio.

The last three characters of the back side of the tablet deal with the "New Land of the Sea-Fowl-god Osiris" which is America. In general, birds are very popular among sea-farers, and it is thought there were many more of them on the water prior to the mechanized fishing of today. Prehistoric boats often show heads of birds on the prow (Figs.4,5,7). According to Jim Bailey, (Ref.9): "the Celts ... and before them the megalith builders, must have followed the skeins of duck and geese in Spring, when they migrated to their breeding grounds in Iceland, and the geese could have led them on to Greenland ... Even some South African birds nest each year in Greenland. Birds can also be a sure sign of the closeness and direction of land. We know that (as early as) the Sumerians (mariners) took birds with them to release when they were lost at sea: if they were beyond reach of land, the birds would come back to the ship, as in the story of Noah in the Bible."

Doubtless there were great numbers of sea-birds in the extensive marshes of

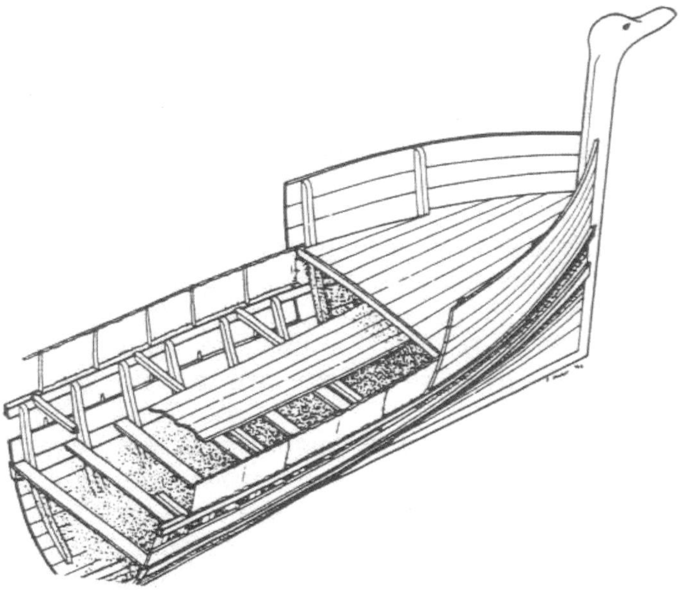

**Fig.4** **Top:** Reconstruction of a Sea People's ship, with bird head on the prow (c.1200 BC, Ref.28). **Below:** Petroglyph of the defeat of the Sea People by Ramses III (Medinet Habu, Egypt, 1176 BC, Ref.28).

Long Island, New York, which would suggest use of the bird ideogram for the Sea-god Osiris. The bird on the Egyptian side of the tablet resembles an American duck, the "European Widgeon" more than any other. According to "The Birds of North America, a Guide to Field Identification" (Ref.11): "This Old World duck is a regular fall visitor to the northern coasts of North America, though never in large numbers". It is probable that the group of Egyptian visitors identified with this bird, and also with Osiris reaching the land of the dead, America!

## The religious meaning

On the front side with the bull, Fig.1, it is striking that inscription 2 is not a Libyan character, but a drawing of a boat. It also might be seen as a double underlining of inscription 1, which is slightly different from inscription 6, both however with the same Libyan meaning: "t(a)" (Ref.1). Inscription 1 is not merely a standard Libyan character, because a small head and two feet are added to it. It is a drawing of a man, meaning a human, or the crew of the boat, as well as a sail. Let us try to formulate a "religious translation" of the tablet, omitting all details which are unimportant for this purpose. If 1 and 2 are translated as "this man", "the crew", or "we", then 3, 4, and 5 should not be translated by "is a vessel", but rather by "is a tool" or "belong", respectively. If we mention "Horus", we don't have to mention "Egyptian", because originally Horus was an Egyptian god. According to the old Egyptian religion, the divine kings were the substitutes of the gods Horus and Osiris. It is the old known sun/moon, Horus/Osiris, bull/bird, father/son mythology. In view of this perspective, the best "religious translation" of the front side, which is different from the literal one, is: **"We belong to the empire of our king, substitute of the god Horus".**

Now we look at the religious meaning of the back side, with the bird (Fig.2). The interpretation of character 9, meaning "region" or "land" should be changed. Compare it again with inscription 13 on the front side, which was an Egyptian hieroglyph (meaning "empire") between Libyan characters. Actually, it may be translated by the more religious "empire" too. Because of the inscriptions 1 and 2 of the front side (now meaning "we"), and the fact that Osiris is an Egyptian god, the translation of "A ship's crew from Upper Egypt" (groups 1&2) can simply be replaced by "We", too. If we replace the geographic word of "expedition" by a more general expression, we arrive at the best "religious translation" of the back side: **"We made this tablet, because we are heading for the empire of our new king, substitute of the god Osiris.**

Again it is clear, that the second Egyptian text with the bird is a continuation of the first Libyan text with the bull, and not the other way around. It confirms that the Libyan text is the front side, and the Egyptian text is the back side.

On the back side of the tablet it is emphasized, that the ship's crew comes from Upper Egypt (character groups 1&2, Fig.2). The front side contains 7

**Fig.5** **Top, left:** Sketch of Olmec pyramid of La Venta, the earliest mound in the Americas, with its fluted contours. **Top, Right:** "Undulating contours of the Mountain of the Far West with the Solar Eagle from a Ramessid Tomb" (Ref.22). Note the 3 gods corresponding with the 3 island groups of the Azores, pointing to 9 rectangles, corresponding with the 9 islands of the Azores. **Below, right:** The Bull-god Horus (right) and the Sea Fowl-god Osiris as prows on the sacred boat Phre (Egypt, c.1000 BC, Ref.27). **Below, left:** Duck-headed papyrus raft inscription from tomb of Ramses II (c.1250 BC, Metropolitan Museum of Art, New York , Ref.28).

groups of characters, equal to 7 Libyan words. Together with the 16 characters of the back side it forms 7+16= 23 units, corresponding to the latitude of the Holy Tropic of Cancer, and the center of the Sunreligion in Upper Egypt, at 23°N. The back side contains 9 groups of characters, equal to 9 Egyptian words. Together with the 14 characters of the front side it forms 9+14= 23 units, which confirms this important latitude. The front side contains 7 and the backside contains 9 groups of characters, together 7+9= 16 units, confirming the latitude of the north coast of Honduras at 16°N. It was the main travel goal, to discuss the Sunreligion at this latitude.

The new meaning of the whole tablet is focused strongly on religion and government. The knowledge and experience of the Egyptians in these issues, which were millennia old, were brought to America by the makers of this tablet, and in particular to Central America, the 16°N region of coastal Guatemala, southern Belize, and northern Honduras (Refs.6-8,15,16). Note that they want to emphasize that these two "empires" are completely different, like day and night. The story about the first empire is written on the front side of the tablet (Fig.1), in Libyan language, from bottom to top (going from left to right), with the bull-god looking along the long side of the rectangle. The story about the second empire is written on the backside (Fig.2), in Egyptian, hieratic language, from left to right and back (in boustrophedon, meaning to be read "as a plowman walks"), going from top to bottom, and the Sea Fowl-god looks along the short side of the rectangle. The two sides of the tablet should be compared with the two volumes of the Bible, which is the oldest book in the world: The Old Testament and the New Testament. The religious text of the Orient Tablet was the reason for their expedition, and the subject of their discussions with the leaders in Honduras and in New England!

Note that this religious meaning has been described in a geographic way. The Old World is considered to be the Old Egyptian Empire of the Bull-god Horus, and the New World is considered to be the New Egyptian Empire of the Sea Fowl-god Osiris (America). Between these empires is a huge ocean (the Atlantic), and one empire is at one side of the planet, and the other empire is at the other side of the planet. The reigning Pharaoh was the living Horus in Egypt, and Osiris ruled the hereafter, in the land in the west: "The realm of the dead is in the west, at the other side of the waters, in the land where the sun sets". When we read the text of the tablet, it appears that the main goal of the visit was not trade, but political and religious consultation.

The second character of group 6 of the Egyptian side of the tablet (Fig.2), which is part of the word "expedition", differs from the hieratic style of Egypt, but matches the hieroglyph still used in modern Micmac hieroglyphic writing in New England (Refs.1,2). The Micmac, a tribe of the Algonquian Indians in the northeast of North America, had learned hieroglyphics from Egyptian visitors, and probably interbred with them. For this reason, Fell thought the Orient Tablet might be an Algonquian copy of

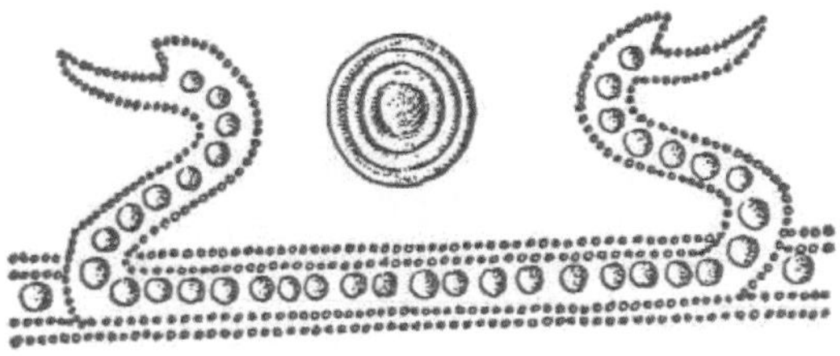

**Fig.6 (Top, Right):** Birdman of Repousse Copper (pounded into a sheet of copper), found in a mound in Georgia, U.S; probably 1100-1300 AD. Mahan reports the conclusion that these "eagle men composites" represented mythological cult-bringers who taught esoteric knowledge of the fire-sun diety religion". (Refs. 33,44).

**Fig.7 (Bottom, Right):** Boat painted on a krater sherd from Tiryns (Greece), with bird heads on both ends of the vessel. In the background may be the 3 island groups of the Azores with megalithic-style distance lines (c.1200 BC, Ref.28). **(Top, Left):** Bronze Bird-Ship with 3-ring Sun symbol and 38 decorations celebrating the 38° latitude of the Azores (Rossin, Pomerania, "Late Bronze Age", Ref.43).

**Fig.8 (Bottom, Left):** Sketch of petroglyph called the "Queen of Punt" (from the temple of Queen Hatsjepsut, c.1500 BC, Der-el-Bari, Egypt, Egyptian Museum, Cairo, Ref.36). This male Central American with a SunGod necklace was probably carved in Egypt from a living person who returned to Egypt. Though Quetzalcoatl has been described as corpulent, and wearing a long robe (Ref.39), he is also described as tall, charismatic and bearded. We may never know for sure who this is.

the original! We do not agree with this idea. We think this unique tablet was an official gift of the Egyptian ship's crew to the religious and political leaders of the local people in New England, done near their departure from America. They wrote the word "expedition" in the Micmac-way, to emphasize that their voyage was not only meant for Honduras at 16°N (the 16 words on the tablet, and the 16 characters on this side of it), but also for their friendly hosts of New England at 41°N (the width of the tablet from its width/length ratio). We think that in the long course of time, this beautiful tablet was stolen, lost, or forgotten about, to end up in the shell midden.

The idea to make a tablet with a geographic and religious text, in which the number of characters correspond to important latitudes, is a continuation of the old megalithic tradition, in which passage graves and other structures were built with the number of menhirs encoding important latitudes. The megalithic tradition ended in Europe between 2000 and 1500 BC, in North America, probably 1000 BC. The Orient Tablet is a late example of stonework with megalithic characteristics. It is our conclusion that around the time this tablet was made, c.850 BC, the meaning of most megalithic monuments around the Atlantic were still understood.

## The Bull-god Horus

Horus was the "Lord of the Sky", and symbol of divine kingship. The pharoah was seen as a manifestation of the "living Horus" on the throne of Egypt. The power of the pharoah was symbolized by a falcon, and by a bull named "Apis". The Apis bull (Fig.3) is black except for a small white triangular patch on his forehead. Between its horns it carries the emblem of the Sun Disk, symbol of the Sunreligion. On it back are the protective wings of the vulture goddess. The pharoah identified closely with the Apis-bull imagery, with its inherent notion of strength and fertility, being an ancient characteristic of the propaganda of the God-King. This is seen in the carved slate palettes (Fig.3), and in one of the mentioned names used in royal protocol, "Victorious Bull". Celebrating his Jubilee Festival, a ceremony concerned with the rejuvenation of the monarch's powers, the pharoah strides briskly alongside the galloping Apis bull. On the death of Apis, Egypt mourned as if for the loss of the monarch himself. The innate power of the bull, its strength and its verility, was conveyed by means of a bull's tail, worn by the king suspended from the back of his kilt, a regular component of his royal dress (Ref.32). According to the Egyptian history writer Manetho, the Apis bull was already officially recognised during the 2nd dynasty of Egypt (c.2900 BC), together with the cult of the heliopolic Memphis bull and the goat of Mendes (Ref.4). The greatest sanctuaries of the Apis bull are the Serapeums in Sakkara and in Alexandria. The well-known petroglyphs of the ritual sacrifice of bulls of the Egyptian medical doctor and writer Achanecht date from the time of the Old Kingdom (c.2780-2263 BC) (Ref.4).

The Palace of Knossos of King Minos on Crete (Greece) has a famous large fresco of a sacred bull, involved in a leaping scene (c.1450 BC, Ref.12). Two red painted sacred bulls of earthenware of the weather-god Hatti (c.1m high, c.1400 BC) were found in the royal palace of the Hittites in their capital Hattusa (Bogazkale) in Anatolia, 160km east of the present capital of Turkey, Ankara (Ref.12). King Assoernasirpal II of Assyria, who reigned from 884 until 859 BC (contemporary with the Egyptian expedition of the Orient Tablet), built a huge palace in the capital Nimrod with two enormous winged bulls, symbolising their main god (Ref.10).

## The Sea Fowl-god Osiris

In Old Kingdom theology, once the ruler of Egypt died, he became Osiris, King of Duat, Ruler of the Underworld in the West. "In the law court Osiris sits on his throne holding the sceptres and supervises the judgement of the new applicants for paradise" (Ref.31). While he is depicted in human form in Egypt, birds flying between primordial waters and earth and heaven and, symbolically between life and death, provided a means for synthesis of myths around the world (see birdman, Fig.6). Figures 4,5,&7 show several examples of bird heads (ducks) on the prows of ancient boats from the Mediterranean Sea.

In his book Sailing to Paradise (Ref.9) author Jim Bailey did a lot of research into the meaning of sacred birds and birdmen in Africa. In the museum of the Ivory Coast capital Abidjan, a casting shows that men wearing bird masks were the first ancestors. The bird masks represent the celestial powers of God in heaven. Other examples are old statuettes of two bronze birdmen from the Senufo tribal area. They are based on an ancient tradition, which is that in the days of the tribe's first ancestors, a race of attractive human birds appeared, possessing all the sciences, which they handed on to mankind. An old wooden statue of the black hornbill from the south of the country shows this divine bird, carrying a disc instead of wings. It has 23 spirals, corresponding to the latitude of the holy Tropic of Cancer, 23°N. Pictures of birdmen occur in various parts of Africa, including the east side, where there was a lot of mining in the prehistorical period. In the ruins of Zimbabwe, these figures carry at their bases circles and water signs which are associated with sea-going peoples. The tradition of the sacred bird and sacred birdman is Bronze Age and Copper Age in origin.

Jairazbhoy (Ref.22) shows photos of a ceramic Olmec duck basin from San Lorenzo, Mexico, incised on the front with a duck with open mouth flying up from a pool, as though it were flying out of the waters of chaos. The Egyptian Book of the Dead describes such a bird, and this bird is also found right next to the sun's boat in an Egyptian temple. It is the "Great Cackler", born of the Primaeval waters. Osiris-Keb is the form of Osiris fused with this Goose-God, who produced the Cosmic Egg. This first event in creation was commemorated in one of the later Mexican festivals in which men jumped into the water and began to cackle like ducks and other birds.

Tortora reed boats with birdmen in attendance are painted on Michicha pottery from the Pacific Coast of Peru (Ref.9). Thor Heyerdahl (Refs.9,20,21) records that bird masks were worn by the captains of ancient ocean going reed boats of which his own boat Ra was a copy. In the Easter Island village of Orongo, the petroglyphs of sungod masks and birdmen are said to have been part of the powerful birdman cult there. The cult featured an awesome cliff dive and swim competition to an offshore rock, to collect and return intact the year's first seabird egg, and earn the position of "birdman" for the next year, as shown in a PBS Special on US television, and previously described by Heyerdahl.

## The Olmec Civilization

When an Egyptian king died, he joined the SunGod Ra in his journey across the Sky during the day, and through the underworld at night. He was called "the great helmsman, who voyaged over the two parts of heaven", and such a journey was described in quite literal terms. There are many indications (Ref.22), that Ramses III (c.1195-1164 BC) yearned passionately to reach the western horizon. He prepared a large navy, and "sent it forth into the great sea of the inverted water". He then referred to the ships arriving in the land of Punt (Central America) safely, and coming back laden. There were 83 vessels attached to the Temple of the SunGod Ammon-Ra at Thebes under his government. A journey across the sky is also represented in his tomb, this time involving seven ships, and Mexican tradition reports their first ancestors came in seven ships. Egyptian papyrus paintings show the sun's ship sailing inverted across the underside of the sky, and an inverted solar boat is shown on an important Olmec relief, with reed clumps on bow and stern, entering the Underworld.

Jairazbhoy (Ref.22) suggests that the colossal Olmec heads with negroid features were the military governors of the new found Egyptian colony on the Gulf of Mexico. Nowhere else in the world at the time were sculptures of this scale being erected other than Egypt. The Olmec heads wear leather helmets, with parallel lines incised on them, like the Egyptian ones from Tanis (c.1250 BC), and just like others on megalithic statues in Columbia (Ref.41) and on an 18" stone figure from Bolivia (Ref.42).

Here is what the respected authority Father Benardino de Sahagun (Ref.22) says in the days after the Conquistadores: "Concerning the origin of this people, the account which the old people give is that they came by sea from toward the north, and it is certain they came in vessels of wood from toward Florida, and then they came along the coast, and disembarked at the port of Panuco (north of Veracruz), which they call Panco, which means, place where those who crossed the water arrived. These people came looking for paradise, and they settled near the highest mountains they found".

The Olmec heartland in southern Mexico (Refs.26-29), was a hot and humid coastal lowland landscape, whose river levees were flooded each year to a depth of about 7 meters (21 feet). Like Egypt, the Olmec civilization was the "gift of the river", leaving behind a deep, rich soil after each year's flooding, conferring an invaluable asset on an agricultural people. It was land that did not have to be cleared before planting, as if ordained by nature to become a Cradle of Civilization. Excavations at La Venta in the 1940s revealed a stockade around the fluted cone of the Great Pyramid of 600 two-ton basalt columns 10 feet tall, moved from 60 miles away. The stockade and the stele of the caucasian-looking Bearded Man had been buried 3,000 years (Ref.40). The National Museum of Mexico owns a famous cache of 22 Olmec 8" jade and sepentine figurines and celts. The 16 male figurines of this group correspond to the latitude of the north coast of Honduras, at 16 N. Together with the 6 celts, the number corresponds with the northern tip of Yucatan, at 16+6= 22 N. Yale archaeologist Coe (Ref.26) says "the Olmec heartland is still barely touched".

The Olmec settlers are thought to have first arrived c.1900 BC, with the culture developed by 1500 BC, and cities flourishing at 1200 BC. Before the Olmec, there were villages and pottery in Mexico, but not much more. With the Olmecs, all kinds of civilized activity appears, including massive organization of labor, a trade network, ceremonial centers with pyramids, colossal sculpture, relief carving, wall painting, orientation of structures, gods and religious symbolism, an obsession with the Underworld, representations of foreign facial types, hieroglyphic writing and scribes, seals and rings, use of iron, and so on. The Olmec sculpture of a jaguar copulating with a woman, is comparable with a text at Medinat Habu (Egypt), in which the god Ptah-Tatenen says to king Ramses III (c.1200 BC), that he assumed the form of a Ram and cohabited with his mother in order to fashion him. Another possible trace of Egyptian royalty in Olmec times is the Olmec use of a double crown, a falcon over another animal. The double crown, symbolizing the unity of the two parts of the country, was an exclusively Egyptian conception (Upper & Lower Egypt, Ref.22). However, most of what appears in this context is not obviously Egyptian in appearance, or it would have been detected at once. A replica of Egyptian civilization was not transported onto Mexican soil as analysis reveals. The migrants were of various ethnic types, and their journey seems to have been motivated by the purpose of discovering the entrance to the Underworld, which was imagined to be in the Far West of the World (Ref.22). It seems reasonable that a substantial number of people would have arrived in the New World after the catastrophe in the eastern Mediterranean at c.1200 BC (Ref.38).

When the Spanish arrived in the New World, hundreds of hieroglyphic books were tossed into bonfires by ardent missionaries. Only four books have made it to the present day. One of these is the Popol Vuh (Refs.24-27). The lords of Quiche consulted their book when they sat in council, and their name for it was Popol Vuh, meaning "Council Book", also titled "The Light That Came From Across the Sea", and "The Dawn of Life". According to this sacred book: "The first people came from across the sea, from the east. They came here in ancient times, black people, white people, people of many faces, people of many languages. Canopy, throne, gourd of tobacco, bone flute, nosepiece, paint of powdered yellow stone, they brought all these. From across the sea, they also brought back writings. They settled countless mountains, giving them epithets and names". Eventually some "advised their sons, we are going back", and they disappeared across the sea (Ref.24).

## Discussion

Today's sailors would probably write a message similar to the Orient Tablet like this: **"We have come on a sailing vessel from Egypt. We wrote this message to commemorate our expedition to America".** In our efforts to decipher the past, we all should take care not to dismiss petroglyphic messages lightly, as though they were ancient graffiti, as has sometimes been suggested. This petroglyph is a good example: while this modern translation sounds nice to us, it does not include the important religious meanings and geographic encodings that were important to those who inscribed the tablet. We should respect that inscribing these petroglyphic messages into rock was time consuming in planning and execution, and we should not be surprised to find the symbolic meanings more complex than they first appear.

# References


1. Fell, B., America BC, Pocket Books, Simon & Schuster, 1994
2. Fell, B., "Occasional Publications", Vol.3, Epigraphic Society, 1975
3. Young, D. A., Heye Foundation, New York (Drawing)
4. Strelocke, H., Egypte, Geschiedenis, kunst en kultuur in het Nijldal, Cantecleer 1979 (Dutch)
5. Jonge, R.M. de, and IJzereef, G.F., De Stenen Spreken, Kosmos Z&K, Utrecht/Antwerpen, 1996 (ISBN 90-215-2846-0) (Dutch)
6. Peterson, F.A., Ancient Mexico, 1959
7. Stuart, G. E., "New Light on the Olmec", National Geographic, Nov. 1993
8. Bernal, I., The Olmec World, University fo California Press, London, 1969 (ISBN 0-520-02891-0)
9. Bailey, J., Sailing to Paradise, Simon & Schuster, 1994
10. Raadselachtige Vondsten uit het Verleden, Aldus Books, Londen, 1979
11. Robbins, C.S., et al, Birds of North America, A Guide to Field Identification, Golden Press, New York, 1966, pg.46
12. Old World Civilizations/The Rise of Cities and States, Weldon Owen Pty Lim., Australia, 1994
13. Casson, L., Ships and Seafaring in Ancient Times, British Museum Press, 1994 (ISBN 0-525-47545-1)
14. Mallery, A. H., and Harrison, M.R., The Rediscovery of Lost America, The Story of the Pre-Columbian Iron Age in America, Dutton, NY, 1979 (ISBN 0-525-47545-1)
15. Thompson, G., American Discovery, Misty Isles Press, Seattle, 1994
16. Zapp, I., and Erikson, G., Atlantis in America, Navigators of the Ancient World, Adventures Unlimited Press, 1998 (ISBN 0-932813-52-6)
17. "Tour Guide Map", America's Stonehenge, POB 84, N. Salem, NH 03073
18. Ferryn, P., "5000 Years Before Our Era: The Red Men of the North Atlantic", NEARA Journal, Vol.XXXI, No.2, pg.59, 1997
19. Lenik, E.J. and Gibbs, N.L., "The Frost Valley Petroglyph, a Catskill Mountains Enigma", NEARA J., Summer, 1999 (ISSN 0149-2551)
20. Heyerdahl, T., The Ra Expeditions, George Allen & Unwin, London, '71
21. Heyerdahl, T., The Tigris Expedition, George Allen & Unwin, London, 1983
22. Jairazbhoy, R.A., Ancient Egyptians and Chinese in America, Rowman & Littlefield, Totowa, N.J., 1974 (ISBN 0-87471-571-1)
23. Stengel, M.K., "The Diffusionists have Landed", Atlantic Monthly, Jan. 2000, pg.35
24. Tedlock, D., trans., Popol Vuh, The Definitive Edition of the Mayan Book of the Dawn of Life and the Glories of Gods and Kings, Simon & Schuster, New York, 1985 (ISBN 0-671-61771-0)
25. Mertz, H., Atlantis, Dwelling Place of the Gods, Box 207 Loop Sta., Chicago, 60690, 1976 (ISBN 0-9600952-3-3)
26. Coe, M.D., (Principal), Mysteries of the Ancient Americas, Reader's Digest General Books, 1986, USA (ISBN 0-89577-183-7)
27. Stacy-Judd, R., Atlantis, Mother of Empires, 1939 (1999 by Adventures Unlimited Press, ISBN 0-932813-69-0)
28. Wachsmann, S., Seagoing Ships and Seamanship in the Bronze Age Levant, College Station, Texas, 1998
29. Kennedy, R.G., Hidden Cities: The Discovery and Loss of Ancient North American Civilization, Penguin Bks, NY 1994 (ISBN 0-14-02.5527-3)

30. Adams. B., and Cialowicz. K., Protodynastic Egypt. Shire Egyptology. Princes Risborough. 1997
31. Hart. G., A Dictionary of Egyptian Gods and Goddesses., Routledge. London. 1986 (ISBN 0-7102-0167-2)
32. Kemp. B.J., Ancient Egypt. Anatomy of a Civilization. London. Routledge. 1991
33. Rolt-Wheeler. F. Ed., The Science-History of the Universe. Current Literature Publishing Co., 1917
34. Nietfeld. P., Collections Manager. Smithsonian Institution. National Museum of the American Indian. Cultural Resources Center. 4220 Silver Hill Road. Suitland. MD 20746-2863.
35. Wallis Budge. E.A., Osiris and the Egyptian Resurrection. 2 Vol., Dover Pub., N.Y., 1973 (ISBN 0-486-22780-4)
36. Ancient Egypt. National Geographic Maps. Supplement to the National Geographic. April. 2001. National Geographic Magazine.
37. Gruener. J. The Olmec Riddle. An Inquiry into the Origin of Pre-Columbian Civilization. Vengreen Publications. 1987. Rancho Santa Fe. Cal. (ISBN 0-9421`85-56-0)
38. Drews. R., The End of the Bronze Age. Changes of Warfare and the Catastrophe. c.1200 BC. Princeton Paperbacks. 1993 (ISBN 0-691-02591-6)
39. Wilkins. H.T., Mysteries of Ancient South America. (1947). Adventures Unlimited Press. 2000. Ill. (ISBN 0-932813-26-7)
40. Lambert. J.D., America's Stonehenge. An Interpretive Guide. Sunrise Publications. Kingston. N.H., 1996 (ISBN 0-9652630-0-2)
41. Milton. J., The Serpent and the Robe. The Pre-Columbian God-Kings. Boston Publishing Co., Boston. 1986 (ISBN 0-06-055159-3)
42. Campbell. J., Historical Atlas of World Mythology. Part III. The Middle and Southern Americas. Harper & Row. New York. 1989 (ISBN 0-06-055159-3)
43. Gelling. P., Davidson. H.E., The Chariot of the Sun. and Other Rites and Symbols of the Northern Bronze Age. Frederick A. Praeger. N.Y., 1969
44. Mahan. J.B., North American Sun Kings. Keepers of the Flame. ISAC Press. Columbus. Ga. , 1992 (ISBN 1-880820-03-x)

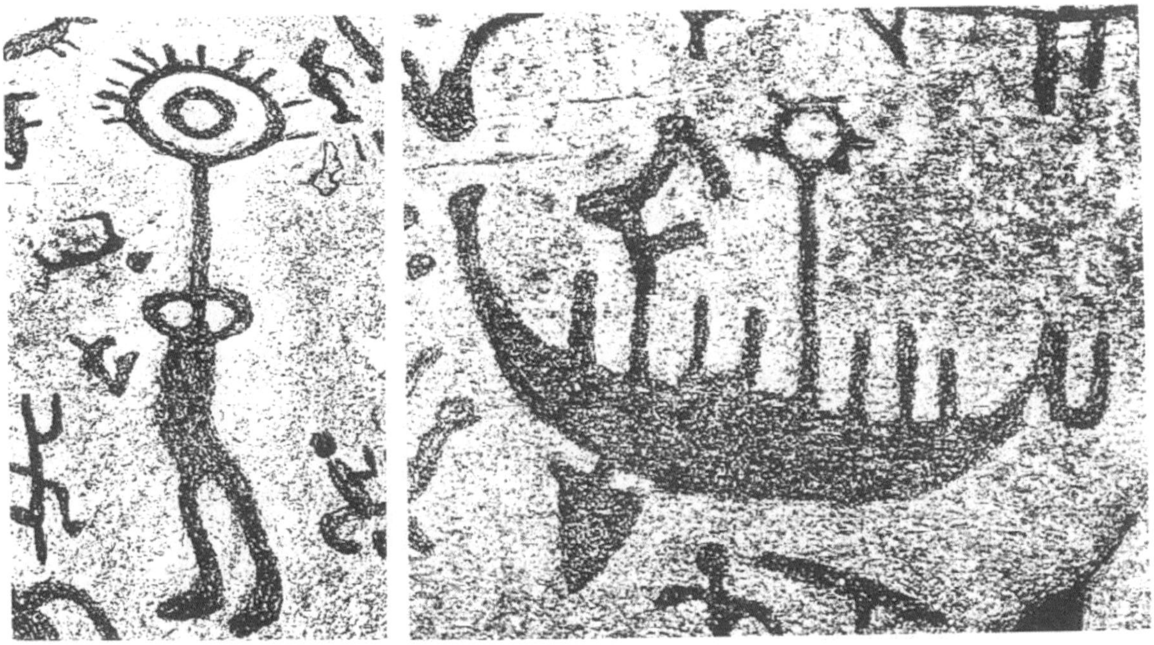

**Fig.9** Scandinavian petroglyphs of the SunGod (left). and a Sunship (right). both at Peterborough. and accompanied by Tifinag script (here omitted) deciphered by Barry Fell. (Peterborough. Ontario. Canada. c.1700 BC. Ch.13. Refs. 4.5.44.45)

# Index

## About The Authors

This book has been written by a Dutchman and an American, who met on the steps of the Archaeological Museum of Carnac, France, while being locked out at lunch.

**Reinoud de Jonge** (1949) is a theoretical physical chemist, and a teacher at an International School in the Netherlands. In his youth, he read Thor Heyerdahl's stories, and learned of heroic discoveries in the Atlantic Ocean that were not taught at school. His interest in megalithic monuments started in 1991, when he was challenged by an article in a Rotterdam newspaper which claimed, that despite an abundance of factual information, Stonehenge was inexplicable. In 1993 de Jonge's first publication "Stonehenge as Sea Chart" (in Dutch) appeared in the periodical BRES (No.158). In 1996 he published De Stenen Spreken (The Speaking Stones), with Professor Gerard IJzereef, which focused on the important petroglyphs of Dissignac, France. In 1998, in cooperation with the Atlantis Foundation, he organized an exposition in the Pinkenberg Museum, outside Arnhem, Holland, which presented an interpretation of these petroglyphs.

**Jay Stuart Wakefield** (1943) is a zoologist who has worked as a newsletter and book editor, and in property management. Like Reinoud, his interest in megalithic prehistory was inspired by childhood reading of Heyerdahl's books. His father Richard's high school graduation address in 1932 entitled The Antiquity of Man in North America began a lifetime of father-son reading, and archaeological exploration. Their coastal and offshore sailing experience has been helpful in this research.

Jay and Reinoud have visited sites together, and worked together on this manuscript for four years in their respective libraries on opposite sides of the world. This is the first time any of this material has been printed in English. The authors are clearly indebted to those who precede them, particularly Mrs. E. Shee Twohig, whose magnificent out-of-print catalogue of European petroglyphs has been indispensable, and her figures frequently used in this work. The authors also wish to acknowledge Dr. Gerard F. IJzereef, Professor of Archaeology of the University of Amsterdam, for his support.

From the data we have derived from these sites, the thinking and achievements of prehistoric man are being brought to light. Now it may be possible to further sort out the profuse stories and legends of the late prehistorical period. There is a lot of work to be done both in literature and in the field. We welcome your help, and suggestions, at our email addresses.

www.ingramcontent.com/pod-product-compliance
Lightning Source LLC
LaVergne TN
LVRC080246220926
848482LV00042B/342

* 9 7 8 1 9 1 7 4 3 8 2 1 6 *